Springer Series in Information Sci

Springer
Berlin
Heidelberg
New York
Barcelona
Hong Kong
London
Milan
Paris
Singapore
Tokyo

Springer Series in Information Sciences

Editors: Thomas S. Huang   Teuvo Kohonen   Manfred R. Schroeder

30 **Self-Organizing Maps**
By T. Kohonen   3rd Edition

31 **Music and Schema Theory**
Cognitive Foundations of Systematic Musicology
By M. Leman

32 **The Maximum Entropy Method**
By N. Wu

33 **A Few Steps Towards 3D Active Vision**
By T. Viéville

34 **Calibration and Orientation of Cameras in Computer Vision**
Editors: A. Gruen and T. S. Huang

35 **Computer Speech**
Recognition, Compression, Synthesis
By M. R. Schroeder

Volumes 1–29 are listed at the end of the book.

B. Roy Frieden

Probability, Statistical Optics, and Data Testing

A Problem Solving Approach

Third Edition
With 115 Figures

Springer

Professor B. Roy Frieden

University of Arizona
Optical Research Center
Tucson, AZ 85721, USA

Series Editors:

Professor Thomas S. Huang

Department of Electrical Engineering and Coordinated Science Laboratory,
University of Illinois, Urbana, IL 61801, USA

Professor Teuvo Kohonen

Helsinki University of Technology, Neural Networks Research Centre, Rakentajanaukio 2 C,
02150 Espoo, Finland

Professor Dr. Manfred R. Schroeder

Drittes Physikalisches Institut, Universität Göttingen, Bürgerstrasse 42-44,
37073 Göttingen, Germany

ISSN 0720-678X
ISBN 3-540-41708-7 3rd Edition Springer-Verlag Berlin Heidelberg New York

ISBN 3-540-63310-9 2nd Edition Springer-Verlag Berlin Heidelberg New York

Library of Congress Cataloging-in-Publication Data

Frieden, B. Roy, 1936–
Probability, statistical optics, and data testing: a problem solving appraoch / Roy Frieden. – 3rd ed.
(Springer series in information sciences, ISSN 0720-678X; 10)
Includes bibliographical references and index.
ISBN 3540417087 (alk. paper)
1. Probabilities. 2. Stochastic processes. 3. Mathematical statistics. 4. Optics–Statistical methods.
I. Title. II. Series
QA273 .F89 2001 519.2–dc21 20001020879

Springer-Verlag Berlin Heidelberg New York
a member of BertelsmannSpringer Science+Business Media GmbH
http://www.springer.de

Printed in Germany

Typesetting: LE-TeX Jelonek, Schmidt & Voeckler GbR, Leipzig
Cover design: *design & production* GmbH, Heidelberg
Printed on acid-free paper SPIN: 10794554 56/3141/YL - 5 4 3 2 1 0

To Sarah and Miriam

Preface to the Third Edition

The overall aim of this edition continues to be that of teaching the fundamental methods of probability and statistics. As before, the methods are developed from first principles, and the student is motivated by solving interesting problems in optics, engineering and physics. The more advanced of these problems amount to further developments of the theory, and the student is guided to the solutions by carefully chosen hints. In three decades of teaching, I have found that a student who plays such an active role in developing the theory gets to understand it more fundamentally. It also fosters a sense of confidence in the analytical abilities of the student and, hence, encourages him/her to strike out into unknown areas of research. Meanwhile, the passage of ten years since the second edition has given the author, as well, lots of time to build confidence and learn more about statistics and its ever-broadening domains of application. Our immediate aim is to pass on this increased scope of information to the student. Important additions have been made to the referencing as well. This facilitates learning by the student who wants to know more about a given effect. Of course, all known typographical errors in the previous editions have also been corrected.

Additional problems that are analyzed range from the simple, such as winning a state lottery jackpot or computing the probability of intelligent life in the universe, to the more complex, such as modelling the bull and bear behavior of the *stock market* or formulating a new central limit theorem of optics. A synopsis of these follows.

A new *central limit theorem of optics* is developed. This predicts that the sum of the position coordinates of the photons in a diffraction point spread function (PSF) follow a *Cauchy probability law*. Also, the output PSF of a relay system using multiply cascaded lenses obeys the Cauchy law. Of course the usual central limit theorem predicts a normal or Gaussian law, but certain of its premises are violated in incoherent diffraction imagery. Other limiting forms are found to follow from a general approach to central limit theorems based upon the use of an *invariance principle*. Other specifically optical topics that are newly treated are the *Mandel formula* of photoelectron theory and the concept of *coarse graining*. The topic of maximum probable estimates of optical objects has been updated to further clarify the scope of application of the *MaxEnt* approach.

The chapter on *Monte Carlo calculations* has been extended to include methods of generating jointly fluctuating random variables. Also, methods of artificially generating *photon depleted*, two-dimensional images are given.

The treatment of *functions of random variables* has been extended to include functions of combinations of random variables, such as quotients or products. For example, it is found that the quotient of two independent Gaussian random variables obeys a Cauchy law. A further application gives the amount by which a probability law is distorted due to viewing its event space from a relativistically moving frame.

Fractal processes are now included in the chapter on *stochastic processes*. This includes the concepts of the *Hausdorff dimension and self-similarity*. The ideas of connectivity by association, and *Erdos numbers*, are also briefly treated.

The subject of *parameter estimation* has been broadened in scope, to include the Bhattacharyya bound, receiver operating characteristics and the problem of estimating multiple parameters.

It is shown that systems of differential equations such as the *Lotka–Volterra* kind are amenable to *probabilistic* solutions that complement the usual analytical ones. The approach also permits *classical trajectories* to be assigned to *quantum mechanical* particles. The trajectories are not those of the physicist D. Bohm, since they are constructed in an entirely different manner from these.

The *Heisenberg uncertainty principle* is closely examined. It is independently derived from two differing viewpoints: the conventional viewpoint that the widths of a function and its Fourier transform cannot both be arbitrarily small; and a measurement viewpoint which states that the mean-square error in estimation of a parameter and the information content in the data cannot both be arbitrarily small. Interestingly, the information content is that of *Fisher*, and not Shannon. The uncertainty principle is so fundamental to physics that its origin in Fisher information prompts us to investigate whether physics in general is based upon the concepts of measurement and Fisher information. The surprising answer is "yes", as developed in the new Sect. 17.3 of Chap. 17.

An alternative statement of the uncertainty principle is *Hirschman's inequality*. This uses the concept of entropy instead of variances of error. It is shown that the entropy of the data and of its Fourier space cannot both be arbitrarily small.

The general use of *invariance principles* for purposes of finding unknown probability laws is extensively discussed and applied. A simple example shows that, based upon invariance to change of units, *the universal physical constants should obey a* $1/x$ *or reciprocal probability law*. This hypothesis is tested by use of a Chi-square test, as an example of the use of this test upon given data. Agreement with the hypothesis at the confidence level $\alpha = 0.95$ is found.

A section has been added on the diverse *measures of information* that are being used to advantage in the physical and biological sciences, and in engineering. Examples are information measures of *Shannon*, *Renyi*, *Fisher*, *Kullback-Leibler*, *Hellinger*, *Tsallis*, and *Gini-Simpson*.

The fundamental role played by Fisher information I, in particular in deriving the Heisenberg uncertainty principle (see above) motivated us to further study I for its mathematical and physical properties. A calculation of I for the case of correlated additive noise shows the possibility of perfect processing of data. Also, I is found to be a measure of the *level of disorder* of a system. It also obeys a property of additivity, and a monotonic decrease with time, $\mathrm{d}I/\mathrm{d}t \leq 0$ under a wide range of conditions. The latter is a restatement of the *Second Law of Thermodynamics* with I replacing the usual entropy term. These properties imply that I may be used in the development of thermodynamics from a non-entropic point of view that emphasizes measurement and estimation error in place of heat. A novel concept that arises from this point of view is a Fisher *temperature of estimation error*, in place of the usual Kelvin temperature of heat.

A remaining property of Fisher information is its *invariance to unitary transformation of whatever type* (e.g. Fourier transformation). This effect is recognized in Chap. 17 to be a *universal invariance principle* that is obeyed by all physically valid probability laws. The universal nature of this invariance principle allows it to be used as a key element in a new *knowledge-based procedure that finds unknown probability laws*. The procedure is called that of "extreme physical information", or EPI for short. The EPI approach is developed out of a model of measurement that incorporates the observer into the observed phenomenon. The observer collects information about the aim of the measurement, an unknown parameter. The information is uniquely Fisher information, and an analysis of its flow from the observed phenomenon to the observer gives the EPI principle. This mathematically amounts to a Lagrangian problem whose output is the required probability law. A zeroing of the Lagrangian gives rise, as well, to a probability law, and *both the extremum and zero solutions have physical significance*. A method of *constructing Lagrangians* is one of the chief goals of the overall approach.

Since EPI is based upon the use of Lagrangians, an added Appendix G supplies the theory needed to understand how Lagrangians form the differential equations that are the solutions to given problems.

EPI is applied to various measurement problems, some via guided exercises, and is shown to give rise to many of the known physical effects that define probability laws: the Schroedinger wave equation, the Dirac equation, the Maxwell–Boltzmann distribution. One of the strengths of the approach is that it also gives rise to valid effects that we ordinarily regard (incorrectly) as *not* describing probability laws, such as Maxwell's equations. Another strength is that the dimensionality of the unknown probability law can be left as a free parameter. The theoretical answer holds for any number of dimensions. However, there is an *effective dimensionality*, which is ultimately limited by that of the user's chosen data space. A third advantage is that EPI rests upon the concept of information, and not a specialized concept from physics such as energy. This means that EPI is applicable to more than physical problems. For example, it applies to genetics, as shown in one of the guided exercises.

An aspect of EPI that is an outgrowth of its thermodynamic roots is its *game aspect*. The EPI mathematical procedure is equivalent to the play of a zero-sum game, *with information as the prize*. The players are the observer and "nature." Nature is represented by the observed phenomenon. Both players choose optimal strategies, and the payoff of the game is the unknown probability law. In that $dI/dt \le 0$ (see above), the observer always loses the game. However, he gains perfect knowledge of the phenomenon's probability law. As an example, the game aspect is discussed as giving rise to the Higgs mass phenomenon. Here the information prize is the acquisition of mass by one of two reactant particles at the expense of the field energy of the other. This is thought to be the way mass originates in the universe.

Tucson, *B. Roy Frieden*
December 2000

Preface to the Second Edition

This new edition incorporates corrections of all known typographical errors in the first edition, as well as some more substantive changes. Chief among the latter is the addition of Chap. 17, on methods of estimation. As with the rest of the text, most applications and examples cited in the new chapter are from the optical perspective. The intention behind this new chapter is to empower the optical researcher with a yet broader range of research tools. Certainly a basic knowledge of estimation methods should be among these. In particular, the sections on likelihood theory and *Fisher information* prepare readers for the problems of optical parameter estimation and probability law estimation. Physicists and optical scientists might find this material particularly useful, since the subject of Fisher information is generally not covered in standard physical science curricula.

Since the words "statistical optics" are prominent in the title of this book, their meaning needs to be clarified. There is a general tendency to overly emphasize the statistics of photons as the *sine qua non* of statistical optics. In this text a wider view is taken, which equally emphasizes the random medium that surrounds the photon, be it a photographic emulsion, the turbulent atmosphere, a vibrating lens holder, etc. Also included are random interpretations of ostensibly deterministic phenomena, such as the Hurter–Driffield (H and D) curve of photography. Such a "random interpretation" sometimes breaks new ground, as in Chap. 5, where it is shown how to produce very accurate raytrace-based spot diagrams, using the *statistical* theory of Jacobian transformation.

This edition, like the first, is intended to be *first and foremost* an introductory text on methods of probability and statistics. Emphasis is on the linear (and sometimes explicitly Fourier) theory of probability calculation, chi-square and other statistical tests of data, stochastic processes, the information theories of both Shannon and Fisher, and estimation theory. Applications to statistical optics are given, in the main, so as to give a reader with some background in either optics or linear communications theory a running start. As a pedagogical aid to understanding the mathematical tools that are developed, the *simplest possible statistical model* is used that fits a given optical phenomenon. Hence, a semiclassical model of radiation is used in place of the full-blown quantum optical theory, a poker-chip model is used to describe film granularity, etc. However, references are given to more advanced models as well, so as to steer the interested reader in the right direction. The listing "Statistical models ..." in the index gives a useful overview of the variety of models used. The reader might, for example, be amazed at how much "mileage" can be obtained from a model as simple as the checkerboard model of granularity (Chaps. 6 and 9).

The reader who needs an in-depth phenomenological viewpoint of a specific topic, with statistical optics as the emphasis and probability theory subsidiary, can consult such fine books as *Photoelectron Statistics* by B. Saleh [Springer Ser. Opt. Sci., Vol. 6 (Springer, Berlin, Heidelberg 1978)] and *Statistical Optics* by J.W. Goodman (Wiley, New York 1985). We recommend these as useful supplements to this text, as well.

The *deterministic* optical theory that is used in application of the statistical methods presented is, for the most part, separately developed within the text, either within the main body or in the exercises (e.g., Ex. 4.3.13). Sometimes the development takes the form of a sequence of exercises (e.g., Exercises 6.1.21–23). The simplest starting point is used – recalling our aims – such as Huygens' wave theory as the basis for diffraction theory. In this way, readers who have not previously been exposed to optical theory are introduced to it in the text.

The goal of this book remains as before: To teach the powerful problemsolving methods of probability and statistics to students who have some background in either optics or linear systems theory. We hope to have furthered this purpose with this new edition.

Tucson,
December 1990

B. Roy Frieden

Preface to the First Edition

A basic skill in probability is practically demanded nowadays in many branches of optics, especially in image science. On the other hand, there is no text presently available that develops probability, and its companion fields stochastic processes and statistics, from the optical perspective. [Short of a book, a chapter was recently written for this purpose; see B.R. Frieden (ed.): *The Computer in Optical Research*, Topics in Applied Physics, Vol. 41 (Springer, Berlin, Heidelberg, New York 1980) Chap. 3]

Most standard texts either use illustrative examples and problems from electrical engineering or from the life sciences. The present book is meant to remedy this situation, by teaching probability with the specific needs of the optical researcher in mind. Virtually all the illustrative examples and applications of the theory are from image science and other fields of optics. One might say that photons have replaced electrons in nearly all considerations here. We hope, in this manner, to make the learning of probability a pleasant and absorbing experience for optical workers.

Some of the remaining applications are from information theory, a concept which complements image science in particular. As will be seen, there are numerous tie-ins between the two concepts.

Students will be adequately prepared for the material in this book if they have had a course in calculus, and know the basics of matrix manipulation. Prior formal education in optics is not strictly needed, although it obviously would help in understanding and appreciating some of the applications that are developed. For the most part, the optical phenomena that are treated are developed in the main body of text, or in exercises. A student who has had a prior course in linear theory, e.g. out of Gaskill's book [J.D. Gaskill: *Linear Systems, Fourier Transforms, and Optics* (Wiley, New York 1978)], or in Fourier optics as in Goodman's text [J.W. Goodman: *Introduction to Fourier Optics* (McGraw-Hill, New York 1968)], is very well prepared for this book, as we have found by teaching from it. In the main, however, the question is one of motivation. The reader who is interested in optics, and in problems pertaining to it, will enjoy the material in this book and therefore will learn from it.

We would like to thank the following colleagues for helping us to understand some of the more mystifying aspects of probability and statistics, and for suggesting some of the problems: B.E.A. Saleh, S.K. Park, H.H. Barrett, and B.H. Soffer. The excellent artwork was by Don Cowen. Kathy Seeley and her editorial staff typed up the manuscript.

Tucson, April 1982 *Roy Frieden*

Preface to the First Edition

[illegible]

Contents

1. Introduction

The landmark subjects in mathematics are those that show us how to look at things in a fundamentally new way. Algebra, calculus, and matrix theory can each be remembered as providing almost a step-function impact upon our way of viewing the physical world.

A course in probability and statistics usually provides the last chance in a person's education for such insights into the workings of the physical world. (These subjects need calculus and matrix theory as prior building blocks.) The probabilistic perspective on the world is truly novel. From this view, the world is a chaotic and unpredictable place, where God *really does* play dice. Nothing is known for sure. Everything must be estimated, with inevitable (and estimable) error. We cannot state that event A will happen, but rather that A will happen with a certain probability.

This view is often forced upon us by nature. Many phenomena cannot be explained *except* in terms of probability. Alternatively, we may *choose* to take this view because of practical considerations such as time or cost (Sect. 1.1.1).

For example, random effects permeate the optical sciences. Images are degraded by such well-known and disturbing (since observable) effects as atmospheric turbulence, laser speckle and computational noise. Electrooptical device outputs suffer from random noise due to photon shot noise, additive resistor noise, random vibration of lenses, etc. Other random effects are *disguised as deterministic effects*, such as the photographic H–D curve (Chap. 6) of an emulsion. All such random effects may be analyzed by employing suitable methods of probability and statistics, including the use of an appropriate *statistical model*. We shall limit attention to some of the simpler, by now "classical," models of statistical optics.

It is our primary aim to derive from first principles the powerful, problem-solving methods of probability and statistics. Problems in statistical optics are mainly used for specific applications of the methods, and not as goals in themselves. In the long run, the reader will make more use of the methods that are developed than of the applications. However, if the reader will thereby gain better insight into statistical-optics phenomena, as we expect he inevitably must, so much the better.

This book evolved out of a course at the Optical Sciences Center, which originally emphasized information theory, with probability theory only playing an introductory role. However, through the years it became apparent that the introductory material was more important to the students than the body of the course! The many applications of probability to optics are more relevant than the single application to information theory. Hence, the shift in emphasis.

While so developing the course, the author chanced upon a book by *Brandt* [1.1] on statistical methods for physics data, which suggested a further avenue of development. This is into the area of statistical data testing. Indeed, physical data of the optical kind are prime candidates for standard statistical tests such as Chi-square, Student t- and Snedecor F-test. This is because they suffer from random noise, and often are insufficient in quality or quantity to permit deterministic answers.

However, to a great extent, the standard statistical tests have not been used in optics. Why? Merely because, it becomes apparent, optical researchers are simply not acquainted with them. By comparison, researchers in medical imagery, who usually have a different educational background, have made good use of the tests, and continue to do so. Accordingly, we have developed the course to include these tests and other techniques of statistics such as principal-components analysis and probability estimation. The derivation of these approaches might, in fact, be regarded as the culmination of the probability theory developed in prior chapters.

Stochastic processes, the other main subject to be treated, have widespread occurrence in optics. All continuous signals with random parts, be it randomness of noise or some other factors, are stochastic processes. In particular, images are stochastic. Most of the interesting applications of probability to optics are actually stochastic in nature. Hence, a general theory of stochastic processes had to be included. In doing so, we have followed many of the ideas set forth by *Papoulis* [1.2] in his landmark text. The latter, and *Parzen* [1.3] constituted this author's introduction to probability theory. We unequivocally recommend them as supplementary texts, as well as the book by *Saleh* [1.4].

Separate mention should also be made of *O'Neill*'s book [1.5], which became a classic in its day, and constituted for many of us in the early 1960's *the* introduction to the world of image research. We recommend it enthusiastically (almost, affectionately) as supplementary reading, not for its statistical methods, which are rather *ad hoc* and sparse, but rather for its excellent exposition on image formation and synthesis.

One really learns probability and statistics by doing problems. We cannot emphasize this point too much. For this reason, many problems are given within the text. These are in the exercises, which are usually at the end of a chapter but sometimes within as well (for the longer chapters). The problems are generally of two types. The first problems in an exercise are direct, usually elementary, applications of the methods developed in that chapter. These are often surprisingly easy to do. (Difficult problems are supplied with hints for steering the student in the right direction.) Following these are usually problems that apply the methods to specifically optical subjects. These are meant to explain some of the more important optical effects that are statistical in nature, such as the Van Cittert–Zernike theorem. Honors problems, of a more optional nature, are also given so as to satisfy the more inquisitive mind. The exercises are demarked, for the reader's convenience, by a vertical line in the margin. Answers to selected problems may be obtained by writing to the author directly, and enclosing a stamped, self-addressed envelope (about $81/2 \times 11$ *in*) for return.

1.1 What Is Chance, and Why Study It?

The study of probability has humble origins, like most sciences. Microbiology originated as the study of fermentation in wines; chemistry and medicine were practiced for millenia as alchemy and witchcraft; astronomy and astrology were originally identical. So it is with probability, which started with the study of the outcomes when a pair of ankle bones of a sheep are tossed. These bones are curiously shaped, like small cubes, and became today's dice. So, probability theory owes its start to an early version of the game of "craps." Such is reality. *Something you've always wanted to know:* the word "craps" traces from the word "crapaud", meaning "toad" or "toad-like" in French, and is attributed to French Cajun settlers in early New Orleans in describing the squatting position taken by the dice players.

Rolling dice is a popular pastime principally because it is a game of "chance". After all, Lady Luck can smile on anyone. This leads us to ask what a "chance" or "random" event is. The definition I like best is that a random event is an *unpredictable event.* Although negative definitions are not usually preferred, this one has some appeal. Everyone knows, after all, what a predictable event is.

1.1.1 Chance vs. Determinism

It is common observation that some events are less chancey, or predictable, than others. There is a chance the sun will not rise tomorrow, but it is very remote. When, then, is a person justified as regarding a given phenomenon as random? Some reflection leads one to conclude that there are two general categories that random events fall into: (i) random events due to natural law, such as the position of a photon in its diffraction pattern; and (ii) random events that exist only in the eye of the beholder, stemming from insufficient data and consequently, a degree of ignorance about the event. Category (i) is not surprising, but (ii) has some important implications. It states that an event can be regarded as random at will, by an observer, if he merely chooses to do so.

This has an interesting implication. Type (ii) randomness is dependent for its existence upon the presence of an observer. Without the observer, the only events that could be considered random (although, by whom?) would be those due to natural law. This is not so startling, however, to a student of communication theory. The observer is an integral part of most communication links. He is the ultimate detector of information about the event, information that must be inevitably limited by whatever channel has been used for its transmission. Hence, in summary, randomness of type (ii) is consistent with, and in fact, demanded by modern communication theory.

The stance we shall take, then, and in fact the stance taken by the majority of modern workers in statistics, is that any event can *at will* be regarded as being either deterministic (predictable), or random. In fact, as a practical matter, every probability law has its deterministic limit, so that one needn't fear using probability indiscriminately. In general, the stance to be taken depends upon what kind of *answers* are being sought. An example will show this.

Suppose, as frequently happens, an ordinary spoon is dropped accidentally and comes to rest on the kitchen floor, perhaps with an intermediate bounce off a chair *en route*. Let us consider its final rest position (x, y) as the event under consideration. As a summary of the preceding discussion, (x, y) can be predicted by either a deterministic or a probabilistic viewpoint. By the former, we have simply a problem in mechanics, albeit a difficult one. By the latter, we have a problem in random scatter to all possible points $(\hat{x}, \hat{y})$ (the karat mark denotes an estimate) on the floor. *The viewpoint to take will depend upon the required accuracy in the answer* $(\hat{x}, \hat{y})$. If an exact (in the sense of Newtonian mechanics) answer is required, then we must take the deterministic view and solve a pair of Lagrange equations. But, if an answer such as "the spoon will land no farther from point $(\hat{x}, \hat{y})$ than 1 cm with 90% probability" is acceptable, then the probabilistic view may be taken.

Frequently, the degree of accuracy demanded of the answer is a matter of economics. It usually costs more money to do a deterministic analysis, with its high requirement upon precision, than to do the corresponding statistical analysis. For example, by merely repeating the spoon-dropping experiment with statistically representative initial conditions, say, a total of 20 times, one can get a good idea of the histogram of occurrences (x, y), and hence of "landing within 1 cm with 90% probability" of an estimated point $(\hat{x}, \hat{y})$.[1] Compare this with the tedious problem in Newtonian mechanics which the exact approach would require.

We end this little discussion of what may be regarded as random with perhaps the most startling example of them all. It turns out that probabilities themselves may often be productively regarded as random quantities. This leads to the somewhat peculiar concept of "the probability of a probability". Nevertheless, it is a useful concept of Bayesian statistics (Chaps. 10 and 16).

1.1.2 Probability Problems in Optics

Where does the need for knowledge of probability and statistics enter into optics? First, let us consider the subject of probability.

Optics is the study, and manipulation, of photons; but not photons that sit still (of course, an impossibility); rather, photons that travel from an airborne laser down through the atmosphere, across the air-ocean boundary, and into the ocean, perhaps to be detected at a sensor many fathoms down. Or, photons that travel from twinkling stars to waiting telescopes. Also, photons that illuminate a microscope specimen, which is then greatly magnified by some ingenious optics, to enter the eye with some hope of being detected. The list is endless. In all cases, the photons are subjected to randomness of many varieties: (a) Being quantum mechanical entities, they have

[1] This is a primitive use of the Monte Carlo approach of statistics. We shall say a lot more about this approach later.

an intrinsic uncertainty in position (as well as time, energy and momentum). This gives rise to the well-known Airy disk diffraction pattern, which actually represents a photon's probability on position. (b) When arising from a weak source, they arrive in time as a "clumpy" progression, so much so that their time spread about equals the root-mean arrival rate. This describes the shot-noise or Poisson nature of photon beams. (c) On the way from source to detector, they may suffer random refraction effects, as in atmospheric turbulence, or random scattering effects by atmospheric particles. (d) At the detector their signal may not be directly sensed, but rather its square may be the registered quantity. And the detector may suffer from random noise of its own because of its electronic circuity (Johnson noise) and other causes. Or, if the detector is photographic, the well-known granularity problems of that medium will arise. (e) Either the source or the detector may be randomly vibrating.

Because of these and other random effects, problems of probabilistic nature arise widely in optics. We shall address many of these, in the first half of the book, Chaps. 2–8.

1.1.3 Statistical Problems in Optics

All of the foregoing may be described as problems of analysis. That is, given such-and-such properties of source and medium and detector, what will the output photon data look like? The observer puts himself in the position of looking forward in time. On the other hand, there is a completely complementary situation that arises when the observer is in the position of knowing only photon data, as some finite set, and from this has to deduce something about the phenomena that went into forming it. This is the general problem area called "statistics." Here, the observer looks backward in time, to what were the *probable* causes of the given data; probable causes, since the data are not of high-enough quality to imply a unique answer. For example, they are either too noisy, or too sparse. As an example, he may want to improve his estimate of c, the speed of light, through averaging the outputs $\{c_n\}$ of N independent experiments. Statistics will tell him what the probable error is in this estimated c, as a function of N. And this can be known even if the probable error in any one determination is *not* known! Or, given a new photographic emulsion of unknown sensitometry, he may want to set up a sequence of experiments whose density data will enable him to decide which development parameters (such as development time, temperature, time in bath, etc.) are the most "significant." Or, given a very noisy image, the observer might want to be able to judge if the picture contains any "significant" structure. Many problems of this type weill be taken up in this text These subjects comprise the second half of the book. Chaps. 9–17.

1.1.4 Historical Notes

Gambling was very popular in the courts of the kings of France: Thus the early domination of the field of probability by French thinkers, among them Laplace, De Moivre, Poisson, etc., names that will appear throughout this book.

Students of optics might be interested to know that one of the world's first opticists was the famous philosopher Baruch (Benedictus) Spinoza. Scientists and philosophers of his era, the 17th century, often depended for their financial support upon the generosity of wealthy patrons. However, Spinoza's ethics required him to be self-sufficient. And so he chose to grind and polish lenses for a living. Unfortunately, he fell victim to his beliefs, dying of lung disease caused by inhaling the optical rouge he used in this work. His book *Calculation of Chances* [1.6], published in 1687, discusses both the probability of rolling a seven in dice and the optical origin of the rainbow.

2. The Axiomatic Approach

The entire edifice of probability theory, and its offshoots statistics and stochastic processes, rests upon three famous axioms of *Kolmogoroff* [2.1]. Indeed, *everything in this book derives from these simple axioms*. The axioms are deceptively simple, in fact so simple that at first they appear to say nothing at all. However, let us recall that axioms are supposed to be that way. The simpler they appear, and the more they imply, the better they are mathematically.

Before addressing these axioms, however, it is necessary to define terms that are mentioned in the axioms. For example, one does not speak abstractly about a "probability," but rather about the probability *of an event*. What, then, is an event? Events, it is apparent, result from "experiments." What is an experiment? We have to become very basic here. The *most* basic approach is to take the rigorous set-theory approach as in, e.g., [2.2]. However, we feel that this unnecessarily drags out the subject. The rigorous approach merely appears to accomplish mathematical definitions of the words "or" and "and," terms which we have an intuitive grasp of anyhow. Hence, we shall take these words in their intuitive sense, and refer the reader who wants rigor to Papoulis' fine book [2.2].

2.1 Notion of an Experiment; Events

An *experiment* may be defined in the broadest sense as any fixed, but repeatable, procedure of steps which lead to an *outcome*. Cooking a meal from a recipe is an experiment. The outcome is whatever comes out of the oven, ranging from a medium-rare roast of beef to an incinerated one. Now, depending upon its success, or other aspects, the experiment might be repeated. Each such repetition is called a *trial*.

Hence, suppose we have a situation where numerous trials are performed, resulting in a set of outcomes. The *outcomes are*, as in the above example, *direct observables*. Why should they differ from one repetition to the next? Well, either out of choice or at the whim of fate, more or less seasoning might be used, the oven temperature might vary, etc. They are assumed to occur in an unpredictable, and random, way.

Each cooking experiment in fact results in one of a set of possible outcomes. Now the user (eater) may choose to cause an associative action or description to accompany each such outcome. These are called *events*. For example, the outcome "medium-rare roast of beef" might cause the event "the roast is rapidly and appreciatively eaten."

In general, each outcome of an experiment will have an event associated with it. For example, suppose the experiment consists of a roll of a die. The outcome of each experiment is a single number on the upturned face. This is the physical observable, or outcome, of the experiment. However, the user may be gambling at an "under or over" table and hence be supremely interested in whether the outcome is exactly, more than, or less than, three. Hence each die outcome will be associated with one of the events "more than three," "less than three," or "exactly three." One might say that *the event describes the physical significance of the outcome to the user*. Hence, the delegation of outcomes to events is completely arbitrary. Sometimes, the events are made to equal the outcomes. Table 2.1 summarizes this terminology.

Table 2.1. Basic terminology

An *experiment* E is a fixed procedure
Each repetition of E is a *trial*
Each trial causes an *outcome*, a direct observable
Each outcome is arbitrarily associated with an *event* A_n
The events $\{A_n\}$ comprise the set whose probabilities $P(A_n)$ are desired

Having defined "events," we are now able to speak of *the probability of an event*. Let us denote the probability P of the event A_n as $P(A_n)$. Before speaking further of the concept of probability, however, we have to distinguish among different types of events.

2.1.1 Event Space; The Space Event

A set of N events which includes all possible events for an experiment defines event space. For example, in the roll of a die, events $\{A_n\} = 1, 2, \ldots, 6$ are elements of a space. The *space event* is defined as event $C = (A_1 \text{ or } A_2 \text{ or } \ldots \text{ or } A_N)$.

However, the event space for a given experiment is not unique. This is because the events are *arbitrarily* assigned to the outcomes (as previously discussed). Therefore, the same experiment may be described by many different possible spaces. For example, in the roll of a die we might instead define as event space

$$\{A_n\} = (\text{any roll less than 3, the roll 3, any roll greater than 3}) .$$
$$n = 1, 2, 3$$

The space to use depends on the probability it is desired to estimate (more on this later in Sect. 2.7).

2.1.2 Disjoint Events

This concept ultimately allows probabilities to be computed. Two events A and B are called mutually exclusive, or disjoint, if the occurrence of A rules out the occurrence of B. For example, if the die turns up two dots, this event rules out the event "five dots." Hence, the events "two dots," "five dots" are disjoint.

Two *non*-disjoint events would be "a roll of less than 4," "a roll greater than 2." Obviously, if the outcome is a "3" both these events occur simultaneously. There is no exclusion. An event can be a vector of numbers, such as the positional event $A = (x, y, z)$ defining a position in space. Is, then, an event $A_1 = (1, 2, 3)$ of this type *disjoint with* an event $A_2 = (1, 2, 4)$? Despite the coordinates $(1, 2)$ in common, the difference in the third coordinate defines two different events A_1, A_2. As an illustration, if a particle is at a position $(1, 2, 3)$ this rules out the event that it is also at another position $(1, 2, 4)$. The two positional events are disjoint.

2.1.3 The Certain Event

Let us consider the event "any number from 1 to 6," when the experiment is the roll of a die. This event is *certain* to occur at each repetition of the experiment. It is consequently called the "certain" event. In general, every experiment has a certain event C associated with it, namely, its entire event space (A_1 or A_2 or ... or A_N).

These event types are summarized in Table 2.2, using rolls of a die as examples.

Table 2.2. Event types

General description	Example using die roll(s)
Space event	1 or 2 or 3 or 4 or 5 or 6
Certain event	ditto
Disjoint events	event 2, event 6

Exercise 2.1

An experiment E consists of imaging a single photon with an optical system. Each trial output is one of the x-coordinate positions $x_1, \ldots, x_N$ in the image plane. Each repetition of E results in a value x from among these $\{x_n\}$. Use this experiment to provide illustrative examples for the concepts in Tables. 2.1 and 2.2.

2.2 Definition of Probability

Associated with each possible event A of an experiment E is its "probability" of occurrence $P(A)$. This number is defined to obey the following axioms [2.1]:

Axiom I $P(A) \geq 0$. That is, a probability is never negative.
Axiom II $P(C) = 1$, where C is the "certain" event. This ultimately fixes a scale of numbers for the probabilities to range over.

Axiom III If events A and B are disjoint, $P(A \text{ or } B) = P(A) + P(B)$. The word "or" is taken in the intuitive sense; point set theory [2.2] defines it mathematically. For example, for the die experiment $P(\text{roll 1 or roll 2}) = P(\text{roll 1}) + P(\text{roll 2})$. This is reasonable, on the basis of frequency of occurrence (see below).

These three axioms, along with a generalization of Axiom III to include an infinity of disjoint events (see below), *imply all the rules of probability, and hence all the methods of statistics* (as will be shown).

2.3 Relation to Frequency of Occurrence

It is more than interesting that the concept of "frequency of occurrence" also obeys the above three axioms. This is no coincidence, since probability theory is basically a way to give the concept of frequency of occurrence a rigorous, mathematical underpinning. The word "probability" is but a mathematical abstraction for the intuitively more meaningful term "frequency of occurrence." In real problems, *it is usual to visualize the unknown probability as a frequency of occurrence*, but then to write of it as a probability. This trick is mathematically justified, however, since in the limit of an infinity of independent occurrences, the frequency of occurrence of an event A equals its probability $P(A)$. This "law of large numbers" (Sect. 2.9) is the exact connection between the two concepts.

The law of large numbers, i.e. frequency of occurrence in an infinite record, is in fact the *Von Mises* [2.8] *definition* of probability. However for various reasons it is not accepted as the basis for an axiomatic theory of probability. Nevertheless the idea is often used to *compute* probabilities, and therein lies its strength.

2.4 Some Elementary Consequences

Building on these axioms, we note the following. Suppose $\overline{A}$ denotes "not A," the complementary event to A. Obviously, A and $\overline{A}$ are disjoint, by Sect. 2.1. Therefore, by Axiom III, $P(A \text{ or } \overline{A}) = P(A) + P(\overline{A})$. Furthermore, the two events, A, $\overline{A}$ together comprise the "certain" event by Sect. 2.1.3. Therefore, by Axiom II, $P(A \text{ or } \overline{A}) = P(A) + P(\overline{A}) = 1$. Combining the latter with Axiom I applied to $\overline{A}$, it must be that $P(A) \leqq 1$. Hence, combining this with Axiom I applied to A,

$$0 \leqq P(A) \leqq 1 \,. \tag{2.1}$$

All probabilities lie between 0 *and* 1. An event having zero probability is termed the "null" or "impossible" event. Hence, all events lie somewhere between the certain and the impossible (a not very profound statement, but certainly necessary for a correct theory!).

2.4.1 Additivity Property

Consider disjoint events A_1 and A_2. Also, suppose that each of A_1 and A_2 is disjoint with an event A_3. Temporarily let $B_1 = (A_2 \text{ or } A_3)$. Then, by definition of "disjoint," A_1 is disjoint with B_1 (see Fig. 2.1). Hence, by Axiom III, $P(A_1 \text{ or } B_1) = P(A_1) + P(B_1)$. Then combining the last two relations

$$P(A_1 \text{ or } A_2 \text{ or } A_3) = P(A_1) + P(A_2 \text{ or } A_3) ,$$

and hence by Axiom III, once again

$$P(A_1 \text{ or } A_2 \text{ or } A_3) = P(A_1) + P(A_2) + P(A_3) .$$

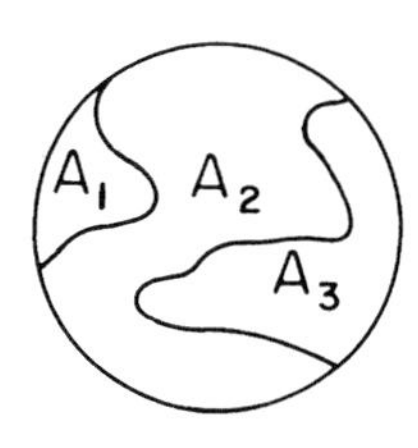

Fig. 2.1. Area representation of events. Each point within area A_1 represents an event of type A_1, etc. for A_2, A_3. A concrete example of this is a dartboard so partitioned into zones. The experiment E consists of throwing a dart at the board. Where the dart sticks in defines an event of either type A_1, A_2 or A_3. Obviously, if it sticks within A_1, it does not stick within A_2. Hence, A_1 and A_2 are disjoint events. Similarly for A_2 and A_3. Finally, by the same reasoning, A_1 is disjoint with the event $(A_2 \text{ or } A_3)$

This process may evidently be continued to N disjoint events. Hence, if the events $A_1, A_2, \ldots, A_N$ are disjoint

$$P(A_1 \text{ or } A_2 \text{ or } \ldots \text{ or } A_N) = P(A_1) + P(A_2) + \cdots + P(A_N) . \tag{2.2}$$

This is called the "additivity" property for N disjoint events. Note that in a rigorous formulation of probability this result cannot be generalized to $N = \infty$. This property, for $N = \infty$, must be regarded as Axiom IV.

2.4.2 Normalization Property

Suppose events $A_1, A_2, \ldots, A_N$ are disjoint and form an event space (see Sect. 2.1). Then

$$(A_1 \text{ or } A_2 \text{ or } \ldots \text{ or } A_N) = C ,$$

the certain event. Then Axiom II may be combined with result (2.2) to produce

$$P(A_1) + P(A_2) + \cdots + P(A_N) = 1 , \tag{2.3}$$

the normalization property for probabilities that form a space. We see, from the preceding, that (2.3) is really a statement that some one of the possible events $A_1, \ldots, A_N$ *must occur* at a given experiment. For example, in Fig. 2.1, the dart must strike the board *somewhere*.

2.5 Marginal Probability

Consider a *joint event* A_m and B_n denoted as $A_m B_n$. Let the $\{B_n\}$ be disjoint and form a space (B_1 or B_2 or ... or B_N). For example, suppose the dartboard in Fig. 2.1 is used in a contest between red and green darts. If A_1, A_2 and A_3 denote the board subareas as before, and now in addition B_1 = event (red dart), and B_2 = event (green dart), then the joint event $A_1 B_2$ denotes the event (green dart strikes area A_1); etc.

Since $A_m B_n$ is an event, it has a probability $P(A_m B_n)$. We now ask, if $P(A_m B_n)$ is known, for $n = 1, 2, \ldots, N$, can $P(A_m)$ be somehow computed from it?

The event A_m is equivalent to the joint event A_m and (B_1 or B_2 or ... or B_N), since the latter is the certain event. For example, dart event A_1 always occurs as A_1 and either (red dart or green dart). This joint event may also be expressed as (A_m and B_1) or (A_m and B_2) or ... or (A_m and B_N). That is, the compound dart event A_1 and (red dart or green dart) is synonymous with the event (A_1 and red dart) or (A_1 and green dart). Furthermore, events (A_m and B_n), $n = 1, \ldots, N$, are disjoint since the $\{B_n\}$ are disjoint (see end of Sect. 2.1.2). Recalling that these are equivalent to event A_m, we may use identity (2.2), which yields

$$P(A_m) = \sum_{n=1}^{N} P(A_m B_n) . \tag{2.4}$$

$P(A_m)$ is called the "marginal" probability, in the sense of summing the mth row of elements across the page to the margin. (This is the way they used to be calculated before the advent of electronic computers.)

As a matter of notation, we shall sometimes denote $P(A_m B_n)$ as $P(A_m, B_n)$, the comma denoting "and."

2.6 The "Traditional" Definition of Probability

Suppose all the probabilities $P(A_n)$, $n = 1, \ldots, N$ in the normalization (2.3) are equal. Then necessarily each $P(A_n) = 1/N$. As an example, in Fig. 2.1 if area $A_1 = A_2 = A_3$, then since $N = 3$ we have $P(A_1) = P(A_2) = P(A_3) = 1/3$.

Suppose further that an event B is defined as a *subset* (A_1 or A_2 or ... or A_n) of event space, with $n \leqq N$. For example, in Fig. 2.1, let event B = event (A_1 or A_3). Then by (2.2),

$$P(A_1 \text{ or } A_2 \text{ or } \ldots \text{ or } A_n) \equiv P(B) = n(1/N) ,$$

or simply

$$P(B) = n/N . \tag{2.5}$$

For our dart problem, then, $P(B) = 2/3$.

This was the traditional definition of probability taught in secondary schools a generation ago. It is usually attributed to Laplace but, in fact, was first published by the mathematician Girolamo Cardan (1663), who also was the first to find the roots of a cubic polynomial equation. The definition (2.5) has a simple interpretation. Recall that n is the number of disjoint events that may each imply the event B. Also, N is the total number of disjoint events possible. Then (2.5) states that $P(B)$ *is simply the ratio of the number of independent ways B can occur, to the total number of permissible events.* The exercise in secondary school consisted of counting up by permutations the unknowns n and N. (This thought still conjures up dreaded memories of red and white balls drawn from urns.)

Equation (2.5) actually allows a probability to be *computed.* Many common experiments permit n and N to be directly (if not easily) computed. The example in Sect. 2.7 below illustrates this point.

Note of interest: According to the "many worlds" view of reality by *Everett* [2.9], *every event* A_n *happens* at each given trial, but in a *different world.* Our world is but one of them. By this interpretation, a probability (2.5) is a ratio of numbers of different worlds!

2.7 Illustrative Problem: A Dice Game

Two "fair" dice are repeatedly rolled. By "fair" is meant equal probability for each of the six possible outcomes of a die roll. The dice fall side by side, so there is always a left-hand die and a right-hand die at each roll. Find the probability of the event $B =$ (left-hand die even, right-hand die greater than 4).

The unknown probability is to be found by the use of identity (2.5). This means that B has to be enumerated as the occurrence of n equi-probable events. Also, to simplify the calculation, B ought to be one member of event space. This point will become clear below.

A frontal attack on the problem is to consider each possible experimental outcome with regard to its evenness or oddness (parity), and size relative to level four. Thus, let e represent the event (left die even), o the event (left die odd), LE the event (right die less than or equal to 4), and G the event (right die greater than 4). In particular, event $B \equiv (e, G)$ in terms of this joint event. Then the following *joint* event space may be constructed for rolls of the two dice:

$$(e, LE), (e, G), (o, LE), (o, G) . \tag{2.6}$$

There are no other possibilities as regards these joint events. Note that we have made B one event (the second) in the event space. This is an important step, as the number of ways n in which B can occur forms the numerator of (2.5).

Now, according to (2.5), we have to express event B in terms of equiprobable events. But directly B can only be formed as die outcomes

$$B \equiv (e, G) = [2, 5] \text{ or } [2, 6] \text{ or } [4, 5] \text{ or } [4, 6] \text{ or } [6, 5] \text{ or } [6, 6] ,$$

where $[2, 5]$ is the joint event (left die outcome 2, right die outcome 5), etc. for the other square brackets. Each of these number couplets is equally probable, by the "fairness" of the dice; see preceding. Therefore, we have but to count them, and this defines an n of 6.

Notice that outcome [2, 5] is of the event type (e, G) only. Each outcome $[m, n]$ belongs to one and only one event listed in (2.6). Hence the latter are disjoint and form an event space, as required.

By the same counting procedure we may compute values n for each of the other events in (2.6), and merely add them to get N since (2.6) form a space. In this way, we find $N = 12 + 6 + 12 + 6 = 36$. Hence by (2.5) $P(B) = 6/36$.

This problem was given merely as an exercise. The next problem has some everyday significance.

2.8 Illustrative Problem: Let's (Try to) Take a Trip

Airlines purposely overbook seats on flights, since in practice a significant number of booked passengers do not show up. Suppose a certain airline books 105 people for 100 seats on a given flight. If all 105 passengers show up, what is the probability that a given person will *not* get a seat?

Analysis: Let $S =$ event that the person *gets* a seat. $\overline{S}$ is the complementary event that he does not. It is easier to compute $P(S)$ first and then use $P(\overline{S}) = 1 - P(S)$ by the exhaustive nature of S and $\overline{S}$.

Consider the 100 seats to be labelled in some order as seat 1, seat 2,. . . , seat 100. *Imagine these 100 seats to be allocated in some random way among the 105 travellers.* The given traveller has a chance of 1/105 of getting seat 1, ditto for seat 2, etc.

Now event S occurs if the given traveller gets *either* seat 1 or seat 2 or . . . or seat 100.

The events are disjoint, there are $n = 100$ of them, and each has the same elemental probability of 1/105 (as found above).

Therefore, by identity (2.2),

$$P(S) = \underbrace{1/105 + 1/105 + \cdots + 1/105}_{\longleftarrow \quad 100 \text{ times} \quad \longrightarrow} = 100/105 \tag{2.7}$$

and the answer is

$$P(\overline{S}) = 5/105\ .$$

Although we have solved the problem, an apparent paradox arises. Suppose contrary to the given problem there were more seats available than passengers. It would appear then from (2.7) that $P(S)$ would now exceed 1, which is of course impossible for a probability. Is the overall approach, then, wrong?

To be specific, suppose there are now 106 seats and 105 passengers. It appears from blind use of (2.7) that now $P(S) = 106/105$. But a more careful use of the approach shows the following: Event S occurs if the traveller gets either seat 1 or seat 2 or ... or seat 106. Again, the events are disjoint, and there are now $n = 106$ of them. But, observe the point of departure from the preceding problem. Since there are now more seats than passengers, then *vice versa the passengers may be pictured as being randomly distributed among the seats*. Hence, the traveller now has a chance of $1/106$ of getting seat 1, ditto for seat 2, etc. Hence, $P(S) = 1/106 + 1/106 + \cdots + 1/106$ a total of 106 times, or simply $P(S) = 1$, as we know must be true.

Other interesting problems for which (2.5) is used to compute a probability are given in Sects. 2.25 and 2.26.

2.9 Law of Large Numbers

We will diverge temporarily from the axiomatic approach here and simply present the result. A proof is given in Sect. 9.6.

Suppose an experiment is devised that gives rise to an event B, along with other events, the totality of events comprising a space. The experiment is carried through a large number of times. What is the *relative number* of times that event B will occur? This is defined as the *absolute number* of times m that B occurs divided by the number N of experiments performed, and is called the "frequency of occurrence" $f(B)$. Hence, $f(B) = m/N$.

The "law of large numbers" states that $f(B)$ asymptotically approaches $P(B)$ as $N \to \infty$,

$$P(B) = \lim_{N\to\infty} (m/N) \ . \tag{2.8}$$

This allows the theoretical $P(B)$ to be determined *experimentally*, for the first time; see Sect. 9.7 for a discussion of the accuracy of this method when N is finite.

The law of large numbers is often the vital link between abstract probability theory and real measurable phenomena. (The law of large numbers also corresponds to the Von Mises definition of probability, mentioned at the end of Sect. 2.3) The following example illustrates this point.

2.10 Optical Objects and Images as Probability Laws

The law of large numbers allows us to compute probabilities on the basis of *measured observables*. This is a very important property that allows probability theory to be applied to the real world. For example, consider the basic imaging system of Fig. 2.2. Photons originate from within intervals $(x_n, x_n + \Delta x)$ of an incoherent object, of intensity profile $o(x)$, and follow unknown paths through the given lens system to the image plane, where they are detected at intervals $(y_m, y_m + \Delta y)$. The image intensity profile is $i(y)$. What is the probability $p(x_n)$ that a randomly chosen photon will be emitted from an object plane interval $(x_n, x_n + \Delta x)$? Also, what is the probability that a randomly chosen photon will be detected within the interval $(y_m, y_m + \Delta y)$ of the image? The photons have wavelengths that are within a narrow band centered upon a mean value λ:

According to the law of large numbers (2.8), $P(x_n)$ may be represented as the number of events (photon is emitted from interval x_n) divided by the total number of emitted photons. This is providing the latter is a very large number, which is, of course, true in the majority of optical cases. But, in this quasi-monochromatic case, the number of events (photon is emitted from interval x_n) is directly proportional to the energy flux $o(x_n)\Delta x$, and likewise the total number of emitted photons is proportional to $\sum_n o(x_n)\Delta x$. Hence, by (2.8)

$$P(x_n) = o(x_n)/\sum_n o(x_n) .$$

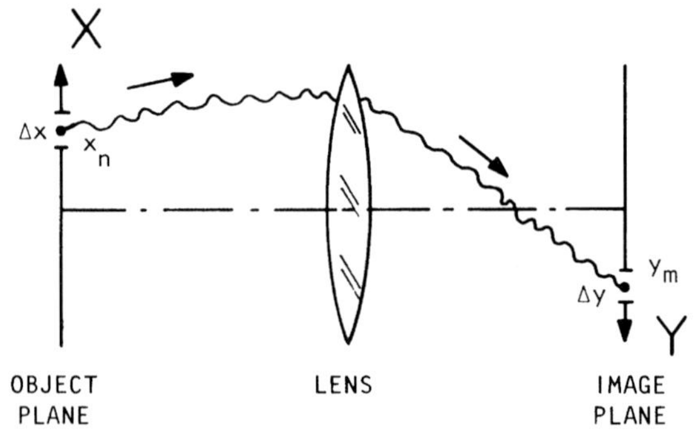

Fig. 2.2. A photon leaves the object plane within interval $(x_n, x_n + \Delta x)$ and strikes the image plane within interval $(y_m, y_m + \Delta y)$

To simplify matters, the denominator may be made to define 1 unit of energy flux. Hence,

$$P(x_n) = o(x_n) \tag{2.9}$$

identically! Or, the probability law on position for a photon in the object plane equals the object intensity distribution. In this manner, the *law of large numbers enables us to infer a probability law from physical observables*. Therein lies the power of the law.

By the same reasoning, the probability $P(y_m)$ for image plane photons obeys

$$P(y_m) = i(y_m) \tag{2.10}$$

if the total image flux is normalized to unity. This would be the case, e.g., if the total object flux is normalized to unity and the lens system is lossless. Equation (2.10) states that the image intensity distribution is equivalent to the probability law on position for a photon in the image plane.

All this has a familiar ring to it, since by quantum mechanics, intensity (actually, energy) patterns statistically represent probability laws *on position* for material particles such as electrons, mesons, etc. However, photons had long been considered *not* amenable to representation in terms of such a *spatial* probability law. Instead, the proper representation was thought to be in *momentum* space. But, on the contrary, photons have recently been found to be representable as energy events in *positional* space [2.12]. The proviso is that the space be partitioned into regions the order of a photon wavelength in size. It must be a "coarse grained" space. This is consistent as well with the modern view that light that has both a wave- and a particle nature. It should be noted that a coarse graining limitation occurs in quantum mechanics as well, where the grain size is the Compton wavelength h/mc of the particle (see Sect. 17.3.11). Here h is Planck's constant, m is the particle mass and c is the speed of light.

2.11 Conditional Probability

Suppose that a trial of experiment E produces an event A. We may ask, what is the probability that it simultaneously produces an event B? That is, what is the conditional probability of event B *if event A has already occurred*? This is defined as

$$P(B|A) \equiv P(AB)/P(A) , \tag{2.11}$$

where $P(B|A)$ denotes the conditional probability of event (B if A). Note: we shall use the word "if" and its notation | interchangeably. Also, as described in Sect. 2.5, $P(AB)$ denotes the *joint probability* of event (A and B).

Figure 2.3 illustrates the conditional phenomenon. Suppose experiment E consists of throwing a dart at the double bullseye shown in the figure. All such darts stick in somewhere within either region A, or region B, or both; the latter denoted as AB (shaded in figure).

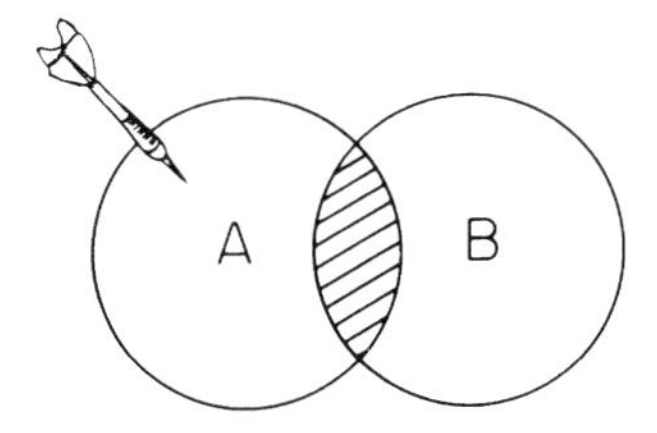

Fig. 2.3. Dartboard consisting of overlapping bullseyes, illustrating the concept of conditional probability

Consider next only those events A, i.e., dart events within region A. Some of these also lie in region AB, while some do not. Those that do, we call events (AB if A). But since these also lie within region B, they are identical with events (B if A) as well, i.e. event $(B|A) = (AB|A)$. Hence, $P(B|A) = P(AB|A)$. *Let us now use* (2.5) *to actually compute $P(AB|A)$.*

The denominator of (2.5) now corresponds to the number of ways an event A can occur, since we are limiting attention to the space of events A by hypothesis.

If a large number M of darts are thrown, this equals $M \cdot P(A)$, by the law of large numbers (2.8). The numerator of (2.5) represents the number of events (AB). This is $M \cdot P(AB)$. Hence, by (2.5), $P(AB|A) = MP(AB)/MP(A) = P(AB)/P(A)$. Finally, by the equality of $P(B|A)$ with $P(AB|A)$ shown previously, we get the result (2.11) once again.

Hence, although (2.11) may be regarded as defining the concept of conditional probability, we see that the concept may alternatively be *derived* from elementary considerations including the law of large numbers.

The relation (2.11) is commonly used in the context of frequency of occurrence. For example, in a certain student body 60% are coeds, and of these, 72% wear jeans daily. What percentage of the student body is both coed and wears jeans? By intuition, the answer is simply the fraction of the student body that is female (i.e., *if* female) times the fraction of these who wear jeans. That is, fraction $0.60 \times 0.72 = 0.432$. This is in direct compliance with the law (2.11), with probabilities replaced by corresponding frequencies of occurrence. Of course, this correspondence is no coincidence; see Sect. 2.3.

Exercise 2.2

2.2.1 Let us return to Fig. 2.2. Use the law of large numbers to relate $P(y_m|x_n)$ for a photon to $s(y_m; x_n)$, the point spread function for the imagery. (Note: the "point spread function" is defined as the image, or spread in photon positions y_m, due to a point-impulse object located at x_n. See Ex. 3.1.23 for further explanation.)[1]

2.2.2 Consider the object plane in Fig. 2.2. The object is incoherent.[2] Let the separation between any successive pair of radiated photons from the object be denoted by t_m. t_m is, of course, a random number. By use of the marginal probability law (2.4) show that $P(t_m)$ is proportional to $\sum_{n=1}^{N} o_n o_{n+m}$, the autocorrelation at lag separation t_m (*Hint:* apply the law (2.4) to the joint probability of event [one photon at position x_n and the other photon distance t_m away, at position $x_{n+m} = x_n + t_m$].)

2.12 The Quantity of Information

Suppose an event B occurs randomly in time. A report about the occurrence of B is generated, and passes along to us by way of a *communication channel* (e.g., a radio link). In turn, we receive a message A which states that event B has occurred. Or, it may state that B *did not* occur. The message may or may not be true. How much "information" is there in message A about the occurrence of event B? Or, to what extent is message A evidence for the occurrence of B?

The situation is illustrated in Fig. 2.4. $P(B)$ is the total probability of occurrence of event B. $P(A)$ is the overall probability of receiving a message A, regardless of event. Conditional probability $P(A|B)$ is a measure of the spread or randomness in

[1] Ex. 3.1.23 is a shorthand for problem 23 of Ex. 3.1. This shorthand will be used throughout the text.

[2] See Ex. 3.1.24.

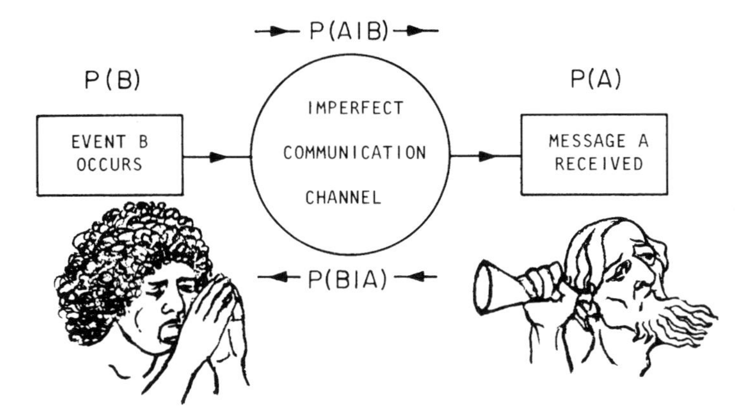

Fig. 2.4. How much information is there in message A about the occurrence of event B, if the channel is imperfect? The channel is specified by either the "forward" conditional probability $P(A|B)$ or by the "backward" equivocation probability $P(B|A)$

the outputs A, given an event B. Conversely, probability $P(B|A)$ specifies the equivocation, i.e., how well we can infer the event B from its given message A. Depending upon circumstance, sometimes $P(A|B)$ is given, and sometimes $P(B|A)$. [Only one need be given because of Bayes' rule (2.23) below.] Either serves to define the channel mathematically.

The quantity of information has been defined differently by *Shannon* [2.3] and by *Fisher* [2.4]. We shall use Shannon's concept in this chapter, as it makes for a wonderful application of the concept of conditional probability. According to Shannon, the information $I(A, B)$ in a message A about an event B obeys

$$I(A, B) \equiv \log_b \frac{P(B|A)}{P(B)} \,. \tag{2.12}$$

The logarithmic base b is arbitrary, and serves to define a unit for I. If base $b = 2$ is used, the unit for I is the "bit." This is by far the commonest unit in use.

By this definition, information I measures simply the probability $P(B|A)$ that event B occurred if message A is received, relative to the probability $P(B)$ that B occurred *regardless of any message*. Does this definition really measure the amount of information contained in message A about the occurrence of event B? It does up to a point, since if the message A contains significant information or evidence about the occurrence of B, once A is known surely the occurrence of B should be more probable than if no message were received. Definition (2.12) has this very property, in fact, since the larger $P(B|A)$ is relative to $P(B)$, the larger is the information I. The following example clarifies this point.

The reliability of weather forecasts may be judged by the use of definition (2.12). Suppose you live in a part of the country where, for the time of year, it rains 25% of the days. The nightly weather forecaster predicts that it will rain the next day. From past performance, it actually does rain when he predicts rain 75% of the time. How much information about the event (occurrence of rain) exists in his forecast of rain?

Analysis: Here message A = (rain forecast), event B = (occurrence of rain), $P(B|A) = 0.75$, while $P(B) = 0.25$. Then, by definition (2.12), $I(A, B) = \log_2(0.75/0.25) = 1.6$ bit. We note that if he were a better forecaster, with $P(B|A) = 1.0$, then $I(A, B) = 2.0$ bit. This is still a finite number, even though the forecaster is perfectly reliable, since the occurrence of rain is fairly common, at 25%. If rain were rarer, say with $P(B) = 1/16$, we would find $I(A, B) = \log(1.0/1/16) = 4.0$ bit, appreciably higher information content.

The last point is very important. The more "unexpected" is an event B, i.e. the smaller $P(B)$ is, the higher is the information content in message A about B. With a perfectly reliable channel, $I(A, B) = -\log_b P(B)$.

We may note that the same analysis may be used to rate any other kind of predictive program. For example, consider medical testing for disease B, where the diagnosis is either A (yes) or $\overline{A}$ (no) (one merely need substitute the word "cancer" for "rain" in the preceding problem to get the idea); or, the identification of weapons site type B from evidence A in an image.

Taking *the ratio* of the probabilities in (2.12) guarantees that $I(A, B)$ will grow unlimitedly with the degree to which $P(B|A)$ exceeds $P(B)$. This unlimited growth also makes sense, since it would be artificial to have $I(A, B)$ approach some asymptote instead.

The logarithm in (2.12) serves to slow down the growth of I with the ratio of probabilities, and more importantly serves to guarantee that the information from "independent" messages add (Sect. 2.16).

The main point of introducing the concept of information here is its intrinsic dependence upon the concept of conditional probability. It makes an interesting, and in fact, useful application of the concept. Some further examples of its use follow.

A "fair" coin [P(head) = P(tail) = 1/2] is flipped by Mr. X. He sends the message "it's a tail" to Mr. Y, over a perfectly reliable channel of communication. How much information about the event "tail" is there in Mr. X's message? Mr. Y knows that the channel is reliable.

Analysis: Since the channel is 100% reliable, P (transmitted "tail" | received "tail") = 1, P (transmitted "head" | received "tail") = 0. Therefore, Mr. Y knows that both event B and message A are "tail," and use of (2.12) yields I (tail, tail) = $\log_2(1.0/0.5) = 1.0$ bit.

In the same way, a binary (2-level) message such as 00111010001100100 where 1's and 0's occur independently with equal randomness and where there is no error present, contains exactly 1.0 bit of information per symbol of the sequence. Binary messages are very commonly used, especially within computers.

2.13 Statistical Independence

If $P(B|A) = P(B)$, then events A and B are termed "independent." This equality describes the situation where the probability of B given the event A is the same as the probability of B when it is *not known* whether event A holds. Hence, event A

really has no effect upon the occurrence of B; they are independent effects. Note that in this case, definition (2.11) becomes

$$P(AB) = P(A)P(B) .$$

This is generalizable to the case where n events $A_1, \ldots, A_n$ are statistically independent. Then

$$P(A_1 A_2 \ldots A_n) = P(A_1)P(A_2)\ldots P(A_n) . \tag{2.13}$$

A note of caution: the converse is not necessarily true. That is, given (2.13) for events $A_1, \ldots, A_n$, it is not necessarily true that these are statistically independent events. To prove their mutual independence from expressions such as (2.13) requires $2^n - (n+1)$ expressions

$$P(A_{k_1} A_{k_2} \ldots A_{k_m}) = P(A_{k_1})P(A_{k_2})\ldots P(A_{k_m}) , \quad m = 2, 3, \ldots, n . \tag{2.14}$$

The concepts of statistical independence and disjointness sound similar. Then if two events A_1, A_2 are disjoint, are they also independent? By the definition of "disjoint" given previously, if event A_2 occurs then A_1 cannot. Or $P(A_1|A_2) = 0$. On the other hand, if A_1 and A_2 are independent $P(A_1|A_2) = P(A_1)$. Hence, unless A_1 is the impossible event (a trivial case) disjoint events *are not* independent events.

2.13.1 Illustrative Problem: Let's (Try to) Take a Trip (Continued)

We found in the illustrative problem of Sect. 2.8 that deliberate overbooking by a certain airline causes a probability $P(\overline{S}) = 1/21$ that a passenger will *not* get his reserved seat when he shows up for a flight. After how many flights is there an even chance that he will have suffered loss of a seat at least once?

Analysis: Consider the two complementary events B and $\overline{B}$. B is the event that the traveller gets his seat on each of n consecutive flights. We want to know n such that $P(\overline{B}) = 1/2$.

Now since $P(\overline{S}) = 1/21$, necessarily $P(S) = 20/21$. Event B is related to event S by

$$B = S \text{ and } S \text{ and } S \text{ and } \ldots \text{ and } S .$$
$$\longleftarrow \quad n \text{ times} \quad \longrightarrow$$

The events S are independent, since the circumstances surrounding each flight are generally different. Hence, independence condition (2.13) holds, and

$$\begin{aligned} P(B) &= P(S)^n \\ &= (20/21)^n . \end{aligned}$$

And so, we have to solve for n

$$P(\overline{B}) = 1 - P(B) = 1 - (20/21)^n \equiv 1/2 .$$

The solution is $n = 14$ flights, to the nearest whole number.

2.14 Informationless Messages

If a received message A is statistically independent of the occurrence of event B, then intuitively message A contains no information about the occurrence of B. Definition (2.12) has this property, since if A and B are independent $P(B|A) = P(B)$ by definition, so that $I(A, B) = \log_b 1.0 = 0$ identically. Here, by the way, is another reason for taking the logarithm.

2.15 A Definition of Noise

The optical layout in Fig. 2.2 may be regarded as defining a communication channel for photon positions. An event x_n gives rise to a message y_m after passing through the communication channel consisting of the lens and intervening medium. Then the quantity[3]

$$I(X, Y) = \log_b \frac{P(y_m|x_n)}{P(y_m)} \tag{2.15}$$

measures the information in observation of an image coordinate y_m about its original, object plane position x_n. We saw from Ex. 2.2.1 that $P(y_m|x_n)$ *physically represents* a spread in image positions y_m about the conjugate point to x_n.

Suppose, now, that the possible positions y_m for a given x_n are the same, irrespective of x_n. That is, $P(y_m|y_n) = P(y_m)$, irrespective of the conjugate point x_n. This is how we defined "statistical independence" in Sect. 2.13. We see from (2.15) that the information is zero. But what, in addition, does the physical situation mean?

If $P(y_m|x_n) = P(y_m)$ this means, by the law of large numbers, that an object source placed at x will after many photon events give rise to an image proportional to $P(y_m)$; an object source at x_2 will give rise to the *same image* $P(y_m)$; etc., across the whole object. This situation describes what we commonly call "noise." That is, the image profile will now have nothing whatsoever to do with the object intensity profile.

This situation could arise if the lens in Fig. 2.2 were of such poor quality that its spread function $s(y; x)$ extends over the entire image detection region. Then the probability law $P(y_m|x_n)$ is effectively $\mathrm{Rect}(y_m/2y_0)$, where $2y_0$ is the image detection region. The observed coordinates y_m would then indeed be independent of their starting points x_n in the object.

In summary, a received message or set of messages (such as photon image positions) which contains *zero information* about a set of events (photon object positions), consists entirely of noise. This is a definition of noise which also shows its arbitrariness: the received message that is independent of one set of events may be dependent upon and describe a different set. For the latter events, the messages are not noise and do contain finite information.

[3] by symmetry condition (2.26)

2.16 "Additivity" Property of Information

A person perusing the morning newspaper, item (message) by item, likes to have the feeling that each new item is *adding* to his storehouse of information about the outside world. This especially makes sense if the individual items are unrelated (statistically independent), and occur in a reliable newspaper. Hence, in general, does information obey a property called "additivity," where the information content in a joint message B_1B_2 equals the information in message B_1 plus that in message B_2?

Since the messages are here assumed to be correct, then events B_1 and B_2 have occurred with unit probability, and from definition (2.12) the information is

$$I(B_1B_2, B_1B_2) \equiv \log_b \frac{1}{P(B_1B_2)} . \tag{2.16}$$

This is the "self information" in message B_1B_2 about the same event B_1B_2. Now, since B_1 and B_2 are given as independent events, the denominator of (2.16) separates into $P(B_1)P(B_2)$, and the logarithm operation produces

$$\begin{aligned} I(B_1B_2, B_1B_2) &= \log_b \frac{1}{P(B_1)} + \log_b \frac{1}{P(B_2)} \\ &= I(B_1, B_1) + I(B_2, B_2) . \end{aligned} \tag{2.17}$$

Hence, the information in the combined message B_1B_2 does equal the information in message B_1 plus that in message B_2.

As an example, given a string of binary numbers 001110010100111010, where the 1's and 0's occur with equal probability and independently of one another, the information in each digit of the sequence is exactly one bit (Sect. 2.12), and the total information in the sequence is simply the number of digits in the sequence times 1 bit, or 18 bits in all.

Hence, statistical independence has a profound effect upon the quantity of information, in two distinct ways: If a message is independent of an event, then the message contains zero information about the event. Or, if a compound message is received describing independent events, and if additionally the message is correct, then the total information is the sum total information in the individual events.

It can be shown that in the more general case, where the messages are not reliably correct, the information from independent messages still adds. An additivity property holds as well for Fisher information. See Ex. 17.3.16.

2.17 Partition Law

Let us start from the marginal probability law (2.4) where events $\{B_n\}$ are disjoint and form a space. Replace $P(A_mB_n)$ by its equivalent product $P(A_m|B_n)P(B_n)$ through definition (2.11) of conditional probability. Then, (2.4) becomes

$$P(A_m) = \sum_{n=1}^{N} P(A_m|B_n)P(B_n) . \tag{2.18}$$

This is called the "partition law," in that event A_m has been partitioned into all its possible component events $\{B_n\}$. It is also sometimes called the "total probability law," in analogy with the total differential law of calculus.

2.18 Illustrative Problem: Transmittance Through a Film

Suppose a light beam composed of three primary colors impinges upon a film. The colors are C_1, C_2, and C_3, or red, blue, and green, respectively. The relative proportions of each color in the beam are $P_1 = 0.4$, $P_2 = 0.5$ and $P_3 = 0.1$ (the overall color is close to purple). Suppose the film energy transmittances for the colors are $t_1 = 0.6$, $t_2 = 0.1$ and $t_3 = 0.3$, respectively. What is the probability that a randomly selected photon will pass through the film?

Analysis: Let B represent the event that a given photon passes through the film. Then we want $P(B)$.

The concept of energy transmittance is intrinsically probabilistic. Suppose a photon to be randomly selected from color types C_1, C_2, or C_3. It has a definite energy that is either $h\nu_1$, $h\nu_2$, or $h\nu_3$ according to its light frequency ν, where h is Planck's constant. For example, let it be $h\nu_1$, corresponding to a red photon (the event C_1). Imagine the photon to be incident upon the film. Since B represents the event that the photon gets through the film, we are interested in $P(B|C_1)$.

By the law of large numbers (2.8) we are to imagine a large number N of *red photons* impinging upon the film, each either causing an event B or not. Then if the total number of these events $(B|C_1)$ is m, the ratio m/N is $P(B|C_1)$ according to the law. But physically, $m/N \equiv mh\nu_1/Nh\nu_1$ is also the net transmitted energy over the total incident energy, or the energy transmittance t_1 for red photons. In summary then,

$$P(B|C_1) = t_1 \ .$$

Likewise,

$$P(B|C_2) = t_2 \ , \quad P(B|C_3) = t_3 \ .$$

Also, by the law of large numbers, the relative number P_1 of red photons must equal $P(C_1)$, the probability of the occurrence of color C_1 for a randomly selected photon. Similarly, $P_2 = P(C_2)$ and $P_3 = P(C_3)$.

Hence, our problem of computing the total probability $P(B)$ of passage through the film has been made conditional upon color. The space of events B has been partitioned into distinct color types. Direct use of the partition law yields

$$P(B) = P(C_1)P(B|C_1) + P(C_2)P(B|C_2) + P(C_3)P(B|C_3)$$

or $P(B) = 0.4(0.6) + 0.5(0.1) + 0.1(0.3) = 0.32$. This is also of course the fraction of all incident photons that pass through the film.

2.19 How to Correct a Success Rate for Guesses

Consider the following experiment in physiological optics. An observer is shown a split field of view, one half of which contains uniform light and the other half a very weak set of fringes. He is to decide which half contains the fringes. The aim of the experiment is to assess the probability p of detecting the fringes. This is to be computed from the experimental detection rate P over many trials.

But, in order to avoid making the observer judge *when* he does and when he does not see the fringes, he is told to always make a decision on their location, *whether he thinks he sees them or not*. (The observer does not always know when he "sees" the fringes, i.e., they are sometimes subliminally detected.) This is called a "forced-choice" experiment.

If the observer's success rate is P, what is the true probability of detection p? Notice that, with many trials, P ought to be about $1/2$ even if the observer never detects the fringes and guesses their location every time. Our objective, then, is to correct the experimental P for guessing.

Analysis: A correct decision can be made because either the fringes are detected or because a correct guess was made. Then by (2.18) the probability P of a correct decision obeys

$$P \equiv P(\text{correct}) = P(\text{correct} \mid \text{detects})\, P(\text{detects}) + P(\text{correct} \mid \text{guesses})\, P(\text{guesses}) \,. \tag{2.19}$$

We may assume that if the observer detects, he will be correct, or P (correct | detects) = 1. Also, in the long run P (correct | guesses) $= 1/2$. Finally, the observer guesses when he doesn't detect, so that $P(\text{guesses}) = 1 - P(\text{detects})$. Then by (2.19)

$$P(\text{detects}) \equiv p = \frac{P - 1/2}{1 - 1/2} = 2P - 1 \,.$$

In the more general case where there are M alternatives to choose from

$$p = \frac{P - 1/M}{1 - 1/M} \,. \tag{2.20}$$

This expression is most commonly used to correct the scores on multiple-choice tests for random guessing, where each question has M possible answers.

Exercise 2.3

2.3.1 For the case of incoherent image formation, as illustrated in Fig. 2.2, we have already identified $P(x_n) = o(x_n)$, $P(y_m) = i(y_m)$ and $P(y_m|x_n) = s(y_m; x_n)$ in terms of physical quantities. Consider arriving photons at image plane positions y_m to be partitioned with respect to possible object plane positions x_n. Then total probability $P(y_m)$ may be computed by use of the partition law (2.18). Show that because of the above physical correspondences this is precisely the law of image formation

$$i(y_m) = \sum_n o(x_n) s(y_m; x_n) \,. \tag{2.21}$$

2.3.2 Consider all the photon arrivals at a given image point y_m. Suppose that fraction f of these occur after passage from the object through the lens, and the remaining fraction $\overline{f} = 1 - f$ occur from an exterior source of photons. The conditional probability of (y_m if x_n) due to the former photons is $s(y_m; x_n)$ as before, while the conditional probability of (y_m if x_n) due to the latter photons is $n(y_m)$. These lack an x_n-dependence since they bypass the object. This constitutes "noise," according to Sect. 2.15. Now, compute the total probability $P(y_m)$, and show that it obeys $P(y_m) = f \sum_n o(x_n)s(y_m; x_n) + \overline{f}n(y_m)$. [Hint: $P(y_m|x_n)$ now is formed disjointly by photons passing through the lens, as one class, and by photons avoiding the lens, as the other class.]

2.20 Bayes' Rule

Intuitively, the word "and" is a commutative operator, so that an event (A and B) is the same as the event (B and A). Then $P(AB) = P(BA)$. Definition (2.11) of conditional probability then yields

$$P(A)P(B|A) = P(B)P(A|B) . \tag{2.22}$$

This is called "Bayes' Rule," after the philosopher and mathematician Thomas Bayes. It is commonly used in the form

$$P(B|A) = \frac{P(B)P(A|B)}{P(A)} \tag{2.23}$$

to compute $P(B|A)$ from a known $P(A|B)$. Why this calculation would be of value is exemplified in Fig. 2.4. The observer is usually at the receiver, not at the transmitter or source of the event. Therefore, upon receiving message A which states that event B has occurred, he really is not certain that event B has occurred. What he has to go on are three valuable pieces of data: the channel specifier $P(A|B)$, the known probability of occurrence of the message A, and the known probability of occurrence of the event B in question. If, on this basis, (2.23) shows that $P(B|A)$ is close to unity, then it is probable that event B has indeed occurred, in agreement with the message. Bayes' rule has many applications, most of which involve "looking backwards" to estimate the probable cause of some observed data.

Bayes' rule is particularly useful under the situation where event B is but one of N disjoint events $B_1, \ldots, B_N$ forming a space, and where A is an event for which one knows the conditional probabilitites $P(A|B_n)$ and also the net $P(B_n)$. Then it takes the form

$$P(B_n|A) = \frac{P(B_n)P(A|B_n)}{P(A)}$$

as in (2.23), but further by use of the partition law (2.18),

$$P(B_n|A) = \frac{P(B_n)P(A|B_n)}{\sum_{m=1}^{N} P(B_m)P(A|B_m)} . \tag{2.24}$$

Notice that the right-hand side only requires knowledge of quantities $P(B_n)$ and $P(A|B_n)$. Optical application (c) below will illustrate such a situation.

2.21 Some Optical Applications

a) We showed in Sect. 2.10 and beyond that the optical system of Fig. 2.2 obeys the physical correspondences $o(x_n) = P(x_n)$, $i(y_m) = P(y_m)$, and $s(y_m; x_n) = P(y_m|x_n)$. Application of Bayes' rule (2.23) then shows that

$$P(x_n|y_m) = \frac{o(x_n)}{i(y_m)} s(y_m; x_n) \; . \tag{2.25}$$

This curious expression gives the backward probability of an object position x_n if the observed image position for the photon is y_m. It measures the "spread" in possible positions x_n, and so represents a kind of "backward looking spread function."[4]

b) Let us return to the photon transmittance problem of Sect. 2.18. B is the event that a photon passes through the film. What fraction of the photons emerging from the film will be spectrally blue?

Analysis: By use of the law of large numbers (2.8), we want quantity $P(C_2|B)$. By Bayes' rule (2.23), $P(C_2|B) = P(C_2)P(B|C_2)/P(B)$, or $P(C_2|B) = P_2 t_2/P(B)$ in terms of given quantities, yielding $P(C_2|B) = 5/32$.

c) A computed tomograph (C. T.) scanner produces x-ray images of the cross-section of a person's brain. These images may be used to diagnose the presence of tumors. Now, such images are highly effective detectors of tumors. For example, suppose $P(Y|T) = 0.95$, where Y is the event that the image discloses a "yes" tumor diagnosis, and T is the event that the subject does have a tumor. Suppose also that $P(\overline{Y}|\overline{T}) = 0.95$ as well, where $\overline{Y}$ is the complementary event to Y, i.e., a "no" diagnosis, and $\overline{T}$ is the complementary event to T, i.e., the absence of a tumor in the subject. These probabilities indicate a quite effective test procedure. Not only are existing tumors detected 95% of the time, but the "false alarm" rate $P(Y|\overline{T})$ of a spurious tumor diagnosis is down to 5% as well. (Note: All probabilities here are hypothetical.)

All well and good, but suppose that a person is unfortunate enough to have the verdict "tumor" from a C. T. scan. What is his actual probability of having a tumor? That is, what is $P(T|Y)$? With a test as effective as described above, we would intuitively expect a high probability for this unfortunate occurrence. But, let us work out the answer by use of Bayes' rule (2.24) and see if the situation is so bleak.

Analysis: Suppose that brain tumors occur in 1% of the population at any one time. Hence $P(T) = 0.01$, and consequently $P(\overline{T}) = 0.99$. Then by (2.24)

$$\begin{aligned} P(T|Y) &= \frac{P(Y|T)P(T)}{P(Y|T)P(T) + P(Y|\overline{T})P(\overline{T})} \\ &= \frac{(0.95)(0.01)}{(0.95)(0.01) + (0.05)(0.99)} = 0.16 \; . \end{aligned}$$

Hence, although the patient has 16 times the chance of having a brain tumor as does the general population, on an absolute basis he still has a small chance of having it!

---

[4] This concept has been used in an image restoration approach [2.6].

This surprising result can be seen to trace from the extremely rare occurrence of the disease, $P(T) = 0.01$, over the general population.

See the related Exs. 2.4.12, 13.

Receiver operating characteristics (roc): Probability $P(Y|T)$ is commonly called the probability of *detection* of the diagnostic system, and is denoted as P_D. Likewise $P(Y|\bar{T})$ is called its false alarm rate, and is denoted as P_F. An ideal system would have P_D close to unity and P_F close to zero. However, in practice these values depend upon the system's operating conditions, such as the quality of the data and the decision rule in use. A plot of P_D vs. P_F under varying operating conditions is often called the *receiver operating characteristic* curve of the diagnostic system; see, for example, *Van Trees* [2.10].

2.22 Information Theory Application

A direct substitution of Bayes' rule (2.23) into definition (2.12) of information casts it in the alternate form

$$I(A, B) = \log_b \frac{P(A|B)}{P(A)} \equiv I(B, A) \, . \tag{2.26}$$

There is a basic symmetry between this result and definition (2.12). In words, the information existing in a message about the occurrence of an event B equals the information in observation of the occurrence of the event about the existence of the message. More simply, information is commutative in the roles played by message and event.

2.23 Application to Markov Events [2.6]

There are many situations where the probability of an event changes from trial to trial. If the probability of an event depends upon the outcome of the preceding trial, the sequence of events is called "Markov," after the Russian mathematician A.A. Markov. An example follows.

Two products A and B are competing for sales. The relative quality of these products are such that if a person buys A at trial n, on his next purchase he will tend to buy A again with high probability, $P(A_{n+1}|A_n) = 0.8$. The trend for B is not nearly as strong, $P(B_{n+1}|B_n) = 0.4$. In the long run, i.e., $n \to \infty$, how often will A be bought, and how often will B be bought?

Analysis: We want

$$\lim_{n\to\infty} P(A_n) \equiv P(A)$$

and

$$\lim_{n\to\infty} P(B_n) \equiv P(B) \, .$$

By the Markov nature of the situation, purchase $n+1$ depends upon what was purchased at n. Hence, by partition law (2.18)

$$\begin{aligned} P(A_{n+1}) &= P(A_{n+1}|A_n)P(A_n) + P(A_{n+1}|B_n)P(B_n) \,, \\ P(B_{n+1}) &= P(B_{n+1}|A_n)P(A_n) + P(B_{n+1}|B_n)P(B_n) \,. \end{aligned} \tag{2.27}$$

We also know from what was given that $P(A_{n+1}|B_n) = 0.6$ and $P(B_{n+1}|A_n) = 0.2$. Then

$$\begin{aligned} P(A_{n+1}) &= 0.8P(A_n) + 0.6P(B_n) \,, \\ P(B_{n+1}) &= 0.2P(A_n) + 0.4P(B_n) \,. \end{aligned} \tag{2.28}$$

This comprises two coupled difference equations whose solution could be found at any n, provided initial conditions $P(A_0)$, $P(B_0)$ were given. However, our problem is even simpler, in that we want the "equilibrium" answer that obtains as $n \to \infty$. Equilibrium means that

$$\begin{aligned} \lim_{n\to\infty} P(A_{n+1}) &= P(A_n) \equiv P(A) \,, \\ \lim_{n\to\infty} P(B_{n+1}) &= P(B_n) \equiv P(B) \,, \end{aligned}$$

Then (2.28) become

$$\begin{aligned} P(A) &= 0.8P(A) + 0.6P(B) \,, \\ P(B) &= 0.2P(A) + 0.4P(B) \,. \end{aligned}$$

Both of these imply that $0.2P(A) = 0.6P(B)$, and we seem to have an impasse. But, of course, in addition

$$P(A) + P(B) = 1 \,.$$

Combining the last two relations produces the solution,

$$P(A) = 0.75 \,, \qquad P(B) = 0.25 \,.$$

Notice that this result is independent of what the purchaser bought initially. After an infinite number of purchases, the original purchase does not matter.

Many physical situations may be modelled in this way. For example, a communication link receives temporally one of two signals, dot or dash, with probabilities dependent upon the preceding received signal. Or, 26 signals A through Z are received according to their predecessors and the conditional probabilities that define average use of the language. Or, 8 gray levels are received in raster scan fashion according to the conditional probabilities that define a given picture.

Another interesting application of Markov events is given in Ex. 2.4.14.

2.24 Complex Number Events

An event A might be the occurrence of a random number $X + \mathrm{j}Y$, $\mathrm{j} = \sqrt{-1}$. How can one speak of the probability of a complex number?

Complex numbers are easily accommodated into the theory by merely regarding them as real number couplets. For example, with $A = X + \mathrm{j}\,Y$, by $P(A)$ we really mean the joint probability $P(X, Y)$ of X and Y. Since X and Y are both real numbers, we have thereby circumvented the problem of a complex number in the argument of P.

Probability $P(X, Y)$ follows all the usual laws for joint randomness. For example, if real and imaginary parts of A are independent, $P(X, Y) = P(X)P(Y)$ as usual.

2.25 What is the Probability of Winning a Lottery Jackpot?

Here are the rules for winning the jackpot of a typical state lottery (that of Arizona). Other state lotteries may be analyzed similarly. 42 balls are contained within a box, each ball with a different number in the range 1–42 painted on it. Six balls are chosen by random from the box, typically by tumbling it until six eventually fall out through a small hole. (Usually the selection is shown on television, so as to keep interest up and to maintain credibility.) A viewer plays the game by choosing six numbers in the same range 1–42. The choice can be either purposeful or by chance. If these six numbers match up with the six that were chosen on TV, regardless of order, then the play wins. What is the probability that a given play will win?

Analysis: Let m_i represent any ball number, a number in the range 1–42. Let the event B = (a win). A play of the game is a six-dimensional, or "sextet" event $(m_1, \ldots, m_6)$ of ball numbers. In a fair game each sextet is equally probable to be the event B. This satisfies the groundrules for the use of (2.5) as the method of computing the probability of winning. In (2.5) the numerator n is to define the number of ways the event B can occur. It is true that there is only one winning sextet. However, the six numbers can be received in any order. The number of such orderings is then n. It may be found as follows (see also Sect. 6.1):

Consider six boxes, each of which can contain one of the winning numbers. The first can contain any one of the 6. This leaves 5 left to allocate. The second box can contain any of the 5 for *each* of the 6 possible in the first box. This amounts to $6 \cdot 5 = 30$ combinations of the two. Similarly, the third box can contain any one of the 4 for each of the $6 \cdot 5$ combinations in the first 2 boxes; etc. The result is a number $n = 6!$ of possible orderings of the six winning numbers. This is the number of different ways it is possible to win.

Now we consider the denominator N. This is the total number of different possible sextets that can be obtained in choosing six different numbers in the range 1–42. The same counting approach as the preceding may be taken. Hence there are again 6 boxes. The first box can contain any one of 42 numbers. With such a number placed in that box, this leaves 41 left, any one of which can be placed in the second box; etc. The result is a number $42 \cdot 41 \cdots 37$ of different possible sextets. This includes all possible sets of *different* numbers as well as all possible orderings of them. All should be counted since all have equal significance in any play of the game.

By (2.5) the result is a probability

$$P(B) = \frac{n}{N} = \frac{6!}{42 \cdot 41 \cdots 37} \approx \frac{1}{5.24 \times 10^6} .$$

Since the population of the state in question is roughly 5.0×10^6 these are appropriate odds for producing a winner about every other play. See Ex. 2.4.17.

2.26 What is the Probability of a Coincidence of Birthdays at a Party?

At a party of M people, what is the probability that two or more people have the same birthday (in the usual sense of the same month and day)? This has a surprisingly positive answer, and can win the reader some money since people at a party are usually willing to bet against the coincidence.

Analysis: Here the event $B =$ (1 or 2 or ... M coincidences of birthday) over the population of M is quite complex to consider directly, since it includes all possible numbers of coincidences *of all types* – pairs, trios, etc. It is obviously easier to work with the unique event $\bar{B}$ of no coincidence. Also, $\bar{B}$ is the event that all birthdays $bd_i, i = 1, \ldots, M$ in the given party are *different*.

Make the simplifying assumption that every day of the year is equally probable to be the birthday of a given person. Then every possible set of birthdays $(bd_1, bd_2, \ldots, bd_M)$ – call it the "birthday vector" – can occur with equal probability. Then we may use (2.5) to compute $P(\bar{B}) = n/N$ where n is the number of birthday vectors that incur no coincidences among its elements, and N is the total number of different possible birthday vectors.

It is easier to compute N first. Here each of the M elements bd_i is simply any number from $1-365$, the number of days in a year (excluding leap years). This amounts to $365 \cdot 365 \cdots 365 = 365^M$ possible party vectors (see Sect. 2.25 preceding). However, this counts a mere reordering of the M birthdays of a given vector as a new birthday vector. Clearly it should not be so counted. This amounts to overcount by a factor $M!$ (see Sect. 2.25). Thus, $N = 365^M/M!$

By the same counting procedure as in Sect. 2.25, there are 365 different possible birthdays for a given person in the party, and for no coincidence 364 possible remaining birthdays for a second, 363 for a third, etc., for a total of $365 \cdot 364 \cdots (365 - M + 1)$ different possible party vectors incurring no coincidence. However, as in the preceding paragraph, this counts mere reorderings of the individual birthdays as counting toward the total number. Therefore the number should again be divided by $M!$ to obtain n.

Using these values of n and N in (2.5), the factors $M!$ cancel and we find that the probability of no coincidence is $P(\bar{B}) = [365 \cdot 364 \cdots (365 - M + 1)]/365^M$. Therefore the probability of any coincidence is its complement,

$$P(B) = 1 - \frac{365 \cdot 364 \cdots (365 - M + 1)}{365^M} .$$

This is surprisingly large even for modest sized parties. As examples, $P(B) = 0.71$ for $M = 30$, $P(B) = 0.80$ for $M = 34$, and even for as few as $M = 23$ people $P(B)$ exceeds 0.50.

Exercise 2.4

2.4.1 The probability of an individual contracting cancer sometime in his life-time has been estimated as 0.24. A group of three people are discussing this fact. What is the probability of at least one of them contracting the disease eventually?

2.4.2 We found in Sect. 2.13.1 that deliberate overbooking by a certain airline will cause a passenger to more likely than not lose his seat at least once if he takes 14 or more flights. After how many flights is there a 75% chance that he will never have suffered loss of a seat?

2.4.3 In Sect. 2.21 c, how much information in bits is there in the message Y about the event T? Can the diagnostic procedure be improved enough, by enlarging $P(Y|T)$, to increase the information content by a factor of two?

2.4.4 The crime rate in Tucson is currently 10%, meaning that the probability of a given resident being a crime victim in a given year is 0.10. After how many years and months is it *likely* ($P > 0.50$) that a given resident will have been a crime victim at least once?

2.4.5 A bag contains 100 coins. Each coin is biased in one of two ways toward obtaining a head when flipped. There are 60 coins for which $P(\text{head}) = 1/2$, and 40 for which $P(\text{head}) = 2/3$. A coin is randomly chosen from the bag. What is its probability of a head on a given flip?

2.4.6 You must decide whether to take an innoculation shot for a certain disease. The shot should ideally consist entirely of dead bacteria. Ninety percent of all shots do, and for them the disease rate among recipients is reduced to 5%. However, 10% of all shots contain enough live bacteria to actually cause the disease: the disease rate is 50% among its recipients.
(Note: these are hypothetical numbers.)

(a) If you take the shot, what is the probability of getting the disease?

(b) Given that the disease rate is 10% among *non* shot-takers, should you opt to take the shot?

2.4.7 The attendees at a certain meeting of the Optical Society of America consist of 10% university professors, 30% industrial researchers, 20% government researchers, 10% students and 30% salesmen. A bellboy finds that the relative number of tippers in these groups are, respectively, 80%, 95%, 70%, 20%, and 90%. What is the probability that he will be tipped for any one job?

2.4.8 In a city where it rains 40% of the days on average, an avid reader of the daily weather forecast in the local newspaper has taken a statistic on how often the forecast is right. He finds that if rain is predicted for a given day it does rain 80% of such days. Then, how much information in bits is contained in a given forecast of rain? (Hint: B is the event that rain occurs, message A is the forecast of rain.)

2.4.9 This exercise will show that film density D represents information I in a certain sense. Suppose the energy transmittance of a certain film is t. Film density D is defined as $D = -\log_{10} t$. The probability $p(B)$ of the event $B =$ (photon passes through the film) is also t, by the law of large numbers. Suppose that a detector notes the arrival of each photon that passes through the film by causing a light to flash. The detector is perfect, in that the light flashes if and only if a photon does pass through. Show then that the information I in the message $A =$ (light flashes) about the event B obeys $I = D$ precisely, if base 10 is used in definition (2.12) of information. I then has the unit of "hartley," in honor of R.V.L. Hartley, the earliest proponent of a logarithmic measure of information [2.5].

2.4.10 A binary communication link is capable of receiving and transmitting 0's and 1's only. It is tested for reliability by repeated observation of its inputs and corresponding outputs. It is observed to output a 1 when 1 is the input 0.90 of the time. Also, a 1 is input to it 0.50 of the time, and it outputs a 1 0.55 of the time. What then is the probability of an output 1 from the link being correct, i.e., what is $P(1 \text{ input} \mid 1 \text{ output})$?

2.4.11 A communication link transmits 5 intensity levels of information per pixel of a picture formed by a spacecraft. These are named, in order of increasing intensity values v_1, v_2, v_3, v_4, v_5. The data are received with random errors, of course. These may be described as the following frequency of occurrences coefficients $P(v_m \text{ received} \mid v_n \text{ sent})$:

		← received →				
	$P(v_m \mid v_n)$	v_1	v_2	v_3	v_4	v_5
↑	v_1	0.4	0.1	0.1	0.1	0.3
	v_2	0.2	0.5	0.2	0.1	0
sent	v_3	0.3	0.3	0.3	0.1	0
	v_4	0	0.1	0.2	0.6	0.1
↓	v_5	0.3	0	0	0	0.7

For the class of pictures transmitted, level v_n occurs $P(v_n)$ of the time as follows:

n	v_n	$P(v_n)$
1	0	0.1
2	0.2	0.2
3	0.4	0.4
4	0.6	0.2
5	0.8	0.1

(a) Are the given $P(v_m \mid v_n)$ mathematically consistent probabilities?
(b) Find the vector $P(v_m \text{ received})$, $m = 1, \ldots, 5$. What does this represent?

(c) Construct the table of "backward" probabilitites $P(\upsilon_n \text{ sent} \mid \upsilon_m \text{ received})$.
(d) If a datum υ_m is received, the quantity

$$\sum_{n=1}^{5} \upsilon_n P(\upsilon_n \text{ sent} \mid \upsilon_m \text{ received})$$

defines an estimate $\hat{\upsilon}_{PM}$ of what was transmitted at the spacecraft. If υ_m(received) $= 0.2$, find $\hat{\upsilon}_{PM}$. This estimate is called the "posterior mean."

(e) If a datum υ_m is received, the value υ_n(sent) which maximizes the curve $P(\upsilon_n \text{ sent} \mid \upsilon_m \text{ received})$ is called the "maximum *a posteriori* estimate of $\hat{\upsilon}_{MAP}$". If υ_m(received) $= 0.2$, find the estimate $\hat{\upsilon}_{MAP}$ in this way. Compare with the answer to (d).

2.4.12 We found in Sect. 2.21 c that with a "yes" diagnosis, a patient has a probability of a tumor which is 16 times that of the public at large. What would be his probability of having a tumor if there were instead a "no" diagnosis?

2.4.13 It is usual to repeat a diagnostic test if at first it gives a positive result. Suppose that the patient of Sect. 2.21c who obtained a "yes" diagnosis is tested again, and once again obtains a "yes" diagnosis. *Now* what is his probability of having a tumor? (Hint: $P(YY|T) = P(Y|T)P(Y|T)$, since the tests are independent.)

On the other hand, the repeated test might be negative. Show that if it is, the patient's probability of having a tumor is reduced to that of the general population once again. In other words, from the standpoint of probabilities, the combination of a positive and then a negative verdict is equivalent to never having been tested. Note that this reasoning may not completely mollify the patient, however.

2.4.14 This exercise deals with the probability of ultimately winning, or, random walk on a finite interval [2.7].

There are many situations where two adversaries compete for a limited resource. For example, A and B have a total capital of N dollars (or other units) between them, and play games of cards with a stake of one dollar per game *until one of them has the entire capital* N. (This is aptly called "gambler's ruin.") If A wins a given game, he gains one dollar and B loses one dollar. If A starts out with n dollars, and his probability of winning a game is p, what is his probability P_n of ultimately winning all N dollars and taking his opponent broke?

A parallel situation arises in the problem of random walk on a finite interval. There are N positions that a particle may occupy on the interval. It starts at position n. Its progress is stepwise governed by the outcome of flipping a biased coin, with p the probability of a "head." With a "head" outcome, the particle proceeds one step to the right; if a "tail," it goes one step to the left. What is its probability P of reaching endpoint N before it reaches the opposite endpoint 0?

Analysis: Let "Wins" denote the event that A takes the entire capital N. Let "wins" denote the event that A wins a given game. A key point to understand is that each game is independent of the others, so that A's probability of "Winning" is P_n if he has n dollars at *any* given game. Consider such a game.

If A wins this game his capital is $n + 1$; if he loses, it is $n - 1$. He wins with probability p, and loses with probability $1 - p \equiv q$. Then by the partition law (2.18)

$$P(\text{Wins}) \equiv P_n = P(\text{Wins} \mid \text{wins})p + P(\text{Wins} \mid \text{loses})q\ .$$

Evidently $P(\text{Wins} \mid \text{wins}) = P_{n+1}$ since his capital is bumped up by 1 unit; and conversely $P(\text{Wins} \mid \text{loses}) = P_{n-1}$ since he has lost 1 unit. Then

$$P_n = P_{n+1}p + P_{n-1}q\ . \tag{2.29}$$

This is a difference equation, which is supplemented by the boundary values

$$P_0 = 0\ , \quad P_N = 1\ . \tag{2.30}$$

These state, respectively, that if A has zero capital, his probability of Winning is 0 (he has already Lost); while if A has capital N, he has all the capital and has Won for sure. States $n = 0$ or N are called "trapping states," since they stop the process.

The solution to the system of Eqs. (2.29, 2.30) is obtained by recognizing that it resembles a homogeneous differential equation. This tempts us to try an exponential solution, of the form

$$P_n = a(1 - b^n)\ , \tag{2.31}$$

with a and b to be determined.

(a) Show that a and b may be found by the use of (2.31) in boundary conditions (2.30) applied to (2.29). In this way, obtain the solution

$$P_n = \frac{1 - (q/p)^n}{1 - (q/p)^N}\ , \quad n = 0, 1, \ldots, N\ . \tag{2.32}$$

(b) Note that solution (2.32) becomes indeterminate if the game is "fair," i.e., $p = q = 1/2$. Then use l'Hopital's rule to show that for the fair game

$$P_n = n/N\ . \tag{2.33}$$

Solution (2.32) shows interesting trends. If $p > 1/2$, A has a very strong chance of Winning provided n is not too small compared to N. This means that it is to his advantage to risk a significant amount of capital n. The latter becomes an even stronger need if $p < 1/2$, where A has a *disadvantage* each game. If A has a large enough fraction n/N of the total capital, *his* P_n *can still exceed* $1/2$. For example, if $p = 0.4$, which stands for a significant disadvantage, if $N = 5$ total, then with $n = 4$ his probability of Winning is still 0.616.

Solution (2.33) for the fair game shows this tendency to favor high capital n even stronger. Directly, the larger n/N is, the higher is A's probability of Winning. "The rich get richer."

2.4.15 Show by the use of definition (2.11) that the probability of the joint event $A_1, A_2, \ldots, A_m$ obeys

$$P(A_1, \ldots, A_m) = P(A_1)P(A_2|A_1)P(A_3|A_1, A_2) \ldots P(A_m|A_1, \ldots, A_{m-1}) \,. \tag{2.34}$$

Show that if $A_1, \ldots, A_m$ is a *Markov events sequence*, (2.34) becomes simply

$$P(A_1, \ldots, A_m) = P(A_1)P(A_2|A_1)P(A_3|A_2) \ldots P(A_m|A_{m-1}) \,. \tag{2.35}$$

2.4.16 ***The a priori probability of intelligent life in the universe.*** This depends critically upon the joint presence of seven factors. This joint probability may be expressed as a product (2.34) of $m = 7$ conditional probabilities. A plausible *smallest value* for each such probability is given below [2.11]:

Fraction of stars having planets	$P = 10^{-2}$
Fraction of these having at least one habitable planet	10^{-1}
Fraction of these that last long enough for life to evolve	10^{-1}
Fraction of these for which life does evolve	10^{-1}
Fraction of these that last long enough for intelligence to evolve	10^{-1}
Fraction of these for which intelligence does evolve	10^{-1}
Fraction of these for which intelligence lasts to the present	10^{-7}

We can use (2.34) only if we can associate events in the table with conditional events in (2.34). *Make the association.*

The product in (2.34) gives $P = 10^{-14}$ for the probability of the present existence of intelligent life associated with a random star. A modest assessment of the number of stars in the universe is 10^{30}. Then by (2.8) the *minimum expected number* of intelligent "events" is about 10^{16}. This is a large number! However, because of the great distances between stars in the universe, the closest such civilization would probably be about 10 million light years away. (See the related Ex. 6.1.27.) This is too far away to be within our galaxy the Milky Way and, so, too distant to communicate with. Intelligent life may exist somewhere out there, but we can't know it or make any use of it, given today's technology.

2.4.17 ***The chance of winning by repeated play.*** The probability of winning a certain game on a given play is $p = 1/N$, where N is an integer. Knowing this, a player guesses that he is guaranteed to win if he plays the game N times. In fact, *what is* the probability P of winning at least once in N plays? What is the value of N for which P is maximized? Show that in the limit as $N \to \infty$, $P \to 1 - 1/e \approx 0.64$. *Hint:* $\ln(1+\epsilon) \approx \epsilon$ for ϵ small.

2.4.18 ***Connectivity by association.*** It has been estimated that each person on earth has knowledge of each other person through at most 6 intermediate people. Thus person A knows person G because (A knows B who knows ... who knows G). Assuming that each person knows randomly 100 others, in 6 links each person connects with $100^6 = 10^{12}$ people. Since this much exceeds the population of earth, assumed for our purposes to be 4×10^9, it is improbable that a given person *will not be* included within that number of people. *What is this probability?* In general, the number of links required for such an association is called the "Erdos number." See [2.13] for further on the concept of connectivity.

3. Continuous Random Variables

Until now, all experiments E have had discrete events $\{A_n\}$ as their outputs, as in rolls of a die. On the other hand, everyday experience tells us that continuously random events often occur, as in the waiting time t for a train, or the position x at which a photon strikes the image plane.

Continuously random variables may also extend to infinite values. Consider, e.g., the waiting time t in a doctor's office.

3.1 Definition of a Random Variable

We therefore define a random variable (RV) x as the outcome of an experiment which can have a continuum of possible values $-\infty \leq x < +\infty$. Let us now consider the event A that $a \leq x \leq b$, written $A = (a \leq x \leq b)$. What is its probability $P(A)$?

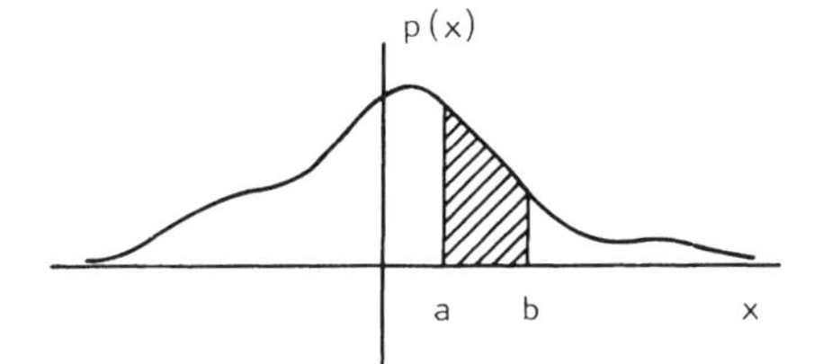

Fig. 3.1. Probability density curve $p(x)$. The shaded area is $P(a \leqq x \leqq b)$

For this purpose it is necessary to introduce the concept of a "probability density" function $p(x)$, defined to obey

$$P(A) = \int_a^b p(x)\,\mathrm{d}x \ . \tag{3.1}$$

This is required to hold true for *all* intervals (a, b) where $a \leqq b$. See Fig. 3.1, where $P(A)$ is the shaded area. Then what properties must $p(x)$ have?

3.2 Probability Density Function, Basic Properties

Taking extreme values for the interval (a, b) unravels the significance of the new function $p(x)$. First, consider the case $a = -\infty$, $b = +\infty$. Now A is the event

$(-\infty \le x \le \infty)$, which is of course the *certain* event (Sect. 2.1). Then, by Axiom II (Sect. 2.2)

$$P(-\infty \le x \le \infty) = P(C) = 1 = \int_{-\infty}^{\infty} \mathrm{d}x\, p(x)\ . \tag{3.2}$$

This is a normalization property analogous to (2.3).

Next consider the opposite extreme, where $b = a + \mathrm{d}x$, $\mathrm{d}x > 0$. Then by (3.1)

$$P(a \le x \le a + \mathrm{d}x) = p(a)\,\mathrm{d}x = p(x)|_{x=a}\,\mathrm{d}x\ , \qquad \mathrm{d}x > 0\ . \tag{3.3}$$

This shows that $p(x)\,\mathrm{d}x$ is *actually a probability*, and that therefore $p(x)$ is a probability *density*. In fact, (3.3) holds true for a general vector $\boldsymbol{x}$ of random variables, where $p(\boldsymbol{x})$ is the probability density function for the joint behavior of $x_1, x_2, \ldots, x_n$, and $\mathrm{d}\boldsymbol{x} = \mathrm{d}x_1\,\mathrm{d}x_2 \ldots \mathrm{d}x_n$. The derivation of the vector form of (3.3) follows an obvious generalization of steps (3.1) through (3.3).

Also, if $z = x_1 + \mathrm{j}x_2$ is a *complex* random variable, then by Sect. 2.24 and the preceding, $p(x_1, x_2)$ is the probability density for the occurrence of z. It is handled as simply the particular case $n = 2$ of the theory. Finally, for joint complex variables $z_1 = x_1 + \mathrm{j}x_2$, $z_2 = x_3 + \mathrm{j}x_4$, etc., the appropriate probability law is $p(x_1, x_2, x_3, x_4, \ldots)$, or $p(\boldsymbol{x})$, with n even.

The consequence of $p(\boldsymbol{x})\,\mathrm{d}\boldsymbol{x}$ being a probability is that *all the preceding laws from the Axioms through (2.24) are also obeyed by a probability density* $p(x)$, with suitable modification (sums replaced by integrals, etc.). Key ones are listed:

$$0 \le p(\boldsymbol{x}) \le \infty\ ; \tag{3.4}$$

$$p(\boldsymbol{x}|\boldsymbol{y}) = p(\boldsymbol{x}, \boldsymbol{y})/p(\boldsymbol{y})\ , \tag{3.5}$$

defining the conditional probability density $p(\boldsymbol{x}|\boldsymbol{y})$ where $(\boldsymbol{x}, \boldsymbol{y})$ means $\boldsymbol{x}$ *and* $\boldsymbol{y}$, $\boldsymbol{y} = y_1, y_2, \ldots, y_m$;

$$p(\boldsymbol{x}|\boldsymbol{y}) = p(\boldsymbol{x}) \quad \text{and} \quad p(\boldsymbol{x}, \boldsymbol{y}) = p(\boldsymbol{x})p(\boldsymbol{y}) \tag{3.6}$$

if $\boldsymbol{x}$ and $\boldsymbol{y}$ are *statistically independent;*

$$p(\boldsymbol{x}) = \int_{-\infty}^{\infty} \mathrm{d}\boldsymbol{y}\, p(\boldsymbol{x}|\boldsymbol{y})p(\boldsymbol{y})\ , \tag{3.7}$$

a *continuous partition law;* and

$$p(\boldsymbol{x}) = \int_{-\infty}^{\infty} \mathrm{d}\boldsymbol{y}\, p(\boldsymbol{x}, \boldsymbol{y})\ , \tag{3.8}$$

the *marginal* probability density $p(\boldsymbol{x})$.

It is interesting that an alternative to (3.1) exists for defining $p(x)$. This is to regard the real line $-\infty \le x \le \infty$ of event space as consisting of disjoint events $x_n \equiv n\,\mathrm{d}x$, $-\infty \le n \le \infty$, and to *define* $p(x)$ to obey $P(x_n) \equiv p(x_n)\,\mathrm{d}x$. Then note that the event $A \equiv (a \le x \le b)$ may be expressed as $x = (a$ or $a + \mathrm{d}x$ or $a + 2\mathrm{d}x$ or $\ldots$ or $b - \mathrm{d}x$ or $b)$. As each of these alternatives is disjoint, by additivity (2.2)

$$P(a \le x \le b) = P(a) + P(a + \mathrm{d}x) \cdots + P(b)$$
$$= \sum_{x_n=a}^{b} p(x_n)\,\mathrm{d}x = \int_a^b p(x)\,\mathrm{d}x\;,$$

which is the previous definition (3.1) but now derived.

3.3 Information Theory Application: Continuous Limit

The concept of information was defined for discrete events and messages, at (2.12). We can now extend it to a continuous situation. Suppose that the message A is a quantity y lying in the range $a \le y \le a + \mathrm{d}y$, and the event B is a quantity x lying in the range $b \le x \le b + \mathrm{d}x$. For example, the event B is a continuously variable intensity value x, and the message A is its reading y from a phototube. These do not have to be equal because of noise and other effects in the phototube. How much information $I(A, B)$ is there in the message A about the event B?

Now $P(A) \equiv P(a \le y \le a + \mathrm{d}y) = p(y)\mathrm{d}y$ by (3.3), and similarly $P(B) \equiv P(b \le x \le b + \mathrm{d}x) = p(x)\mathrm{d}x$. Also, $P(A|B) \equiv P(A, B)/P(B) = p(x, y)\,\mathrm{d}x\,\mathrm{d}y/p(x)\,\mathrm{d}x$. But by identity (3.5) $p(x, y) = p(y|x)p(x)$. Then combining results,

$$P(A|B) = p(y|x)\,\mathrm{d}y\;.$$

Then by (2.26),

$$I(A, B) \equiv \log \frac{P(A|B)}{P(A)} = \log \frac{p(y|x)}{p(y)}\;, \tag{3.9}$$

the common factor $\mathrm{d}y$ having cancelled.

That (3.9) is devoid of differentials $\mathrm{d}x$ or $\mathrm{d}y$ is important. Quantities such as $\log \mathrm{d}y$ or $\log \mathrm{d}x$ are unbounded and would be difficult to accommodate in the theory. We note that the differentials do not exist in (3.9) because of the ratio of probabilities taken in the definition (2.12). This is yet another good property of this particular definition of information.

In summary, the continuous version (3.9) of information is strictly analogous to the discrete definition (2.12), continuous probability densities having replaced discrete probabilities.

3.4 Optical Application: Continuous Form of Imaging Law

It was found in Ex. 2.3.1 that the discrete law of image formation $i(y_m) = \sum o(x_n)s(y_m; x_n)$ follows the identification of probability law $P(y_m)$ with the image $i(y_m)$, probability $P(x_n)$ with the object $o(x_n)$, and probability $P(y_m|x_n)$ with the point spread function $s(y_m; x_n)$. This assumed, as in Fig. 2.2, a discrete subdivision of points y_m, with separation Δy in image space, and a discrete subdivision of points

x_n, with separation Δx in object space. What happens to the law of image formation as both $\Delta x \to 0$ and $\Delta y \to 0$?

Evidently, all probabilities will approach probability densities, according to result (3.3). Moreover, by similar arguments to those in Sect. 2.10, the probability density $p(x)$ for object photon position x must now equal the object density of flux (intensity) law $o(x)$; $p(y)$ should equal the image density law $i(y)$; and $P(y|x)$ should equal $s(y; x)$, the point spread function. Therefore, we may use the continuous version (3.7) of the partition law with these quantities, to find that

$$i(y) = \int_{-\infty}^{\infty} \mathrm{d}x\, o(x)\, s(y; x)\ , \tag{3.10}$$

which is the well-known imaging law for a continuous object.

This continuous law is the usual one quoted, since it includes the discrete law as a special case. If the object is discrete, it may be represented as a sum of weighted Dirac delta functions, $o(x) = \sum_n o_n \delta(x - x_n)$. Substitution into (3.10) produces, by the sifting property of the delta functions, a discrete law once again.

The latter operations are a particular example of the transition from continuous probability laws to discrete ones, as taken up in Sect. 3.12 below.

3.5 Expected Values, Moments

Suppose the outcome of each experiment E is operated upon by a fixed function f. First consider discrete experiments. If $\{A_n\}$ are the elements of event space, an outcome event A_n gives rise to quantity $f(A_n)$. Then the mean or expected value of f is defined by

$$\langle f \rangle \equiv \sum_{n=1}^{N} f(A_n) P(A_n)\ . \tag{3.11}$$

By analogy, for continuous RV's $\boldsymbol{x} \equiv (x_1, \ldots, x_n)$,

$$\langle f \rangle = \int_{-\infty}^{\infty} \mathrm{d}\boldsymbol{x}\, f(\boldsymbol{x})\, p(\boldsymbol{x}) \tag{3.12}$$

defines the mean value of a function f of $\boldsymbol{x}$.

The function f may be any well-defined function. Consider the special cases where $f(\boldsymbol{x}) = x^k$ for a single random variable x, or $f(A_n) = x_n^k$ for discrete events $A_n = x_n$ with k an integer. Then the expected value is the kth moment m_k,

$$\left\langle x^k \right\rangle \equiv m_k \equiv \int_{-\infty}^{\infty} \mathrm{d}x\, x^k\, p(x) \tag{3.13a}$$

or

$$\left\langle x_n^k \right\rangle \equiv m_k \equiv \sum_{n=0}^{\infty} x_n^k P_n\ , \quad k = 1, 2, \ldots\ . \tag{3.13b}$$

The particular number $m_1 \equiv \langle x \rangle$ is usually called the "mean-value of x" or the "mean" of the probability density $p(x)$.

The *central* moments μ_k are defined as

$$\mu_k \equiv \int_{-\infty}^{\infty} \mathrm{d}x (x - \langle x \rangle)^k p(x) \tag{3.14}$$

and similarly for the discrete case. These are the moments of $p(x)$ about axes located at the mean of $p(x)$. The particular moment μ_2 is termed the *variance* and is denoted as σ^2. The square root of this quantity, σ, is called the *standard deviation.*

Direct use of (3.14) for $k = 2$ establishes the important relation

$$\sigma^2 = \langle x^2 \rangle - \langle x \rangle^2 . \tag{3.15}$$

In practice, σ^2 is usually computed from this relation, rather than from (3.14).

3.6 Optical Application: Moments of the Slit Diffraction Pattern [3.1]

The point spread function for a diffraction-limited slit aperture obeys (Ex. 3.1.23)

$$s(y) = \frac{\beta_0}{\pi} \operatorname{sinc}^2(\beta_0 y) , \quad \operatorname{sinc}(x) \equiv \sin(x)/x . \tag{3.16}$$

This is also called the slit "diffraction pattern." Parameter β_0 is related to the light wavelength λ, slit width $2a$ and image conjugate distance F through

$$\beta_0 = ka/F , \quad k = 2\pi/\lambda . \tag{3.17}$$

According to Sect. 3.4, *$s(y)$ also represents the probability density function for position y of a photon in the image plane,* if it originates at a point located at the origin in the object plane. What are the moments of $s(y)$?

By the recipe (3.13a)

$$m_1 = \int_{-\infty}^{\infty} \mathrm{d}y \, y \frac{\beta_0}{\pi} \operatorname{sinc}^2(\beta_0 y) = 0 \tag{3.18}$$

since $s(y)$ is an even function so that the integrand is odd. Note that the mean value is at the peak of the curve $s(y)$, meaning that the average position for a photon in the diffraction pattern is identical with its single most-likely position as well.

The second moment is, by (3.13a)

$$m_2 = \int_{-\infty}^{\infty} \mathrm{d}y \, y^2 \frac{\beta_0}{\pi} \operatorname{sinc}^2(\beta_0 y) = \infty , \tag{3.19}$$

since the y^2 factors cancel, leaving the integral of an all-positive function $\sin^2(\beta_0 y)$. This is a curious situation. Not only is the second moment unbounded, but so are all higher even moments, as further use of (3.13a) will show.

Hence, insofar as m_2 measures "average spread," a photon suffers infinite average spread in the slit diffraction pattern. This is somewhat unsettling, but does it somehow disqualify $s(y)$ from being a bona fide probability density function? No, since $s(y)$ obeys normalization and also is positive everywhere, the only two really necessary conditions; see Eqs. 3.2 and 3.4.

On the other hand, this effect has an important *practical* consequence: see Sect. 9.14.

Further insight into the reason for the bizarre behavior (3.19) may be had by considering the Fourier transform of $s(y)$, called the "optical transfer function" and denoted $T(\omega)$; see (3.64). This turns out to be a triangle function

$$T(\omega) = \text{Tri}(\omega/2\beta_0)\ , \quad \text{Tri}(x) = \begin{cases} 1-\ |x| & \text{for } |x| \le 1 \\ 0 & \text{for } |x| > 1 \end{cases} . \tag{3.20}$$

It can easily be shown (Sect. 4.2) that $m_2 = -T''(0)$, the second derivative at the origin of the triangle function. However, note that a triangle function has a discontinuous slope at the origin, and hence a second derivative there of $-\infty$. Hence, $m_2 = -(-\infty) = \infty$ once again. In general, optical transfer functions which have a discontinuous slope, or cusp, at the origin will have spread functions which have infinite second, and higher-order, moments.

Finally, it should be noted that the infinite second moment is actually demanded by Heisenberg's uncertainty principle. This states that if a photon (or other particle) has a y-component of momentum μ at the instant that it is at position y in its diffraction pattern, then the spreads in y and μ obey (17.117),

$$\sigma_Y \sigma_\mu \ge h/4\pi\ .$$

Quantity h is Planck's constant. Now the photons issue from a point source and pass through a slit whose width is tiny compared with distance F to the image plane (Ex. 3.1.23). Then the transverse momentum values μ can only be 0. Hence the spread $\sigma_\mu = 0$. Then in order for product $\sigma_Y \sigma_\mu$ to exceed $h/4\pi$, σ_Y must be infinite.

3.7 Information Theory Application

The information $I(A, B)$ given by (3.9) represents the information about an event x obtained by observing a *single* datum y. But the event x is usually unknown to the observer. How then could he compute $I(A, B)$? The answer is that he can't. Instead, the *average* of information $I(A, B)$ over all possible x and y is computed. This is called the "trans-information," and is defined as

$$I(X, Y) \equiv \iint_{-\infty}^{\infty} \mathrm{d}x\,\mathrm{d}y\, p(x, y) \log \frac{p(y|x)}{p(y)} . \tag{3.21}$$

This new quantity measures the information in all data values y about all possible events x, and so does not require the particular x corresponding to a given y. This law somewhat simplifies if the logarithm of the quotient is expanded out. Then using the marginal probability law (3.8) we get

$$I(X, Y) = \iint_{-\infty}^{\infty} \mathrm{d}x\,\mathrm{d}y\, p(x, y) \log p(y|x) - \int_{-\infty}^{\infty} \mathrm{d}y\, p(y) \log p(y) . \tag{3.22}$$

The last integral has the form of an "entropy,"

$$H(Y) = -\int_{-\infty}^{\infty} \mathrm{d}y\, p(y) \log p(y) , \tag{3.23}$$

so that analogously the first integral is made to define an entropy

$$H(Y|X) \equiv -\iint_{-\infty}^{\infty} \mathrm{d}x\, \mathrm{d}y\, p(x, y) \log p(y|x) . \tag{3.24}$$

Then (3.22) reads

$$I(X, Y) = H(Y) - H(Y|X) . \tag{3.25}$$

This states that *the trans-information is the entropy of the observed values y as reduced by the conditional entropy characterizing the channel of communication.*

The concept of entropy has other uses as well. Maximizing the entropy of an unknown probability law $p(y)$ is a means of producing a good estimate of $p(y)$ (Sect. 10.4.2).

The Shannon information (3.25) is by no means a unique measure of information. It all depends upon the aims of the user. Other information measures are listed in Sect. 3.18.

3.8 Case of Statistical Independence

If the $\{x_n\}$ are statistically independent, then by separation effect (2.13), identity (3.12) becomes

$$\langle f \rangle = \int_{-\infty}^{\infty} \cdots \int \mathrm{d}x_1 \ldots \mathrm{d}x_n f(x_1, \ldots, x_n) p(x_1) \ldots p(x_n) . \tag{3.26}$$

Finally, if in addition f is itself a product function $f_1(x_1) \ldots f_n(x_n)$ we get complete separation of the integrals, so that

$$\langle f \rangle = \langle f_1(x_1) \rangle \ldots \langle f_n(x_n) \rangle . \tag{3.27}$$

3.9 Mean of a Sum

Consider a set of random variables $\{x_n\}$ which are generally dependent. We wish to know what the mean-value of their weighted sum is,

$$\left\langle \sum_{i=1}^{n} a_i x_i \right\rangle .$$

The $\{a_i\}$ are any set of constant numbers.

This calculation falls within the realm of (3.12). Hence,

$$\left\langle \sum_{i=1}^{n} a_i x_i \right\rangle = \int_{-\infty}^{\infty} \cdots \int_{-\infty}^{\infty} \mathrm{d}x_1\, \mathrm{d}x_2 \ldots \mathrm{d}x_n \left(\sum_{i=1}^{n} a_i x_i \right) p(x_1, x_2, \ldots, x_n) .$$

The multiple integrals may be easily evaluated by noting that the term a_1x_1 is constant with respect to integration variables $x_2, \ldots, x_n$, so that these only integrate out $p(x_1, \ldots, x_n)$ to produce $p(x_1)$. Hence the term a_1x_1 contributes

$$\int_{-\infty}^{\infty} \mathrm{d}x_1 a_1 x_1 p(x_1) = a_1 \langle x_1 \rangle$$

to the answer. In the same way, term a_2x_2 contributes $a_2\langle x_2\rangle$, etc. for the n terms of the sum. The answer is then

$$\left\langle \sum_{i=1}^{n} a_i x_i \right\rangle = \sum_{i=1}^{n} a_i \langle x_i \rangle \ . \tag{3.28}$$

In words, *the mean of a weighted sum equals the weighted sum of the means.* This is a general result, quite independent of the state of joint dependence of the individual $\{x_n\}$.

3.10 Optical Application

Image formation often occurs where either the object $o(x_n)$ or the spread function $s(y_m; x_n)$ are randomly varying quantities. What then is the mean image $i(y_m)$?

We start with the imaging equation (2.21)

$$i(y_m) = \sum_{n=1}^{N} o(x_n) s(y_m; x_n) \ . \tag{3.29}$$

Take the mean of both sides when the $o(x_n)$ are random variables. Then by identity (3.28)

$$\langle i(y_m) \rangle = \sum_{n=1}^{N} \langle o(x_n) \rangle \, s(y_m; x_n) \ .$$

An object may be considered a random variable at each of its points x_n when it is a randomly selected member of a set of objects, such as letters of the alphabet. If the observer cannot know which member of the set will be imaged, he must regard the object as random.

When instead the $s(y_m; x_n)$ are random variables, the result is

$$\langle i(y_m) \rangle = \sum_{n=1}^{N} o(x_n) \, \langle s(y_m; x_n) \rangle \ .$$

If a given object is imaged repeatedly through short-term atmospheric turbulence, then the spread function for each position couplet $(y_m; x_n)$ is a random variable over the class of images so obtained. Spread function s is called a "speckle" pattern [3.2]. Then the last equation shows that the average image so formed is equivalent to convolving the object with an average spread function. This latter quantity is equivalent to one "long exposure" average spread function.

3.11 Deterministic Limit; Representations of the Dirac δ-Function

If a random variable x can take on only one value x_0, then $p(x)$ can still describe this situation by taking on a Dirac delta (δ) function form

$$p(x) = \delta(x - x_0) . \tag{3.30a}$$

The Dirac $\delta(x)$ function is defined as obeying

$$\delta(x) = 0 \quad \text{for} \quad x \neq 0 \tag{3.30b}$$

and

$$\int_{-\infty}^{\infty} \mathrm{d}x\, \delta(x) = 1 . \tag{3.30c}$$

These almost conflicting properties can be satisfied only if $\delta(x)$ is imagined to have *an infinitely narrow* [in view of (3.30b)] *spike* centered on the origin, *with infinite height* so as to yield unit area in (3.30c).

The significance of representation (3.30a) for $p(x)$ is that $p(x)$ is then infinitely narrow and concentrated about the event $x = x_0$. Hence, no other event but x_0 can occur. This is, of course, what we mean by *determinism.*

Dirac $\delta(x)$ has a further "sifting" property,

$$\int_{-\infty}^{\infty} \mathrm{d}x\, f(x)\delta(x - x_0) = f(x_0) , \tag{3.30d}$$

which follows simply from definition (3.30b,c). Hence in the deterministic limit, by (3.12)

$$\langle f \rangle = f(x_0) ,$$

as would be expected since $p(x)$ has no spread about the point x_0.

Use of the correspondence (3.30a) with $x_0 = 0$, in particular, allows us to form functional representations for $\delta(x)$. Every probability density $p(x)$ has a free parameter, call it q, which governs the degree of concentration in $p(x)$ about its mean. For example, with a Gaussian law, $q = \sigma$, the standard deviation. In general, with $q \to 0$, the law becomes ever-more concentrated about its mean of zero, approaching $\delta(x)$. Hence, *we may represent* $\delta(x)$ *as simply*

$$\delta(x) = \lim_{q \to 0} p(x) \tag{3.30e}$$

where $p(x)$ *is any probability density.* Note that (3.30e) has unit area under it regardless of the value of q. Hence $\delta(x)$ so formed automatically obeys normalization property (3.2).

As an example of such use, we may choose the "slit diffraction" probability law (3.16). There, parameter β_0 governs the concentration of the law about its mean or zero, with $q = \beta_0^{-1}$. As $\beta_0 \to \infty$ the law becomes more concentrated. Hence

$$\delta(x) = \lim_{\beta_0 \to \infty} \frac{\beta_0}{\pi} \operatorname{sinc}^2(\beta_0 x) . \tag{3.30f}$$

An infinity of such representations for $\delta(x)$ may be formed in this way.

Finally, we note the "Fourier representation" for $\delta(x)$, which is formed by taking the Fourier transform of an arbitrary function $f(x)$ and then inverse transforming this. If the result is to once again be $f(x)$, by the sifting property (3.30d) it must be that

$$\delta(x) = (2\pi)^{-1} \int_{-\infty}^{\infty} \mathrm{d}x' \, \mathrm{e}^{\mathrm{j}xx'} \, . \tag{3.30g}$$

3.12 Correspondence Between Discrete and Continuous Cases

A discrete law P_n may be regarded as an extremely spikey version of a corresponding continuous law

$$p(x) = \sum_{n=0}^{N} P_n \delta(x - x_n) \, . \tag{3.31}$$

Notice that because of the Dirac delta functions, the right-hand sum is zero except at the yield points $x = x_n$. It resembles an uneven "picket fence", as in Fig. 3.2.

This relation is very useful, since it allows any relation involving a probability density to be expressed in the proper form for a discrete probability case. One merely substitutes (3.31) into the relation in question and performs any necessary integration by use of the handy sifting property (3.30d) for $\delta(x)$. For example, the normalization property (3.2) for density becomes, after substitution of (3.31)

$$1 = \int_{-\infty}^{\infty} \mathrm{d}x \sum_{n=0}^{\infty} P_n \delta(x - x_n) = \sum_{n=0}^{\infty} P_n \int_{-\infty}^{\infty} \mathrm{d}x \, \delta(x - x_n) = \sum_{n=0}^{\infty} P_n \, ,$$

which is, of course, the normalization property for the discrete case.

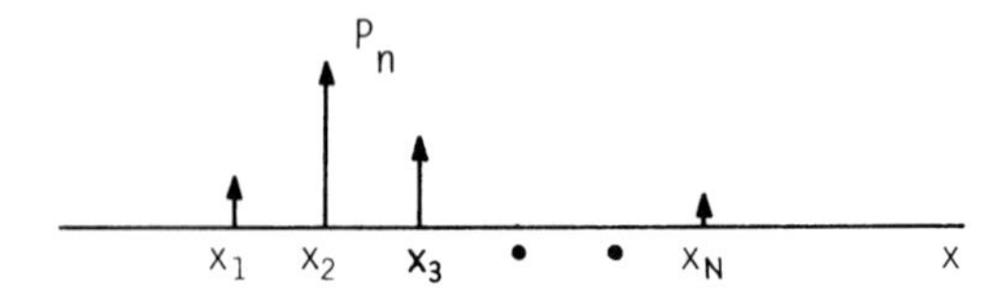

Fig. 3.2. Picket-fence probability law

Hence, *we may now consider the continuous situation to be the general one; interested readers can take the discrete limit (3.31) of a continuous law as required.* Most of the development that follows therefore emphasizes the continuous situation.

3.13 Cumulative Probability

Here we ask, what is the probability that a RV x' is less than a value x? Directly from definition (3.1) of density,

$$P(-\infty < x' < x) = \int_{-\infty}^{x} dx' p(x') . \tag{3.32}$$

This probability is also simply denoted as $F(x)$, and is commonly called the *cumulative probability* or *distribution function* for random variable x. Since $p(x') \geq 0$, F is a monotonically increasing function of x. Note that (3.32) has the same mathematical form as an optical edge spread function. There are many parallels between the mathematics of image formation and that of probability "formation." This will become particularly evident in Chap. 4.

3.14 The Means of an Algebraic Expression: A Simplified Approach

Suppose a random variable y is formed from two other random variables x_1 and x_2 according to

$$y = ax_1 + bx_2 , \quad a, b \text{ constants} . \tag{3.33}$$

The expression in x_1 and x_2 (and any other number of variables) may be of higher-order power than this, but we take the linear case for simplicity. The problem we address is how to find all moments of y. Here we shall only seek moments $\langle y \rangle$, $\langle y^2 \rangle$ and σ_Y^2. Higher-order moments may be found by an obvious extension of the method.

There are (at least) two ways of attacking the problem. First we show the cumbersome way, by direct use of definition (3.12). This requires setting up a multidimensional integral, e.g., for the first moment

$$\langle y \rangle = \iint_{-\infty}^{\infty} dx_1 \, dx_2 (ax_1 + bx_2) p(x_1, x_2) . \tag{3.34}$$

The multidimensional nature of the integral can become quite awkward as the number of random variables $x_1, x_2, \ldots$ increases.

Rather than take the approach (3.34), we may instead directly take the mean-value of both sides of (3.33): if the two sides of (3.33) are equal then their means must also be equal. Then by identity (3.28)

$$\langle y \rangle = a \langle x_1 \rangle + b \langle x_2 \rangle . \tag{3.35}$$

This shows directly that $\langle y \rangle$ depends upon knowledge of two quantities, $\langle x_1 \rangle$ and $\langle x_2 \rangle$. Now we can see why this approach is simpler than by (3.34). Means $\langle x_1 \rangle$ and $\langle x_2 \rangle$ may be evaluated as one-dimensional integrals such as

$$\langle x_1 \rangle = \int_{-\infty}^{\infty} dx_1 x_1 p(x_1) , \tag{3.36}$$

compared with the two-dimensional, or higher-order nature of (3.34). This follows since by direct use of (3.12)

$$\langle x_1 \rangle = \iint_{-\infty}^{\infty} dx_1 \, dx_2 x_1 p(x_1, x_2) = \int_{-\infty}^{\infty} dx_1 x_1 p(x_1)$$

by the marginal law (3.8). The *reduction in dimensionality* is the desired effect. It occurs for higher order moments as well, such as $\langle x_1^2 \rangle$, $\langle x_2^2 \rangle$, etc.

Cross-terms such as $\langle x_1^2 x_2^3 \rangle$ would still require a two-dimensional integration in general, except if x_1 and x_2 are given as statistically independent (a common occurrence in problems). Then directly $\langle x_1^2 x_2^3 \rangle = \langle x_1^2 \rangle \langle x_2^3 \rangle$ by identity (3.27), and each of the right-hand means involves again a *one*-dimensional integration.

As an example of the method, suppose that in (3.33) x_1 and x_2 are independent and each is sampled from a uniform probability law over interval (0, 1). Now what are $\langle y \rangle$, $\langle y^2 \rangle$ and σ_Y^2? We already have (3.35) for $\langle y \rangle$. To form $\langle y^2 \rangle$ in the same fashion we merely square both sides of (3.33) and then take the mean value of both sides. Then, invoking identity (3.28) yields

$$\left\langle y^2 \right\rangle = a^2 \left\langle x_1^2 \right\rangle + b^2 \left\langle x_2^2 \right\rangle + 2ab \left\langle x_1 x_2 \right\rangle \ ,$$

or further

$$\left\langle y^2 \right\rangle = a^2 \left\langle x_1^2 \right\rangle + b^2 \left\langle x_2^2 \right\rangle + 2ab \left\langle x_1 \right\rangle \left\langle x_2 \right\rangle \tag{3.37}$$

by identity (3.27).

Equations (3.35) and (3.37) show that the averages $\langle x_1 \rangle$, $\langle x_2 \rangle$, $\langle x_1^2 \rangle$ and $\langle x_2^2 \rangle$ need only be computed. These are one-dimensional,

$$\langle x_1 \rangle = \langle x_2 \rangle = \int_0^1 \mathrm{d}x\, x = 1/2 \ , \quad \text{and} \quad \left\langle x_1^2 \right\rangle = \left\langle x_2^2 \right\rangle = \int_0^1 \mathrm{d}x\, x^2 = 1/3$$

by definition (3.13a).

Direct substitution into (3.35) and (3.37) produces the answers

$$\langle y \rangle = a/2 + b/2 \ , \quad \left\langle y^2 \right\rangle = a^2/3 + b^2/3 + ab/2$$

and by identity (3.15)

$$\sigma_Y^2 = a^2/12 + b^2/12 \ .$$

3.15 A Potpourri of Probability Laws

We show below some of the commonest probability laws that arise in optics. The list is by no means complete; for others, see the Probability law listing in the Subject Index.

3.15.1 Poisson

The most well known discrete probability law $P(A_n) \equiv P_n$ is the Poisson, defined as

$$P_n = \mathrm{e}^{-a} a^n / n! \ , \quad n = 0, 1, 2, \ldots \ , \tag{3.38}$$

with $a > 0$. This curve is illustrated in Fig. 3.3 for particular values of a. Parameter a is the sole determinant of curve shape. For small values of a the curves are strongly

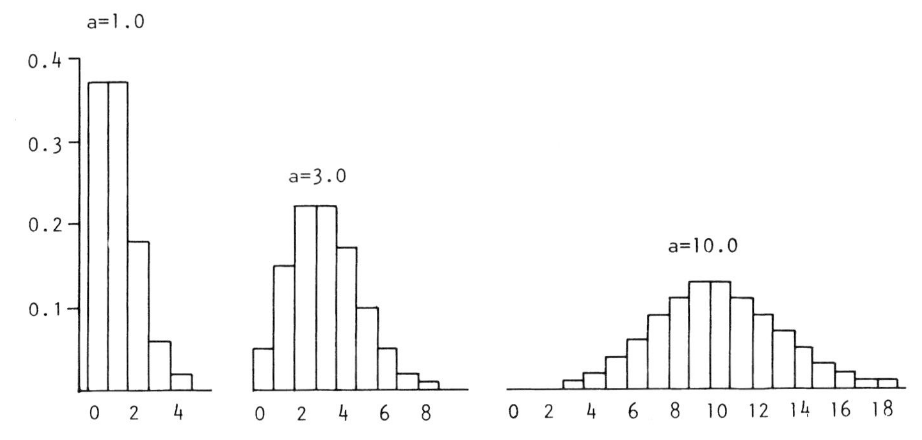

Fig. 3.3. Poisson probability law for three different mean-values a

skewed to the right, but approach a symmetric bell shape as $a \to \infty$. (Note the shape for $a = 10$.) In fact the bell shaped curve is the *normal curve*: see Sect. 6.5.3.

By the use of definition (3.13b) it can directly be shown that a is both the mean and the variance, and even the third central moment, of the curve.

The Poisson law arises optically as the probability law for n photon arrivals over a time interval T, if the photons arrive independently with uniform randomness in time, and if there is a sufficiently small time interval dt such that not more than one photon arrives within each such interval (Sect. 6.6).

3.15.2 Binomial

Another discrete probability law that arises commonly is the binomial law

$$P_n = \binom{N}{n} p^n q^{N-n}, \quad \binom{N}{n} \equiv \frac{N!}{n!(N-n)!} \tag{3.39}$$

where $p+q = 1$ and brackets denote the binomial coefficient. Direct use of definition (3.13b) shows that $\langle n \rangle = Np$ and $\sigma^2 = Npq$.

The binomial law arises under the same circumstances that produce the Poisson law. However, it does not require rare events, as does the Poisson, and is of more general validity (Sect. 6.1). The Poisson law is often an approximation to the binomial law.

3.15.3 Uniform

The most elementary probability *density* is the uniform law

$$p(x) = b^{-1}\mathrm{Rect}\left(\frac{x-a}{b}\right), \quad \mathrm{Rect}(x) \equiv \begin{cases} 1 & \text{for } |x| \le 1/2 \\ 0 & \text{for } |x| > 1/2 \end{cases} \tag{3.40}$$

This is shown in Fig. 3.4. Use of (3.13a) and (3.14) shows that $\langle x \rangle = a$, the midpoint, and $\sigma^2 = b^2/12$. Use of (3.13a) shows that moment $m_k = [(a + b/2)^{k+1} - (a - b/2)^{k+1}]/b(k + 1)$.

A uniform law arises optically as the probability on position x for a photon in a uniformly bright object (Sect. 2.10). It will be seen that a uniform probability law also characterizes optical phase in a model for laser speckle (Sect. 5.9.5).

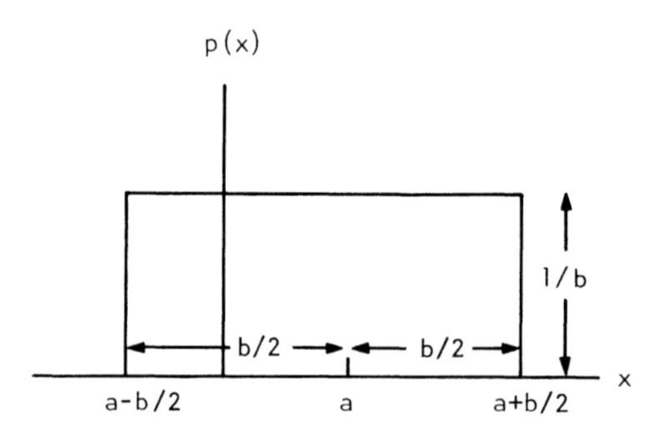

Fig. 3.4. Uniform probability law $\frac{1}{b}\text{Rect}\left(\frac{x-a}{b}\right)$

The RANF$(\cdot)$ computer function generates a random variable whose fluctuations follow the uniform law (3.40) with $a = \frac{1}{2}$, $b = 1$. This corresponds to uniformity over interval (0, 1), and is the only probability law that may be directly accessed on many computers. Yet Monte Carlo calculations (Chap. 7) and other simulations usually require random variables that follow other probability laws. Luckily, a mere mathematical transformation on the outputs from the RANF$(\cdot)$ function will make these outputs fluctuate according to any prescribed law. This is one of the subjects we take up later (Chap. 5).

3.15.4 Exponential

This is of the form (Fig. 3.5)

$$p(x) = \begin{cases} a^{-1}\exp(-x/a) & \text{for } x \geqq 0 \\ 0 & \text{for } x < 0 \end{cases} \tag{3.41}$$

Use of (3.13a) for computing means shows that

$$\langle x \rangle = a\,, \quad \langle x^2 \rangle = 2a^2\,, \quad \text{and} \quad \sigma^2 = a^2\,.$$

[The latter by (3.15).] Hence, the variance equals *the square* of the mean. Compare this with the Poisson case (3.38) where the variance equals the mean. See also Ex. 3.1.14 for the general moment m_k.

This law arises in laser speckle as the probability density for an intensity level x, when the mean intensity is a (Sect. 5.9.8). It also arises as the probability law for time between successive Poisson events (Ex. 6.1.16). Most generally it arises as the probability density for a random variable whose mean is fixed, where the probability density is required to have *maximum entropy* (Sect. 10.4.11).

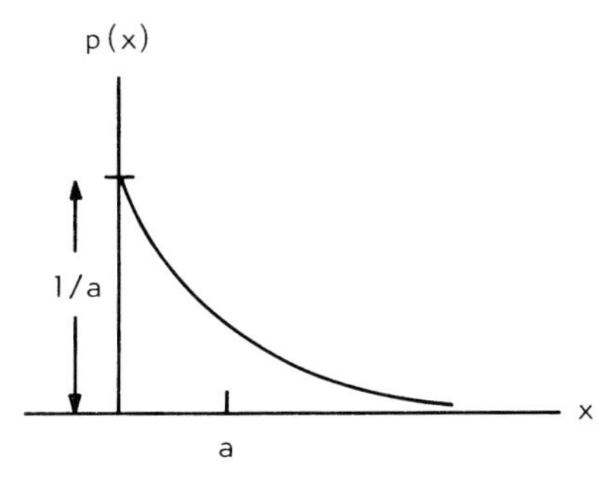

Fig. 3.5. Exponential probability law

3.15.5 Normal (One-Dimensional)

The most frequently *used* probability law is the famous normal law

$$p(x) = (2\pi\sigma^2)^{-1/2}\exp[-(x - \langle x\rangle)^2/2\sigma^2]\,. \tag{3.42}$$

Its moments μ_k are given in Ex. 3.1.15. The phrase "x is a normal random variable with mean $\langle x\rangle$ and variance σ^2" is used so often that it now has its own notation,

$$x = N(\langle x\rangle\,, \sigma^2)\,. \tag{3.43}$$

Particular values for $\langle x\rangle$ and σ^2 are usually inserted. In the particular case $\langle x\rangle = 0$, the normal law is called the "Gaussian" probability law. This is illustrated in Fig. 3.6.

The reason for the popularity of the normal law is its common occurrence in physical problems, this due to the *"central limit" theorem* (Sect. 4.25).[1] For example, optical phase after passing through atmospheric turbulence obeys a normal (in fact, Gaussian) law (Sect. 8.6.1).

3.15.6 Normal (Two-Dimensional)

The most common bivariate law (i.e., describing *two* random variables) in use is the *normal* bivariate one,

$$p(x, y) = \frac{1}{2\pi\sigma_1\sigma_2(1-\rho^2)^{1/2}}\exp\left\{-\frac{1}{2(1-\varrho^2)}\right.$$
$$\left.\cdot\left[\frac{(x-m_x)^2}{\sigma_1^2} + \frac{(y-m_y)^2}{\sigma_2^2} - \frac{2\varrho(x-m_x)(y-m_y)}{\sigma_1\sigma_2}\right]\right\}\,. \tag{3.44}$$

For example, this law describes the *joint* fluctuation of pupil phase values x and y at two points, caused by atmospheric turbulence (Sect. 8.6.1).

Remarkably, the *marginal* density $p(x)$, computed by use of identity (3.8), is independent of parameters ϱ and σ_2^2, obeying

$$x = N(m_x, \sigma_1^2)\,. \tag{3.45}$$

[1] This theorem applies so extensively that an aura has developed about it. The assertion "x must be normal because of the central limit theorem," is usually considered proof enough, especially if said in an authoritative voice.

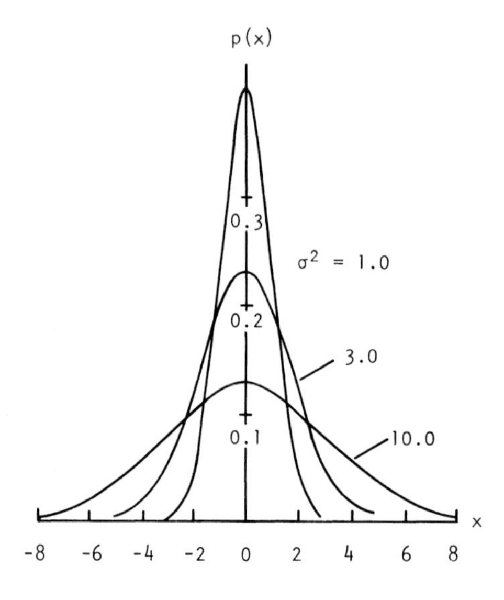

Fig. 3.6. Gaussian probability law for three different variances σ^2

A similar case exists for random variable y.

The free parameters $m_x, m_y, \sigma_1, \sigma_2$, and ϱ of this law may be related to various moments of the law. For these purposes, we must use the mean value form (3.12) with $n = 2$,

$$\langle f \rangle \equiv \iint_{-\infty}^{\infty} \mathrm{d}x\,\mathrm{d}y\, f(x, y)p(x, y) \,. \tag{3.46}$$

For example,

$$\langle (x - \langle x \rangle)(y - \langle y \rangle) \rangle \equiv \iint_{-\infty}^{\infty} \mathrm{d}x\,\mathrm{d}y (x - \langle x \rangle)(y - \langle y \rangle) p(x, y) \,.$$

Substitution of (3.44) for $p(x, y)$ in this integral yields

$$\langle (x - \langle x \rangle)(y - \langle y \rangle) \rangle = \varrho \sigma_1 \sigma_2 \,,$$

so that

$$\varrho = \langle (x - \langle x \rangle)(y - \langle y \rangle) \rangle / \sigma_1 \sigma_2 \,. \tag{3.47}$$

ϱ is called the "correlation coefficient" for fluctuations in x and y. If $\varrho = 0$, from (3.44) $p(x, y) = p(x)p(y)$, or x and y are statistically independent. ϱ thereby measures the degree to which x and y fluctuate in a mutually dependent manner. ϱ is a coefficient, that is, a pure number lying between 0 and 1. This traces from the division by $\sigma_1 \sigma_2$ in (3.47); see Ex. 3.2.8.

Regarding other means, from (3.46)

$$\langle x \rangle = \iint_{-\infty}^{\infty} \mathrm{d}x\,\mathrm{d}y\, x p(x, y) = \int_{-\infty}^{\infty} \mathrm{d}x\, x p(x)$$

by (3.8), so that

$$\langle x \rangle = m_x$$

by identity (3.13a). Likewise $\langle y \rangle = m_y$. Finally, by computing moments $\langle x^2 \rangle$ and $\langle y^2 \rangle$, by (3.15) it is easily shown that

$$\sigma_1^2 = \langle x^2 \rangle - m_x^2 , \quad \sigma_2^2 = \langle y^2 \rangle - m_y^2 .$$

3.15.7 Normal (Multi-Dimensional)

As a direct generalization of the preceding, the n-dimensional normal density function is of the form

$$p(x_1, x_2, \ldots, x_n) \equiv p(\boldsymbol{x}) = \frac{|A|^{1/2}}{(2\pi)^{n/2}} \exp\{-2^{-1}(\boldsymbol{x} - \langle \boldsymbol{x} \rangle)^T [A](\boldsymbol{x} - \langle \boldsymbol{x} \rangle)\} \tag{3.48}$$

where T denotes transpose and vector

$$(\boldsymbol{x} - \langle \boldsymbol{x} \rangle) \equiv \begin{pmatrix} x_1 - \langle x_1 \rangle \\ \cdot \\ \cdot \\ \cdot \\ x_n - \langle x_n \rangle \end{pmatrix} .$$

Matrix $[A]$ is $n \times n$ and *positive definite*. The latter means that all its principal minors are positive

$$A_{11} > 0 , \quad \det \begin{bmatrix} A_{11} & A_{12} \\ A_{21} & A_{22} \end{bmatrix} > 0, \ldots, \det[A] > 0 . \tag{3.49}$$

The n-dimensional generalization to the correlation coefficient (3.47) is the *covariance coefficient*

$$\varrho_{jk} \equiv \frac{\langle (x_j - \langle x_j \rangle)(x_k - \langle x_k \rangle) \rangle}{\sigma_j \sigma_k} . \tag{3.50a}$$

Use of (3.48) in the n-dimensional average (3.12) yields

$$\varrho_{jk} = \frac{\text{cofactor } A_{jk}}{(\text{cofactor } A_{jj} \text{cofactor } A_{kk})^{1/2}} , \quad j \neq k \tag{3.50b}$$

with

$$\sigma_j^2 = \frac{\text{cofactor } A_{jj}}{\det[A]} . \tag{3.51}$$

In the special case where the $\{x_n\}$ are mutually independent, matrix $[A]$ is diagonal, so the exponent in (3.48) becomes purely a sum of squares of the $(x_n - \langle x_n \rangle)$. The effect is to make $p(x_1, \ldots, x_n)$ a product of *one-dimensional* normal density functions (3.42). Likewise, with $[A]$ diagonal the cofactors A_{jk}, $j \neq k$, are zero so that by (3.50b) all coefficients ϱ_{jk}, $j \neq k$, are zero.

3.15.8 Skewed Gaussian Case; Gram–Charlier Expansion

A real probability law is often close to Gaussian, but significantly different to require a separate description. A very useful way to describe such a probability law is by the "Gram–Charlier" expansion

$$p(x) = p_0(x)\left[1 + \sum_{n=1}^{N} \frac{c_n}{n!} \mathrm{H}_n(x)\right] , \tag{3.52}$$

where $p_0(x)$ is the closest Gaussian to $p(x)$, obtained say by least-squares fit. Coefficients $\{c_n\}$ measure the departure from a pure Gaussian $p_0(x)$ case, and are to be determined. The $\{\mathrm{H}_n(x)\}$ are Hermite polynomials. The latter obey an orthogonality relation

$$\int_{-\infty}^{\infty} \mathrm{d}x \frac{1}{\sqrt{2\pi}} \mathrm{e}^{-x^2/2} \mathrm{H}_m(x)\mathrm{H}_n(x) = n!\delta_{mn} \tag{3.53}$$

with respect to a Gaussian weight function. The first few are

$$\mathrm{H}_0(x) = 1 , \quad \mathrm{H}_1(x) = x , \quad \mathrm{H}_2(x) = x^2 - 1 , \quad \mathrm{H}_3(x) = x^3 - 3x . \tag{3.54}$$

The size of coefficients $\{c_n\}$ determines closeness to the pure Gaussian $p_0(x)$ case. The odd coefficients c_1, c_3, etc., multiply odd polynomials $\mathrm{H}_n(x)$ and hence describe a "skewness" property for the $p(x)$ curve. The even coefficients in the same way define an even contribution to the curve which is often called "peakedness," in that it may cause $p(x)$ to be more peaked (or tall) than a pure Gaussian. With $p(x)$ observed experimentally as a histogram (frequency of occurrence data), the unknown $\{c_n\}$ may be determined in the usual manner for a series of orthogonal functions. Combining (3.52) and (3.53),

$$c_n = (2\pi)^{-1/2} \int_{-\infty}^{\infty} \mathrm{d}x[p(x)/p_0(x) - 1]\mathrm{e}^{-x^2/2}\mathrm{H}_n(x) . \tag{3.55}$$

3.15.9 Optical Application

The observation of glitter patterns reflected from ocean waves has been proposed as a means of remotely determining wind velocity at the ocean surface [3.3]. Glitter patterns are formed by specular reflection of sunlight. To infer wind speed from them requires knowledge of the *probability law for wave slope* under different wind conditions.

Cox and *Munk* [3.4] established this law. They determined the joint probability $p(x, y)$ linking wave slope x *along* wind direction and slope y *transverse to* wind direction. We describe this work next.

First, note that $p(x)$ should differ from $p(y)$. Along x, waves tend to lean away from the wind, and hence statistically should have gentler slopes on the windward side than on the leeward side. This should cause the histogram $p(x)$ for such slopes to be skewed away from the origin (zero slope) toward steep leeward (negative on plots) slope. The histogram $p(y)$ for slope *transverse* to the wind *should not show such bias*, since waves in this direction are not directly formed by the wind, but rather by a leakage of energy from the longitudinal wave motion *per se*.

The latter also suggests that x and y are generally *dependent* variables, so that $p(x, y)$ should not separate into a simple product. Should, then, $p(x, y)$ be a normal bivariate form (3.44)? More generally, Cox and Munk used a two-dimensional Gram–Charlier expansion of the type (3.52),

$$p(x, y) = p_0(x)p_0(y)\left[1 + \frac{c_1 d_2}{2!}\mathrm{H}_1(x)\mathrm{H}_2(y) + \frac{c_3}{3!}\mathrm{H}_3(x) + \frac{d_4}{4!}\mathrm{H}_4(y) + \frac{c_2 d_2}{2!2!}\mathrm{H}_2(x)\mathrm{H}_2(y) + \frac{c_4}{4!}\mathrm{H}_4(x)\right]. \tag{3.56}$$

Series (3.56) is actually *selected terms* in the direct product $p(x)p(y)$ of two series (3.52). However, since not all terms of the product are included, the expression is *not* factorable into $p(x) \cdot p(y)$, as was required above.

The coefficients in (3.56) were determined using linear regression and least-square fits (Chap. 14) to the experimental data. Results are shown in Fig. 3.7 for the two wind directions.

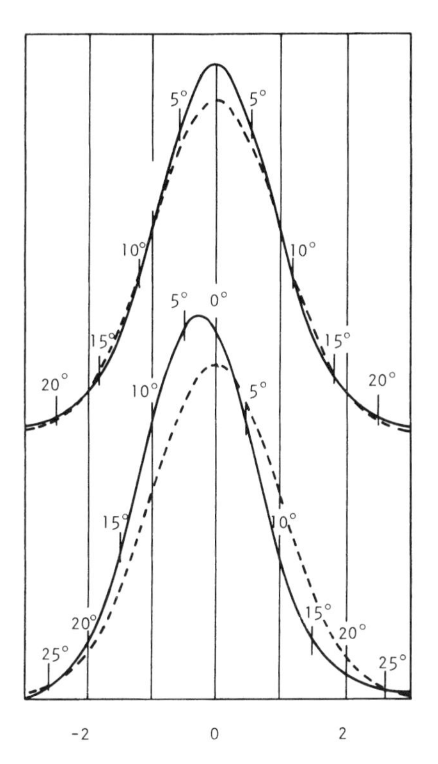

Fig. 3.7. Principal sections through the Cox-Munk wave slope probability law: (top) cross-wind direction, (bottom) along-wind. Dashed curves are the closest Gaussians $p_0(x)$, $p_0(y)$. Vertical stripes show the slopes relative to the standard deviations for slope in the two directions. Angles indicated are the actual slope angles, in degrees, for a wind speed of 10 m/s. Note the skewness from the closest Gaussian in the bottom curve [3.4]

3.15.10 Geometric Law

A probability law arising in quantum optics is the geometric (or Pascal) law (Fig. 3.8)

$$P_n = \frac{a^n}{(1+a)^{n+1}} . \tag{3.57}$$

In a thermally generated light field, this describes the probability of a photon having energy level $(n + \frac{1}{2})h\nu$, when the light source is at temperature T (Ex. 10.1.17). Parameter a is the "Bose factor"

$$a = \frac{1}{\exp(h\nu/kT) - 1} , \tag{3.58}$$

where h is Planck's constant, ν is the light frequency, and k is the Boltzmann constant.

It can be shown that $\langle n \rangle = a$ and $\sigma_n^2 = a(a+1)$ (Ex. 4.1.11). Comparison with Fig. 3.5 shows that the geometric law much resembles the exponential law. Both have a maximum at the origin, regardless of mean value, and both have a variance going as the *square* of the mean.

The geometric law also arises in the theory of reliability (Ex. 6.1.24).

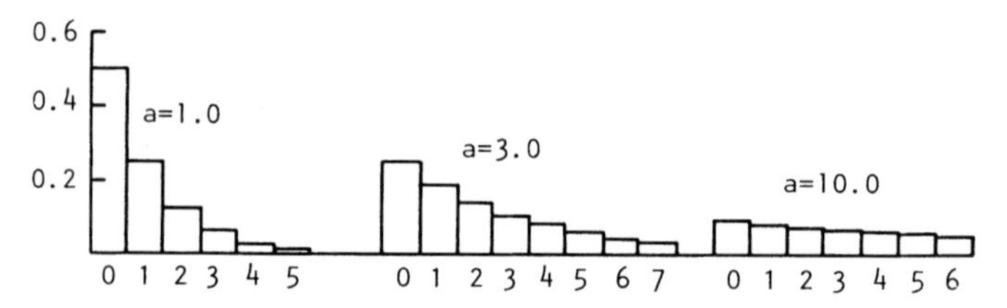

Fig. 3.8. Geometric probability law for different means a

3.15.11 Cauchy Law

The Cauchy (or Lorentz) law

$$p(x) = \frac{1}{\pi} \frac{a}{a^2 + x^2} \tag{3.59}$$

is plotted in Fig. 3.9. This law governs the distribution of light flux along a line in a plane containing a point source. Its unique self-convolution properties verify Huygen's principle (Ex. 5.1.18).

Although the mean of the law is zero, its second and higher even moments are infinite because it does not fall off fast enough with x.

3.15.12 sinc² Law

It has been shown that images are probability laws. The most commonly occurring image is probably (3.16), that of a point source as formed by diffraction-limited slit optics (Ex. 3.1.23). This probability law is shown in Fig. 3.9b.

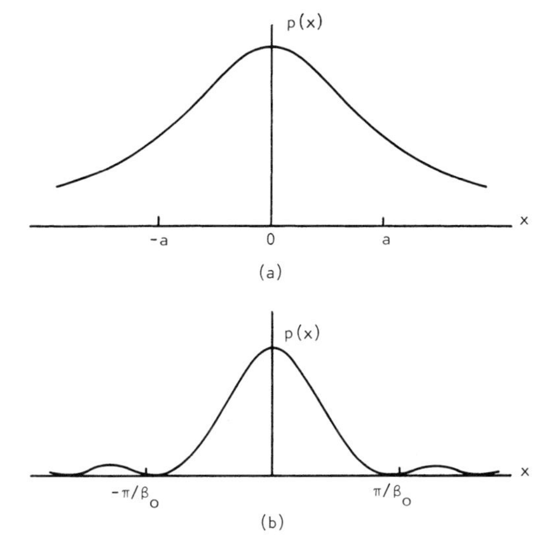

Fig. 3.9. Two probability laws having infinite second moment. (a) Cauchy probability law. (b) $\text{sinc}^2(\beta_0 x)$ probability law

As discussed in Sect. 3.6, the sinc^2 law has infinite second moment, a consequence of the Heisenberg uncertainty principle. Despite this, 0.77 of the events obeying the law occur within interval $|x| \leqq \pi/2\beta_0$; or, 0.90 occur within interval $|x| \leqq \pi/\beta_0$, the central core width. This may be established by numerical integration of the sinc^2 function.

Exercise 3.1

3.1.1 Name five optical experiments whose events are discrete; name five whose events are continuous random variables.

3.1.2 Photographic emulsions suffer from signal-dependent noise whereby the variance of transmittance t is proportional to its mean. This may be represented by a situation $t = \langle t \rangle + \sqrt{a\langle t \rangle}\, n$, a constant, where n is a RV of zero mean and unit variance.

a) Show this, i.e., show that $\sigma_T^2 = a\langle t \rangle$ if t is represented this way.

It is proposed to remove the signal-dependent nature of the noise, at least to second order statistics, by forming a new RV defined by $s = \sqrt{t}$.

b) Show that this will be accomplished if ratio $\sigma_T/\langle t \rangle \ll 4$. That is, show that $\sigma_S^2 \simeq a/4$, independent of $\langle t \rangle$ as required. (*Hint:* The smallness of $\sigma_T/\langle t \rangle$ means that $a/16 \ll \langle t \rangle$, and also that n is sufficiently small to be expanded out in Taylor series wherever it occurs.)

3.1.3 Using (3.5) compute $p(y|x)$ if $p(x, y)$ is Gaussian bivariate as in (3.44).

3.1.4 a) Can the function $\sin(x)/x$ represent a probability density? Explain.

b) The function $1/x$ has a pole at the origin. Can it represent a probability density? Yet function $\delta(x)$ has a pole at the origin and can represent a probability density. Explain why.

3.1.5 Derive relation (3.15) from definition (3.14) of the central moments.

3.1.6 Suppose that x and y are Gaussian bivariate random variables. Then their fluctuations correlate, and intuitively observation of values x imply values y close to x. Therefore, the values x contain information about the values y. Show that this information $I(X, Y)$ for the Gaussian bivariate case is $-2^{-1}\ln(1-\varrho^2)$. [*Hint:* Use (3.21), (3.44) with $m_x = m_y = 0$, and the answer to Ex. 3.1.3 above. For simplicity, also assume that $\sigma_1 = \sigma_2 \equiv \sigma$.]

3.1.7 Let random variables x and y be statistically independent and chosen from the same probability density $(1/a)\,\text{Rect}(x/a)$, $a > 0$. Compute $\langle x^2 y^2 \rangle$.

3.1.8 Establish a representation for Dirac $\delta(x)$ from the probability density $(1/a)\,\text{Rect}(x/a)$.

3.1.9 Show that the definition (3.13a) of moment m_k for a random variable x goes over into definition (3.13b) for discrete random events n, under the "picket fence" substitution (3.31) for $p(x)$.

3.1.10 Find the cumulative probability law $F(x)$ for a random variable x whose probability density is $a^{-1}\text{Rect}(x/a)$, $a > 0$.

3.1.11 Show by the use of definition (3.13b) that the variance of a Poisson variable equals its mean.

3.1.12 Show by the use of definition (3.13b) that for the binomial law $\langle n \rangle = Np$ and $\sigma^2 = Npq$.

3.1.13 Show by the use of definition (3.13a) that for the uniform probability density (3.40) the variance obeys $\sigma^2 = b^2/12$.

3.1.14 The exponential law (3.41) has an interesting property. When multiplied by x^k and integrated from 0 to ∞ in order to find moment m_k, the integral takes on the form of the gamma function. Show in this way that a simple formula exists for the kth moment, $m_k = a^k k!$

3.1.15 Show that the kth central moment (3.14) for the *normal* case (3.42) obeys $\mu_k = \pi^{-1/2} 2^{k/2} \sigma^k \Gamma[(k+1)/2]$. The integral to be evaluated in (3.14) is found in standard tables.

3.1.16 Find $\langle e^{kx} \rangle$, k constant, if random variable $x = N(a, \sigma^2)$.

3.1.17 Find $\langle e^{kx} \rangle$, k constant, if random variable x obeys the uniform law (3.40).

3.1.18 Find $\langle e^{y} \rangle$, if random variable y is formed from independent random variables x_1, x_2 according to a function $y = ax_1 + bx_2 + c$, with a, b, c constants. Assume that each of x_1 and x_2 are uniformly random over interval $(0, 1)$.

3.1.19 Suppose random variable x has a uniform probability density over interval $(0, 1)$. Form a new random variable $y = ax + b$. What do constants a and b have to be in order for the mean $\langle y \rangle$ and variance σ_Y^2 of y to be specified numbers? (*Hint:* Use the approach of Sect. 3.14.)

3.1.20 Suppose RV y obeys $y = ax_1 + bx_2$, where x_1 and x_2 are independent random variables which each obey the same probability law $x = N(c, \sigma^2)$. Find moments $\langle y \rangle$, $\langle y^2 \rangle$ and σ_Y^2. (See preceding Hint.)

3.1.21 Suppose RV's y_1 and y_2 obey $y_1 = x_1 + x_2$ and $y_2 = x_1 + x_3$, where RV's x_1, x_2, and x_3 are all independently sampled from the same, uniform probability law over interval (0, 1). What is the correlation coefficient ϱ_{12}, defined at (3.47), between y_1 and y_2?

3.1.22 ***Honors Problem***. A two-point image i_1, i_2 is repeatedly formed from a two-point object o_1, o_2 according to the imaging equations

$$i_1 = o_1 s_1 + o_2 s_2 + n_1 \, ,$$
$$i_2 = o_1 s_2 + o_2 s_1 + n_2 \, .$$

The object o_1, o_2 is fixed, as are the spread function values s_1, s_2. Quantities n_1, n_2 are noise values which are independent, of zero mean, and with standard deviation σ_N.

For each set of noisy image data i_1, i_2, the above equations are inverted for an object estimate $\hat{o}_1, \hat{o}_2$. Because of the random noise present, the object estimates will have errors

$$e_1 = o_1 - \hat{o}_1 \, ,$$
$$e_2 = o_2 - \hat{o}_2 \, .$$

Find:

(a) $\langle e_1 \rangle$, $\langle e_2 \rangle$. (If these are zero, estimates $\hat{o}_1, \hat{o}_2$ are called "unbiased.")
(b) $\sigma_{e_1}, \sigma_{e_2}$, the standard deviations of error in the estimates.
(c) $\varrho_{e_1 e_2}$, the correlation in the error between the two estimates.

Observe the tendencies in quantities (b) and (c), (i) as σ_N^2 grows, and (ii) as higher resolution is required in the outputs, i.e., as $s_1 \to s_2$. Would an effect $\varrho_{e_1 e_2} \to -1$ bode well for the estimation procedure? (See also Ex. 5.1.21.)

3.1.23 ***The optical transfer function and related quantities.*** This exercise is meant to lay the groundwork for future analyses of image formation. Here we shall derive some basic deterministic relations governing image formation. They may be generalized to two dimensions in an obvious way.

In Fig. 3.10 a point source of light, of wavelength λ, is located at infinity along the optical axis (OA) of the given lens. Spherical waves emanate from it, but after travelling the infinite distance to the lens they appear plane, as illustrated at the far left. The lens has a coordinate y which relates to a "reduced" coordinate β by

$$\beta = ky/F \tag{3.60}$$

where $k = 2\pi/\lambda$, and F is the distance from the lens to the image plane. The lens aperture size β_0 is assumed to be small enough to allow Huyghen's scalar theory of light to be used, and to permit the approximations in Sect. (a) below to be made. At each coordinate β of the pupil there is a phase distance error Δ from a sphere whose radius is F (i.e., which would perfectly focus light in the image plane).

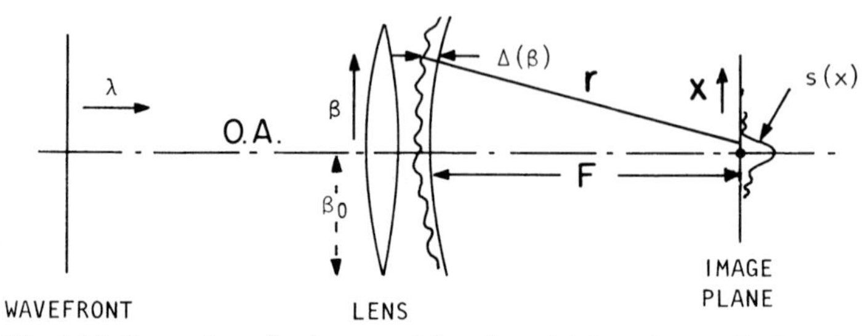

Fig. 3.10. Formation of point spread function $s(x)$ from lens suffering phase errors $\Delta(\beta)$

Huygen's principle [3.1] states that each point β of the wavefront just to the right of the lens acts like a new source of spherical waves. This defines a complex "pupil function"

$$e^{jk\Delta(\beta)} \tag{3.61}$$

at each point β of the pupil.

The focussed image has an intensity profile $s(x)$ called the "point spread function" which is located in the image plane. This may be represented in terms of a point amplitude function $a(x)$ as

$$s(x) = |a(x)|^2 . \tag{3.62}$$

Quantity $a(x)$ relates to the pupil function (3.61) by Fraunhofer's integral [3.5]

$$a(x) = \int_{-\beta_0}^{\beta_0} d\beta \exp\{j[k\Delta(\beta) - \beta x]\} , \tag{3.63}$$

which is basically a finite Fourier transform of the pupil function. An inconsequential multiplicative constant has been dropped.

a) Prove (3.63) by representing $a(x)$ as a superposition integral over the pupil of Huygen's waves $r^{-1}\exp(jkr)$, r the distance from a general point β on the wavefront in the pupil to a general point x in the image plane. [Hint: $r^2 = (y-x)^2 + (F+\Delta)^2$, where $(F+\Delta) \gg |y-x|$, so that $r \simeq F + \Delta +$ small terms, by Taylor series. Assume that the lens has a compensating phase $ky^2/2F$ and that x is close to the optical axis.]

Finally, the optical transfer function $T(\omega)$ is defined as basically the Fourier transform of $s(x)$,

$$T(\omega) \equiv K^{-1} \int_{-\infty}^{\infty} dx\, s(x) e^{-j\omega x} , \tag{3.64}$$

where

$$K \equiv \int_{-\infty}^{\infty} \mathrm{d}x\, s(x) \ .$$

b) By combining the last three equations, show that $T(\omega)$ *is also the autocorrelation of the pupil function,*

$$T(\omega) = (2\beta_0)^{-1} \int_{\omega-\beta_0}^{\beta_0} \mathrm{d}\beta \exp\{\mathrm{j}k[\Delta(\beta) - \Delta(\beta - \omega)]\} \ . \tag{3.65}$$

[*Hint:* Perform the integration dx, using representation (3.30g) of the Dirac delta function, then its sifting property (3.30d).]

Result (3.65) will be used extensively in statistical analyses of atmospheric "seeing" effects in Chap. 8.

The two-dimensional version of result (3.65) is

$$T(\omega) = A^{-1} \int_{\substack{\text{overlap}\\ \text{region}}} \mathrm{d}\boldsymbol{\beta} \exp\{\mathrm{j}k[\Delta(\boldsymbol{\beta}) - \Delta(\boldsymbol{\beta} - \omega)]\} \tag{3.66}$$

where

$$\omega \equiv (\omega_1, \omega_2) \tag{3.67}$$

are the two Cartesian components of frequency,

$$\boldsymbol{\beta} \equiv (\beta_1, \beta_2) \tag{3.68}$$

are the two reduced pupil coordinates, and

$$A \equiv \text{area of the pupil} \ . \tag{3.69}$$

The "overlap region" is the area common to pupil function $\exp[\mathrm{j}k\Delta(\beta)]$ and its shifted version $\exp[-\mathrm{j}k\Delta(\boldsymbol{\beta} - \boldsymbol{\omega})]$, with $\boldsymbol{\omega}$ the shift.

A *three-dimensional* optical transfer function also exists. It defines the diffraction image of a three-dimensional object [3.6].

c) The result (3.65) implies a value of $\omega \equiv \Omega$ beyond which $T(\omega) = 0$ regardless of the form of $\Delta(\beta)$. This Ω is called the "cutoff frequency" for the optical system. Show that

$$\Omega = 2\beta_0 \ . \tag{3.70}$$

d) Show by the use of (3.62) and (3.63) that if $\Delta(\beta) = 0$ the slit aperture diffraction pattern (3.16) results. This also has been made to have unit area, and to thus represent a probability law, by use of the proper multiplicative constant.

3.1.24 ***Van Cittert-Zernike theorem.*** In radio astronomy, extended images of distant objects are produced without the use of a lens. Instead, the correlation in amplitude is measured at pairs of points in a receiving plane. The Fourier transform of this correlation function then produces the object intensity pattern. We are in a position to derive this interesting result, shown next in one dimension for simplicity.

In Fig. 3.11, an extended, radiating object consists of elements of length $\mathrm{d}y$, one of which is centered on point y shown. The light amplitude at y is $a(y)$. The elements radiate incoherently, by which is meant

$$\langle a(y)a^*(y')\rangle = o(y)\delta(y - y') , \tag{3.71}$$

the average over time. The asterisk denotes complex conjugate. Radiance $o(y)$ is the required object profile.

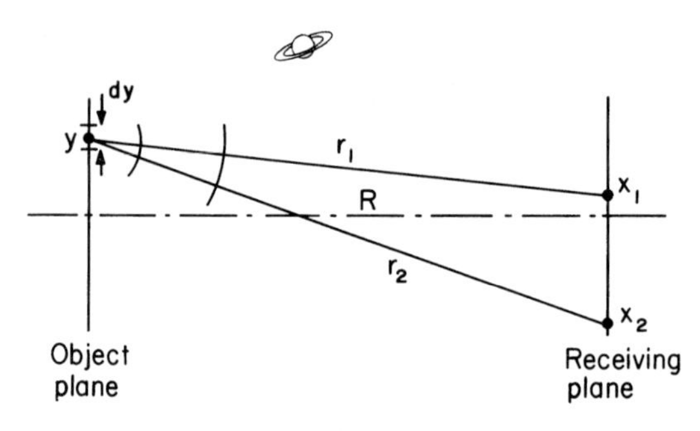

Fig. 3.11. Geometry for the Van Cittert-Zemike theorem. Note the absence of a lens

The time-average correlation in amplitude $\varrho_{12} \equiv \langle a(x_1)a^*(x_2)\rangle$ in the receiving plane is to be computed, both amplitudes measured at the same time. By Huygen's principle, amplitude $a(x_1)$ is a simple sum of spherical waves from each element $\mathrm{d}y$ of the object. Because distance R is so great, these are essentially plane waves, so that

$$a(x_1) = \int_{-\infty}^{\infty} \mathrm{d}y\, a(y) \exp[\mathrm{j}(kr_1 - 2\pi\nu t)] \tag{3.72}$$

where $k = 2\pi/\lambda$, λ is the light wavelength, and ν is its frequency. We have suppressed an inconsequential time delay in the argument y of $a(y)$. Amplitude $a(x_2)$ obeys a similar expression.

(a) By the use of (3.71) and (3.72), show that

$$\varrho_{12} = \int_{-\infty}^{\infty} \mathrm{d}y\, o(y) \exp[\mathrm{j}k(r_1 - r_2)] . \tag{3.73}$$

[*Hint:* The sifting property (3.30d) of Dirac $\delta(x)$ must be used, along with identity (3.28).] Hence, the observable correlation ϱ_{12} is a known transformation of the desired $o(y)$. This simplifies even further.

(b) Show by the use of

$$r_1 = [R^2 + (y - x_1)^2]^{1/2} \simeq R - x_1 y/R + (x_1^2 + y^2)/2R$$

and similarly for r_2, that

$$\varrho_{12} = \mathrm{e}^{\mathrm{j}b} \int_{-\infty}^{\infty} \mathrm{d}y\, o(y) \exp[-\mathrm{j}k(x_1 - x_2)y/R] . \tag{3.74}$$

Or, the observable correlation at separation $x_1 - x_2$ is the Fourier transform of the unknown object. Show that for ordinary stellar objects b is negligible compared to the rest of the exponent.

In radio astronomy, correlation ϱ_{12} is ideally observed at all possible separations $x_1 - x_2$, and this is inverse Fourier transformed to produce $o(y)$.

For later use, a quantity

$$\langle a(x_1, t_1)\, a^*(x_2, t_2)\rangle \equiv \Gamma_{12}(t_2 - t_1)$$

is called the "mutual coherence function." Evidently $\Gamma_{12}(0) = \varrho_{12}$. It is easy to show that

$$\Gamma_{12}(t) = \mathrm{e}^{-2\pi \mathrm{j} \nu t} \varrho_{12} \tag{3.75}$$

for this case, so that the mutual coherence function as well is linked to the Fourier transform of the distant object. Also, the concept of a "coherence time" $\Delta\tau$ is defined as

$$\Delta\tau^2 \equiv \int_{-\infty}^{\infty} \mathrm{d}t\, t^2 \Gamma_{11}(t) \Big/ \int_{-\infty}^{\infty} \mathrm{d}t\, \Gamma_{11}(t)\ . \tag{3.76}$$

This is the second moment of the mutual coherence where the observation points x_1, x_2 coincide spatially.

3.1.25 ***Maximum possible variance.*** Consider the RV x confined to an interval $a \leqq x \leqq b$.

(a) Show that regardless of $p(x)$

$$\sigma^2 \leqq (b-a)^2/4 \equiv \sigma^2_{\max}\ . \tag{3.77}$$

Hint: Use

$$\begin{aligned}\sigma^2 &\equiv \int_a^b \mathrm{d}x\, p(x)(x - \langle x\rangle)^2 \\ &= \int_a^b \mathrm{d}x\, p(x) \left[\left(x - \frac{b-a}{2}\right) + \left(\frac{b-a}{2} - \langle x\rangle\right)\right]^2 .\end{aligned}$$

Square out and evaluate integrals, use $[x - (b-a)/2]^2 \leqq (b-a)^2/4$.

(b) Show that $\sigma_{\max}$ is attained by the particular density

$$p(x) = \frac{1}{2}[\delta(x-a) + \delta(x-b)]\ , \tag{3.78}$$

with events concentrated at each of the extreme values of x only.

(c) Find the *minimum* value of σ^2 and the $p(x)$ attaining it, if RV x is confined to $a \leqq x \leqq b$.

3.1.26 ***Irradiance of an underwater object*** [3.7]. In Fig. 3.12, a light source is irradiating a screen located distance x away. The entire system is under water. Then the total light reaching the screen consists of two distinct components: direct light and scattered light due to interaction with large, suspended particles and plankton. In particular, the scattered radiation causes an irradiance $I(x)$ at the screen obeying

$$I(x) = 2.5 I_0 K(1 + 7\mathrm{e}^{-Kx})\mathrm{e}^{-Kx}/4\pi x\ , \tag{3.79}$$

where $x_0 \leqq x \leqq \infty$. [Assume that $I(x) = 0$ for $x < x_0$.] This is a semiempirical formula, based on diffusion theory. K is the attenuation coefficient for scattered light, I_0 is a known constant.

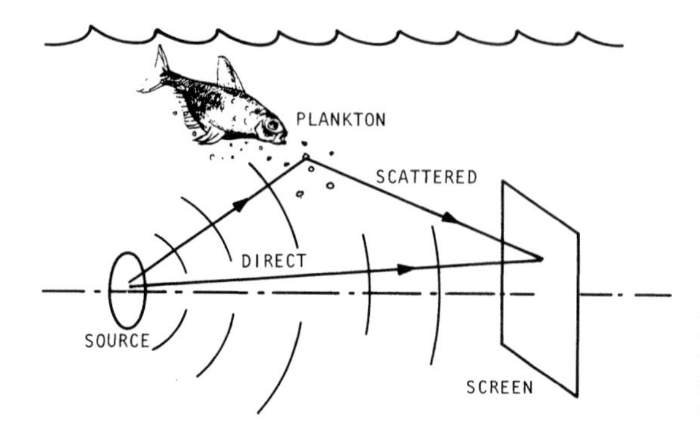

Fig. 3.12. Total illumination on a screen under water has a direct component and a scattered component

Alternatively, (3.79) may be regarded as the probability density for scattered photons at the screen. Suitably normalized, it represents the probability that a randomly selected source photon will reach the screen distance x away.

(a) Find the normalization constant that transforms (3.79) into a probability law.

(b) Find its moments by direct integration. The following integrals will be of use:

$$\int_0^\infty dx\, x^m\, e^{-x} = m!\,, \quad m = 0, 1, \ldots$$

$$\int_0^\infty dx \frac{e^{-ax}}{x+b} = -e^{ab}\mathrm{Ei}(-ab)\,, \quad a > 0,\ b > 0\,,$$

where Ei is the "exponential integral", defined as

$$\mathrm{Ei}(z) \equiv \int_{-\infty}^{z} dx\, x^{-1} e^{x}\,.$$

3.1.27 ***Forced-choice detection in physiological optics*** [3.8]. The "psychometric curve" for an individual is his probability of detecting a given stimulus as a function of its brightness above background. This is obtained as a success rate $P(C)$, C = correct identification of the stimulus. By "correct" is meant correct identification of *where* in a split (say) field the stimulus actually is.

One model of detection assumes that if a uniform brightness b is present, it will be perceived as a random brightness x from a probability law $p_X(x)$ *whose mean is* b. Suppose the observer looks at a split field, one of which contains a uniform brightness level b and the other a brightness level $b + d$. By the detection model, what will be the dependence of the psychometric curve $P(C)$ upon stimulus level d?

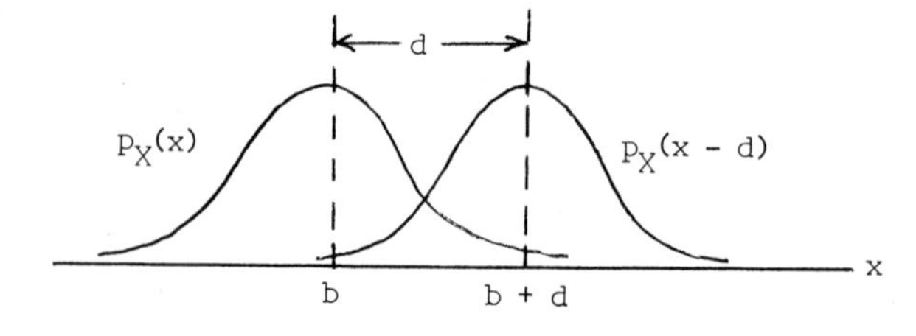

Fig. 3.13. Probability law $p_X(x)$ with mean level b, and its shifted version $p_X(x-d)$. What is the probability that a value y sampled from the latter will exceed a value x sampled from $p_X(x)$?

A correct identification is now the identification of the field that contains the incremental mean level d. Of course, the viewer does not perceive the fields to be at levels b and $b + d$, but rather at levels x and y as respectively sampled from the law[2] $p_X(x)$ centered on mean b and the law $p_X(y)$ centered on mean $b + d$. Hence, a correct identification is the event that $y \geqq x$. This is illustrated in Fig. 3.13. What is $P(y \geqq x)$ or, equivalently, $P(x \leqq y)$?

Analysis: By definition (3.1)

$$P(C) \equiv P(x \leqq y) = \int_{-\infty}^{\infty} \mathrm{d}y \int_{-\infty}^{y} \mathrm{d}x\, p(x, y) .$$

But the viewer perceives the two fields independently, so that

$$P(x, y) = p_X(x) p_Y(y) .$$

Also, $p_Y(y)$ is merely a shifted version of $p_X(x)$. By Fig. 3.13[3]

$$p_Y(y) = p_X(y - d) .$$

Then

$$P(C) = \int_{-\infty}^{\infty} \mathrm{d}y \int_{-\infty}^{y} \mathrm{d}x\, p_X(x) p_X(y - d) .$$

(a) Show that the latter collapses into *the convolution*

$$P(C) = \int_{-\infty}^{\infty} \mathrm{d}y\, F_X(y) p_X(y - d) \tag{3.80}$$

of the cumulative probability law for perceived brightness with its density law.

(b) More generally, there may be $m + 1$ fields present, where m fields are at brightness level b and one is at level $b + d$. Then a correct decision occurs when $y \geqq x_1, y \geqq x_2, \ldots,$ *and* $y \geqq x_m$, where the $\{x_n\}$ are the perceived brightnesses in the fields at mean level b. Show that then

$$P(C) = \int_{-\infty}^{\infty} \mathrm{d}y [F_x(y)]^m p_X(y - d) , \tag{3.81}$$

still in convolution form.

It may be noted that this procedure may be generalized to where the functional form of $p_X(x)$ changes with its mean. The result (3.81) would simply be amended by allowing $F(y)$ to be the cumulative law for a different function than $p_X(x)$.

It can be seen from (3.81) that if $p_X(x)$ is a normal curve with unit variance, $P(C)$ is only a function of two parameters, m and d. This allows it to be tabulated, as in [3.9].

3.1.28 ***Persistence of the Poisson law*** [3.10]. X-rays are being emitted by a medical specimen, at an average rate λ photons/time, during an exposure time T. The probability law for m photons so emitted is Poisson (3.38) with parameter $a = \lambda T$ (Sect. 6.6).

---

[2] By convention, the subscript to p names the RV in question. It is usually capitalized.

[3] And keeping in mind that y is a value of x.

A detector located some distance away has probability p of detecting a photon. And if m photons are emitted by the source, the probability $P(n|m)$ that n will be detected is binomial (3.39) with index $N = m$ (Sect. 6.1).

Show that the total probability $P(n)$ that n photons will be detected is *again* Poisson, now with parameter $a = \lambda Tp$. *Hint:* Use partition law (2.18) with a summation index m, $n \leq m \leq \infty$. Also, change variable $m - n \equiv i$.

3.16 Derivation of Heisenberg Uncertainty Principle

Physically, this principle states that certain pairs of physical attributes, such as position and momentum, cannot both be known simultaneously with arbitrarily small errors. If one error is small the other must be large, a property commonly called *reciprocity*. Such reciprocity follows because *the product* of the two errors must exceed a universal constant. There are at least two different ways to establish this product form. One way is to start from Eqs. (17.18) and (17.19), which states that *there is such a product* linking the mean-square error ϵ^2 in any estimated quantity with the Fisher information level I in its measurement. When this is applied to the particular measurement of a material particle's location, the result is the Heisenberg uncertainty principle [Ex. 17.3.6].

Since the uncertainty principle is so basic to physics, and results from consideration of measurement and information content, this suggests that maybe all of physics can be derived from such considerations. This topic is taken up again in Sect. 17.3.

The preceding derives the Heisenberg principle from the standpoint of measurement. Another, more standard, route to the principle is purely analytical in nature: as simply the statement that *a function and its Fourier transform cannot both be arbitrarily narrow.* This route is taken next. (See also Ex. 17.3.18.) First a preliminary relation must be stated.

3.16.1 Schwarz Inequality for Complex Functions

The Schwarz inequality, Eq. (17.21), is the statement that the dot product of two *vectors*, of whatever dimension, is a maximum when there is zero angle between them. This may be readily generalized to *complex functions* $G = G(x)$, $H = H(x)$ as a statement

$$\left| \int \mathrm{d}x G^* H \right|^2 \leq \int \mathrm{d}x |G|^2 \int \mathrm{d}x |H|^2 . \tag{3.82}$$

We now apply this inequality to the problem at hand.

3.16.2 Fourier Relations

Suppose that two complex functions ψ, ϕ are Fourier transforms of one another,

$$\psi(x) = \frac{1}{\sqrt{2\pi}} \int \mathrm{d}\omega\, \phi(\omega) \exp(-\mathrm{j}\omega x) \tag{3.83}$$

and also are normalized in the sense of

$$\int \mathrm{d}x\psi\psi^* = \int \mathrm{d}\omega\,\phi\phi^* = 1\,. \tag{3.84}$$

This normalization property, according to (3.2), holds when products $\psi\psi^*$and $\phi\phi^*$ are *probability densities*. Then the individual functions ψ, ϕ are called probability *amplitudes*. Of course this is precisely the case in quantum mechanics, where coordinate x is the position of a particle and ω is its momentum coordinate (in 'natural units' $\hbar = h/2\pi = 1$, h Planck's constant). We continue with the quantum mechanical application. Both coordinates x and ω are now fluctuations from their mean values.

By (3.13a) the mean-square fluctuations in position and momentum obey

$$\langle x^2 \rangle = \int \mathrm{d}x x^2\,\psi\psi^*, \quad \langle \omega^2 \rangle = \int \mathrm{d}\omega\,\omega^2\phi\phi^*\,. \tag{3.85}$$

We next work toward replacing the second integral by one over x. Differentiating (3.83) $\mathrm{d}/\mathrm{d}x$ shows that $\psi'(x)$ is the Fourier transform of $-\mathrm{j}\omega\phi(\omega)$. Using these, respectively, for functions $f(x)$ and $F(\omega)$ in Parseval's theorem (8.55) gives

$$\int \mathrm{d}x\psi'\psi'^* = \int \mathrm{d}\omega\,\omega^2\phi\phi^*\,. \tag{3.86}$$

3.16.3 Uncertainty Product

Using (3.86) in the 2nd Eq. (3.85) and multiplying by the 1st Eq. (3.85) gives

$$\langle x^2 \rangle \langle \omega^2 \rangle = \int \mathrm{d}x(x\psi)(x\psi)^* \int \mathrm{d}x\psi'\psi'^*\,. \tag{3.87}$$

Using $G = x\psi$, $H = \psi'$ in the Schwarz inequality (3.82) shows that the right side of (3.87) exceeds a certain quantity, so that the left side does as well,

$$\langle x^2 \rangle \langle \omega^2 \rangle \geq |A|^2, \quad A = \int \mathrm{d}x x\psi^*\psi'\,. \tag{3.88}$$

We may note that $A + A^* = \int \mathrm{d}x x(\psi^*\psi' + \psi\psi'^*) = \int \mathrm{d}x x p'$, where $p = \psi\psi^*$ is a probability density [see below (3.84)]. Integrating this by parts gives $\int \mathrm{d}x x p' = -\int \mathrm{d}x p = -1$, since p evaluated at the infinite limits is zero by the normalization requirement (3.84). Thus, $A + A^* = -1$, so that $2\,\mathrm{Re}(A) = -1$ or $2|A|\cos\theta = -1$ where θ is the phase of A. Squaring the last equality gives $4|A|^2 = 1/\cos^2\theta$, showing that $|A|^2 \geq 1/4$. Using the latter in (3.88) gives the desired result

$$\langle x^2 \rangle \langle \omega^2 \rangle \geq 1/4\,. \tag{3.89}$$

This is the Heisenberg uncertainty principle, stating reciprocity between fluctuations in position x and momentum ω. The right hand side $1/4$ is replaced more generally by $\hbar^2/4$ if natural units $\hbar = 1$ are not used.

3.17 Hirschman's Form of the Uncertainty Principle

The fundamental analytical effect in the Heisenberg uncertainty principle is that the probability densities for x and ω cannot both be *arbitrarily narrow*. Eq. (3.89) is but one way of expressing this fact. Entropy is another measure of the effective width of a probability density. For example, for a normal random variable x of variance σ^2, the entropy is $H(X) = \beta + \log \sigma$, with β a pure number (Ex. 3.2.5) so that *the entropy increases with the width of the probability law.*

On this basis, *Hirschman* [3.11] and *Beckner* [3.12] express the uncertainty principle in terms of entropies,

$$H(Y) + H(\Omega) \geq 1 + \ln(h/2) \ . \tag{3.90}$$

This states that the entropy (3.23) of $p(y)$ plus that of $p(\omega)$ must exceed a universal constant. Thus, by the above correspondence between entropy and width of probability law, it is impossible for the effective widths of $p(y)$ and $p(\omega)$ *to both* be arbitrarily narrow. Again, this implies that a narrow range of fluctuations *in both* y and ω is impossible. In the all-Gaussian case, Hirschman's inequality (3.90) precisely goes over into Heisenberg's (3.89) (Ex. 3.2.1).

The conceptual importance of Hirschman's inequality (3.90) lies in expressing Heisenberg uncertainty in terms of information quantities rather than the usual "spreads" $\langle \omega^2 \rangle$ and $\langle x^2 \rangle$. Are there other information quantities that are candidates for this purpose?

3.18 Measures of Information

The Shannon measure of information (3.21) is actually a measure of the "distance" between two probability densities $p(y|x)$ and $p(y)$. For example, notice that if the two are the same the "distance" is zero. There are other measures of the distance between two densities, and these are likewise called "information" measures. Let the two densities be generally denoted as $p(x)$ and $r(x)$. The latter is often called a "reference" density or function. The following is a list of the more important alternative information measures $I(p, r)$:

3.18.1 Kullback–Leibler Information

$$I_{KL}(p, r) = -\int \mathrm{d}x \, p(x) \log[p(x)/r(x)] \ . \tag{3.91}$$

See *Kullback* [3.13]. This is also called the "cross entropy". In the particular case $r(x) = 1$ it reverts back to (3.23), the entropy. The integral (3.91) may be generalized in the obvious way to two-dimensional cases of variables (x, y). In the particular two-dimensional case of $p = p(y)$, $r = p(y|x)$ the Kullback–Leibler information becomes the Shannon information (3.21). Hence I_{KL} is actually the more general of the two.

The form (3.91) may be used to estimate an unknown density p in the presence of a known reference function r and input data. The latter can be, for example, known moments of p. (See Chap. 10.) The answer for p is then always proportional to r. Hence, the estimate p follows the hills and valleys of the input function r, except as the data constrain it away from these. Then r truly operates as a "reference" function for p. See Ex. 3.2.2.

The cross-entropy also arises as the maximum-probable estimate of an empirical rate; see (10.50) and (10.56).

3.18.2 Renyi Information

$$I_R(p, r; \alpha) = (1-\alpha)^{-1} \log \int \mathrm{d}x\, p(x)^\alpha r(x)^{1-\alpha} . \tag{3.92}$$

See *Renyi* [3.14]. This has a free parameter $\alpha > 0$. Its significance depends upon the application.

In the limit as $\alpha \to 1$ the Renyi information approaches the Kullback–Leibler information (3.91). See Ex. 3.2.4.

For a special choice of reference function $r(x)$, information I_R becomes proportional to the Fisher information [(3.97) below]. See Ex. 3.2.9.

3.18.3 Wootters Information

$$I_W(p, r) = \cos^{-1} \int \mathrm{d}x \sqrt{p(x) r(x)} . \tag{3.93a}$$

See *Wootters* [3.15]. For a special choice of reference function $r(x)$, information I_W becomes proportional to the *Fisher* information [(3.97) below]. See Ex. 3.2.10.

A related information is the simple *inner product* between two probabilities,

$$I_{IP}(p, r) = \int \mathrm{d}x\; p(x) r(x) . \tag{3.93b}$$

It can easily be shown (Ex. 3.2.14) that

$$I_{IP}(p, r) \le \sqrt{I_{GS}(p) I_{GS}(r)} . \tag{3.93c}$$

The new information I_{GS} is called the *Gini-Simpson information* [3.21], defined as $I_{GS}(p) \equiv I_{IP}(p, p) = \int \mathrm{d}x p^2 = \langle p \rangle$, the average of the probability density. In the case of a normal law $p(x)$, information

$$I_{GS}(p) = \frac{1}{\sqrt{4\pi}} \frac{1}{\sigma} . \tag{3.93d}$$

This is to be compared with the Fisher information (3.99) for the same case, $I_F = 1/\sigma^2$.

The complement $1 - I_{GS}(p)$ to the Gini-Simpson information is called the Gini-Simpson index of "discrimination" or "diversity", and is often used in the fields of population biology and anthropology. It is easily shown (Ex. 3.2.15) that the index of discrimination is maximized by a uniform probability law $p(x)$.

3.18.4 Hellinger Information

$$I_H(p, r; \alpha) = \int \mathrm{d}x [p(x)^\alpha - r(x)^\alpha]^{1/\alpha} . \tag{3.94}$$

This is another measure with a free parameter $\alpha > 0$. Often the value $\alpha = 1/2$ is used [3.19].

There are other information measures which are not distance measures to a reference function $r(x)$ but, instead, are *absolute measures* of the density p. Some of these are as follows:

3.18.5 Tsallis Information

In the special case where the reference function in (3.94) is $r(x) = p(x)^{1/\alpha}$, if one ignores the exponent $1/\alpha$ and adds an appropriate denominator, (3.94) becomes the Tsallis information [3.16]

$$I_T(p; \alpha) = \frac{1}{\alpha - 1} \int \mathrm{d}x [p(x) - p(x)^\alpha] . \tag{3.95}$$

Again, this has a free parameter $\alpha > 0$. In the limit as $\alpha \to 1$ the information $I_T \to H(x)$, the entropy (3.23). See Ex. 3.2.3.

3.18.6 Fisher Information

$$I_F(p_a; \alpha) = \int \mathrm{d}x \, p_a \left(\frac{\partial \log p_a}{\partial a} \right)^2 , \quad p_a = p(x|a) . \tag{3.96}$$

See *Fisher* [3.17], also (17.19). Density p_a is here a *likelihood* law (see examples (17.1), (17.2)), which is a probability law that is conditional upon a parameter α. Fisher information is a measure of how well a parameter can be estimated from a measurement; see (17.18).

In the particular case of shift invariance, where $p_a(x|a) = p(x - a)$ and p is the probability density on the shift (e.g., the system noise), a change of integration variable in Eq. (3.96) gives

$$I_F \equiv I_F(p) = \int \mathrm{d}x \, \frac{(dp/dx)^2}{p} , \quad p = p(x) . \tag{3.97}$$

We have redefined x to be the new integration variable. *This information provides a basis for deriving many probability laws of physical origin*; see Chap. 17, Sect. 17.3.

The Fisher information (3.97) is actually proportional to a cross entropy (3.91) with special choice of reference function,

$$I_F = \lim_{\Delta x \to 0} \left(\frac{2}{\Delta x^2} \right) \int \mathrm{d}x \, p(x) \log \frac{p(x)}{p(x + \Delta x)} . \tag{3.98}$$

See Ex. 3.2.5. The reference function is an infinitesimally shifted version $p(x+\Delta x)$ of the probability law $p(x)$. Thus, Fisher information may be thought of as the "distance" between a density $p(x)$ and its slightly shifted version $r(x) \equiv p(x+\Delta x)$, a kind of Kullback–Leibler *self-distance*.

Fisher information is likewise a self-distance of many other information measures, including Renyi information and Wootters' information. See Exs. 3.2.9 and 3.2.10. In this regard Fisher information is a "mother information".

In the case of a normal density function $p(x)$ of variance σ^2

$$I_F = \frac{1}{\sigma^2} \; . \tag{3.99}$$

See Ex. 3.2.7. The Fisher information is the reciprocal of the variance for a host of probability laws, those belonging to the *exponential family* of laws. Examples are the normal, exponential, binomial, geometric and χ^2 laws. Also, compare (3.99) with (3.93d), the Gini-Simpson information for the same case.

Fisher information has definite units, in contrast with Shannon information. Notice in (3.99) that the units are those of the reciprocal of a squared data value. However, a units-free Fisher information may be defined. See Ex. 3.2.11.

3.18.7 Fisher Information Matrix

Fisher information is probably most well-known in its *matrix form*. This arises naturally in the context of the following problem of estimation.

Suppose you are given data $y_1, \ldots, y_N \equiv \boldsymbol{y}$ about *some parameters* $a_1, \ldots, a_K \equiv \boldsymbol{a}$ *that are desired to be known*. You also know the conditional probability law $p(\boldsymbol{y}|\boldsymbol{a})$ of all possible data $\boldsymbol{y}$ in the presence of the parameters. This is called *the likelihood law*. Given only the data, the estimates $\hat{\boldsymbol{a}}$ of $\boldsymbol{a}$ can only be formed out of them, as functions $\hat{\boldsymbol{a}}(\boldsymbol{y})$. How good can such estimates be, in the sense of incurring smallest possible mean-square errors

$$\left\langle (\hat{a}_k(\boldsymbol{y}) - a_k)^2 \right\rangle \equiv \int \mathrm{d}\boldsymbol{y} (\hat{a}_k(\boldsymbol{y}) - a_k)^2 p(\boldsymbol{y}|\boldsymbol{a}) \equiv \epsilon_k^2, \; k = 1, \ldots K? \tag{3.100}$$

(Notice that averaging $\langle \cdots \rangle$ is always with respect the likelihood law $p(\boldsymbol{y}|\boldsymbol{a})$.)

Assume that any worthwhile estimator will be unbiased, that is, on average will give the correct answer,

$$\left\langle \hat{\boldsymbol{a}}(\boldsymbol{y}) \right\rangle = \boldsymbol{a} \; . \tag{3.101}$$

Next, we prove a later-needed identity:

$$\left\langle \hat{a}_j(\boldsymbol{y}) \frac{\partial \ln p}{\partial a_k} \right\rangle \equiv \int \mathrm{d}\boldsymbol{y} \, \hat{a}_j \, p \frac{1}{p} \frac{\partial p}{\partial a_k} = \int \mathrm{d}\boldsymbol{y} \, \hat{a}_j \frac{\partial p}{\partial a_k} \tag{3.102}$$
$$= \frac{\partial}{\partial a_k} \int \mathrm{d}\boldsymbol{y} \, \hat{a}_j \, p = \frac{\partial \langle \hat{a}_j \rangle}{\partial a_k} = \frac{\partial a_j}{\partial a_k} = \delta_{jk} \; ,$$

the Kronecker delta function, where $p \equiv p(\boldsymbol{y}|\boldsymbol{a})$. The first equality follows from the definition of the expectation and the formula for the derivative of a logarithm, the second from a cancellation, the third because $\hat{a}_j$ depends upon $\boldsymbol{y}$ and not a_k, the fourth from the definition of the expectation, and the fifth as an elementary property of partial derivatives.

The route to finding ϵ_k^2 is as follows. First do the case $k = 1$. Form a vector

$$\boldsymbol{x} \equiv \begin{bmatrix} \hat{a}_1 - a_1 \\ \partial \ln p/\partial a_1 \\ \cdot \\ \cdot \\ \cdot \\ \partial \ln p/\partial a_K \end{bmatrix} \tag{3.103}$$

From this we may form a mean "outer product" matrix

$$\langle \boldsymbol{x}\boldsymbol{x}^{\mathrm{T}} \rangle = \begin{bmatrix} \langle(\hat{a}_1 - a_1)^2\rangle & \langle(\hat{a}_1 - a_1)\frac{\partial \ln p}{\partial a_1}\rangle & \langle(\hat{a}_1 - a_1)\frac{\partial \ln p}{\partial a_2}\rangle & \cdots \\ \langle(\hat{a}_1 - a_1)\frac{\partial \ln p}{\partial a_1}\rangle & \langle(\frac{\partial \ln p}{\partial a_1})^2\rangle & \langle\frac{\partial \ln p}{\partial a_1}\frac{\partial \ln p}{\partial a_2}\rangle & \cdots \\ \langle(\hat{a}_1 - a_1)\frac{\partial \ln p}{\partial a_2}\rangle & \langle\frac{\partial \ln p}{\partial a_2}\frac{\partial \ln p}{\partial a_1}\rangle & \langle(\frac{\partial \ln p}{\partial a_2})^2\rangle & \cdots \\ \cdot & \cdot & \cdot & \\ \cdot & \cdot & \cdot & \\ \cdot & \cdot & \cdot & \end{bmatrix} \tag{3.104}$$

By definition the mean is to be of each matrix element. The matrix is symmetric. The Fisher information matrix $[I]$ is defined as the matrix *excluding the top row and column.* Thus, $[I]$ has general elements

$$I_{ij} \equiv \int \mathrm{d}\boldsymbol{y}\, p \frac{\partial \ln p}{\partial a_i}\frac{\partial \ln p}{\partial a_j}, \quad p \equiv p(\boldsymbol{y}|\boldsymbol{a}) \tag{3.105}$$

for $i, j = 1, \ldots, K$. We next find a very important significance to this matrix.

The elements of the top row (and column, by symmetry) of matrix (3.104) may be readily evaluated. Element (1, 1) is just ϵ_1^2 by (3.100). Element $(1, 2) = (2, 1)$ is evaluated in steps as

$$\begin{aligned}(1, 2) &= \langle \hat{a}_1 \partial \ln p/\partial a_1 \rangle - \langle a_1 \partial \ln p/\partial a_1 \rangle = 1 - a_1 \langle \partial \ln p/\partial a_1 \rangle \\ &= 1 - a_1 \int \mathrm{d}\boldsymbol{y}\, p(1/p)\partial p/\partial a_1 = 1 - a_1 \partial/\partial a_1 \int \mathrm{d}\boldsymbol{y}\, p = 1.\end{aligned} \tag{3.106}$$

The first equality is trivial, the second by (3.102) and the constancy of the parameter a_1, the third by the formula for differentiating a logarithm, the fourth by a cancellation and since p is not a function of a_1, and the fifth is by the constancy of a normalization integral. Element $(1, 3) = (3, 1) = 0$, by the same steps, since the first right-hand term of (3.106) is now $\langle \hat{a}_1 \partial \ln p/\partial a_2 \rangle = 0$ by (3.102). In the same way, elements $(1, 4), (1, 5), \ldots, (1, K + 1)$ and the corresponding column elements are likewise zero. The result is that (3.104) becomes

$$\langle \boldsymbol{x}\boldsymbol{x}^{\mathrm{T}} \rangle = \begin{bmatrix} \epsilon_1^2 & 1 & 0 & \cdots & 0 \\ 1 & I_{11} & I_{12} & \cdots & I_{1K} \\ 0 & I_{21} & I_{22} & \cdots & I_{2K} \\ \cdot & \cdot & \cdot & & \cdot \\ \cdot & \cdot & \cdot & & \cdot \\ 0 & I_{K1} & \cdot & \cdots & I_{KK} \end{bmatrix} \tag{3.107}$$

in terms of the error ϵ_1^2 and Fisher information matrix elements I_{ij}.

Because of its construction as an outer product $\langle \boldsymbol{x}\boldsymbol{x}^{\mathrm{T}} \rangle$, the matrix (3.107) is *positive definite*. This means in particular that its determinate must be positive or zero. Taking the determinant by expanding by cofactors along the top row gives

$$\epsilon_1^2 \det[I] - 1 \cdot \det \begin{bmatrix} 1 & I_{12} & \cdots & I_{1K} \\ 0 & I_{22} & \cdots & I_{2K} \\ \cdot & \cdot & & \cdot \\ \cdot & \cdot & & \cdot \\ \cdot & \cdot & & \cdot \\ 0 & I_{K2} & \cdots & I_{KK} \end{bmatrix} \geq 0\,. \tag{3.108}$$

Expanding the right-hand matrix by cofactors along its leftmost column gives directly $Cof\ I_{11}$. The result is that

$$\epsilon_1^2 \geq \frac{Cof\ I_{11}}{\det[I]} \equiv [I]_{11}^{-1}\,, \tag{3.109}$$

the $(1, 1)$ element of the inverse matrix $[I]^{-1}$ to $[I]$.

An inequality for the k'th error ϵ_k^2 may be developed in the same way. See Ex. 3.2.12. The result is

$$\epsilon_k^2 \geq [I]_{kk}^{-1}\,, \quad k = 1, \ldots, K\,. \tag{3.110}$$

When does the equality sign hold in this expression? That is, when is the lowest possible error achievable by the k'th estimator? This is when the likelihood law $p(\boldsymbol{y}|\boldsymbol{a}) \equiv p$ obeys a relation

$$\hat{a}_k(\boldsymbol{y}) - a_k = \sum_{j=1}^{K} B_j \frac{\partial \ln p}{\partial a_j}\,, \qquad B_j = \text{const}\,. \tag{3.111}$$

The estimator $\hat{a}_k(\boldsymbol{y})$ is then termed *efficient*. See Ex. 3.2.13.

Exercise 3.2

3.2.1 ***Equivalence of two uncertainty principles.*** Show that when probability densities $\psi(x)$ and $\phi(\omega)$ are Gaussian, Hirschman's inequality (3.90) becomes Heisenberg's (3.89). Show also that in this case both inequalities become equalities, so that the system obeys minimum uncertainty overall. In quantum mechanics the system is said to obey a "minimum *uncertainty product*". We see by (3.90) that it also obeys a minimum *entropy sum*.

3.2.2 ***Maximum cross entropy solutions.*** Show that when Eq. (3.91) is extremized in choice of density p subject to data constraints F_k as in Eq. (10.25), the answer is

$$\hat{p}(x) = q(x)\exp[-1 + \mu + \sum_k \lambda_k f_k(x)] \,. \tag{3.112}$$

Hint: Follow the procedure of Sect. 10.4.7. This relation shows that the reference function q also acts like a *bias function* for $\hat{p}$. At values x for which q is high, $\hat{p}$ will likewise tend to be high, predicting that these x occur with high probability; and conversely low for x for which q is low. This allows a high degree of prior knowledge to be inserted into the approach, and is generally beneficial. The exception is when the bias function $q(x)$ is misregistered from the true $p(x)$. Then the entire approach suffers badly. See [3.18].

3.2.3 ***Entropy as limit of Tsallis information.*** Show that in the limit as parameter $\alpha \to 1$ the Tsallis information (3.95) obeys $I_T(p;\alpha) \to H(X)$, the entropy. *Hint:* Use l'Hopital's rule and the identity $\frac{\mathrm{d}}{\mathrm{d}x}(b^x) = b^x \ln b,\ \ b = \text{const.}$

3.2.4 ***Cross entropy as limit of Renyi information.*** Show that in the limit as $\alpha \to 1$ the Renyi information (3.92) approaches the cross entropy (3.91). *Hint:* As in the preceding problem.

3.2.5 ***Fisher information as a limiting cross entropy.*** Show that in the indicated limit in (3.98) the latter becomes proportional to the Fisher information (3.97). *Hint:* In the indicated limit, (3.98) becomes an indeterminate form 0/0. Therefore l'Hopital's rule must be used. In view of the second-order factor Δx^2 in the denominator, this rule must be used twice.

3.2.6 ***Entropy for a normal probability law.*** Show by direct substitution of the normal form (3.42) into the entropy formula (3.23) that the entropy for this case obeys

$$H = \frac{1}{2} + \ln(\sqrt{2\pi}\sigma) \,. \tag{3.113}$$

Hint: Evaluate H as the expectation $-\langle \log p \rangle$, avoiding the explicit evaluation of integrals by identifying the expectations as moments of p.

3.2.7 ***Fisher information for a normal law.*** By direct substitution of a normal form (3.42) for p in Eq. (3.97) show that the resulting Fisher information I_F is the reciprocal of the variance. *Hint*: Represent I_F as an expectation $\langle (d \ln p/\mathrm{d}x)^2 \rangle$. Avoid the explicit evaluation of integrals by simply identifying the expectations as moments of p.

3.2.8 ***Bounded nature of correlation coefficient.*** Show that the definition (3.47) of ϱ guarantees that it obeys boundedness

$$0 \le |\varrho| \le 1 \,. \tag{3.114}$$

Hint: Use the Schwarz inequality (3.82). Notice that the coefficient can go negative. What does this mean? Discuss the cases $\varrho = -1, 0, +1$.

3.2.9 ***Fisher information as a limit of Renyi information.*** Show that with the choice $q(x) = \sqrt{p(x+\Delta x)}$ in (3.92), in the limit as $\Delta x \to 0$, $I_R(p,q) \to \Delta x^2 2^{-1} \alpha I_F(p)$. *Hint:* Expand out $p(x+\Delta x)$ in Taylor series to second order in Δx, performing the needed integrations using the identities $\int \mathrm{d}x\, p'(x) = \int \mathrm{d}x\, p''(x) = 0$.

3.2.10 ***Fisher information as a limit of Wootters' information.*** Show that with the choice $q(x) = \sqrt{p(x+\Delta x)}$ in (3.93), in the limit as $\Delta x \to 0$, $I_W(p,q) \to \Delta x^2 4^{-1} I_F(p)$. *Hint:* As in previous problem.

3.2.11 ***Units-free Fisher information.*** A form of Fisher information that lacks units is [3.20]

$$F(p_a;\alpha) = \int \mathrm{d}x (x-a)^2\, p_a \left(\frac{\partial \log p_a}{\partial a}\right)^2, \quad p_a = p(x|a)\,. \tag{3.115}$$

Using dimensional analysis of the integrand, show the lack of units. Show that in the shift-invariant case $p_a(x|a) \equiv p(x-a)$, Eq. (3.115) becomes simply

$$F(p) = \int \mathrm{d}x\, x^2 \frac{(\mathrm{d}p/\mathrm{d}x)^2}{p}, \quad p = p(x)\,. \tag{3.116}$$

This may be compared with (3.97). The extra factor x^2 cancels the units.

3.2.12 Generalize the inequality (3.109) for ϵ_1^2 to the result (3.110) for ϵ_k^2, using an analogous approach. *Hint*: define the helper vector $\boldsymbol{x}$ as in (3.103) except with the top element replaced by $\hat{a}_k(\boldsymbol{y}) - a_k$.

3.2.13 By considering the elements of the matrix formed by the outer-product

$$\boldsymbol{x}\boldsymbol{x}^{\mathrm{T}} \equiv \begin{bmatrix} x_1 \\ x_2 \\ \cdot \\ \cdot \\ \cdot \\ x_N \end{bmatrix} \begin{bmatrix} x_1\ x_2 \cdots x_N \end{bmatrix}, \tag{3.117}$$

show that if element x_1 is a linear combination of the other elements then the top row of the product matrix is a linear combination of the other rows. Then the determinant of the matrix is zero. Hence, with an $\boldsymbol{x}$ formed as in (3.103) show that the efficiency condition (3.111) results for the case $k = 1$.

3.2.14 Prove that the inner-product information $I_{IP}(p,r)$ is bounded by the geometric mean between Gini-Simpson informations as in (3.93c), $I_{IP}(p,r) \le \sqrt{I_{GS}(p) I_{GS}(r)}$. *Hint:* Use the Schwarz inequality.

3.2.15 Show that in the case of a finite range $a \le x \le b$ of x the Gini-Simpson index of discrimination $1 - \langle p \rangle$ is maximized by a uniform probability law $p(x)$. *Hint:* Use the Euler-Lagrange equation (G.14) of Appendix G, with a constrained Lagrangian (G.19) containing an added normalization term.

4. Fourier Methods in Probability

There is probably no aspect of probability theory that is easier to learn than its Fourier aspect. All of the linear theory [4.1] involving convolutions, Dirac delta functions, transfer theorems, and even sampling theorems has its counterparts in probability theory.

This is fortunate, since the majority of problems the opticist encounters that are statistical in nature seem to fall within the Fourier regime. In other words, they are *not* of the "permutation and combination" variety encountered in secondary school problems, or in Chap. 2. To this author, at least, those were much more difficult problems. No, in practice the typical problem encountered by an opticist is the calculation of a mean value of a quantity of interest, that is, evaluation of an integral. It may take some ingenuity to find the proper $p(x)$ function to place within the integral, but then again the task is not often that difficult, thanks to Fourier methods and, sometimes, the devising of a simple physical model (more on this later).

The connection with Fourier theory arises from the concept of a "characteristic function," defined next.

4.1 Characteristic Function Defined

Consider the operation of taking the (inverse) Fourier transform of the probability density function $p(x)$,

$$\varphi(\omega) \equiv \int_{-\infty}^{\infty} \mathrm{d}x\, p(x)\, \mathrm{e}^{\mathrm{j}\omega x} \,. \tag{4.1}$$

Parameter ω is conjugate to x, in the sense that the product ωx must be unitless, the latter because of term $\exp(\mathrm{j}\omega x)$. Hence, *the unit for* ω must be inverse to that of x. We shall sometimes call ω a "frequency," in analogy to the quantity in communications theory. It will not always, however, have units of frequency (time^{-1} or length^{-1}).

Quantity $\varphi(\omega)$ is called the "characteristic function" for random variable x. What significance does $\varphi(\omega)$ have? First of all, noting definition (3.12) of the expectation, we see that in actuality

$$\varphi(\omega) = \left\langle \mathrm{e}^{\mathrm{j}\omega x} \right\rangle \,. \tag{4.2}$$

That is, it represents the average response to a sinusoid, which is in the spirit of linear theory (Fig. 4.1). But of what use is $\varphi(\omega)$?

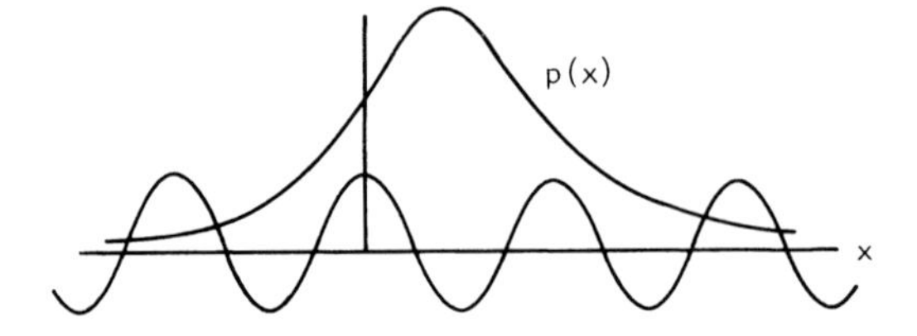

Fig. 4.1. The integrated product of the two curves is the characteristic function (real part)

4.2 Use in Generating Moments

By taking successive derivatives of (4.1) at $\omega = 0$ we see that

$$\varphi^{n}(0) = \mathrm{j}^{n} \int_{-\infty}^{\infty} \mathrm{d}x\, x^{n} p(x) = \mathrm{j}^{n} \left\langle x^{n} \right\rangle , \quad n = 1, 2, \ldots \tag{4.3}$$

with the particular value $\varphi^{(0)}(0) \equiv \varphi(0) = 1$.

Hence, the behavior of $\varphi(\omega)$ at the origin defines all the moments of $p(x)$. This has practical use with regard to *computing* the moments, since it is usually easier to differentiate $\varphi(\omega)$ than to integrate $x^{m} p(x)$ (provided $\varphi(\omega)$ is known analytically).

4.3 An Alternative to Describing RV x

This suggests that $\varphi(\omega)$ offers an alternative to $p(x)$ for describing the random nature of x. Of course it does, since (4.1) may be inverted, to give the Fourier transform

$$p(x) = (2\pi)^{-1} \int_{-\infty}^{\infty} \mathrm{d}\omega\, \varphi(\omega)\, \mathrm{e}^{-\mathrm{j}\omega x} \tag{4.4}$$

Hence, either $p(x)$ or $\varphi(\omega)$ suffices to describe RV x completely.[1]

4.4 On Optical Applications

A point worth noting about $\varphi(\omega)$ is that *optical amplitude* (3.61) is precisely of the form $\exp(\mathrm{j}\omega x)$, if x is regarded as optical phase and frequency $\omega \equiv k = 2\pi/\lambda$. Hence by (4.2) $\varphi(\omega)$ actually has a *physical* interpretation, namely, the average amplitude at a point in the field of a randomly fluctuating phase front.

Another optical application arises from the fact that the spread function $s(x)$ is a probability law on photon position (Sect. 3.4). Now by (3.64) the Fourier transform of $s(x)$ defines the optical transfer function $T(\omega)$. Hence, $T(\omega)$ is both the optical transfer function and the characteristic function for probability law $s(x)$. This identification is taken advantage of in the theory of atmospheric turbulence (Sect. 8.6.2).

[1] Note, however, that if the RV is discrete, the use of (4.4) to construct $p(x)$ will give rise to Dirac delta functions as in (3.31). The *multipliers* of these will be the required $\{P_n\}$.

4.5 Shift Theorem

It is well known in linear theory that a linear phase term in the Fourier domain is equivalent to a shift in the space domain. This has its counterpart here. Suppose RV x undergoes a linear shift, to form RV

$$y = ax + b\,, \qquad a, b\,\text{constants}\,.$$

What is the characteristic function $\varphi_Y(\omega)$ for RV y? (By convention, the subscript to φ or to p denotes the RV in question.)

Using definition (4.2),

$$\varphi_Y(\omega) = \left\langle e^{j\omega y}\right\rangle = \left\langle e^{j\omega(ax+b)}\right\rangle = e^{j\omega b}\left\langle e^{j(a\omega)x}\right\rangle\,,$$

or

$$\varphi_Y(\omega) = e^{j\omega b}\varphi_X(a\omega)\,. \tag{4.5}$$

This is called the "shift" theorem.

Characteristic functions for the commonest probability laws are found next. For other laws, see the Characteristic function listing in the Subject Index.

4.6 Poisson Case

The Poisson probability law is a *discrete* one (3.38), whereas the definition (4.1) assumes a continuous $p(x)$ in use. But, by using the picket-fence identity (3.31), we may now define a characteristic function for this discrete case. Definition (4.1) becomes

$$\varphi(\omega) = \int_{-\infty}^{\infty} dx\, e^{j\omega x}\sum_{n=0}^{\infty} P_n\delta(x - x_n) = \sum_{n=0}^{\infty} P_n\, e^{j\omega x_n} \tag{4.6}$$

by the sifting property. Note that this is $\langle \exp(j\omega x_n)\rangle$ which directly corresponds to version (4.2).

To evaluate $\varphi(\omega)$ for the Poisson case we merely substitute (3.38) for P_n into (4.6), and let $x_n = n$. Then

$$\varphi(\omega) = \sum_{n=0}^{\infty}(e^{-a}a^n/n!)\,e^{j\omega n} = e^{-a}\sum_{n=0}^{\infty}(a\,e^{j\omega})^n/n!\,,$$

or

$$\varphi(\omega) = e^{-a}\,e^{a\,e^{j\omega}}\,, \tag{4.7}$$

having recognized the series for $\exp(\cdot)$. It is not often that we see a *double exponential* in actual use. By successively differentiating (4.7) we easily find the successive moments of the Poisson RV, via identity (4.3). The mean, variance σ^2, and (remarkably) third central moment all equal a.

4.7 Binomial Case

The binomial law (3.39) is discrete, so that we once again use the discrete form (4.6) of the characteristic function. With in addition $x_n = n$, it becomes

$$\varphi(\omega) = \sum_{m=0}^{N} \mathrm{e}^{\mathrm{j}\omega m} \binom{N}{m} p^m q^{N-m} .$$

To evaluate this sum write it as

$$\varphi(\omega) = \sum_{m=0}^{N} \binom{N}{m} (p\mathrm{e}^{\mathrm{j}\omega})^m q^{N-m} ,$$

and compare it with a mathematical identity called the *binomial theorem,*

$$(a+b)^N \equiv \sum_{m=0}^{N} \binom{N}{m} a^m b^{N-m} .$$

Directly the answer is

$$\varphi(\omega) = (p\mathrm{e}^{\mathrm{j}\omega} + q)^N . \tag{4.8}$$

The moments may be found as usual by successively differentiating (4.8). In this manner we obtain the mean and variance

$$\langle m \rangle = Np \tag{4.9}$$

and

$$\sigma_M^2 = Np(1-p) . \tag{4.10}$$

4.8 Uniform Case

Direct integration of (4.1) for the case (3.40) of uniform probability yields

$$\varphi(\omega) = \mathrm{e}^{\mathrm{j}a\omega} \operatorname{sinc}\left(\frac{b\omega}{2}\right) . \tag{4.11}$$

Regarding moments, this is one case where it is easier to integrate directly in definition (3.13a) than it is to differentiate (4.11) many times over.

4.9 Exponential Case

Substitution of exponential law (3.41) into definition (4.1) yields

$$\varphi(\omega) = \frac{1}{1 - \mathrm{j}a\omega} \tag{4.12}$$

as the characteristic function.

Successive differentiation of (4.12) shows that moment $m_k = a^k k!$ (see also Ex. 3.1.14).

4.10 Normal Case (One Dimension)

Substitution of normal law (3.42) into definition (4.1) yields

$$\varphi(\omega) = \exp(\mathrm{j}\omega \langle x \rangle - \sigma^2\omega^2/2) \tag{4.13}$$

after completing the square.

The first few moments are easily obtained by differentiation, but things rapidly get out of hand for higher orders. See Ex. 3.1.15 for the general answer.

4.11 Multidimensional Cases

In complete analogy with definition (4.1), we may define the joint characteristic function $\varphi(\omega_1, \omega_2, \ldots, \omega_n) \equiv \varphi(\boldsymbol{\omega})$ for a joint probability density $p(\boldsymbol{x})$ as

$$\varphi(\boldsymbol{\omega}) \equiv \int_{-\infty}^{\infty} \mathrm{d}\boldsymbol{x}\, p(\boldsymbol{x})\, \mathrm{e}^{\mathrm{j}\boldsymbol{\omega}\cdot\boldsymbol{x}} = \left\langle \mathrm{e}^{\mathrm{j}\boldsymbol{\omega}\cdot\boldsymbol{x}} \right\rangle \tag{4.14}$$

by (3.12).

The properties of $\varphi(\boldsymbol{\omega})$ are in complete analogy with one-dimensional version $\varphi(\omega)$. For example, regarding the generation of moments, we have

$$\left.\frac{\partial^{(k)}\varphi(\boldsymbol{\omega})}{\partial\omega_r^k}\right|_{\omega=0} = \mathrm{j}^k \int_{-\infty}^{\infty} \mathrm{d}x_r\, p(x_r) x_r^k \equiv \mathrm{j}^k \left\langle x_r^k \right\rangle , \tag{4.15}$$

where identity (3.8) was used. Hence, the kth moment of RV x_r relates simply to the kth derivative with respect to ω_r of the joint characteristic function. Similar identities hold for *mixed* moments of the form $\langle x_r^k x_s^m \rangle$, etc.

4.12 Normal Case (Two Dimensions)

Substitution of the normal bivariate law (3.44) into definition (4.14) with $n = 2$ yields

$$\begin{aligned}\varphi(\omega_1, \omega_2) = {} & \exp[\mathrm{j}(\omega_1 \langle x \rangle + \omega_2 \langle y \rangle) \\ & -2^{-1}(\sigma_1^2\omega_1^2 + \sigma_2^2\omega_2^2 + 2\sigma_1\sigma_2\varrho\omega_1\omega_2)] .\end{aligned} \tag{4.16}$$

This result applies directly to the problem of finding the long-term average optical transfer function due to turbulence. The latter is characterized by a normal bivariate phase distribution (Sect. 8.6.2).

4.13 Convolution Theorem, Transfer Theorem

The probability density function $p(x)$ is conceptually quite similar to that of the point impulse response in linear systems theory. That is, $p(x)$ defines the "spread" permissible in x at a single event (impulse occurrence). Continuing the comparison, $\varphi(\omega)$ is then analogous to the transfer function of the linear system since it is the Fourier transform of $p(x)$.

Will a convolution operation occur? In linear systems theory, a convolution occurs when the input to the system is an extended function of x, and the output function is sought. In probability theory, a convolution arises when the sum of two independent RV's is sought. This is shown next.

4.14 Probability Law for the Sum of Two Independent RV's

Suppose random variables y and z are statistically independent. Define a new RV

$$x = y + z\,. \tag{4.17}$$

We seek the probability density for x. Call this $p_X(x)$. (By convention, the subscript to p names the RV in question.)

Let us first try to find $\varphi_X(\omega)$, the characteristic function for RV x. By definition

$$\begin{aligned}\varphi_X(\omega) &\equiv \left\langle \mathrm{e}^{\mathrm{j}\omega x}\right\rangle \\ &= \left\langle \mathrm{e}^{\mathrm{j}\omega(y+z)}\right\rangle\end{aligned}$$

by (4.17),

$$= \left\langle \mathrm{e}^{\mathrm{j}\omega y}\right\rangle \left\langle \mathrm{e}^{\mathrm{j}\omega z}\right\rangle$$

by independence effect (3.27),

$$= \varphi_Y(\omega)\varphi_Z(\omega) \tag{4.18}$$

by definition (4.2).

This may be considered a *transfer theorem for statistics*, when two independent RV's add. We may now use the inverse transform (4.4) to find the required $p_X(x)$. The transform of product (4.18) yields a convolution

$$p_X(x) = \int_{-\infty}^{\infty} \mathrm{d}x'\, p_Y(x')\, p_Z(x - x') \equiv p_Y(x) \otimes p_Z(x) \tag{4.19}$$

where $\otimes$ denotes "convolution" (Fig. 4.2).

The result (4.19) makes sense intuitively. Since by (3.4) both p_Y and p_Z are nonnegative, their convolution in (4.19) can only be a broader and smoother curve then either of them. That is, RV x must be more random than either y or z. Or, the sum of two RV's is more random than either RV. Certainly this is a reasonable effect. As an example, consider the following problem.

Suppose y and z are both Gaussian RV's, with respective variances σ_Y^2 and σ_Z^2. What is the probability law for their sum? Although convolution expression (4.19) may be used, it is in fact easier to use the transfer theorem (4.18). In conjunction with (4.13), this shows that

$$\varphi_X(\omega) = \exp[-(\sigma_Y^2 + \sigma_Z^2)\omega^2/2]\,. \tag{4.20}$$

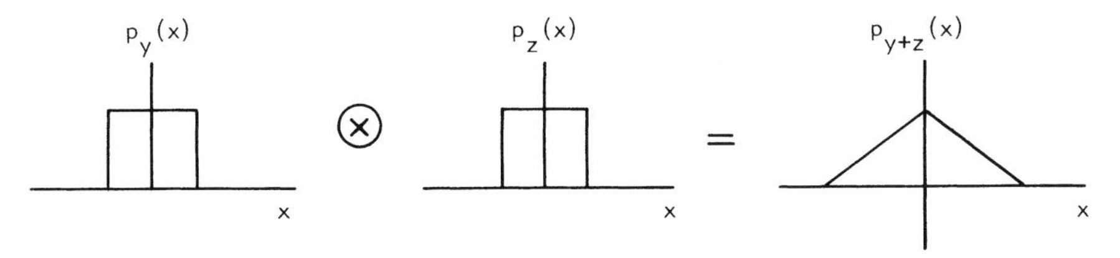

Fig. 4.2. The probability density for the sum of two RV's is formed as the convolution of the two probability densities. The particular case of two uniform probability densities is shown

This has the form of the characteristic function for a *Gaussian* $p(x)$ with net variance $\sigma_Y^2 + \sigma_Z^2 \equiv \sigma_X^2$. This again shows that $p_X(x)$ can only be a broader function than either p_Y or p_Z.

4.15 Optical Applications

4.15.1 Imaging Equation as the Sum of Two Random Displacements

Consider a typical photon in Fig. 2.2, leaving the object plane at coordinate position x and proceeding to coordinate position y in the image plane. For simplicity the imaging is at 1:1 conjugate distances, so that the magnification is unity. Now position y may be expressed as

$$y = x + (y - x) . \tag{4.21}$$

This states that y is x, plus a perturbation $(y - x)$. Let us examine the constituents of the right-hand side. The term x is a RV denoting object position. It follows a probability law $o(x)$ (Sect. 3.4). The term $(y - x)$ is a RV denoting an incremental "kick" to the side according to the probability law $s(y; x)$ for the lens. Now, for each x, $s(y; x) = s_{Y|X}(y|x) = s(y - x|x)$; see (8.47b). But $(y - x)$ is independent of x if the range of x (object size) is sufficiently small. Then $s(y; x) = s(y - x)$. In imaging theory, this is called a condition of "isoplanatism" [4.2]. In probability theory the probability law $s(y - x)$ is called "strict-sense stationary" (Sect. 8.3). Hence, (4.21) is really of the form (4.17). Therefore, identity (4.19) follows, from which

$$i(y) = o(x) \otimes s(x) . \tag{4.22}$$

We again get the imaging equation, but from a new viewpoint. We see that *the image $i(y)$ is really built up from random events y that are the result of the independent addition of transverse positions x in the object plane and $y - x$ in the image plane.* In fact, this principle may be used to artificially build up an image photon by photon (Sect. 7.4).

4.15.2 Unsharp Masking

We noted in Sect. 4.14 that probability law $p_X(x)$ must be broader than either law $p_Y(x)$ or law $p_Z(x)$. What does this mean in the context of image formation? By the correspondence of (4.22) with (4.19), the image $i(y)$ is simply more blurred than either the spread function (of course) or the object.

This *broadening* effect is contingent upon the positivity of $s(x)$ in particular. In fact, the image $i(y)$ could be made locally *narrower* than $o(x)$ if $s(x)$ were allowed to have negative regions. Indeed, this is one of the ways images are enhanced or restored [4.3]. Now negative-going spread functions $s(x)$ are possible to effect by various optical tricks. And by the law of large numbers (2.8) this must have a probabilistic counterpart.

Can, then, a method be found for making a random variable x *less random* by adding to it other independent random variables? At first the mind recoils at this idea, which is certainly not intuitive. But, let us see.

Consider a RV x. Suppose $x = N(\langle x \rangle, \sigma_X^2)$. By sampling x many times over, say N times, we want to estimate its mean value $\langle x \rangle$. Of course, we may simply take the arithmetic mean of the N samples (Sect. 9.1). But, let us see if we can approach the goal by a different route, *by adding random numbers z to the random samples of x.*

Therefore, form a new RV

$$y = x + z \tag{4.23}$$

for each x. The random variable z is chosen to obey a law $z = N(0, \sigma_Z^2)$, independent of samples x. Histograms $p_X(x)$ and $p_Y(y)$ are formed. Now the resulting law $p_Y(y)$ must be broader than $p_X(x)$, as we found in Sect. 4.14. What, then, is the advantage to forming $p_Y(y)$?

Let us now subtract the two histograms, that is, form

$$f(x) = p_X(x) - p_Y(x) \; . \tag{4.24}$$

What properties does $f(x)$ have?

As shown in Fig. 4.3, the function $f(x)$ is centered upon the unknown mean-value $\langle x \rangle$. Moreover, the region over which $\langle x \rangle$ could exist, shaded in Fig. 4.3c, is quite narrow, in fact *much narrower than by direct observation of $p_X(x)$* in Fig. 4.3a. The narrowness of the shaded region depends upon the crossover values x for which $p_X(x) = p_Y(x)$, and hence ultimately upon the choice of probability law $p_Z(x)$. This narrowing effect also occurs in "unsharp masking," where a purposely blurred version of a given image is subtracted from the image. The output difference image has narrower and sharper line profiles than the input.

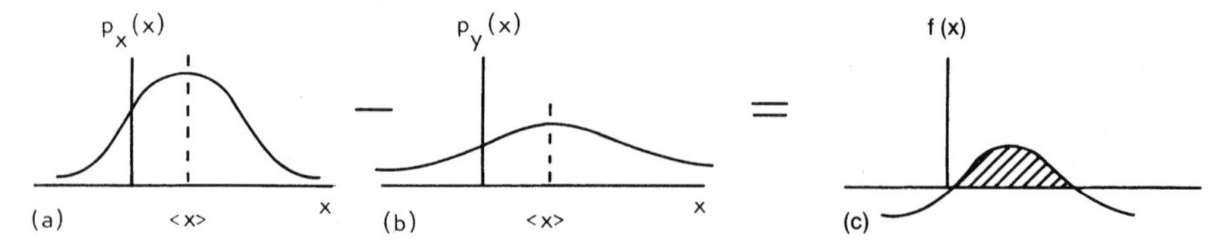

Fig. 4.3. An "unsharp masking" procedure for histograms. Taking their difference yields a narrower curve, whose mean may accordingly be better defined

Hence, oddly enough by adding random numbers to a given set of random numbers x, it is possible to substantially narrow the range of possibilities for the unknown mean behavior $\langle x \rangle$ of x. The extent to which this method may be pushed is currently unknown.

4.16 Sum of *n* Independent RV's; The "Random Walk" Phenomenon

Suppose RV's $y_1, \ldots, y_n$ are statistically independent. By the same methods as above, it can easily be shown that the new RV

$$x \equiv y_1 + \cdots + y_n \tag{4.25}$$

has a characteristic function φ_X that is formed as a multiple-product

$$\varphi_X(\omega) = \varphi_{Y_1}(\omega) \ldots \varphi_{Y_n}(\omega) \,, \tag{4.26}$$

with a density function p_X formed as a multiple-convolution

$$p_X(x) = p_{Y_1}(x) \otimes p_{Y_2}(x) \otimes \cdots \otimes p_{Y_n}(x) \,. \tag{4.27}$$

In the special case where each RV y_m is binary, e.g. with values -1 or $+1$ only, the resulting RV x is called a "random walk" variable. The usual model for this situation is to imagine a drunkard trying to walk a straight line from a point A toward a point B, in a series of n steps. Each step has the magnitude 1, but is independently random as to sign (+ for forward, − for backward). His progression could just as well be accomplished by flipping a coin at each step in order to determine whether it is a step forward (+1) or backward (−1). The distance he has traversed after n such randomly chosen steps is described by x in (4.25). Hence, the expression "random *walk*."

A random walk situation is commonly used to model physical processes. The best known of these is due to *Einstein* [4.4], in modeling Brownian motion.

Exercise 4.1

4.1.1 Random variable x has probability law $p_X(x)$ and characteristic function $\varphi_X(\omega)$, with $-\infty \leqq x \leqq \infty$. What if now the allowed range of x is truncated to $|x| \leqq b$? Calling the truncated RV y, what will $\varphi_Y(\omega)$ be in terms of $\varphi_X(\omega)$?

4.1.2 Random variable x has probability law $p_X(x)$ and characteristic function $\varphi_X(\omega)$. A new probability density $p_Y(x)$ is formed obeying $p_Y(x) = p_X(x) \otimes g(x)$, with $g(x) \geqq 0$ but otherwise arbitrary. Find (a) $\varphi_Y(\omega)$, and (b) moments $\langle y \rangle$, $\langle y^2 \rangle$, and $\langle y^3 \rangle$ in terms of the old moments $\langle x \rangle$, $\langle x^2 \rangle$, and $\langle x^3 \rangle$.

4.1.3 The convolution law (4.19) holds even for the sum of *mixed* continuous and discrete RV's. For example, $x = y + z$ where outcome y is obtained by flipping a fair coin, with $y = +1$ if a head and $y = -1$ if a tail. On the other hand, z is a continuous RV, independent of outcome y, obeying a law $p_Z(x)$. What is $p_X(x)$? Hint: Use $p_Y(x) = 2^{-1}[\delta(x+1) + \delta(x-1)]$, the method of Sect. 3.12.

4.1.4 Do problem 4.1.3 if instead RV y obeys Poisson statistics and RV z is Gaussian. In this way, derive the unusual expression for the probability law of the sum of a Poisson RV with an independent Gaussian RV. Physically, when could such a situation arise?

4.1.5 By use of the moment generating principle (4.3), show that the third central moment of a Poisson RV equals its mean.

4.1.6 By use of principle (4.3), verify that the mean and variance of the normal law (4.13) are $\langle x \rangle$ and σ^2.

4.1.7 In the same way, verify (4.9), (4.10) for the mean and variance of a binomial law.

4.1.8 A RV x is formed as $x = y + z$, with y, z independent, and with $p_Y(x)$ uniform over interval $(0, a)$ and $p_Z(x)$ uniform over interval $(0, 2a)$. What is $p_X(x)$?

4.1.9 Suppose the "unsharp masking" technique (4.24) is tried upon a RV x which is uniformly random over some finite interval (a, b). Would the technique work? That is, would $f(x)$ exhibit a narrower region about the mean than does $p_Y(x)$? Explain.

4.1.10 (a) What does the defining expression become for the characteristic function $\varphi(\omega_1, \omega_2)$ in terms of its density $p(x, y)$ if $p(x, y) = p(r)$, $r = (x^2 + y^2)^{1/2}$?

(b) Evaluate $\varphi(\omega_1, \omega_2)$ when in particular $p(r) = \begin{cases} 1/\pi a^2 & \text{for } 0 \leqq r \leqq a \\ 0 & \text{for } r > a \end{cases}$

(c) Evaluate $\langle r \rangle$ and σ_r^2 for the case (b). Discuss the results.

4.1.11 Show that the characteristic function (4.6) with $x_n = n$ for the geometric law (3.57) is

$$\varphi(\omega) = \frac{1}{1 + a - a\,e^{j\omega}} \,. \tag{4.28}$$

(*Hint:* the summation terms form a geometric series.) By differentiating this, show that $\langle n \rangle = a$ and $\langle n^2 \rangle = 2a^2 + a$.

4.1.12 (a) Show that the characteristic function for the Cauchy law (3.59) is the two-sided exponential

$$\varphi(\omega) = \exp(-a\,|\omega|) \,. \tag{4.29}$$

(b) Using this result, show that if x is Cauchy with parameter a, and y is independently Cauchy with parameter b, then the sum of x and y is Cauchy with parameter $(a + b)$.

4.17 Resulting Mean and Variance: Normal, Poisson, and General Cases

First, consider the special case where each of the $\{y_m\}$ is normal, with generally different means and variances,

$$y_m = N(\langle y_m \rangle, \sigma_m^2) .$$

Then it can be proved from (4.26) that p_X is also normal, and with a variance

$$\sigma_X^2 = \sum_{m=1}^{n} \sigma_m^2 \tag{4.30a}$$

and a mean

$$\langle x \rangle = \sum_{m=1}^{n} \langle y_m \rangle . \tag{4.30b}$$

Next, consider the special case where each of the $\{y_m\}$ is Poisson with mean and variance parameter a_m. Direct use of the Poisson characteristic function (4.7) in the product theorem (4.26) shows that the new RV x is also Poisson, and with a parameter

$$a_X = \sum_{m=1}^{N} a_m . \tag{4.31}$$

Hence, in either the Gaussian case or the Poisson case, the sum of n RV's is a RV whose variance is the sum of the n constituent variances, and whose mean is the sum of the n constituent means.

Finally, results (4.30a, b) also follow in the general case of $\{y_m\}$ independent and following *any distributions*. This may be shown by squaring both sides of (4.25) and taking the expectation termwise (method of Sect. 3.14). See Ex. 4.3.4.

4.18 Sum of n Dependent RV's

Suppose, most generally, that RV's $y_1, y_2, \ldots, y_n$ are generally *dependent*, and the probability density for their sum x is required. Then the characteristic function for x is

$$\varphi_X(\omega) \equiv \left\langle \mathrm{e}^{\mathrm{j}\omega x} \right\rangle = \langle \exp[\mathrm{j}(\omega y_1 + \omega y_2 + \cdots + \omega y_n)] \rangle \tag{4.32a}$$

$$\equiv \varphi_Y(\boldsymbol{\omega})|_{\boldsymbol{\omega}=\omega} \tag{4.32b}$$

by (4.14).

This is the n-dimensional characteristic function of $\varphi_Y(\boldsymbol{\omega})$ evaluated where it intersects a hyperplane $\boldsymbol{\omega} = \omega$.

By definition the required density is

$$p_X(x) \equiv (2\pi)^{-1}\int_{-\infty}^{\infty} d\omega\, \varphi_X(\omega)\, e^{-j\omega x}$$
$$= (2\pi)^{-1}\int_{-\infty}^{\infty} d\omega\, \varphi_Y(\boldsymbol{\omega})|_{\boldsymbol{\omega}=\omega}\, e^{-j\omega x}$$

by (4.32b),

$$= (2\pi)^{-1}\int_{-\infty}^{\infty} d\omega \left[\int_{-\infty}^{\infty}\cdots\int_{-\infty}^{\infty} d\boldsymbol{y}\, p_{\boldsymbol{Y}}(\boldsymbol{y})\, e^{j\boldsymbol{\omega}\cdot\boldsymbol{y}}\right]_{\boldsymbol{\omega}=\omega} e^{-j\omega x}$$

by (4.14)

$$= (2\pi)^{-1}\int_{-\infty}^{\infty}\cdots\int_{-\infty}^{\infty} d\boldsymbol{y}\; p_{\boldsymbol{Y}}\;(\boldsymbol{y})$$
$$\int_{-\infty}^{\infty} d\omega\, \exp[j\omega(y_1 + \cdots + y_n) - j\omega x]$$
$$\longleftarrow\quad 2\pi\;\; \delta(y_1 + \cdots + y_n - x) \quad\longrightarrow$$

$$= \int_{-\infty}^{\infty}\cdots\int_{-\infty}^{\infty} dy_1 \ldots dy_{n-1}\, p_{\boldsymbol{Y}}(y_1, \ldots, y_{n-1}, x - y_1 - \cdots - y_{n-1}) \tag{4.33}$$

by the sifting property (3.30d) of $\delta(x)$.

Hence Fourier methods have allowed us to solve the general problem as well.

Note that (4.33) goes over into the convolution result (4.27), when the $\{y_n\}$ in (4.33) are *independent*. This follows by the separation effect (3.6) for independent RV's.

4.19 Case of Two Gaussian Bivariate RV's

In actual calculations, it is often easier to use (4.32b), with later Fourier transformation, than to use the direct answer (4.33). For example, consider the case where $x = y_1 + y_2$, with y_1, y_2 *jointly Gaussian bivariate*. What is $p_X(x)$? To find $\varphi_X(\omega)$ first, the prescription (4.32b) says to merely take the characteristic function $\varphi_{Y_1Y_2}(\omega_1, \omega_2)$ and replace $\omega_1 = \omega_2 = \omega$. Then by (4.16)

$$\varphi_X(\omega) = \exp[-2^{-1}\omega^2(\sigma_1^2 + \sigma_2^2 + 2\sigma_1\sigma_2\varrho)]\,.$$

By (4.13), this is simply the characteristic function for a Gaussian RV whose variance is

$$\sigma_X^2 = \sigma_1^2 + \sigma_2^2 + 2\sigma_2\sigma_2\varrho\,.$$

Hence $x = N(0, \sigma_1^2 + \sigma_2^2 + 2\sigma_1\sigma_2\varrho)$ in this case.

4.20 Sampling Theorems for Probability

A common way of finding the probability density $p(x)$ for a RV x is to use the product theorem (4.26) to first establish $\varphi_X(\omega)$, and then to Fourier transform this to form $p(x)$. However, $\varphi(\omega)$ is often too complicated a function for its Fourier transform to be known analytically. In this case the Fourier transform integral

$$p(x) = (2\pi)^{-1} \int_{-\infty}^{\infty} \mathrm{d}\omega\, \varphi(\omega)\, \mathrm{e}^{\mathrm{j}\omega x} \tag{4.34}$$

must be evaluated numerically. Of course, the digital computer can only evaluate this integral as a discrete sum, and this usually means a loss of precision. However, there is a situation where no precision is lost. This is the case where a "sampling theorem" holds true, as described next.

4.21 Case of Limited Range of *x*, Derivation

In some cases it is a priori known that RV x cannot take on values beyond a certain value, i.e., $|x| \leqq x_0$. For example, a quantum mechanical particle is confined to a box of length $2x_0$. Or a RV $x \equiv x_0 \cos\theta$, θ a RV. Many such situations exist.

Then by Fourier's basic theorem $p(x)$ may be represented within the limited interval as a series

$$p(x) = \sum_{n=-\infty}^{\infty} a_n\, \mathrm{e}^{-\mathrm{j}n\pi x/x_0}\,, \quad |x| \leqq x_0 \tag{4.35a}$$

with

$$p(x) = 0\,, \quad |x| > x_0\,. \tag{4.35b}$$

The coefficients obey

$$a_n = (2x_0)^{-1} \int_{-x_0}^{x_0} \mathrm{d}x\, p(x)\, \mathrm{e}^{\mathrm{j}n\pi x/x_0}\,. \tag{4.36}$$

This expression may be compared with the one for $\varphi(\omega)$,

$$\varphi(\omega) = \int_{-x_0}^{x_0} \mathrm{d}x\, p(x)\, \mathrm{e}^{\mathrm{j}\omega x}\,, \tag{4.37}$$

using (4.35b). These show that

$$a_n = (2x_0)^{-1} \varphi(n\pi/x_0)\,.$$

Substitution of this a_n back into (4.35a) shows that

$$p(x) = (2x_0)^{-1} \sum_{n=-\infty}^{\infty} \varphi(n\pi/x_0)\, \mathrm{e}^{-\mathrm{j}n\pi x/x_0}\,, \quad |x| \leqq x_0\,. \tag{4.38}$$

This is half of the sampling theorem. The other half is produced by substituting (4.38) back into (4.37), switching orders of summation and integration, and integrating termwise,

$$\varphi(\omega) = \sum_{n=-\infty}^{\infty} \varphi(n\pi/x_0)\,\mathrm{sinc}(x_0\omega - n\pi)\,, \quad \text{all } \omega\,. \tag{4.39}$$

This states that $\varphi(\omega)$ for any ω may be computed as a superposition of $\mathrm{sinc}(x_0\omega)$ functions, weighted by values $\varphi(n\pi/x_0)$ of φ sampled at a discrete subdivision of frequencies. The frequency spacing is $\Delta\omega = \pi/x_0$ (Fig. 4.4). Equation (4.39) shows the interesting property that for $\omega = l\pi/x_0$, all the sinc functions are zero except for $\mathrm{sinc}(x_0 l\pi/x_0 - l\pi)$, which is, of course, unity. Hence, for ω at a sampling point, only the sinc function centered on that point contributes to $\varphi(\omega)$ (Fig. 4.4).

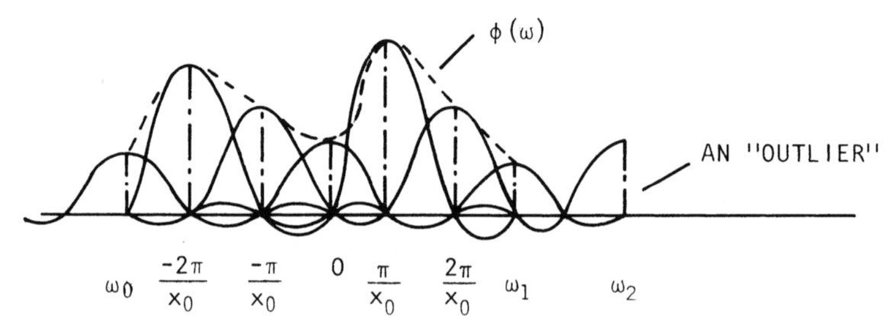

Fig. 4.4. Formation of a characteristic function $\varphi(\omega)$ from its sampled values $\varphi(n\pi/x_0)$. These modulate sinc functions, which are centered on each sampling point $n\pi/x_0$. If it is required to represent $\varphi(\omega)$ in this way over a *finite interval* $\omega_0 \leqq \omega \leqq \omega_1$ (shown), the "outlier" indicated will contribute some of its tails inside the interval

4.22 Discussion

Recall that our original aim was to numerically integrate the right-hand side of (4.34). Compare this with the sampling theorem (4.38), which exactly equals (4.34). We see that the integral has been replaced by a summation over discrete samples of the known integrand and, furthermore, there is no error in making the replacement. This permits $p(x)$ to be evaluated with *arbitrary accuracy* on a digital computer.

The *proviso* has been, of course, that x be limited to a finite interval of values. This is not generally true, of course, and depends on the case under study. Notice that the weaker is the confinement of x, i.e. the larger is the interval size $2x_0$, by (4.38) the closer together the sampling of φ must be, in fact, to the limit where the sum *again becomes* the integral.

Barakat [4.5] has used sampling theorem (4.38) to compute probability densities in the case of laser speckle.

The second sampling theorem (4.39), is actually the Whittaker–Shannon interpolation formula. (See [4.6] for a classic discussion of the analogous sampling theorem for optical imagery.) This formula states that $\varphi(\omega)$ may be perfectly computed anywhere in the continuum $-\infty \leqq \omega \leqq +\infty$, using as data an infinity of sampled values $\varphi(n\pi/x_0)$ of $\varphi(\omega)$. Satisfying as this may be, some of us may object to the need for an infinity of data inputs. There is, however, a way out. Let us instead ask for a *finite interval* $\omega_0 \leqq \omega \leqq \omega_1$ of required values $\varphi(\omega)$. In fact, (4.39) shows that then the continuum of values $\varphi(\omega)$ over the finite interval may be computed from a *finite number* of sampled values $\varphi(n\pi/x_0)$, provided the computed $\varphi(\omega)$ may be slightly in error. The samples $\varphi(n\pi/x_0)$ lie within the finite interval $\omega_0 \leqq \omega \leqq \omega_1$ with a few additional lying just outside. These "outliers" contribute within the interval by means of the tails of their sinc functions. Hence their contributions to accuracy within the interval diminish, the farther away from the interval they lie (Fig. 4.4). Then, the greater the precision required of $\varphi(\omega)$, the higher the number of outliers that must be included within the sum (4.39).

4.23 Optical Application

Point spread functions $s(x)$ in practice have limited support. Although theoretically they have diffraction rings that extend out to infinity [4.2], the ultimate presence of noise in any image makes these diffraction rings impossible to detect. On the other hand, $s(x)$ also represents $p(x)$, a probability density law on photon position within the spread function (Sect. 3.4). Since the Fourier transform of $s(x)$ is the optical transfer function $T(\omega)$, the results (4.38), (4.39) apply, with $s(x)$ substituted for $p(x)$ and $T(\omega)$ for $\varphi(\omega)$:

$$s(x) = (2x_0)^{-1} \sum_{n=-\infty}^{\infty} T(n\pi/x_0)\, e^{jn\pi x/x_0}\,, \quad |x| \leqq x_0 \tag{4.40a}$$

$$T(\omega) = \sum_{n=-\infty}^{\infty} T(n\pi/x_0)\, \mathrm{sinc}(x_0\omega - n\pi)\,, \quad \text{all } \omega\,. \tag{4.40b}$$

We further note that $T(\omega)$ always has limited support. That is, optical systems exhibit sharp cutoff; see [4.2] or (3.65). Call the cutoff frequency Ω. Then (4.40a, b) may be modified to finite sums,

$$s(x) = (2x_0)^{-1} \sum_{n=-N}^{N} T(n\pi/x_0)\, e^{jn\pi x/x_0}\,, \quad |x| \leq x_0 \tag{4.41a}$$

$$T(\omega) = \sum_{n=-N}^{N} T(n\pi/x_0)\, \mathrm{sinc}(x_0\omega - n\pi)\,, \quad \text{all } \omega \tag{4.41b}$$

where N equals the nearest integer to $x_0\Omega/\pi$.

4.24 Case of Limited Range of ω

Sampling theorems may also be established if, instead of x being limited in range, ω is. This would be the case if $p(x)$ is a band-limited function, such as $\text{sinc}^2(ax)$, or $\text{sinc}^2(ax) \otimes g(x)$ with $g(x) \geq 0$ arbitrary, etc. In this event, ω is limited to a finite interval $|\omega| \leq \omega_0$, and sampling theorems of the form (4.38, 39) hold, where now the roles of φ and p are reversed, and similarly for x and ω, and x_0 and ω_0. These new formulas would have similar computational advantages to the preceding.

We also have a third sampling formula. Recall that corresponding to any $p(x)$ is a distribution function $F(x)$ defined by (3.32). A relation exists which expresses $\varphi(\omega)$ in terms of sampled values of $F(x)$,

$$\varphi(\omega) = \mathrm{j}\pi(\omega/\omega_0) \sum_{n=-\infty}^{\infty} F(n\pi/\omega)\, \mathrm{e}^{-\mathrm{j}n\pi\omega/\omega_0} , \quad |\omega| \leq \omega_0 . \tag{4.42}$$

This sampling theorem was invented by *Tatian* [4.7] in order to compute the *optical transfer function* in terms of sampled values of the *edge response*.

Equation (4.42) is useful if the distribution function $F(x)$ is the basic data at hand, rather than the usual $p(x)$. Were $F(x)$ known, $p(x)$ could only be found by numerically differentiating $F(x)$, which is a rather noise sensitive operation. It often does not pay to use a sampling theorem that uses such error-prone $p(x)$ values as inputs. Under the circumstances of given $F(x)$ data, then, (4.42) is a much more accurate sampling theorem to use since it is based upon the *direct* data only.

Tatian also showed that if $p(x)$ is bandlimited, as is assumed here, so is $F(x)$. Therefore, $F(x)$ obeys its own sampling theorem

$$F(x) = \sum_{n=-\infty}^{\infty} F(n\pi/\omega_0)\, \text{sinc}(\omega_0 x - n\pi) , \quad \text{all } x . \tag{4.43}$$

This would permit the distribution function to be constructed at any continuous x value from a set of discrete samples of itself. The more accuracy that is demanded in the construction, the higher the number of discrete samples that would be required.

4.25 Central Limit Theorem

In Sect. 4.16, we found how to form the probability density $p_X(x)$ due to the sum of n independent RV's $\{y_m\}$. The answer (4.26) was that the characteristic function $\varphi_X(\omega)$ is the product

$$\varphi_X(\omega) = \varphi_{Y_1}(\omega) \dots \varphi_{Y_n}(\omega) \tag{4.44}$$

of the individual characteristic functions.

Thus it must be that the functional form of the answer $\varphi_X(\omega)$ will depend upon the form of the individual $\varphi_{Y_m}(\omega)$, $m = 1, 2, \dots, n$. In fact, this is true only to a very limited extent. Practically *regardless* of the form of the $\varphi_{Y_m}(\omega)$, once a value $n \geqq 4$ is used, $\varphi_X(\omega)$ greatly resembles one fixed distribution: the normal curve! This remarkable fact is derived next.

4.26 Derivation [4.8]

As will be seen, the basis for this derivation lies in the fact that a function convolved with itself many times over approaches a Gaussian curve shape.

Suppose a RV x is formed as the sum of n independent RV's $\{y_m\}$

$$x = y_1 + y_2 + \cdots + y_n \,. \tag{4.45}$$

Suppose the $\{y_m\}$ to also have zero-mean (this simplifies the proof), and to be identically distributed according to one density function $p_Y(y)$ and its corresponding characteristic function $\varphi_Y(\omega)$. The variance σ_Y^2 must be finite. (Note that Cauchy RV's, sinc^2 RV's and others originating in diffraction do not obey this requirement.) Then by (4.44) the characteristic function for x obeys

$$\varphi_X(\omega) = [\varphi_Y(\omega)]^n \,. \tag{4.46}$$

We next seek the limiting form for this $\varphi_X(\omega)$ as n becomes large. Accordingly, expand $\varphi_Y(\omega)$ by Taylor series. Then

$$\varphi_X(\omega) = [\varphi_Y(0) + \omega\varphi_Y'(0) + 2^{-1}\omega^2\varphi_Y''(0) + \mu\omega^3]^n \,.$$

But by (4.3), $\varphi_Y(0) = 1$, $\varphi_Y'(0) = j\langle y\rangle = 0$ since the $\{y_m\}$ are zero-mean, and $\varphi_Y''(0) = -\sigma_Y^2$; μ is some constant that need not be determined. Hence

$$\varphi_X(\omega) = (1 - \sigma_Y^2\omega^2/2 + \mu\omega^3)^n \,.$$

Let us now define a new RV

$$s = x/\sqrt{n}$$

which is the required sum of RV's $\{y_m\}$, divided by $\sqrt{n}$. Whereas with $n \to \infty$ the variance in x will increase unlimitedly [see (4.30a)], the new RV s will have a *finite* variance. This is preferable, and we therefore seek the probability for s instead.

By the shift theorem (4.5), $\varphi_S(\omega) = \varphi_X(\omega/\sqrt{n})$, so that

$$\varphi_S(\omega) = (1 - \sigma_Y^2\omega^2/2n + \mu\omega^3/n^{3/2})^n \,.$$

It is convenient now to take logarithms of both sides,

$$\ln \varphi_s(\omega) = n \ln(1 - \sigma_Y^2\omega^2/2n + \mu\omega^3/n^{3/2}) \,.$$

But as n is to be regarded as large, the (...) term is very close to 1. Hence we may use the Taylor expansion

$$\ln(1 + b) = b + \eta b^2 \,,$$

η constant. Then

$$\lim_{n\to\infty} \ln \varphi_S(\omega) = n[-\sigma_Y^2\omega^2/2n + \mu\omega^3/n^{3/2} + \eta(-\sigma_Y^2\omega^2/2n + \mu\omega^3/n^{3/2})^2] \,.$$

Now evaluating the right-hand side in the limit $n \to \infty$, *we see that only the first term in* [...] *remains*. Thus

$$\lim_{n\to\infty} \ln \varphi_S(\omega) = -\sigma_Y^2\omega^2/2 \,,$$

or

$$\lim_{n\to\infty} \varphi_S(\omega) = e^{-\sigma_Y^2\omega^2/2} . \tag{4.47}$$

As φ_S is Gaussian, by identity (4.13) so is RV s, with

$$s = N(0, \sigma_Y^2) . \tag{4.48}$$

This is what we set out to prove. The fact that $p_S(x)$ has this fixed *limiting* form is the statement of the "central limit" theorem.

The central limit theorem also has its "frequency-space" version (4.47). As this follows from the multiple-product (4.46), it states that the product (not convolution now) of n functions tends toward a Gaussian shape, as $n \to \infty$, under rather general conditions.

The derivation also holds when RV's $\{y_m\}$ have a finite mean, as the reader can easily verify. Furthermore, the central limit theorem holds under even more general conditions than assumed above. The $\{y_m\}$ do not have to all obey the same probability law, and they do not all have to be independent. *Lyapunov's condition* [4.9] describes the most general situation for independent RV's. (Also see Ex. 4.2.)

Finally, RV's $\{y_m\}$ do not have to be continuous, as here assumed. The above derivation would hold for *discrete* RV's as well. This fact by itself vastly broadens the scope of application of the central limit theorem.

However, it is important to keep in mind that "the" central limit theorem does not always hold. Adding independent RVs does not always result in a normal law. For example, Cauchy laws often result instead. See Sect. 4.28.1. *A class of central limit theorems* that have as their output a Cauchy law is derived in Ex. 9.2.1 by the use of an invariance principle.

Exercise 4.2

We saw how the situation (4.45) led naturally to a normal law for s, in (4.48). What if, instead of (4.45), x is a *weighted sum* of the independent RV's y_m,

$$x = a_1 y_1 + a_2 y_2 + \cdots + a_n y_n , \tag{4.49}$$

with the $\{a_m\}$ known, nonzero coefficients? Again defining a RV $s = x/\sqrt{n}$, and going through the same steps as in the derivation of result (4.48), show that

$$\lim_{n\to\infty} s = N(0, \sigma_S^2) ,$$

where

$$\sigma_S^2 = \sigma_Y^2 \sum_{m=1}^{n} a_m^2/n , \tag{4.50}$$

provided each $a_m^2 \ll n$, $m = 1, \ldots, n$. We shall have occasion to use result (4.50) in Sect. 5.9.4, to establish the probability law on intensity for laser speckle.

4.27 How Large Does n Have To Be?

As a case in point, consider the $n = 3$ situation where y_1, y_2, and y_3 are identically distributed via $p_Y(y) = \text{Rect}(y - 1/2)$.

The central limit answer and the exact answer (4.27) of $p_X(x) = \text{Rect}(\cdot) \otimes \text{Rect}(\cdot) \otimes \text{Rect}(\cdot)$ are both shown in Fig. 4.5. The curve $p_X(x)$ is here three connected parabolic arcs, but their closeness to the normal curve (shown dashed) defined by (4.48) is remarkable.

Since a Rect function is about as far from resembling a Gaussian curve as any *continuous* $p(x)$ function, we may conclude from this exercise that when the $\{y_m\}$ are each characterized by a continuous density function *it will not take more than* $n = 4$ *of them to make their sum appear normal to a good approximation.* The central limit theorem is a strong statement.

For a sum of *discrete* RV's, $p_X(x)$ tends toward a sequence of equispaced impulses whose *amplitude envelope* is a normal curve. When the spacing between impulses is much less than the variance of the normal curve, for all practical purposes $p_X(x)$ acts like a normal curve.

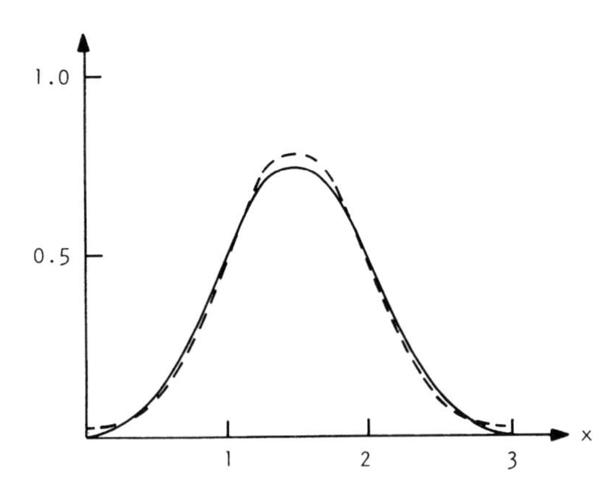

Fig. 4.5. Comparison of exact answer (solid curve) with approximating Gaussian curve (dashed) due to the central limit theorem

4.28 Optical Applications

The central limit theorem applies fairly widely in optics. These applications are not all explicitly statistical in nature. Some arise as purely a multiple-convolution phenomenon.

4.28.1 Cascaded Electro-Optical Systems

Linear electro-optical systems sometimes operate in "cascade", where the output image of one system acts as an incoherent input to the next (Fig. 4.6). Under these

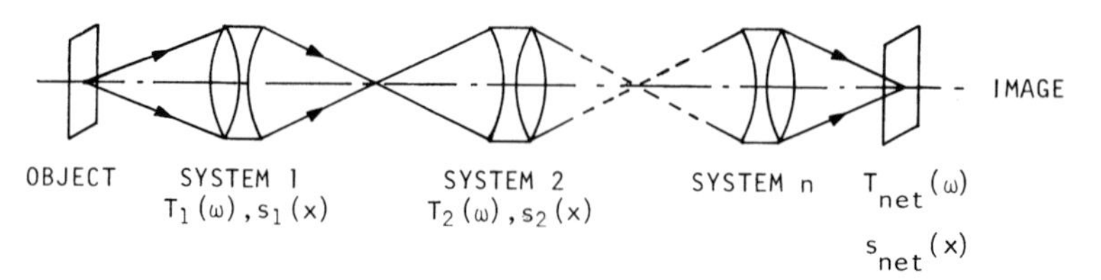

Fig. 4.6. Optical systems operating "in cascade." The output optical transfer function $T_{\text{net}}(\omega)$ and point spread function $s_{\text{net}}(x)$ approach, respectively, an exponential law and a Cauchy law, as the number of lenses n becomes large

conditions the net optical transfer function $T_{\text{net}}(\omega)$ of the entire system is simply a product

$$T_{net}(\omega) = T_1(\omega)T_2(\omega) \cdots T_n(\omega) . \tag{4.51}$$

of those of the sub systems, $T_m(\omega)$, $m = 1, 2, \ldots, n$. Notice that this chain product is of the form (4.44) assumed in derivation of the central limit theorem. Then if the number n of sub systems is three or more, and if the individual $T_m(\omega)$ are somewhat similar functions, by Sects. 4.26–4.27 it would seem that $T_{\text{net}}(\omega)$ should have approximately a Gaussian form, with likewise a Gaussian form for its spread function $s_{\text{net}}(x)$. However, this is not true: Because of the *finite apertures* of the optical sub systems, the underlying variance σ_Y^2 is infinite, violating a premise of the derivation in Sect. 4.26. In fact the form $s_{\text{net}}(x)$ is actually *Cauchy* in the limit as $n \to \infty$. See (9.88). This approximates a Gaussian only in the central core region of the point spread function. See (9.96).

4.28.2 Laser Resonator

In the laser resonator, optical waves bounce back and forth between the resonator's end mirrors. Let these be rectangular, with separation b and width a. Let $v(x_2)$ denote the amplitude at coordinate x_2 on mirror 2, due to amplitude $v(x_1)$ on mirror 1 on the previous bounce. Then [4.10]

$$v(x_2) = \int_{-a}^{a} \mathrm{d}x_1 v(x_1) K(x_2 - x_1) . \tag{4.52}$$

Here, the imaging kernel $K(x)$ is a Fresnel expression [4.2]

$$K(x) = (\lambda b)^{-1/2} \exp(-\mathrm{j}\pi x^2/\lambda b) , \tag{4.53}$$

and λ is the light wavelength.

Equation (4.52) shows that the amplitude on mirror 2 is the amplitude on mirror 1 convolved with kernel $K(x)$. On the return bounce, the roles of x_1 and x_2 are reversed, amounting to a truncation of $v(x_2)$ for $|x_2| > a$ in (4.52). Therefore, the output wave after n propagations (4.52) is the result of n successive pairs of operations (convolution, truncation).

If it were not for the truncation at each step we would have purely n successive convolutions, and again the central limit theorem would hold. The truncations alter the result, however, just as they changed the cascaded systems spread function of Fig. 4.6 from a Gaussian to closer to a Cauchy law. Each truncation cuts off the tails of the previous convolution output. This excites extra oscillations in each successive convolution, so we do not end up with the smooth Gaussian that the central limit theory gives. Such oscillatory functions, of course, ultimately find their way into the mode shapes for the laser.

However, if ratio a/b is large enough *the effect of truncation will be small* in (4.52), since $v(x_1)$ will effectively fall to zero before the rim values $x_1 = \pm a$ are attained. *Now* the resonator output after n reflections is purely due to n convolutions with kernel $K(x)$, and hence should follow a Gaussian law. Fox and Li carried out these operations on a digital computer with the results shown in Fig. 4.7. Cases $n = 1$ and $n = 300$ of such "digital" propagation are shown. The $n = 300$ result is indeed close to a Gaussian shape.

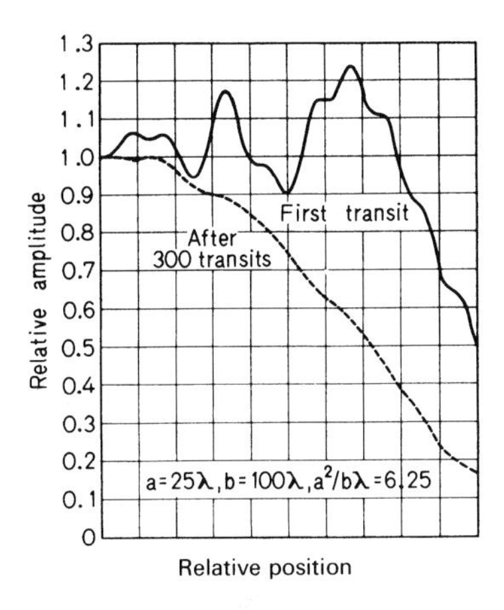

Fig. 4.7. Laser amplitude profile after one transit of the resonator, and after 300 transits. The initially launched wave had a flat profile [4.10]

4.28.3 Atmospheric Turbulence

The other type of application of the central limit theorem is purely statistical in nature. It arises from the addition of n random numbers.

For example, by one model of atmospheric turbulence due to *Lee* and *Harp* [4.11], the turbulent atmosphere between light source and detector is imagined to consist of planes of timewise random optical phase (Fig. 4.8). These planes are oriented with their normals parallel to the line of sight connecting source and detector. Then the net instantaneous optical phase Φ_{net} at the detector is a sum

$$\Phi_{\text{net}} = \Phi_1 + \Phi_2 + \cdots + \Phi_n \tag{4.54}$$

over the individual phases Φ_m due to the planes. Here $\Phi_m = n_m \Delta t_m$, where n_m is the random index of refraction of the mth plane and Δt_m is its random thickness. As the turbulence ought to act *independently* from one plane to the next (or at least from one group of planes to the next), and as the phase fluctuations within each plane should follow about the same physical law, Φ_{net} satisfies the central limit and hence should be normally distributed. This is, in fact, the usual presumption regarding $p(\Phi_{\text{net}})$ at a single point. More generally, for the joint $p(\Phi_{\text{net}}, \Phi'_{\text{net}})$ at two points, a Gaussian bivariate assumption is made (Sect. 8.6.1).

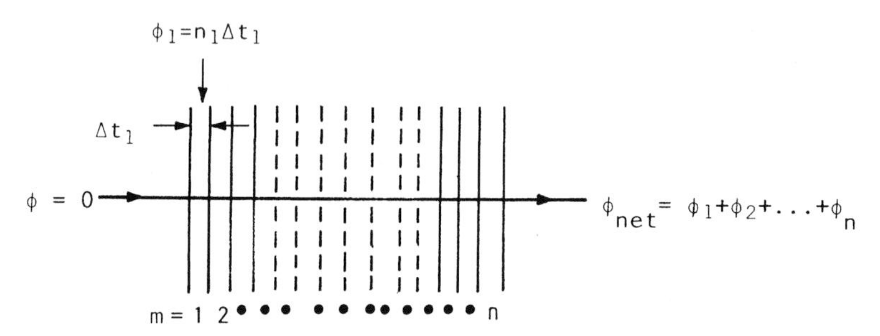

Fig. 4.8. The atmospheric-turbulence model due to Lee and Harp

4.29 Generating Normally Distributed Numbers from Uniformly Random Numbers

In numerical simulation of noisy data, and in other types of Monte Carlo calculations, it is often necessary to generate independent, *normally random* numbers. For example, image noise is commonly modeled as being Gaussian random (normal, with zero mean), and uncorrelated from point to point.

However, often the only random number generator that is easily accessed on a computer instead generates *uniformly random* numbers over (0,1). An example is the function RANF(·) of the CDC Scope library. Now, these numbers have the desirable property of being uncorrelated, a property that takes some care in satisfying, and that is half the requirements of our noise-generation problem. The question arises, then, as to whether it is possible to use these output numbers from RANF and somehow *transform them* into normally distributed numbers (with a required mean and required variance). In fact, there are a few different approaches to this problem, and to the more general one of transformation from one probability distribution to *any* other (Chaps. 5, 7)

Here we content ourselves with the simplest such approach, a mere addition of a few RANF outputs. This has the advantage of being a very fast operation, but has the disadvantage of *approximating* the normal case (as we have seen).

Suppose we add together every three successive outputs y_1, y_2, and y_3 from RANF(·) to form a new RV,

$$x = y_1 + y_2 + y_3 \, . \tag{4.55}$$

This is recognized as simply the case $n = 3$ of the derivation in Sect. 4.26. In particular, each of the $\{y_m\}$ here follows the *uniform* probability law. Then Fig. 4.5, in fact, gives the ensuing probability law for x, a sequence of three parabolic arcs that approximate a normal distribution. Hence, the new RV x will be approximately normal, as was required.

However, merely taking the sum as in (4.55) will not yield a prescribed mean m and variance σ^2 for RV x. For example, by taking the mean of both sides of (4.55) we see that $\langle x\rangle = \langle y_1\rangle + \langle y_2\rangle + \langle y_3\rangle = 3\langle y\rangle = 3/2$.

This leads us to replace (4.55) by a slightly more general form

$$x = a(y_1 + y_2 + y_3) + b\ , \tag{4.56}$$

with constants a, b to be found such that x has the required two moments. The rest is algebra, using the approach of Sect. 3.14. Taking the mean of both sides and using $\langle y\rangle = 1/2$, requires

$$m = 3a/2 + b\ . \tag{4.57}$$

Squaring both sides of (4.56) and taking the mean produces requirement

$$\left\langle x^2\right\rangle \equiv \sigma^2 + m^2 = a^2(3\left\langle y^2\right\rangle + 6\left\langle y\right\rangle \left\langle y\right\rangle) + b^2 + 6ab\left\langle y\right\rangle$$

by identity (3.15), or

$$\sigma^2 + m^2 = (5/2)a^2 + b^2 + 3ab \tag{4.58}$$

since $\langle y^2\rangle = 1/3$.

Equations (4.57, 58) may be solved for unknowns a, b. The solution is $a = 2\sigma$, $b = m - 3\sigma$. Hence, the required transformation is

$$x = 2\sigma(y_1 + y_2 + y_3) + m - 3\sigma\ . \tag{4.59}$$

A random variable x formed in this way from every three successive outputs of RANF($\cdot$) will be very close to a normal RV and will (exactly) have mean m and variance σ^2.

A measure of the degree to which x approximates a normal RV is the range of values that x defined by (4.59) can take on. A true normal RV can, of course, vary from $-\infty$ to $+\infty$. Using the extreme values for y of 0 and 1, (4.59) shows that x is confined to an interval

$$m - 3\sigma \leqq x \leqq m + 3\sigma\ .$$

In most simulation use, this artificial constraint should not be serious, since the normal curve itself has 99.7% of its events x confined to this interval. Exceptions would be where events far out in the tails are very important, as in studies of "noise outliers."

If the degree of approximation is nevertheless unacceptable, the above procedure may be carried through for four successive outputs of RANF($\cdot$) instead; or indeed, for any number n required. The higher n is made, by the central limit, the closer density $p(x)$ will approximate the normal case.

4.30 The Error Function

We have considered many circumstances for which a RV x is normal. It is often important to be able to calculate $P(a \leqq x \leqq b)$, the probability that RV x lies in the interval (a, b). What is this quantity when x is normal?

By (3.1) and (3.42),

$$P(a \leqq x \leqq b) = (2\pi\sigma^2)^{-1/2} \int_a^b dx \exp[-(x - \langle x \rangle)^2/2\sigma^2] . \tag{4.60}$$

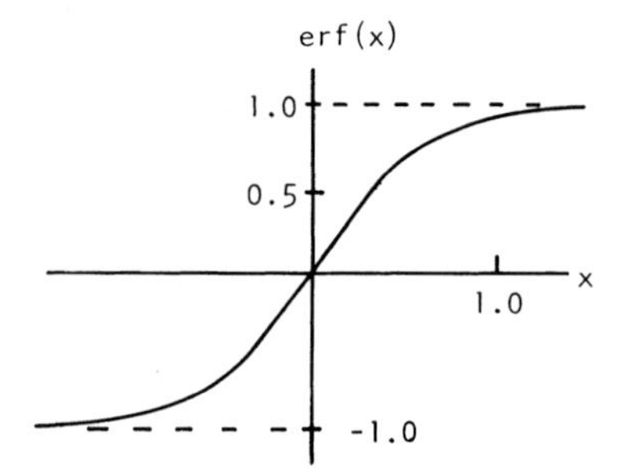

Fig. 4.9. Sketch of erf(x), showing its asymptotes (dashed lines)

This is almost in the form of the error function, defined as [4.12]

$$\mathrm{erf}(z) = 2(\pi)^{-1/2} \int_0^z dt\, e^{-t^2} . \tag{4.61}$$

In fact, by the change of integration variable $y = (x - \langle x \rangle)/\sqrt{2}\sigma$ in (4.60), we easily find that

$$P(a \leqq x \leqq b) = \frac{1}{2}\left[\mathrm{erf}\left(\frac{b - \langle x \rangle}{\sqrt{2}\sigma}\right) - \mathrm{erf}\left(\frac{a - \langle x \rangle}{\sqrt{2}\sigma}\right)\right] . \tag{4.62}$$

Values of the error function exist in most books of mathematical tables. We have reproduced one such set in Appendix A. The function is sketched in Fig. 4.9.

As an example of the use of (4.62), let us consider the basic question: what is the probability that a normal RV lies within $\pm 1\sigma$ of its mean? Then in (4.62), $b = \langle x \rangle + \sigma$, $a = \langle x \rangle - \sigma$, and

$$P = \frac{1}{2}[\mathrm{erf}(1/\sqrt{2}) - \mathrm{erf}(-1/\sqrt{2})] = \mathrm{erf}(1/\sqrt{2}) = 0.68 .$$

Exercise 4.3

4.3.1 A coin is independently flipped n times, forming events $y_1, \ldots, y_n$, where an event $y_m = 1$ if a head occurs, or $y_m = -1$ if a tail occurs. The coin is biased, with $P(\text{head}) = 3/4$, $P(\text{tail}) = 1/4$. The sum $x = y_1 + y_2 + \cdots + y_n$ of events is formed.

(a) What is $\langle x \rangle$?
(b) What is σ_X^2?

Hint: Use results (4.30).

4.3.2 An optical amplitude A is formed as the sum of "phasors"

$$A = \sum_{m=1}^{n} \exp(\mathrm{j}\varphi_m) ,$$

where the φ_m are independent phases identically distributed as uniform over interval $(-\pi, \pi)$. Let $A = x + \mathrm{j}y$, with x and y real.

(a) Show that $\langle x \rangle = \langle y \rangle = 0$.

(b) Find σ_X^2, σ_Y^2. Show that $\sigma_X^2 + \sigma_Y^2 \equiv \langle |A|^2 \rangle = n$, i.e., the "intensity" of the net disturbance simply equals the number of phasors. This shows that a light bulb will glow, despite emitting light of *random phase* from the atoms in its filament.

4.3.3 Prove the assertions (4.30) on the mean and variance for a sum of normal RV's, by use of the product theorem (4.26).

4.3.4 Using the algebraic method of Sect. 3.14, show that the results (4.30) hold regardless of the form of probability laws $p_{Y_m}(x)$, $m = 1, \ldots, n$.

4.3.5 Prove (4.31) for the sum of n Poisson RV's, using identity (4.26).

4.3.6 A RV x has a characteristic function $\varphi(\omega)$ which exhibits sharp cutoff, $\varphi(\omega) = 0$ for $|\omega| > \omega_0$. This results in a sampling theorem of the type (4.39) with the roles of $p(x)$ and $\varphi(\omega)$ interchanged.

(a) What is the spacing Δx of sampling points $x_n = n\Delta x$, $n = 0, \pm 1, \pm 2, \ldots$?

(b) The points x_n are called the "degrees of freedom" of $p(x)$. How many degrees of freedom does $p(x)$ have within a fixed length L of x?

(c) If $p(x)$ is a point spread function, what does quantity $\omega_0 L$ in the previous expression represent physically?

4.3.7 Two optical systems are operating in "cascade" (Sect. 4.28.1). Each is diffraction-limited, with transfer function $T_n(\omega) = 1 - |\omega| / \Omega$, Ω the cutoff frequency, $n = 1, 2$. Calculate $T_{\mathrm{net}}(\omega)$ and compare this with the closest Gaussian.

4.3.8 Carry through the derivation (4.55–59) for the simpler case $x = y_1 + y_2$. To what interval of values about the mean m will this x be confined? What probability law $p_X(x)$ is actually obeyed by this RV?

4.3.9 We showed in Fig. 4.8 the laminar phase model of optical turbulence of *Lee* and *Harp*. Suppose each of the $\{\Phi_m\}$ are chosen independently from the same probability law $p_{\Phi_m}(x) = (1/\Delta\Phi)\mathrm{Rect}(x/\Delta\Phi)$. We found that under these circumstances the net Φ at the output is normal. How will its mean and variance depend upon $\Delta\Phi$ and n, the latter the number of laminar planes (length of the turbulent medium)?

4.3.10 A picture is being transmitted by "delta-modulation," which means that the intensity of the n'th pixel on a line is formed from the $(n-1)$'st by

$$i_n = i_{n-1} + \Delta i_n , \quad n = 2, 3, \ldots, N$$

for a picture N pixels across. First i_1 is transmitted, then differences $\Delta i_2, \Delta i_3, \ldots, \Delta i_N$.

Suppose each of the $\{\Delta i_n\}$ suffers random error according to Gaussian statistics, $\Delta i_n = N(0, \sigma^2)$. Then the error i_n grows with n (across the picture). Derive an expression for σ_n^2, the variance of error at the n'th pixel.

Suppose the picture is required to have at least a certain level of quality, defined in the following way. Let e represent the error at value i_N. It is required that the probability $P(e < 0.1) = 0.9$. If $N = 100$ how small must σ be for this to be true? *Hint:* Use (4.62).

4.3.11 Given that a RV $x = N(0, \sigma^2)$, it is desired to find $P(|x| \leqq 2\sigma)$. Do this in two different ways: (a) by use of the error function formula (4.62), and (b) by use of the quadratic probability law for the three-RANF approximation to the normal law, defined at (4.59).

4.3.12 ***Intrinsic fluctuation in light from a star.*** It is known in astronomy that if star A has a larger diameter than star B, the total light from A will fluctuate less than that from B. This effect is intrinsic to the stars, and independent of fluctuations due to atmospheric turbulence. It may be explained as follows.

Model a star as being composed of M mutually incoherent surface elements of constant area. Within the m'th element, a constant (but independently random) radiance i_m is emitted. Then element m contributes an intensity at the observation point 0 on earth that obeys a simple inverse-square law on distance r_m to 0,

$$I_m = i_m / r_m^2 \ .$$

The total intensity at 0 is merely the sum over these incoherent sources,

$$I = \sum_m^M i_m / r_m^2 \simeq R^{-2} \sum_m^M i_m \ ,$$

the latter because to a very good approximation, all $r_m \simeq R$, the distance to the star. The observable I is an RV, since each i_m is. We want its mean $\langle I \rangle$ and variance σ_I^2.

Assuming that each i_m is selected from one probability law with mean $\langle i \rangle$ and variance σ^2, show that the *relative fluctuation* $\sigma_I / \langle I \rangle$ obeys

$$\sigma_I / \langle I \rangle = M^{-1/2} \sigma / \langle i \rangle \ . \tag{4.63}$$

As M is proportional to the area of the star, the relative fluctuation goes inversely with its diameter.

See the related Ex. 13.1.3.

4.3.13 ***Log-normal probability law for stellar irradiance.*** The irradiance at a point in the image of a star fluctuates randomly due to atmospheric turbulence provided the exposure time is brief enough, on the order of 0.01 s or less. What probability law do these fluctuations follow?

It has often been observed, and assumed in theoretical work, that the law is log normal.[2] A justification for this law was in fact found by *Strohbehn* [4.13]. Its essence lies in the *central limit theorem*, as described next.

Model the turbulent medium as the parallel slabs in Fig. 4.10, each of which scatters away a random amount of light amplitude from an incident beam. Then if t_n is the amplitude transmittance of the nth slab, the amplitude u at the receiver obeys a simple product

$$u = u_0 \prod_{n=1}^{N} t_n \;, \tag{4.64}$$

where u_0 is the initial amplitude and N is the total number of slabs. This is called, for good reason, a "multiplicative" or "multiple-scattering" model of turbulence. Note that by this model the receiver lies within the scattering medium. The complementary situation of an "additive" or "single-scattering" model is taken up in Ex. 5.1.24.

(a) By taking the natural logarithm of (4.64), establish conditions on the t_n and N for which $\ln |u|$ is a normal RV.

(b) The irradiance I obeys $I = |u|^2$. Show that I is log normal as well.

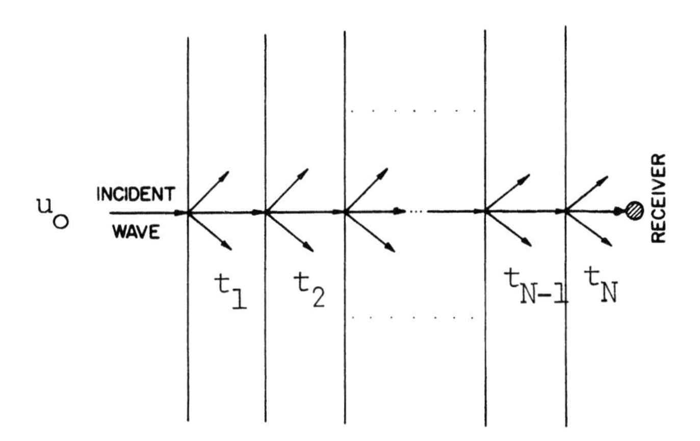

Fig. 4.10. Multiple-scattering model of turbulence [4.13]

4.3.14 ***Persistence of diffusion.*** Suppose that a random variable x is formed as the sum of independent RV's y and z, where y obeys an arbitrary probability law while z obeys a Gaussian one,

$$x = y + z, \quad p_Y(y) \text{ arbitrary}, \quad z = N(0, \sigma^2) \;. \tag{4.65}$$

Define a "time" parameter $\tau = \sigma^2$. Then by (3.42) and (4.19) the probability density for x obeys the convolution

$$p_X(x|\tau) = \frac{1}{\sqrt{2\pi\tau}} \int \mathrm{d}y \, p_Y(y) \exp[-(x-y)^2/2\tau] \;. \tag{4.66}$$

The left-side explicitly indicates that p_X depends conditionally upon the size of the parameter τ. By alternately differentiating this $\partial/\partial\tau$ and $\partial^2/\partial x^2$ under the integral sign, show that p_X obeys [4.14]

$$\frac{\partial p_X}{\partial \tau} = \frac{1}{2}\frac{\partial^2 p_X}{\partial x^2}, \quad p_X = p_X(x|\tau) \;. \tag{4.67}$$

This has the form of *a diffusion equation* in p_X. It becomes an actual diffusion equation if τ is proportional to the time t. This will be the case if z obeys diffusion, since then σ^2 in the 2nd Eq. (4.65) is proportional to the time [1.2]. Hence we have proven the following theorem: if a particle undergoing spatial diffusion is additionally displaced randomly according to *any probability law*, then the resulting total displacement still obeys diffusion.

4.3.15 $\mathbf{p(x)}$ ***in cases*** $\mathbf{p(x|a)}$ ***where parameters*** $\mathbf{a}$ ***are random.*** Any probability law has parameters $a_1, \ldots, a_N = \mathbf{a}$ associated with it. For example, the exponential law (3.41) has a single parameter $\mathbf{a} = a$. A probability law is actually conditional upon particular values of the parameters, as indicated by the notation $p(x|a_1, \ldots, a_N)$. Suppose the parameters are not fixed, but rather obey their own probability law $p_A(a_1, \ldots, a_N)$. What is the statistical behavior of x? This is sometimes called the law of *total* fluctuation.

Analysis: By the partition law (3.7)

$$p_X(x) = \int da_1 \cdots da_N \; p(x|a_1, ..., a_N)\, p_A(a_1, ..., a_N) \,. \tag{4.68}$$

Problem: If x is exponential (3.41) with parameter a, and the latter obeys a law $p_A(a) = K/a$, $B \le a \le C$, with K, B, C constants, find the law of total fluctuation $p_X(x)$.

4.3.16 ***Mandel's formula.*** Einstein [4.4] found that when photons impinge upon a substrate a random number n of electrons (called *photoelectrons*) are ejected. What statistics do these electrons obey? Let w denote the total integrated energy of the photons over a detection interval T, with $p_W(w)$ its probability density. It turns out that for a detection interval T much less than the coherence time, and with polarized thermal radiation, $p_W(w)$ is exponential (3.41) with a mean $\langle w \rangle \equiv w_0$. Also, the statistics of n are conditional upon the value of w, obeying a law $P(n|w)$ which is Poisson with a mean $\langle n \rangle \equiv \alpha w$. Parameter α is called the quantum efficiency. Show via Eq. (4.68) that under these circumstances the law of total fluctuation in photoelectron number n is

$$P(n) = \frac{1}{1 + \alpha w_0} \left(\frac{\alpha w_0}{1 + \alpha w_0} \right)^n . \tag{4.69}$$

This geometric law is called "Mandel's formula".

5. Functions of Random Variables

Consider the notion of a "square-law" detector: If x is an input to the detector, then $y = x^2$ is its output or detected value. Consider next the case where x is a random variable with probability law $p_X(x)$. Then output y is also random. If so, what is its probability law $p_Y(y)$? Certainly this should depend in some way upon $p_X(x)$, since x and y are closely related by the transformation $y = x^2$. The general question of how to find the law $p_Y(y)$ for a transformed or output RV is the subject of this chapter.

The above relation $y = x^2$ describes a "transformation of a random variable," from RV x to RV y. There are many instances where a transformation of an RV occurs in optics. Consider, for example, the geometrical imaging equation

$$1/y + 1/x = 1/F ,$$

where x is the object conjugate position, y is the image conjugate position, and F is the focal length. If x is random, this defines a transformation from an RV x to RV y. Then, what if a probability law is known for x? For example, suppose the object to be vibrating longitudinally. Can we find the resulting probability law on state of focus y for the image? This problem is taken up below.

An even more fundamental problem occurs when simply dealing with the intensity resulting from the complex squaring of an amplitude function. For example, if $z = x + \mathrm{j}y$, where x and y are the real and imaginary parts of the complex amplitude z, suppose the joint probability $p_{XY}(x, y)$ is known. Now, intensity $i \equiv |z|^2 = x^2 + y^2$. This is a known transformation from RV's x and y to a new RV i. Then, can $p_I(i)$ be found? This problem occurs in analyzing laser speckle, as is shown below.

5.1 Case of a Single Random Variable

The problem we first consider is how to find the probability density $p_Y(y)$ for a single RV y, when it is defined in terms of another RV x by a functional relation

$$y = f(x) . \tag{5.1a}$$

The law $p_X(x)$ is assumed known. We might consider f a general "system function," with RV x the input to the system and RV y the output.

Let the inverse relation to (5.1a) be

$$x = f^{-1}(y) . \tag{5.1b}$$

Now, for a given y there might be a unique root x, or a multiplicity of roots $x_1, \ldots, x_n$. This depends upon the form of function f. For example, if f is the squaring operation, then $y = x^2$ and $x = \pm\sqrt{y}$ for two roots. It is simplest to first consider the unique-root case.

5.2 Unique Root

The situation is most easily analyzed graphically, as in Fig. 5.1a. The vertical y axis consists of all permissible disjoint events

$$A_Y = (y \leqq Y \leqq y + \mathrm{d}y) , \quad \mathrm{d}y > 0, -\infty \leqq y \leqq \infty .$$

Note that $\mathrm{d}y > 0$ by definition (3.3).

Consider one such event A_Y (dot in the figure). Because to each y there is a unique x given by (5.1b), the same event A_Y may alternatively be described as event

$$A_X = (x \leqq X \leqq x + \mathrm{d}x) , \quad \mathrm{d}x > 0 ,$$

with x determined by (5.1b).

Therefore, the probability $P(A_Y)$ may be described alternatively as

$$P(A_Y) = P(y \leqq Y \leqq y + \mathrm{d}y) = P(x \leqq X \leqq x + \mathrm{d}x) . \tag{5.2}$$

Intuitively, what this means is that since an event in y has a unique correspondence to one in x, *the relative number of times a value y will occur equals the relative number of times the corresponding value of x will occur.*

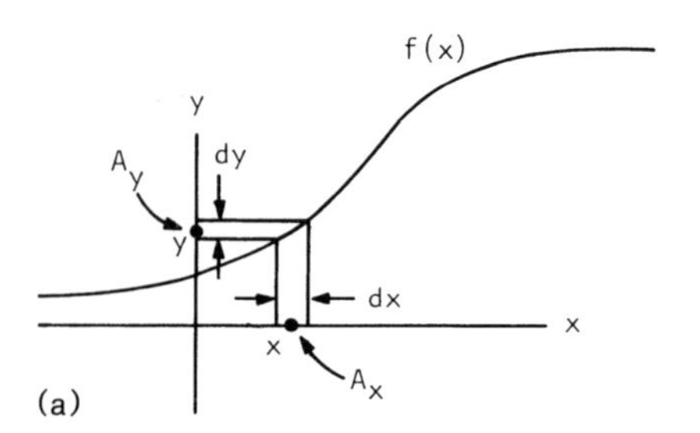

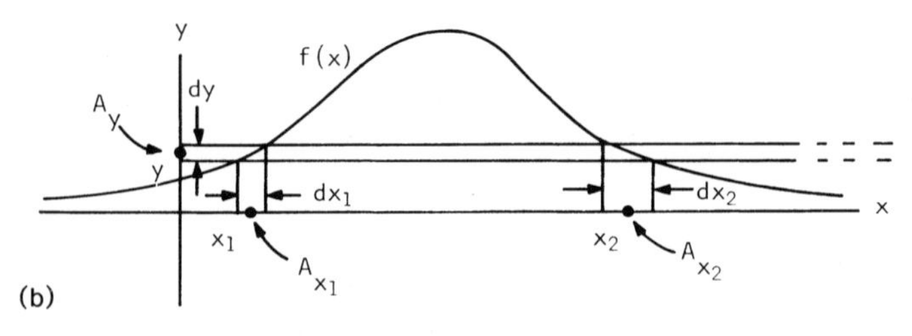

Fig. 5.1. Transformation of a random variable, (a) Unique-root case. For each y there is one root x. (b) Multiple-root case. For each y, there are two or more roots $x_1, x_2, \ldots, x_r$

Then by definition (3.3) of probability density, (5.2) becomes

$$p(x)|_{x=X}\,\mathrm{d}x = p(y)|_{y=Y}\,\mathrm{d}y\ . \tag{5.3}$$

But, we decided before (Sect. 4.5) to specify *the name* of the particular RV whose density is $p(x)$, by a subscript, as in $p_X(x)$. This done, the notation $x = X$ in (5.3) becomes redundant. Similarly for RV y. Then more simply

$$p_X(x)\,\mathrm{d}x = p_Y(y)\,\mathrm{d}y\ . \tag{5.4}$$

Hence by (5.1b),

$$p_Y(y) = p_X[f^{-1}(y)]/\left|y'\right|\ . \tag{5.5}$$

This is the required probability. The y' is the derivative $\mathrm{d}y/\mathrm{d}x$. The absolute value is taken because $\mathrm{d}y$ and $\mathrm{d}x$ are both always positive, by definition (3.3). Note that since $p_Y(y)$ must be expressed purely in terms of y, then y' must be expressed in terms of y as well.

5.3 Application from Geometrical Optics

As an example, consider the geometrical optics relation

$$1/y + 1/x = 1/F\ , \tag{5.6}$$

with F a lens focal length and x, y conjugate distances from the lens. If $p_X(x)$ is known for the object conjugate distance, what is $p_Y(y)$ for the image conjugate distance? According to (5.5) we need to know $f^{-1}(y)$ and $y'(y)$. These are

$$f^{-1}(y) \equiv x = \frac{Fy}{y-F}\ ,$$

and

$$y' = -\frac{(y-F)^2}{F^2}\ .$$

Then by (5.5),

$$p_Y(y) = \frac{F^2}{(y-F)^2}\,p_X\left(\frac{Fy}{y-F}\right)\ . \tag{5.7}$$

This is the general answer for arbitrary $p_X(x)$.

To get the flavor of the distortion of object space manifest in (5.7), consider the particular case of near-equal conjugates, where all x values $2F - F/2 \leqq x \leqq 2F + F/2$ are equally likely. This may be described by a probability

$$p_X(x) = \begin{cases} F^{-1} & \text{for } 2F - F/2 \leqq x \leqq 2F + F/2 \\ 0 & \text{for all other } x\ . \end{cases} \tag{5.8}$$

Using this in (5.7) yields

$$p_Y(y) = \begin{cases} \frac{F}{(y-F)^2} \text{ for } 5F/3 \leqq y \leqq 3F \\ 0 \text{ elsewhere .} \end{cases} \tag{5.9}$$

This law shows a much higher concentration of image events near $y = 5F/3$ than near $3F$. See Fig. 5.2.

A higher significance may be given to result (5.7). By the "law of large numbers" arguments of Sect. 2.10, $p_Y(y)$ actually represents the image intensity distribution $i(y)$ due to an object $o(x) = p_X(x)$. However, in particular, perfect (stigmatic) imagery is assumed to take place since (5.6) allows for no "spread" $s(y; x)$ in events y due to an event x. Thus if a uniformly glowing light rod were placed on the x interval (5.8) in object space, image intensity *along* y would follow the law (5.9) in the stigmatic limit. In particular, the point $y = 5F/3$ would be *nine* times as bright as the point $3F$. This situation is illustrated in Fig. 5.2.

The case of nonstigmatic imagery, where a definite spread $s(y; x)$ exists, is taken up in Ex. 5.1.23. Now the image along y will be (5.9) convolved with a spread function.

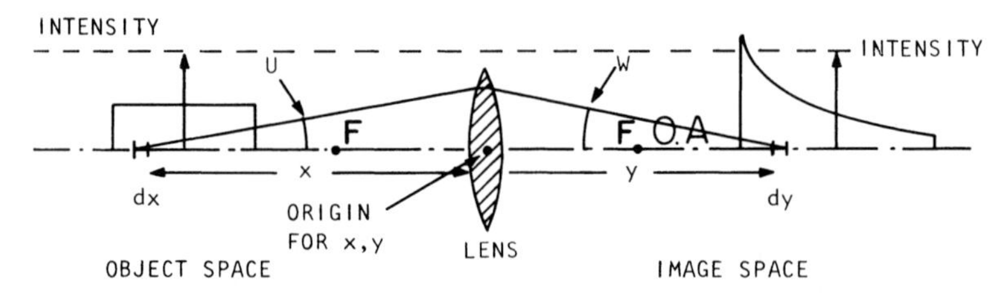

Fig. 5.2. Probability laws as intensity curves in object and image spaces. Both curves describe light intensity along the optical axis OA

5.4 Multiple Roots

Suppose the inverse relation (5.1b) to have many roots, $x_1, \ldots, x_r$, for a given y. For example, this is the case if $y = x^2$, so that $x = \pm\sqrt{y}$ with $r = 2$. The situation is illustrated in Fig. 5.1b.

As in the one-root case, we consider an event y to occur. According to Fig. 5.1b, this may be alternatively viewed as occurrence of the event x_1 or the event x_2 or ... or x_r. Now these r events are disjoint [see paragraph beneath (3.8)]. Hence by the additivity property (2.2),

$$P(y \leqq Y \leqq y + \mathrm{d}y) = P(x_1 \leqq X \leqq x_1 + \mathrm{d}x_1) + \ldots + P(x_r \leqq X \leqq x_r + \mathrm{d}x_r) .$$

Note the need for subscripting the $\mathrm{d}x$'s, since as shown in Fig. 5.1b *they are in general of different length depending upon the local value* y'.

Then by definition (3.3) of probability density,

$$p_Y(y)\,\mathrm{d}y = p_X(x_1)\,\mathrm{d}x_1 + \cdots + p_X(x_r)\,\mathrm{d}x_r\ . \tag{5.10}$$

Finally, recalling that all $\mathrm{d}y$, $\mathrm{d}x_1, \ldots, \mathrm{d}x_r$. are positive by hypothesis,

$$p_Y(y) = p_X(x_1)/\left|y'(x_1)\right| + \cdots + p_X(x_r)/\left|y'(x_r)\right|\ . \tag{5.11}$$

This is the r-root generalization we sought.

5.5 Illustrative Example

As an example, we return to the case $y = x^2$, $x_1 = -\sqrt{y}$, $x_2 = \sqrt{y}$. Here $y' = 2x$, $y'(x_1) = -2\sqrt{y}$, $y'(x_2) = 2\sqrt{y}$, so that by (5.11)

$$p_Y(y) = [p_X(-\sqrt{y}) + p_X(+\sqrt{y})]/2\sqrt{y}\ . \tag{5.12}$$

If, in particular, $x = N(0, \sigma^2)$, (5.12) yields

$$p_Y(y) = \begin{cases} \frac{1}{2\sqrt{y}}\frac{2}{\sqrt{2\pi}\sigma}\exp(-y/2\sigma^2) & \text{for}\ \ y \geq 0 \\ 0 & \text{for}\ \ y < 0\ . \end{cases} \tag{5.13}$$

The zero region for y originates of course in the fact that if $y = x^2$ with x real, y can never be negative.

5.6 Case of *n* Random Variables, *r* Roots

Here we consider the general situation, of a vector $\boldsymbol{x}$ of n RV's transformed to a new vector of n RV's $\boldsymbol{y}$ through a functional relation

$$\boldsymbol{y} = f(\boldsymbol{x})\ . \tag{5.14}$$

If density $p_X(\boldsymbol{x})$ is known, what is the new $p_Y(\boldsymbol{y})$?

To be most general, we shall assume that each RV x_m obtained from the inverse relation

$$\boldsymbol{x} = f^{-1}(\boldsymbol{y}) \tag{5.15}$$

has r roots $x_{m1}, \ldots, x_{mr}$, r general.

In order to avoid cumbersome notation we first consider the particular case of $n = 2$ RV's $\{y_1, y_2\}$, with $r = 2$ roots $\{x_{11}, x_{12}\}$ for x_1, and $\{x_{21}, x_{22}\}$ for x_2, in (5.15). The equivalence of events in $\boldsymbol{x}$ or $\boldsymbol{y}$ space implies

$$\begin{aligned} p_{\boldsymbol{Y}}(y_1, y_2)\,\mathrm{d}y_1\,\mathrm{d}y_2 = {} & p_{\boldsymbol{X}}(x_{11}, x_{21})\,\mathrm{d}x_{11}\,\mathrm{d}x_{21} + p_{\boldsymbol{X}}(x_{11}, x_{22})\,\mathrm{d}x_{11}\,\mathrm{d}x_{22} \\ & + p_{\boldsymbol{X}}(x_{12}, x_{21})\,\mathrm{d}x_{12}\,\mathrm{d}x_{21} + p_{\boldsymbol{X}}(x_{12}, x_{22})\,\mathrm{d}x_{12}\,\mathrm{d}x_{22} \end{aligned} \tag{5.16}$$

In other words, the sum is now over all permutations of the r roots, four terms here or r^n in general.

Since also $\mathrm{d}x_1\,\mathrm{d}x_2 = |\mathrm{J}(x_1, x_2/y_1, y_2)|\,\mathrm{d}y_1\,\mathrm{d}y_2$, where J is the Jacobian

$$J(x_1, x_2/y_1, y_2) \equiv \det \begin{bmatrix} \partial x_1/\partial y_1 & \partial x_1/\partial y_2 \\ \partial x_2/\partial y_1 & \partial x_2/\partial y_2 \end{bmatrix}, \tag{5.17}$$

from (5.16) the answer is

$$\begin{aligned} p_{\mathbf{Y}}(y_1, y_2) = \; & p_{\mathbf{X}}(x_{11}, x_{12})\,|J(x_{11}, x_{21}/y_1, y_2)| \\ & + p_{\mathbf{X}}(x_{11}, x_{22})\,|J(x_{11}, x_{22}/y_1, y_2)| \\ & + p_{\mathbf{X}}(x_{12}, x_{21})\,|J(x_{12}, x_{21}/y_1, y_2)| \\ & + p_{\mathbf{X}}(x_{12}, x_{22})\,|J(x_{12}, x_{22}/y_1, y_2)|\,. \end{aligned} \tag{5.18}$$

The notation

$$J(x_{11}, x_{21}/y_1, y_2) \equiv J(x_1, x_2/y_1, y_2)\Big|_{\substack{x_1 = x_{11} \\ x_2 = x_{21}}}$$

is used, and so forth for the other three terms.

The reader can easily generalize the result (5.18) to any n and r, although the *calculation* of r^n such terms may be tedious to say the least.

5.7 Optical Applications

There are many applications to optics of the principle of transformation of a random variable. These occur in analyzing such diverse phenomena as laser speckle, geometrical ray trace (to establish spot diagrams) and optical turbulence. We shall present below some applications from these problem areas. Although the transformation aspect of the problems will be emphasized, these problems inevitably draw upon other aspects of RV theory as well, and hence will serve to tie these all together.

5.8 Statistical Modeling

It will be seen that each problem, in addition, requires certain physical assumptions as to the actual processes going on. These assumptions comprise a "statistical model," the general aims of which are (a) to be simple, so as to produce a tractable solution, and yet (b) to be realistic enough to approximate reality. This dual aim is often difficult to achieve. On the pragmatic level, a model need only be as complicated as is required to produce agreement with experimental results. This is of course the "empirical method" of science, which is the criterion used in *deterministic* problems of physics as well. Hence, it shall be assumed that *if two competing models equally – well account for some random data, then the simpler model is preferred.* (This is the principle called "Occam's razor," after William of Occam.)

An example of statistical modeling is to regard the diffuser in a laser speckle setup as characterized by a "scatter spot" model, consisting of a large number of phasewise uncorrelated, contiguous areas, within each of which phase is perfectly correlated (Sect. 5.9.3). The same model will be seen to apply in other problem areas as well.

The ability to produce workable models is an acquired skill, which grows with successive problem work. After a while, a researcher amasses a "library" of models, which may be freely drawn upon in attempts to fit new problems. The aim of the following applications, and others beyond, is to start this process going.

5.9 Application of Transformation Theory to Laser Speckle

Anyone who has seen a laser beam in reflection off a wall, or in transmission, has seen laser speckle. It has a granular appearance, whereby the laser light appears to be broken up into a random pattern of grains (Fig. 5.3).

The effect is particularly bothersome because (a) the frequency of occurrence of random speckle intensity falls off rather slowly (single exponential) with intensity, so that large intensity excursion are common; (b) the standard deviation of these fluctuations *increases linearly* with mean intensity level in the light beam, so that increasing the signal level does not buy a loss in speckle effect; and (c) all phase values from 0 to 2π in the beam are *equally* likely, which is about as random as phase can be.

It is an interesting exercise in probability to derive properties (a) to (c). The central limit theorem and the theory of transformation of a RV will play major roles in the analysis below, which generally follows *Dainty*'s approach [5.2].

5.9.1 Physical Layout

As illustrated in Fig. 5.4, let a uniformly bright, collimated light beam of wavelength λ be incident upon a diffuser. Let the light be linearly polarized, for simplicity. Then

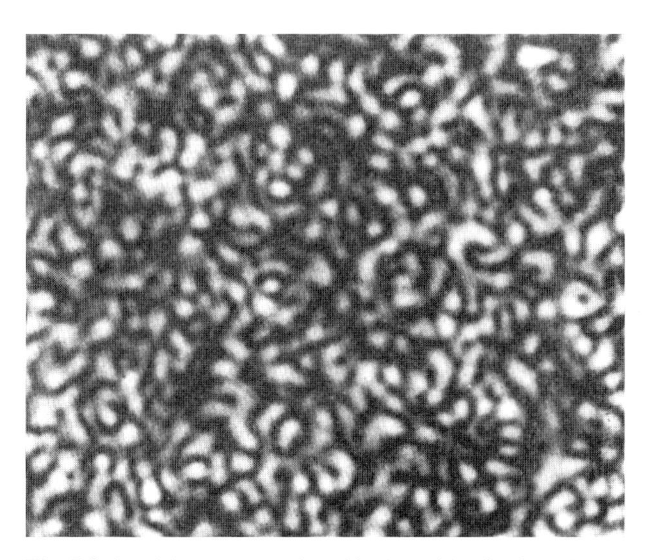

Fig. 5.3. Speckle pattern produced by laser illumination [5.1]

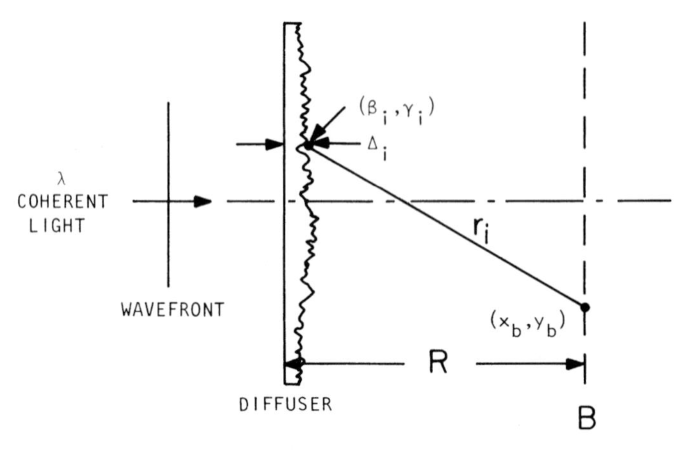

Fig. 5.4. Layout of the laser speckle phenomenon

a speckle pattern will be formed in all planes beyond the diffuser. In particular, we will examine the statistics of the pattern in the plane B located distance R from the diffuser. This speckle pattern could be observed by simply imaging plane B onto a screen, by use of any convenient lens.

5.9.2 Plan

As discussed above, we want to establish the probability laws $p_I(x)$ and $p_\Phi(x)$ for intensity and phase of the speckles in the plane B. To accomplish this, we will make use of the transformation relations

$$I = U_{\mathrm{re}}^2 + U_{\mathrm{im}}^2$$
$$\Phi = \tan^{-1}(U_{\mathrm{im}}/U_{\mathrm{re}}) \tag{5.19}$$

where U_{re}, U_{im}, respectively, are the real and imaginary parts of the complex light amplitude U at a point (x_b, y_b) in plane B. If joint probability $p_{U_{\mathrm{re}}U_{\mathrm{im}}}(x, y)$ can be established first, *the law* $p_{I\Phi}(v, w)$ *follows by a transformation (5.18) of random variables.* With $p_{I\varphi}$ known, the required laws p_I, p_Φ are found by marginal integrations (3.8) of $p_{I\Phi}$. Hence, we set out to find $p_{U_{\mathrm{re}}U_{\mathrm{im}}}(x, y)$.

5.9.3 Statistical Model

We will show here that by a reasonable and yet simple statistical model U_{re} and U_{im} each obey the central limit theorem. Hence they are normally distributed. The assumptions of the model are denoted by • in the following. Two additional assumptions are made below, in Sect. 5.9.4 and 5.9.5.

- If R is not too small, the light amplitude U at any point (x_b, y_b) in plane B is the integral $\mathrm{d}r_i$ over all spherical waves $\exp(\mathrm{j}kr_i)/r_i$ from the diffuser. These are Huygens' celebrated wavelets.

- To keep the analysis simple, assume that the diffuser consists of $N \gg 1$ uncorrelated, contiguous areas called "scattering spots." Within each scattering spot the amplitude is perfectly correlated. This is a statistical model which has been used with success in other problems as well, e.g., in Sect. 5.10.1 or in [5.3].
- Let r_i connect the image point (x_b, y_b) with the center of the ith spot. Then if the spot area Δa is sufficiently small, the integral over Huygens' wavelets becomes well approximated by a discrete sum

$$U = (\Delta a/\lambda R) \sum_{i=1}^{N} \mathrm{e}^{\mathrm{j}kr_i} \,. \tag{5.20}$$

5.9.4 Marginal Probabilities for Light Amplitudes U_{re}, U_{im}

We next want to show that U obeys the central limit theorem (4.48) in each of its real and imaginary parts. This will establish that U_{re} and U_{im} are each normal.

From the geometry of Fig. 5.4,

$$r_i = [(R - \Delta_i)^2 + (\beta_i - x_b)^2 + (\gamma_i - y_b)^2]^{1/2} \tag{5.21}$$

where Δ_i is the diffuser thickness at point (β_i, γ_i). Of course, $\Delta_i \ll R$. Then (5.21) is well-approximated by

$$r_i = (R - \Delta_i) + (2R)^{-1}[(\beta_i - x_b)^2 + (\gamma_i - y_b)^2] \tag{5.22}$$

or simply

$$r_i = A_i - \Delta_i \tag{5.23}$$

where A_i is all the deterministic terms in (5.22).

The substitution of (5.23) into (5.20) gives

$$U = c \sum_{i=1}^{N} \exp[\mathrm{j}k(A_i - \Delta_i)] \,, \quad c = \Delta a/\lambda R \,. \tag{5.24}$$

Does U obey the central limit theorem?

By hypothesis, each scattering spot i has an independent phase Δ_i. Therefore, each term in the sum (5.24) is an independent RV. Let us assume further that the statistics of Δ across the diffuser are invariant with position i. Then, except for the deterministic factor $\exp(\mathrm{j}kA_i)$, each term of the sum (5.24) is identically distributed as well as independent. Further, the factors $\exp(\mathrm{j}kA_i)$ become real weights a_i in the real and imaginary parts U_{re}, U_{im} of sum (5.24). Since in addition N is large, the ground rules described in Exercise 4.2 for satisfying the central limit theorem have been satisfied. Hence, U_{re} and U_{im} are each normal RV's.

Having established that the marginal probabilities $p_{U_{\mathrm{re}}}$ and $p_{U_{\mathrm{im}}}$ are each normal, we now ask what the *joint* probability $p_{U_{\mathrm{re}}U_{\mathrm{im}}}$ is. This quantity is not, of course, uniquely fixed by the marginal laws. For example, if U_{re} and U_{im} are *correlated*, the joint probability in question might be the normal bivariate one. If they are not correlated, we must assume the joint law to be a simple product of the two marginal laws. Therefore, we have to determine the state of correlation.

5.9.5 Correlation Between U_{re} and U_{im}

Directly from (5.24), the correlation in question obeys

$$\begin{aligned}\langle U_{\mathrm{re}} U_{\mathrm{im}} \rangle &= c^2 \sum_{m,n} \langle \cos k(A_m - \Delta_m) \sin k(A_n - \Delta_n) \rangle \\ &= c^2 \sum_{m \neq n} \langle \cos k(A_m - \Delta_m) \rangle \langle \sin k(A_n - \Delta_n) \rangle \\ &\quad + c^2 \sum_{n} \langle \cos k(A_n - \Delta_n) \sin k(A_n - \Delta_n) \rangle \end{aligned} \tag{5.25}$$

according to whether $m = n$ or not. The sum $m \neq n$ separates into the given product of expectations because Δ_m and Δ_n are independent, by the statistical model. The second sum was for $m = n$.

- In order to evaluate the averages in (5.25), we have to fix the statistics of Δ. It is reasonable to assume that a physical diffuser has *uniform randomness* for its total phase $k\Delta$. All this presumes is that $p(\Delta)$ has a $\sigma \gg \lambda$, since then $\langle (k\Delta)^2 \rangle = (2\pi)^2 \langle \Delta^2 \rangle / \lambda^2 = (2\pi)^2 \sigma^2 / \lambda^2 \gg (2\pi)^2$, so that many cycles of phase are a common occurrence. In summary, the final assumption of our statistical model is that

$$p_\Delta(\theta) = (2\pi)^{-1} \mathrm{Rect}(\theta / 2\pi) \,. \tag{5.26}$$

Then identically

$$\begin{aligned} \langle \cos k(A - \Delta) \rangle &= (2\pi)^{-1} \int_{-\pi}^{\pi} \mathrm{d}\theta \cos(kA - \theta) = 0 \,, \\ \langle \sin k(A - \Delta) \rangle &= 0 \end{aligned}$$

and

$$\langle \cos(kA - \theta) \sin(kA - \theta) \rangle = 0 \,.$$

Therefore, U_{re} and U_{im} do not correlate.
These results also imply that

$$\langle U_{\mathrm{re}} \rangle \equiv c \sum_{n=1}^{N} \langle \cos k(A_n - \Delta_n) \rangle = 0$$

and

$$\langle U_{\mathrm{im}} \rangle = 0 \,.$$

Therefore, each of U_{re} and U_{im} is a *Gaussian* RV, in particular.

5.9.6 Joint Probability Law for U_{re}, U_{im}

We have thereby established that the joint law is a simple product of Gaussian marginal laws

$$p_{U_{\mathrm{re}}U_{\mathrm{im}}}(x, y) = (2\pi\sigma^2)^{-1} \exp[-(x^2 + y^2)/2\sigma^2] . \tag{5.27}$$

Here

$$\begin{aligned} 2\sigma^2 &\equiv \langle |U|^2 \rangle = \langle I \rangle = \langle U_{\mathrm{re}}^2 + U_{\mathrm{im}}^2 \rangle = \langle U_{\mathrm{re}}^2 \rangle + \langle U_{\mathrm{im}}^2 \rangle \\ &= c^2 \sum_{n=1}^{N} \langle \cos^2 k(A_n - \Delta_n) \rangle + c^2 \sum_{n=1}^{N} \langle \sin^2 k(A_n - \Delta_n) \rangle \end{aligned}$$

by (5.24), or

$$2\sigma^2 = Nc^2 = N(\Delta a/\lambda R)^2 . \tag{5.28}$$

5.9.7 Probability Laws for Intensity and Phase; Transformation of the RV's

Continuing the plan of Sect. 5.9.2, we want to find the joint law $p_{I\Phi}(v, w)$ through the transformation (5.19) of old RV's U_{re} and U_{im} to new variables I and Φ. This is a unique ($r = 1$ root) transformation. By the approach (5.18) with $r = 1$, we obtain

$$p_{I\Phi}(v, w) = p_{U_{\mathrm{re}}U_{\mathrm{im}}}(x, y) \left| \mathrm{J}\left(\frac{x, y}{v, w}\right) \right| . \tag{5.29}$$

Now

$$\mathrm{J}\left(\frac{x, y}{v, w}\right) \equiv \det \begin{bmatrix} \partial x/\partial v & \partial x/\partial w \\ \partial y/\partial v & \partial y/\partial w \end{bmatrix} ,$$

where by the transformation (5.19)

$$U_{\mathrm{re}} = I^{1/2} \cos \Phi \quad \text{or} \quad x = v^{1/2} \cos w$$

and

$$U_{\mathrm{im}} = I^{1/2} \sin \Phi \quad \text{or} \quad y = v^{1/2} \sin w .$$

Thus

$$\begin{aligned} \mathrm{J}\left(\frac{x, y}{v, w}\right) &= \det \begin{bmatrix} 2^{-1} v^{-1/2} \cos w & -v^{1/2} \sin w \\ 2^{-1} v^{-1/2} \sin w & v^{1/2} \cos w \end{bmatrix} \\ &= 2^{-1} \cos^2 w + 2^{-1} \sin^2 w = 1/2 . \end{aligned}$$

Using this result in (5.29),

$$p_{I\Phi}(v, w) = 2^{-1} p_{U_{\mathrm{re}}U_{\mathrm{im}}}(v^{1/2} \cos w, v^{1/2} \sin w) . \tag{5.30}$$

This would be true for any joint law $p_{U_{\mathrm{re}}U_{\mathrm{im}}}$. In the particular case (5.27), Eq. (5.30) becomes

$$p_{I\Phi}(v, w) = \frac{1}{4\pi\sigma^2} \exp(-v/2\sigma^2) \mathrm{Rect}(w/2\pi) \tag{5.31}$$

where $v \geq 0$.

5.9.8 Marginal Laws for Intensity and Phase

Following the prescription (3.8), Eq. (5.31) integrates dw over interval $(-\pi, \pi)$ to give

$$p_I(v) = (\sqrt{2}\sigma)^{-2} \exp(-v/2\sigma^2) = \langle I \rangle^{-1} \exp(-v/\langle I \rangle) , \tag{5.32a}$$

by (5.28), or it integrates dv over interval $(0, \infty)$ to give

$$p_\Phi(w) = (2\pi)^{-1} \mathrm{Rect}(w/2\pi) . \tag{5.32b}$$

These are what we set out to find.

In summary, under the assumptions of the given statistical model, *speckle intensity follows a negative exponential law (5.32a), and speckle phase follows a uniform law (5.32b).*

The fact that (5.32a) falls off so slowly with intensity v accounts for the very random and grainy appearance in a speckle pattern. Quantitatively, this slow fall off makes the speckle fluctuations comparable with the mean intensity level in the pattern, as shown next.

5.9.9 Signal-to-Noise (S/N) Ratio in the Speckle Image

The signal-to-noise ratio, denoted as S/N, describes the extent to which an image is corrupted by noise. It is defined as the mean intensity value $\langle I \rangle$ over the standard deviation for intensity σ_1. These quantities may easily be found from the known law (5.32a) for $p_1(v)$. Directly,

$$\langle I \rangle = (\sqrt{2}\sigma)^{-2} \int_0^\infty dv\, v \exp(-v/2\sigma^2) = 2\sigma^2 , \tag{5.33}$$

after an integration-by-parts. Further,

$$\langle I^2 \rangle = (\sqrt{2}\sigma)^{-2} \int_0^\infty dv\, v^2 \exp(-v/2\sigma^2) = 8\sigma^4 , \tag{5.34}$$

again integrating by parts. Then by identity (3.15)

$$\sigma_1^2 \equiv \langle I^2 \rangle - \langle I \rangle^2 = 4\sigma^4 . \tag{5.35}$$

Using (5.33) and (5.35),

$$\mathrm{S/N} \equiv \langle I \rangle / \sigma_1 = (2\sigma^2)/(2\sigma^2) = 1 . \tag{5.36}$$

This result shows specifically how noisy a process speckle is. *On the average, its fluctuations equal the mean signal itself!* This accounts for the extremely grainy appearance of a speckle image.

It is possible to reduce the noise level by averaging adjacent image points. Many such "speckle reduction" schemes have been proposed; see, e.g., the survey in [5.4]. One of the simplest of these approaches is to scan the speckle image with a finite aperture. The output of such a scan is the convolution of the noisy speckle image with a "pillbox" spread function, which performs a running average or smoothing effect upon the image. It is a useful exercise in probability to analyze this technique, and this is done next. The probability law which will arise, the chi-square, will later be used for the completely different application of "significance" testing.

5.10 Speckle Reduction by Use of a Scanning Aperture

One direct way of reducing speckle is by picking up the speckled image with a scanning aperture of finite area A [5.5]. The aperture automatically averages the speckle falling within it at each aperture position. Of course, resolution is lost in this manner but it is often a worthwhile tradeoff with speckle reduction.

The ultimate aim of using a scanning aperture is an increase in the S/N ratio. This will be found by determining the probability density for total intensity within the aperture.

5.10.1 Statistical Model

As usual, a statistical model must be devised, and this is denoted by dots •

• Assume the speckle image to consist of discrete cells. Within each cell the light amplitude does not spatially vary, and the amplitudes do not correlate from cell to cell. This model coincides with the "scatter spot" model of Sect. 5.9.3 for the diffuser phase profile. Let aperture A contain M of these cells, as illustrated in Fig. 5.5.

• The total amplitude U (complex) inside the aperture then obeys

$$U = \sum_{m=1}^{M} U_m \, , \tag{5.37}$$

where U_m represents the complex amplitude from the mth cell.

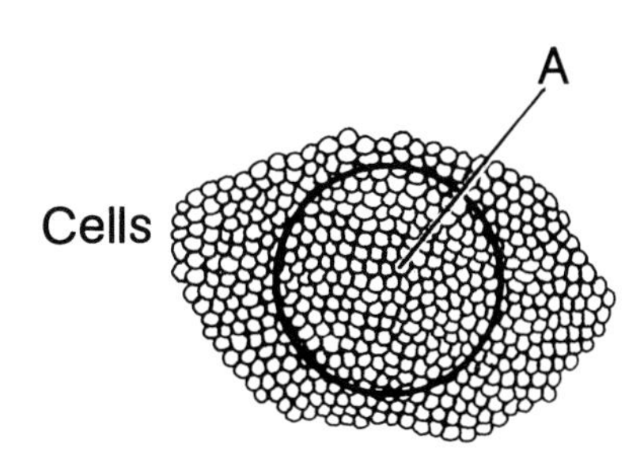

Fig. 5.5. Scanning aperture A encloses a fixed number M of cells containing independent amplitudes. Each cell generally contains many speckles

By (5.37), the total intensity v and net phase w obey

$$v = \sum_{m=1}^{M} (x_m^2 + y_m^2) \, , \quad w = \tan^{-1}\left(\sum_m y_m / \sum_m x_m\right) \tag{5.38}$$

where x_m, y_m are U_{re}, U_{im} for cell m. We are particularly interested in establishing $p_1(v)$, the probability for output intensity from the aperture.

5.10.2 Probability Density for Output Intensity $p_I(v)$

The first of (5.38) may be rewritten as

$$v = \sum_{m=1}^{M} x_m^2 + \sum_{m=1}^{M} y_m^2 \,. \tag{5.39}$$

Now, by (5.27), the marginal probabilities at each cell m obey

$$p_{U_{\rm re}}(x) = p_{U_{\rm im}}(x) = (2\pi\sigma^2)^{-1/2} \exp(-x^2/2\sigma^2) \,. \tag{5.40}$$

Since σ is independent of m, by (5.28), *these probabilities are the same at each m.*

We note the quadratic dependencies in (5.39) and therefore seek the probability law for $U_{\rm re}^2$ and for $U_{\rm im}^2$. By transformation law (5.12) for the square of an RV,

$$p_{U_{\rm re}^2}(x) = p_{U_{\rm im}^2}(x) = (2\pi\sigma^2)^{-1/2} x^{-1/2} \exp(-x/2\sigma^2) \,, \quad x \geqq 0 \,, \tag{5.41}$$

at each cell m. By (5.39), intensity v is the sum of $2M$ such RV's. Also, by the model, these are independent.

In summary to this point, intensity v is the sum of $2M$ identically distributed and independent RV's. Then according to (4.26), the unknown $p_1(v)$ has a characteristic function which is a product of those for $U_{\rm re}^2$ and $U_{\rm re}^2$.

Accordingly, we form

$$\varphi_{U_{\rm re}^2}(\omega) = \varphi_{U_{\rm im}^2}(\omega) = (2\pi\sigma^2)^{-1/2} \int_0^\infty \mathrm{d}x x^{-1/2} \exp(\mathrm{j}\omega x - x/2\sigma^2) \,. \tag{5.42}$$

This is a special case of the more general integral

$$I(\nu) \equiv \int_0^\infty \mathrm{d}x\, x^{\nu-1} \exp(\mathrm{j}\omega x - x/2\sigma^2) \,. \tag{5.43}$$

In fact, comparison shows that

$$\varphi_{U_{\rm re}^2}(\omega) = \varphi_{U_{\rm im}^2}(\omega) = (2\pi\sigma^2)^{-1/2} I(1/2) \,. \tag{5.44}$$

To evaluate $I(v)$, use the change of variable

$$t = -(\mathrm{j}\omega x - x/2\sigma^2) \,.$$

Then

$$\begin{aligned} I(\nu) &= (1/2\sigma^2 - \mathrm{j}\omega)^{-\nu} \int_0^\infty \mathrm{d}t\, t^{\nu-1} \mathrm{e}^{-t} \\ &= (2\sigma^2)^\nu (1 - 2\mathrm{j}\omega\sigma^2)^{-\nu} \Gamma(\nu) \,, \end{aligned} \tag{5.45}$$

in terms of $\Gamma(\nu)$, the gamma function.

Using $\Gamma(1/2) = \pi^{1/2}$ and (5.44), we find

$$\varphi_{U_{\rm re}^2}(\omega) = \varphi_{U_{\rm im}^2}(\omega) = (1 - 2\mathrm{j}\omega\sigma^2)^{-1/2} \,. \tag{5.46}$$

This is the characteristic function for $U_{\rm re}^2$ or $U_{\rm im}^2$ at each of the M independent speckles.

We are now ready to form $\varphi_I(\omega)$. By (5.39) and the independence of random variables x_m and y_m, $\varphi_1(\omega)$ obeys a product (4.26) of the form

$$\varphi_1(\omega) = \prod_{m=1}^{M} \varphi_{U_{\text{re}}^2}(\omega) \prod_{m=1}^{M} \varphi_{U_{\text{im}}^2}(\omega) \, . \tag{5.47}$$

Then, using (5.46) for every term in the products,

$$\varphi_I(\omega) = (1 - 2\mathrm{j}\omega\sigma^2)^{-M} \, . \tag{5.48}$$

The probability density having this characteristic function is the sought-after result. By (5.43), $I(M)$ defines the characteristic function for a probability $x^{M-1}\exp(-x/\sigma^2)$. By result (5.45) this characteristic function is

$$(2\sigma^2)^M \Gamma(M)(1 - 2\mathrm{j}\omega\sigma^2)^{-M} \, . \tag{5.49}$$

Compare this with (5.48). It is $(2\sigma^2)^M \Gamma(M)\varphi_I(\omega)$, almost what we want. Hence a probability density

$$p_I(x) = (2\sigma^2)^{-M}[\Gamma(M)]^{-1} x^{M-1} \exp(-x/2\sigma^2) \tag{5.50}$$

would have precisely (5.48) as its characteristic function. This is our required probability for total intensity.

This probability density is called the "chi-square" distribution. As a check on the result, for $M = 1$ it goes over into the result (5.32a) for *single point* observation of the speckle image.

For further remarks on the degree of accuracy in the result (5.50), and a more exact approach to the subject, see the chapter by *Goodman* in [5.4]. Also see Exercise 5.1.25 for the probability law on the *number* of speckles.

5.10.3 Moments and S/N Ratio

The moments of (5.50) are best obtained by successive differentiation of its characteristic function (5.48). To directly use (5.50) instead in moment integrals would be very cumbersome.

Directly,

$$\varphi_I'(\omega) = 2\mathrm{j}M\sigma^2(1 - 2\mathrm{j}\omega\sigma^2)^{-M-1}$$

and

$$\varphi_I''(\omega) = 4\mathrm{j}^2 M(M+1)\sigma^4(1 - 2\mathrm{j}\omega\sigma^2)^{-M-2} \, .$$

Then

$$\langle I \rangle = 2M\sigma^2 \text{ and } \left\langle I^2 \right\rangle = 4M(M+1)\sigma^4 \, . \tag{5.51}$$

Using identity (3.15), we get

$$\sigma_\mathrm{I}^2 = 4M\sigma^4 \, . \tag{5.52}$$

As might be expected, σ_I^2 for the aperture varies directly with the number M of speckles it encloses. However, so does the mean signal level $\langle I \rangle$, by (5.51). Again defining an S/N ratio (Sect. 5.9.9) by

$$\text{S/N} \equiv \langle I \rangle / \sigma_I ,$$

from (5.51) and (5.52) we get

$$\text{S/N} = M^{1/2} . \tag{5.53}$$

This shows the advantage of aperture averaging. The output S/N directly increases as the square root of M, and hence as the square root of aperture area.

Of course, aperture averaging is not entirely beneficial, since it is equivalent to convolution of the image with a square window function, and this must entail a *loss of spatial resolution*. Work on speckle reduction is still going on.

5.10.4 Standard Form for the Chi-Square Distribution

The chi-square distribution (5.50) was found to arise out of a particular physical effect, speckle averaging. From the derivation, a chi-square distribution in a random variable z can be seen to hold whenever

$$z = \sum_{n=1}^{N} x_n^2 , \quad x_n = N(0, \sigma^2) , \tag{5.54}$$

N finite,[1] where the $\{x_n\}$ are independently chosen from the same, Gaussian, probability law. The $\{x_n\}$ are called the "degrees of freedom" for the problem. There are N degrees of freedom in (5.54).

This *standard form* of the problem corresponds to the speckle problem we considered if one makes replacement $2M \to N$ in (5.48–53). The results are

$$p_Z(z) = (\sqrt{2}\sigma)^{-N}[\Gamma(N/2)]^{-1} z^{N/2-1} \, \mathrm{e}^{-z/2\sigma^2} , \tag{5.55a}$$

$$\varphi_Z(\omega) = (1 - 2\mathrm{j}\omega\sigma^2)^{-N/2} , \tag{5.55b}$$

$$\langle z \rangle = N\sigma^2 , \quad \sigma_Z^2 = 2N\sigma^4 , \quad \text{S/N} = \sqrt{N/2} , \tag{5.55c}$$

Some chi-square probability curves are shown in Fig. 5.6, for $\sigma = 1$ and the indicated number of degrees of freedom.

For small N the curves are extremely skewed to the right, but become less so as N increases. In fact they approach a symmetric, *normal* shape as $N \to \infty$. This is implied by (5.54), which shows that z is formed as the sum $\sum x_n^2 \equiv \sum z_n$ of N identically distributed random variables z_n. By the central limit theorem z is normally distributed as $N \to \infty$.

The standard problem (5.54) arises in many physical problems. It also characterizes the probability law for the "sample variance" and is commonly used in a test of "significance" of experimental data. These are shown in later chapters.

---

[1] The number N and the function N in equations (5.54) are of course different. We apologize for the possible confusion.

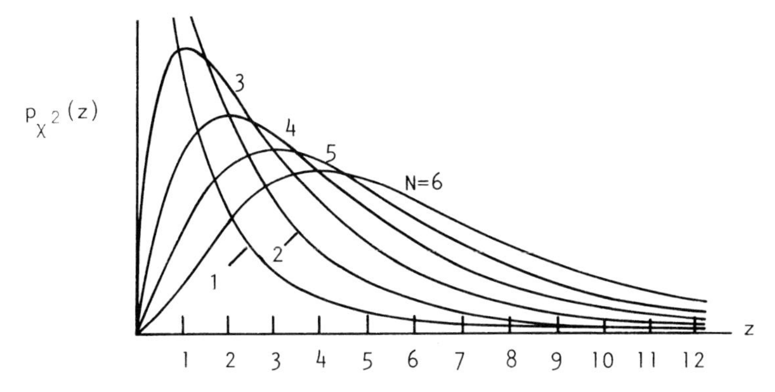

Fig. 5.6. Chi-square probability curves for $\sigma = 1$ and the indicated number of degrees of freedom N

5.11 Calculation of Spot Intensity Profiles Using Transformation Theory

We showed in Sect. 5.3 and in Fig. 5.2 an example of how to calculate a geometrical intensity curve using basic principles of probability. As that particular problem has very limited practical use, we now want to broaden the view as much as possible.

Consider Fig. 5.7, where object and image planes are in fixed positions, and the spot intensity profile in the image due to a single object point P is required. The lens system in between is a general one. Let any ray emanating from P be specified by its *polar and azimuthal angles* x_1 and x_2, respectively. Let the ray strike the image plane at *rectangular coordinates* y_1 and y_2. The density of such ray-strikes as a function of (y_1, y_2) is called the "geometrical spot intensity" profile $p_{Y_1Y_2}(y_1, y_2)$. It is a common figure of merit for lens system performance.

Now *the density of the ray-strikes is*, by the law of large numbers (2.8), *precisely the probability density for these events*. Furthermore, the coordinates (y_1, y_2) relate to the angles (x_1, x_2) by a *known (ray-trace) relation*

$$(y_1, y_2) = f(x_1, x_2) , \tag{5.56a}$$

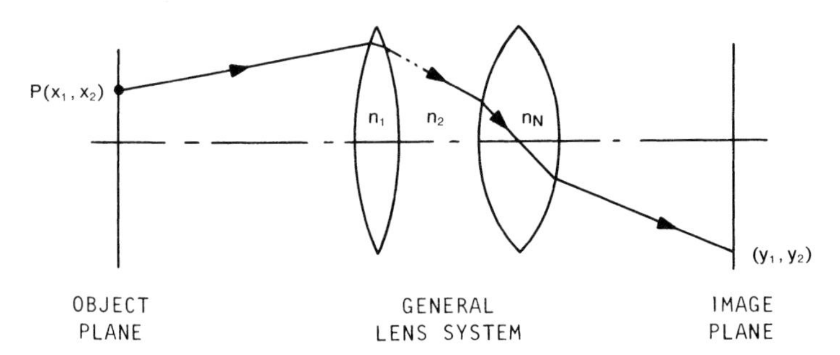

Fig. 5.7. The spot profile in the image plane is a probability law resulting from transformation of a random variable

and also the probability density $p_{X_1X_2}(x_1, x_2)$ for angles (x_1, x_2) is known. (For example, a uniform fan of rays is equivalent to a Rect-function on x_1 and on x_2.) Then our problem of estimating the spot profile $p_{Y_1Y_2}(y_1, y_2)$ function *is fully equivalent to establishing the probability density function* $p_{y_1y_2}(y_1, y_2)$ *due to a transformation* (5.18) of RV's. Aside from the beauty of the correspondence, this is a problem whose answer is already known.

According to the recipe (5.18), we simply have to know the Jacobian of the transformation at each of the roots (x_{1r}, x_{2r}) corresponding to output coordinates (y_1, y_2). In the vast majority of cases there will be only one such root.[2] The answer then is

$$p_{Y_1Y_2}(y_1, y_2) = \begin{cases} \text{Abs det} \begin{bmatrix} \partial x_1/\partial y_1 & \partial x_1/\partial y_2 \\ \partial x_2/\partial y_1 & \partial x_2/\partial y_2 \end{bmatrix} \text{ for } \begin{matrix} a_1 \leqq y_1 \leqq b_1 , \\ a_2 \leqq y_2 \leqq b_2 , \end{matrix} \\ 0 \text{ for } y_1, y_2 \text{ outside above intervals .} \end{cases} \tag{5.56b}$$

This is the Jacobian alone, since a uniform or Rect-function ray fan was supposed for $p_{X_1X_2}(x_1, x_2)$. Limiting coordinate values (a_1, b_1), (a_2, b_2) for y_1, y_2 are determined by (5.56a) where x_1, x_2 take on their extreme values defined by the pupil size.

Hence, the solution centers on knowing the partial derivatives in (5.56b). These could be found, and exactly, if the *functional form* for f in the ray-trace relation (5.56a) were analytically known. We consider such a case first. Later we treat the more general case.

5.11.1 Illustrative Example

To illustrate use of the approach, we have contrived an analytic example where the functional form f of ray-trace relation (5.56a) is known. Consider the one-dimensional system shown in Fig. 5.8. The lens system is a single cylindrical surface of radius R, index n, and paraxial focal length F. The image plane is located at distance D from the surface. The object point is at infinity. We want to find the spot intensity profile $p_Y(y)$ due to a uniform fan of rays $p_X(x) = \text{Rect}(x/3)$ (ignoring an irrelevant multiplicative constant). In this simple case, ray-trace relation (5.56a) becomes

$$y = x(1 - D/F) + (x^3/2F^2)n(n-1)^{-1}[1 - n(n-1)^{-1}D/F] \tag{5.57}$$

to third order in x. Thus, the single derivative $\mathrm{d}x/\mathrm{d}y$ here defining the Jacobian is known analytically, no ray-trace *subroutines* being needed.

With (5.57) known, the procedure is to define a required subdivision in y, invert the cubic polynomial (5.57) for the corresponding x values, evaluate $\mathrm{d}x/\mathrm{d}y$ at these x, and then use (5.56b) to compute $p_Y(y)$, which is here simply $|\mathrm{d}x/\mathrm{d}y|$. Resulting spot profiles are shown in Fig. 5.9, for three relative positions D/F of the image plane. Note that the closer the image plane comes to the focal point the more

[2] Multiple roots occur when the caustic bundle of intersecting rays also intersects the image plane. This does not ordinarily occur, but if it does, it is only at isolated points (y_1, y_2) of the plane.

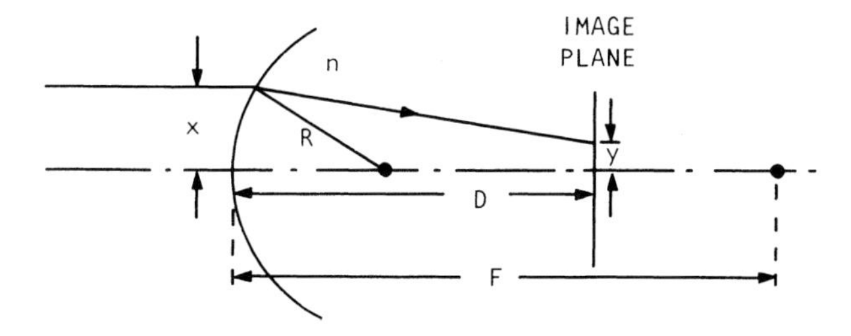

Fig. 5.8. Application to a simple cylindrical system

concentrated is the spot, right to the limit where the central maximum becomes infinite when the image plane is located at the paraxial focus. (This infinite value is, of course, an approximation of the geometrical ray approach taken, and would not occur in a diffraction analysis.)

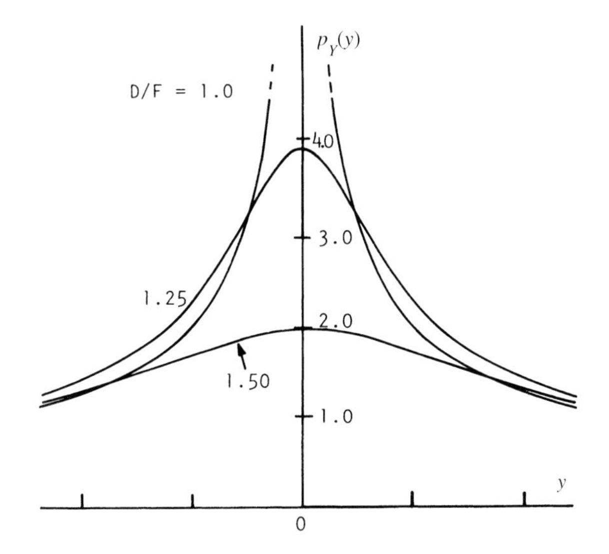

Fig. 5.9. Resulting spot profiles at three different image plane positions D indicated by ratio D/F

5.11.2 Implementation by Ray-Trace

We have shown how the approach (5.56a, b) may be used in the simplest possible case of a single surface. In that special case, the functional form of f in (5.56b) is known analytically. But what about the real situation involving many surfaces? The tracing of rays through a complex lens system does not ordinarily lend itself to such analytic expression. It would seem then that this elegant approach must be abandoned in practice.

There is a way out, however. This is to *numerically* trace through the lens system each ray (x_1, x_2) and its differential counterpart $(x_1 + dx_1, x_2 + dx_2)$; these give rise to ray-strikes (y_1, y_2) and $(y_1 + dy_1, y_2 + dy_2)$. All these quantities are now known

– as *numbers*. Therefore, ratios $\partial x_1/\partial y_2$, etc., may now be *numerically* formed by finite differences, permitting (5.56b) to be used once again.

In practice, the calculation would start by choosing a uniform subdivision of *input angles* (x_1, x_2) and their differentials. Given these, a raytrace would be *directly* used to define the corresponding (y_1, y_2) in image space. [Note that in this way we avoid the horrific problem of somehow *inverting* the ray-trace procedure to find unknown inputs (x_1, x_2) in terms of known outputs (y_1, y_2).] The only price to pay for this ray-trace approach is determination of the required curve $p_{Y_1Y_2}(y_1, y_2)$ *at an uneven spacing* of points (y_1, y_2), since we cannot guarantee even spacing in the output points (y_1, y_2).

However, this uneven spacing is not a severe price. It may be overcome by suitably interpolating between the points, or simply by using more rays.

The novel feature of the approach is that *each pair* (ray, differential ray) *of traced rays establishes with arbitrary precision a point on the spot intensity curve*. This is to be compared with the conventional approach of building up the spot curve as a histogram of ray strikes within narrow zones of the image plane. The latter is intrinsically an approximation because of the finite size of the zones, and also requires hundreds of rays be to traced *to* each zone in order to achieve acceptable error in each *single* estimated spot value.

This approach may also be derived from a deterministic point of view, using conservation of energy along each ray [5.6].

5.12 Application of Transformation Theory to a Satellite-Ground Communication Problem

This problem involving transformation of a RV was encountered by the author while doing research upon a proposed communication channel for the US Navy. It is not simply a problem of obtaining a density $p_Y(y)$ from knowledge of a $p_X(x)$, but like most real life problems (which offer real payment for their solution), involved a slightly higher degree of complexity.

A satellite-ground communication link consists of a laser at the satellite and a detector of light on the ground. A sequence of laser pulses, either "on" or "off" in a binary coded message, are being transmitted from the satellite to the detector. The signals pass through the whole earth's atmosphere, and hence are temporally random at the detector on the ground. They "twinkle." (Fig. 5.10).

Superimposed upon the detected intensity values over time will be randomly fluctuating light due to background sources such as ambient skylight. Now the aim of the detector logic is to correctly identify each transmitted symbol as being either "on" or "off." One strategy for accomplishing this aim is to check whether the received intensity value I exceeds a threshold value. Making the threshold high enough will clip out most of the random background radiation. However, it will have the drawback of erroneously missing those "on" signals that were received with fluctuations far enough *below* their mean to not exceed threshold. These will

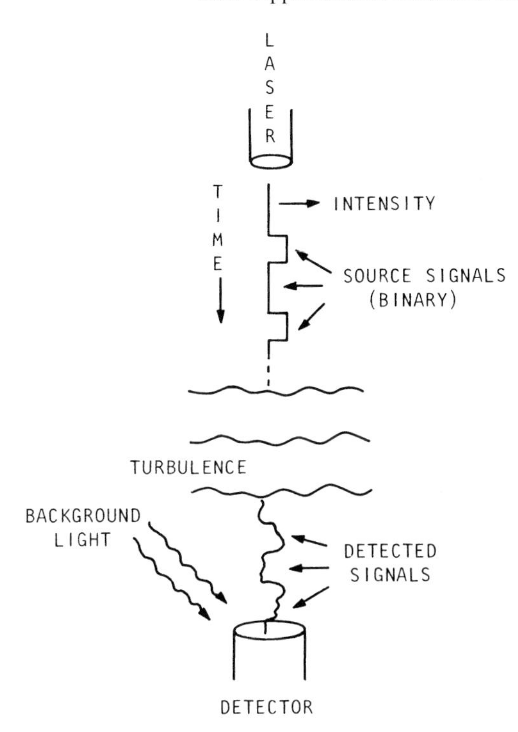

Fig. 5.10. Layout of a satellite-ground communication link

be erroneously identified as "off" signals. We therefore will leave the threshold as variable in the analysis that follows.

Let us also make the reasonable assumption that the background radiation does not fluctuate nearly as strongly as the laser light twinkles. Instead, it "drifts" slowly over time. Hence, virtually all fluctuations in the received signal I are due to laser scintillation. However, a finite threshold still must be used because of incomplete knowledge of the drifting background value and other factors that are outside the scope of this discussion.

Then *the probability that an "on" symbol will be correctly identified by the detector is the probability that the received laser intensity I exceeds threshold.* Let the latter be rm_I, where r is an arbitrary multiplier and m_I is the mean detected laser intensity. Hence, we want to compute $P(I > rm_I)$. (The analysis for the "off" case would follow similar steps.)

Let σ_I^2 denote the variance in intensity I. Now the ratio σ_I/m_I was measured by *Burke* [5.7] for filtered starlight. The latter is a close approximation to laser light as well. He found that for stars near zenith (straight up) viewing, the relative fluctuation in irradiance due to scintillation is approximately

$$\sigma_I/m_I = 0.4\ . \tag{5.58}$$

This is in fact the only data we have to go on! Can the required $P(I > rm_I)$ be found on this basis alone? Now

$$P(I > rm_I) \equiv \int_{rm_I}^{\infty} \mathrm{d}x\, p_I(x) \tag{5.59}$$

where $p_I(x)$ is the probability density law for scintillation. The calculation cannot proceed unless the form of $p_I(x)$ is known.

Actually, what is known is that $\ln I$ follows a normal law; see Ex. 4.3.13. This suggests that we merely transform the known law in $\ln I$ to one in I by using (5.4), and then integrating it in (5.59). However, this is not necessary. By (5.4),

$$\mathrm{d}x\, p_I(x) = \mathrm{d}y\, p_{\ln I}(y) \tag{5.60a}$$

when

$$y = \ln x\ . \tag{5.60b}$$

Hence (5.59) becomes

$$P(I > rm_I) = \int_{\ln rm_I}^{\infty} \mathrm{d}y\; p_{\ln I}(y)\ , \tag{5.61}$$

directly in terms of the *logarithmic* RV. This would allow us to use the known normal law for $p_{\ln I}(y)$ to compute the integral, provided its mean $m_{\ln I}$ and variance $\sigma^2_{\ln I}$ were known numerically. But the only piece of numerical data we have is (5.58), and this is for direct intensity and not log intensity. What can be done? This is the real-life complication we referred to at the outset.

In fact, it is possible to relate the known to the unknown quantities.[3] By a direct evaluation of m_I,

$$m_I \equiv \int_0^{\infty} \mathrm{d}x x\; p_I(x) \tag{5.62a}$$

so that by transformation (5.60a, b)

$$m_I = \int_{-\infty}^{\infty} \mathrm{d}y \mathrm{e}^y\, p_{\ln I}(y)\ . \tag{5.62b}$$

We may now plug in the normal law for $p_{\ln I}(y)$ and carry through the integration, to find

$$m_I = \exp(m_{\ln I} + \sigma^2_{\ln I}/2)\ . \tag{5.63}$$

Remember we want to find $m_{\ln I}$ and $\sigma_{\ln I}$. Equation (5.63) is then one relation in the two unknowns.

The other needed relation is gotten by similarly evaluating

$$\langle I^2 \rangle = m_I^2 + \sigma_I^2 = \int_0^{\infty} \mathrm{d}x x^2\, p_I(x)\ . \tag{5.64}$$

Identity (3.15) was used. Again using (5.60a, b) we get

$$m_I^2 + \sigma_I^2 = \int_{-\infty}^{\infty} \mathrm{d}y \mathrm{e}^{2y}\, p_{\ln I}(y)\ . \tag{5.65}$$

[3] We thank Dr. R.P. Bocker for collaboration on this problem.

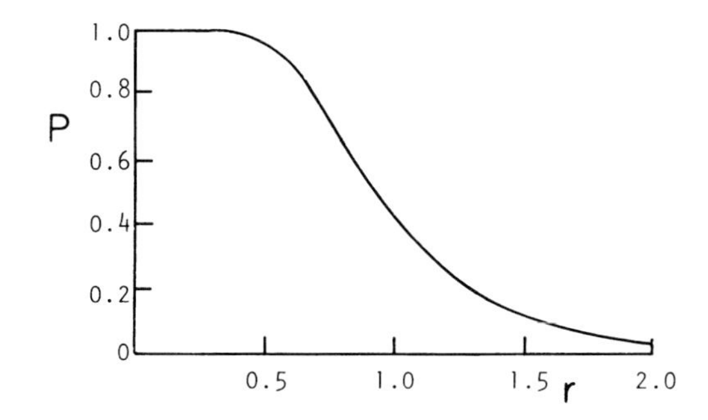

Fig. 5.11. Probability P of exceeding threshold level rm_I

Once again plugging in the normal law for $p_{\ln I}(y)$ and doing the integration, we find

$$m_I^2 + \sigma_I^2 = \exp(2m_{\ln I} + 2\sigma_{\ln I}^2) \ . \tag{5.66}$$

Equations (5.63) and (5.66) are now regarded as two equations in the two unknowns $m_{\ln I}$ and $\sigma_{\ln I}^2$. Their simultaneous solution is

$$m_{\ln I} = \ln m_I - 2^{-1}\ln[1 + (\sigma_I/m_I)^2] \tag{5.67a}$$

and

$$\sigma_{\ln I}^2 = \ln[1 + (\sigma_I/m_i)^2] \ . \tag{5.67b}$$

We think it rather fortuitous that the only data at hand, σ_I/m_I in (5.58), uniquely fixes the unknown $\sigma_{\ln I}^2$ and nearly uniquely fixes $m_{\ln I}$. Even better, $m_{\ln I}$ will completely drop out of the expression for $P(I > rm_I)$, leaving the answer solely dependent upon the only given data, σ_I/m_I! This is shown next.

We now use the known $m_{\ln I}$, $\sigma_{\ln I}^2$ in the evaluation of integral (5.61). This establishes

$$P(I > rm_I) = 1/2 - \text{erf}[(\ln r)/\sigma_{\ln I} + \sigma_{\ln I}/2] \ , \tag{5.68}$$

where $\sigma_{\ln I}$ relates to the datum through (5.67b).

Equation (5.68) was evaluated as a continuous function of parameter r defining threshold. The datum (5.58) was used. Results are shown in Fig. 5.11. The curve is virtually at 1.0 for thresholds $r \lesssim 0.40$. This means that the "twinkling" effect of a star or laser is rarely so severe that it drops to less than 0.4 of its mean level. However, the slightly higher threshold of $r = 0.50$ causes P to significantly drop below 1.0, to level 0.95, amounting to 95% assurance that the "on" pulse will be correctly identified. Once this r is exceeded, P falls off rapidly. However, note that P is finite even for r out to value 2.0. This shows that a twinkle is more apt to be uncharacteristically bright ($r \gtrsim 1.5$) than dark ($r \lesssim 0.5$). Perhaps this is why the effect is so visible, inspiring poets through the ages.

Exercise 5.1

5.1.1 The Newtonian form of the stigmatic imaging equation is $xy = f^2$, where x is the object distance to the front focal plane and y is the image conjugate distance beyond the back focal plane. Suppose the object is vibrating randomly back and forth in x, according to a known probability law $p_X(x)$. What is the corresponding probability law $p_Y(y)$ for image position?

5.1.2 The "lens-makers" formula for the focal length f of a thin lens of refractive index n and front, back radii r_1, r_2 respectively, is $1/f = (n-1)(1/r_1 - 1/r_2)$. What is the probability law $p_F(f)$ if refractive index n follows a law $p_N(n)$, with r_1, r_2 constant? What is the probability law $p_F(f)$ if radius r_1 follows a law $p_R(r_1)$, with n, r_2 constant?

5.1.3 More realistically in the previous problem, *both* radii r_1 and r_2 will randomly vary (n held fixed). Suppose the joint probability law $p_{R_1R_2}(r_1, r_2)$ is known. What is the resulting probability law $p_F(f)$? [*Hint:* To use the Jacobian approach one must have as many old RV's as new ones, two each in this case. Therefore, we have to *invent* a second new RV conjugate to f. By symmetry with f, let this be g defined as $1/g = (n-1)(1/r_1 + 1/r_2)$. Then by transformation theory, one may find the joint law $p_{FG}(f, g)$. Finally, integrating this through in g leaves the required marginal law $p_F(f)$].

5.1.4 The irradiance I at the Earth's surface due to a star is log-normal, i.e., $\ln I$ is a normal RV. Then what is the probability law for I itself?

5.1.5 A "phasor" $a = \exp(j\varphi)$. Suppose φ is a RV confined to the primary interval $|\varphi| \leqq \pi$, and $p_\Phi(\varphi)$ is uniform over the interval. Let $a = a_{\text{re}} + ja_{\text{im}}$. Find the probability laws for a_{re} and for a_{im}.

5.1.6 A sinusoidal pulse leaves a laser at time $t = 0$, and returns from an unknown target located distance r away at time $2r/c$, with c the speed of light. Thus, the acquired amplitude as a function of time is $a(r, t) = \sin(t - 2r/c)$. With t a constant, the detected amplitude depends randomly upon the random distance r of the source. If r is uniformly random over interval (r_1, r_2), what is $p_A(a)$?

5.1.7 Snell's law of refraction of a ray at a plane surface is $n_2 \sin\theta_2 = n_1 \sin\theta_1$ where n_1, n_2 are the indices of refraction on each side of the surface and θ_1, θ_2 are the respective ray angles from the normal to the surface. With n_1, n_2 held fixed, suppose that angular direction θ_1 for the ray is unknown with uniform probability over the interval $|\theta_1| \leqq \Delta\theta$. What is the ensuing probability law on θ_2?

5.1.8 Show that, in Sect. 5.12, if instead of a log-normal law I obeyed an exponential law (3.41) with mean m_I, then the probability $P(I > rm_I)$ that an "on" symbol would be correctly identified is simply $\exp(-r)$.

5.1.9 In holography, the image amplitude is the sum of speckle amplitude $U = (U_{\text{re}}, U_{\text{im}})$ *and a constant amplitude* U_0. Suppose the latter to be purely real. Then the holographic amplitudes $U'_{\text{re}}, U'_{\text{im}}$ obey transformation $U'_{\text{re}} = U_{\text{re}} + U_0$, $U'_{\text{im}} = U_{\text{im}}$ from the speckle amplitudes. Find the probability law on irradiance. *Hint:* Apply this transformation to (5.27) to get the new expression for probability density $p_{U'_{\text{re}}U'_{\text{im}}}(x, y)$ for the new amplitudes. Use this in (5.30) to get the new joint $p_{I\varphi}(v, w)$. Integrate this through in w to get the new marginal $p_I(v)$ for intensity fluctuation.

Show that the answer is

$$p_I(v) = \sigma^{-2} I_0(2\sigma^{-2}\sqrt{i_0 v}) \exp[-\sigma^{-2}(v + i_0)] ,$$

where $i_0 \equiv U_0^2$ and the modified Bessel function $I_0(x)$ is defined as

$$I_0(x) \equiv (2\pi)^{-1} \int_{-\pi}^{\pi} \mathrm{d}w \exp(x \cos w) .$$

5.1.10 Given that $x = N(0, \sigma_I^2)$ and $y = 1/x$, find $p_Y(y)$.

5.1.11 Given that $p_X(x) = x_0^{-1}\mathrm{Rect}(x/x_0)$ and $y = x^2 - 3x + 2$, determine $p_Y(y)$.

5.1.12 Given that $x_1 = N(0, \sigma_1^2)$, $x_2 = N(0, \sigma_2^2)$, x_1, x_2 are independent RV's and $y_1 = x_1 - x_2$, $y_2 = x_1 + x_2$. Find

(a) joint probability $p_{Y_1 Y_2}(y_1, y_2)$.

(b) marginal $p_{Y_1}(y_1)$.

(c) marginal $p_{Y_2}(y_2)$.

5.1.13 Suppose $y = \ln x$ and $y = N(0, \sigma_1^2)$. Find $p_Z(z)$, where $z = 1/x$.

5.1.14 Photographic density D relates to exposure E by the Hurter-Driffield H–D expression $D = \gamma \log(E/E_0)$, γ, E_0 constants.

(a) Find $p_D(x)$ in terms of a given $p_E(x)$.

(b) Suppose E is a *uniformly* random variable between limits E_1 and E_2. Find $p_D(x)$, $\varphi_D(\omega)$, $\langle D \rangle$ and σ_D^2.

5.1.15 Given the H–D expression $D = \gamma \log(E/E_0)$, γ, E_0 constants, suppose RV E fluctuates by small amounts about a mean background level E_b. Then the corresponding fluctuations in D from its background level D_b are also small.

(a) Using differentials, express a small fluctuation $\mathrm{d}D$ about D_b in terms of the corresponding fluctuation $\mathrm{d}E$ about E_b.

(b) Use this expression to get σ_D^2 in terms of σ_E^2. Is this result consistent with the answer to problem 5.1.14?

5.1.16 Many physical problems involving the mutual interaction of particles invoke the use of a spring-like potential $V = 2^{-1}kx^2$ to characterize each interaction. Suppose the particles have random (one-dimensional) positions $\{x_n\}$, so that the net potential V_N at a point, obeying

$$V_N = 2^{-1}k \sum_{n=1}^{N} x_n^2 ,$$

is itself a random variable. Make some reasonable assumptions about the statistics of the $\{x_n\}$ so as to produce a chi-square law for V_N. Find the mean and standard deviation of V_N.

If each of the k-values were allowed to be different, but constant, so that now

$$V_N = 2^{-1} \sum_{n=1}^{N} k_n x_n^2 ,$$

show that under the same assumptions on the statistics for the $\{x_n\}$ the random variable V_N is still chi-square, although with different parameters.

5.1.17 Consider an ideal gas, consisting of M particles of masses $m_1, \ldots, m_M$. Let u_n, v_n, w_n represent the x-, y-, and z-components of velocity for the nth particle at a certain time. Velocity is assumed random in time, obeying a Boltzmann probability law overall,

$$\begin{aligned} &p(u_1, v_1, w_1, \ldots, u_M, v_M, w_M) \\ &= B \exp\left\{-(2kT)^{-1}\left[\sum_{n=1}^{M} m_n(u_n^2 + v_n^2 + w_n^2)\right]\right\} . \end{aligned}$$

This is merely a product of Gaussian densities in each of the u_n, v_n, and w_n separately. But the variances for these Gaussian laws are seen to depend on n, as

$$\sigma_n^2 = kT/m_n$$

so that the problem is not quite in the form (5.39). Instead, regard as RV's

$$\begin{aligned} x_n &= (m_n/2)^{1/2} u_n , \\ y_n &= (m_n/2)^{1/2} v_n , \\ t_n &= (m_n/2)^{1/2} w_n , \end{aligned}$$

and form the total kinetic energy

$$E = \sum_{n=1}^{M} (x_n^2 + y_n^2 + t_n^2)$$

for the gas.

(a) Show that RV E satisfies the conditions (5.54) for being a chi-square random variable. Find the probability law for E.

(b) Physics students are taught that E is a constant for the ideal gas. The S/N ratio for E is a measure of the constancy of E. Find how S/N varies with M, the total number of particles.

5.1.18 ***Verification of Huygens' principle*** [5.8]. A point source is located at S in Fig. 5.12. It radiates uniformly in all directions, within the plane of the figure, so that the probability density $p_\Theta(\theta)$ for direction θ of a photon is $p_\Theta(\theta) = (2\pi)^{-1}\text{Rect}(\theta/2\pi)$.

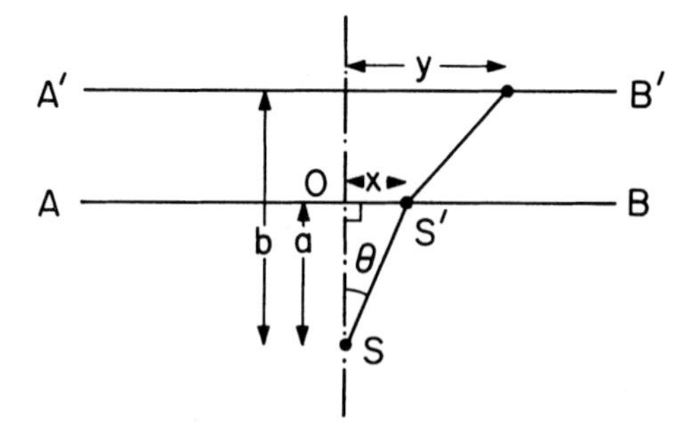

Fig. 5.12. Huygens' principle from properties of the Cauchy probability law (Ex. 5.1.18). The light distribution along $A'B'$ due to the point source S is attained alternatively by a line source along AB with the proper intensity profile. A downward view of light paths in the ballistic galvanometer (Ex. 5.1.19)

(a) Show that the probability density $p_X(x)$ for photons crossing the line AB located distance a from S obeys a Cauchy law (3.59) with parameter a (after suitable renormalization, since half the photons never cross AB).

(b) Suppose the source S is removed, replaced with a line source along AB whose intensity/length obeys the previously found $p_X(x)$. Each point S' on the line source radiates uniformly in all directions, as did S. Let line $A'B'$ be distance b from the former source S. Show that the intensity $p_Y(y)$ along $A'B'$ is exactly Cauchy with parameter b, as if there were instead a real point source at S once again. *Hint:* $p_Y(y) = \int_{-\infty}^{\infty} \mathrm{d}x\, p(y|x) p_X(x)$.

This verifies Huygens' principle, which states that each point on the wavefront of a propagating light wave acts as a new source of secondary light waves, from which the light wave at later times may be constructed.

5.1.19 In the ballistic galvanometer, a mirror rotates on a vertical axis by an angle $\theta/2$ proportional to the current (which it is desired to measure) passing through a coil. Figure 5.12 is a downward view of the apparatus. A horizontal light beam is directed from a source point O to the mirror axis S. It bounces back off the mirror at angle θ and onto a nearby wall scale AB, where its horizontal position x is observed.

Find the probability law for position x, if all currents (and hence angles θ) are equally likely. (Compare with previous problem.)

5.1.20 An incoherent light beam illuminates the surface of a pool of water of depth D and refractive index n (Fig. 5.13). The beam has a narrow but finite lateral extension. It is refracted by the water surface, and its center strikes the pool floor at coordinate x. The pool surface may be modelled as a randomly tilting plane that remains at about a constant distance D from the pool floor. The probability law $p_\theta(x)$ for tilt angles θ is known. The beam has a transverse intensity distribution $g(x)$. We want to predict the probability density $p_X(x)$ for photon positions on the pool floor.

(a) Show that for small enough tilt angles

$$p_X(x) = g(x) \otimes p_0(x) ,$$

where $\otimes$ denotes a convolution, and $p_0(x)$ is the density $p_X(x)$ that would result if the beam were infinitely narrow. [*Hint:* An event x on the pool floor arises disjointly

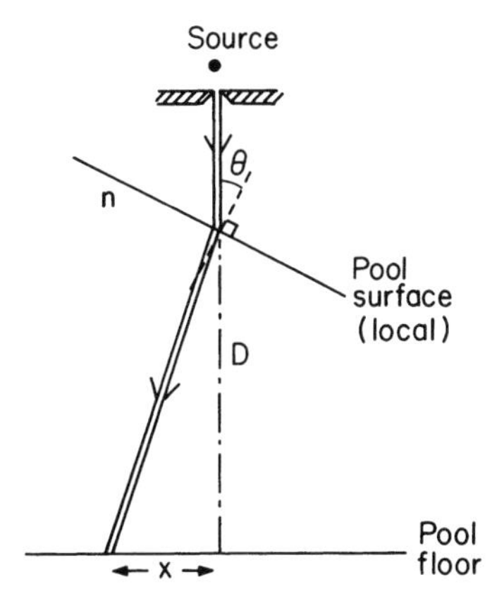

Fig. 5.13. Geometry for problem of light distribution on the bottom of a pool

from all possible lateral positions in the beam, implying use of the partition law (3.7)].

(b) Find $p_0(x)$, first assuming small tilt angles θ, and then allowing for finite sized θ.

(c) What does probability $p_X(x)$ represent physically?

More realistically, the beam will suffer random scattering as well, from suspended particles *within* the water. This more complex problem is treated in Sect. 8.15.12.

5.1.21 ***Propagation of image noise errors into an object estimate.*** Suppose that an image $\{i_m\}$ is formed from an object $\{o_n\}$ and noise $\{n_m\}$ as

$$i_m = \sum_{n=1}^{N} o_n s_{mn} + n_m \,, \quad m = 1, \ldots, N \,, \tag{5.69a}$$

$$n_m = N(0, \sigma^2) \,, \quad \langle n_m n_k \rangle = 0 \,, \quad m \neq k \,. \tag{5.69b}$$

Object, image, and noise are all sampled at a constant spacing. Quantity $\{s_{mn}\}$ is the spread function. The noise is Gaussian and uncorrelated.

An estimate $\{\alpha_m\}$ of this object is formed by making it obey

$$i_m = \sum_{n=1}^{N} \alpha_n s_{mn} \,, \quad m = 1, \ldots, N \,. \tag{5.70}$$

This estimate is called a direct-inverse solution, since in matrix form the solution to (5.70) is

$$\boldsymbol{\alpha} = [S]^{-1}\boldsymbol{i} \tag{5.71}$$

where $\boldsymbol{\alpha} = \{\alpha_m\}^T$, $\boldsymbol{i} = \{i_m\}^T$ and matrix $[S]$ consists of elements $\{s_{mn}\}$.

Because of the noise in (5.69a), $\boldsymbol{i}$ will be random in (5.71), causing the output $\boldsymbol{\alpha}$ to also be random. We want to know how random $\boldsymbol{\alpha}$ will be, for a given σ of randomness in the $\{n_m\}$. Hence, we seek the probability density $p(\boldsymbol{\alpha})$ for the outputs.

The key observation is that the processing formula (5.71) also represents a transformation of RV's, *from* $\boldsymbol{i}$ *to* $\boldsymbol{\alpha}$.

(a) Show that the required probability density $p(\boldsymbol{\alpha})$ obeys

$$p(\boldsymbol{\alpha}) = |\det[S]| \, p_I([S]\boldsymbol{\alpha}) \,, \tag{5.72}$$

where $p_I(\boldsymbol{i})$ is the probability law for the image.

Because of (5.69a, b)

$$\begin{aligned} p_I(\boldsymbol{i}) &= p_{\text{noise}}(\boldsymbol{i} - \langle \boldsymbol{i} \rangle) \\ &= \left(\frac{1}{\sqrt{2\pi}\sigma} \right)^N \exp\left[-\sum_m (i_m - \langle i_m \rangle)^2 / 2\sigma^2 \right] \,, \end{aligned} \tag{5.73}$$

where

$$\langle i_m \rangle = \sum_n o_n s_{mn} = \sum_n \langle \alpha_n \rangle s_{mn}$$

(the latter from (5.70)).

(b) Show that when (5.73) is substituted into (5.72), the result is

$$p(\boldsymbol{\alpha}) = |\det[S]| \, (\sqrt{2\pi}\sigma)^{-N} \exp\left[-(2\sigma^2)^{-1} \sum_{m,n} \Delta\alpha_m \Delta\alpha_n \varrho_{mn}\right] \tag{5.74}$$

where

$$\Delta\alpha_n \equiv \alpha_n - \langle \alpha_n \rangle$$

and

$$\rho_{mn} \equiv \sum_k s_{km} s_{kn} \tag{5.75}$$

is an autocorrelation across the spread function. Comparison with (3.48) shows that *$\boldsymbol{\alpha}$ is distributed normal multivariate.*

(c) Using result (3.51) show that the variance σ_n^2 in α_n obeys

$$\sigma_n^2/\sigma^2 = \text{cofactor}\, \varrho_{nn} / \det[\rho] \equiv G_n \,. \tag{5.76}$$

Quantity G_n measures the noise gain due to processing the data.

(d) Show that if matrix $[\varrho]$ is close to diagonal, G_n is close to

$$G = 1/\varrho_{nn} = 1/\sum_k s_{nk}^2 \,. \tag{5.77}$$

Note that this is essentially independent of n, since for any n the sum is always across the entire spread function.

(e) If the image points are close enough together, the sum in (5.77) may be approximated by an integral across the spread function. In this case, show that if s is a Gaussian blur function of sigma σ_{blur}, the noise gain becomes

$$G = 2\sqrt{\pi}\sigma_{\text{blur}} \,. \tag{5.78}$$

This indicates that the more blurred the image was, the noisier will be the estimated object.

5.1.22 ***Propagation of input data errors into output estimates.*** Any processing formula whatsover represents a transformation from input random variables (the data) to output random variables (the estimate). On this basis, we may solve a very general class of problems. Suppose inputs $\{i_m\} = \boldsymbol{i}$ suffer from zero-mean noise $\{n_m\} = \boldsymbol{n}$, so that

$$i_m = \langle i_m \rangle + n_m \,, \quad m = 1, \ldots, N \,. \tag{5.79}$$

Suppose processing formula

$$\alpha_m = f_m(\boldsymbol{i}) \,, \quad m = 1, \ldots, M \tag{5.80}$$

exists for producing outputs $\{\alpha_m\}$. We want to know what probability law the $\{\alpha_m\}$ will jointly follow.

Show by the Jacobian approach that the answer is

$$p(\boldsymbol{\alpha}) = p_N(\boldsymbol{i} - \langle \boldsymbol{i} \rangle)\, |\mathrm{J}(\boldsymbol{i}/\boldsymbol{\alpha})| \tag{5.81}$$

where p_N is the joint probability law for the noise. Both the argument $(\boldsymbol{i} - \langle \boldsymbol{i} \rangle)$ and the Jacobian $\mathrm{J}(\boldsymbol{i}/\boldsymbol{\alpha})$ are known in terms of $\boldsymbol{\alpha}$ from inverse relations to (5.80)

$$i_m = g_m(\boldsymbol{\alpha}) \,. \tag{5.82}$$

5.1.23 In Sect. 5.3 it was shown that a uniformly glowing light rod placed upon the x-interval (5.8) in object space will give rise to an image with intensity (5.9) along y in image space. The proviso was that each photon departure event x give rise *uniquely* to a photon arrival event y in the image. This is a situation of *stigmatic imagery*. Suppose we lift this restriction and allow a spread in possible positions y' about the unique y corresponding to an x. Let this obey a known probability law $p(y'|y)$, and further assume that

$$p(y'|y) = s(y' - y) \,,$$

a spread function. Now the imagery is *non*stigmatic.

(a) What does the latter equation mean physically? Using the partition law (3.7), show that the nonstigmatic image $p_{\mathrm{NS}}(y')$ obeys simply a convolution

$$p_{\mathrm{NS}}(y') = \int_{-\infty}^{\infty} \mathrm{d}y\, p_Y(y) s(y' - y)$$

of the stigmatic image $p_Y(y)$ with the spread function $s(y)$.

5.1.24 ***Rice-Nakagami probability law for stellar light fluctuation.*** The conditions under which atmospheric turbulence causes a log-normal probability law for stellar irradiance were given in Ex. 4.3.13. We describe here the complementary situation, where the light amplitude at the observer is the sum of a great many amplitudes, each of which has been randomly scattered *but once* by the medium (Fig. 5.14).

The net amplitude u at the observation point will then be the sum

$$u = a + x + \mathrm{j}y \,, \tag{5.83}$$

where a is the average or direct-light amplitude, and x and y are component random amplitudes due to the scattering. We want the probability law for $|u| \equiv t$.

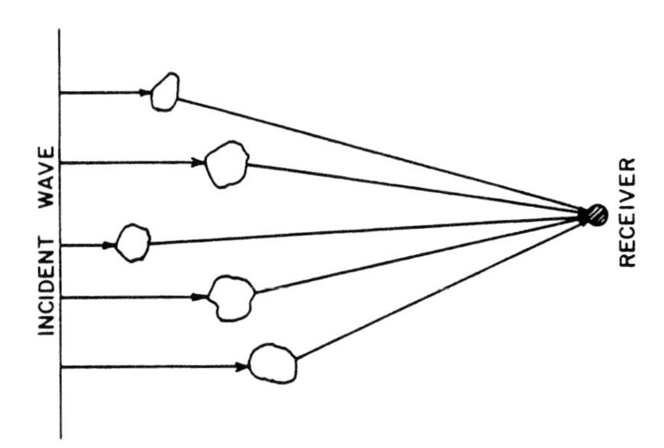

Fig. 5.14. Single-scattering model of turbulence [5.9]

(a) Quantities x and y are the result of a large number of random scatterings. Then it is asserted that x and y are each normal RV's. Why?

(b) Assume that x and y each have the same variance σ^2 and a mean of zero. Further assume that x and y are independent RV's. Then show that the probability law for net amplitude is the Rice-Nakagami [5.9]

$$p_T(t) = (t/\sigma^2)\exp\left(-\frac{t^2+a^2}{2\sigma^2}\right) I_0(at/\sigma^2) . \tag{5.84}$$

Function I_0 is the zero-order modified Bessel function

$$I_0(z) = \pi^{-1}\int_{-1}^{1} dx(1-x^2)^{-1/2}\exp(zx) . \tag{5.85}$$

Hint: Since x and y have known statistics, $p_{XY}(x, y)$ may easily be formed. Then from (5.83), a quantity $z = |u|^2$ is known in terms of RV's x and y. Regard this as a transformation of a RV from the old y to the new z, x held constant. Then form $p_{XZ}(x, z)$ from the known $p_{XY}(x, y)$ using the Jacobian approach. By $t = \sqrt{z}$, transform this to the law for $p_{XT}(x, t)$, again by the Jacobian approach. Finally, integrate this through dx, where $-t - a \leqq x \leqq t - a$, and use identity (5.85).

Note that for $a = 0$, (5.84) goes over into a "Rayleigh" probability law (6.58).

(c) From the Rice-Nakagami law, show that *irradiance* obeys a law

$$p_I(i) = \sigma^{-2}\exp\left(-\frac{i+a^2}{2\sigma^2}\right) I_0(a\sqrt{i}/\sigma^2) . \tag{5.86}$$

Coincidentally, this is the same law as for the holography problem (5.1.9). Also, for $a = 0$, (5.86) goes over into a simple exponential law.

(d) Show that for $a \neq 0$ the law (5.86) falls off slower with i than the laser speckle law, and hence is more random.

The case $a = 0$ physically corresponds to u being formed purely from "diffuse" light, i.e., without direct light from the input beam. Recently, it was suggested [5.10] that instead of being Rayleigh, the amplitude in this case should more generally follow an *m distribution*,

$$p_T(t) = \frac{2m^m t^{2m-1}}{\Gamma(m)\langle i\rangle^m}\exp(-mt^2/\langle i\rangle) , \quad \langle i\rangle = 2\sigma^2 . \tag{5.87}$$

Parameter m is related to the reciprocal of the variance of t and is at the user's discretion. The case $m = 1$ corresponds to a Rayleigh law; compare (5.87) with (5.84) when $a = 0$.

5.1.25 ***The Polya (Negative binomial) probability law*** [5.5] is given by

$$P(n) = \frac{(n+M-1)!}{n!(M-1)!} p^n q^M \,, \quad p = \langle n \rangle / (\langle n \rangle + M) \,, \quad q = 1 - p \,. \tag{5.88}$$

It arises in the study of laser speckle and of photon statistics. We treat the speckle case here. The photon case is given in Exercise 10.1.19.

Return to Sect. 5.10 on speckle reduction by use of a scanning aperture. The aperture contains M fixed cells. A cell may contain any number of speckles, each of energy I_0. We found that the total intensity I within the aperture follows a law $p(I)$ which is χ-square (5.50) with a variance $4M\sigma^2$.

What probability law $P(n)$ governs the *number n* of speckles within the aperture? It is shown in Sect. 6.6 that for a fixed average (or total) intensity I, $P(n|I)$ is Poisson, with mean I/I_0. Hence, both $p(I)$ and $P(n|I)$ are known. This allows $P(n)$ to be found via partition law (3.7). Show that (5.88) results, with $p = 2\sigma^2/(I_0 + 2\sigma^2)$. *Hint:* integrate using (5.45).

5.1.26 ***Probabilistic solutions to first-order differential equations.*** Suppose that a particle travels along a one-dimensional trajectory $x = x(t)$, t the time. Let the trajectory obey a given first-order differential equation

$$\frac{\mathrm{d}x}{\mathrm{d}t} = f(x, t) \tag{5.89}$$

where f can generally be a non-linear function of its arguments. If x is instead a vector this equation becomes a vector equation of Lotka-Volterra type [5.17]. The following theory can easily be generalized to this more general problem.

Can a probability density $p_X(x)$ for x be found? By the law of large numbers, this is simply the histogram of occurrences x sampled from the trajectory $x(t)$ over a required span of time values. What density function can be used for the time?

Analysis: Among the four coordinates (x, y, z, t) time is special. Regardless of problem, each time value occurs once and only once, thereby obeying a density function

$$p_T(t) = 1/T \quad \text{for } 0 \leqq t \leqq T \,, \tag{5.90}$$

T a required time span. Notice that $p_T(t)$ is a density function for *deterministic* values t. Hence it is not a *probability* density per se.

In the presence of a definite trajectory $x(t)$ there is one and only one x value for each value of t [actually one *interval* $(x, x + |\mathrm{d}x|)$ for each interval $(t, t + |\mathrm{d}t|)$]. With $p_T(t)$ known, Sect. 5.2 suggests use of the Jacobian relation

$$p_X(x)|\mathrm{d}x| = p_T(t)|\mathrm{d}t| \,. \tag{5.91}$$

to obtain the required $p_X(x)$. This can be used, despite $p_T(t)$ not being a probability density, because Jacobian relations hold for any density functions, regardless of their nature. In the derivation of Sect. 5.2, it is nowhere assumed that events x and y are unpredictable; the essential ingredient is the known relation (5.1a) between them.

Then using (5.90) we get the formal answer [5.11]

$$p_X(x) = \frac{1}{T}\frac{1}{|\mathrm{d}x/\mathrm{d}t|} . \tag{5.92}$$

Notice that this makes sense. Where the particle speed is low it tends to "linger". Hence these x-values should occur more frequently. The final solution $p_X(x)$ must of course be solely a function of x. It is found by solving the differential equation (5.89) for $x = g(t)$, inverting this to $t = g^{-1}(x)$, and then using $\mathrm{d}x/\mathrm{d}t = g'(g^{-1}(x))$ in (5.92).

Problem: Obtain the solution $p_X(x)$ for the particular case $f(x,t) = at + b$, a, b constant.

5.1.27 ***Classical trajectories for quantum particles.*** In the previous problem we used (5.92) to solve for $p_X(x)$ when $x(t)$ is of a required form. Here we do the inverse problem: With $p_X(x)$ of a required form, what is $x(t)$?

Analysis: From (5.92) it obeys

$$\left|\frac{\mathrm{d}x}{\mathrm{d}t}\right| = \frac{1}{T}\frac{1}{p_X(x)} . \tag{5.93}$$

Substituting in a required $p_X(x)$ results in a first-order differential equation in $x(t)$. Because of the absolute value signs we have to choose a sign for $\mathrm{d}x/\mathrm{d}t$. Since the trajectory $x(t)$ must be continuous one sign holds once and for all over all x and t. Arbitrarily choose the $+$ sign. Since $p_X(x) \geq 0$, (5.93) shows that x will be a monotonically increasing function of t. (With the minus sign it would monotonically decrease.)

Then (5.93) becomes a simple differential equation $x'(t) = 1/[Tp_X(x)]$. This has an elementary solution [5.11]

$$\int \mathrm{d}x\, p_X(x) = \frac{1}{T}(t - t_0) \tag{5.94}$$

The integral in (5.94) is an indefinite one, a known function $h(x)$. Also, t_0 is an additive (and unknowable) constant. Hence the formal solution is

$$x = h^{-1}(\frac{1}{T}(t - t_0)) . \tag{5.95}$$

An interesting application is to quantum mechanics. Let $\psi(x)$ be a stationary solution to the Schroedinger wave equation. Then $p_X(x) = |\psi(x)|^2$. Using this in Eq. (5.94) gives a specific classical trajectory $x(t)$ for the particle. Note that a classical trajectory so constructed is entirely different from one of D. Bohm [5.12] for the same case. The two theories are fundamentally different.

For a free particle trapped in a box of length L, the lowest-order solution is

$$\psi(x) = \sqrt{2/L}\cos(\pi x/L) , \quad -L/2 \leqq x \leqq L/2 . \tag{5.96}$$

Problem: Carry through with this case to obtain the classical trajectory $x(t)$. This will obey a transcendental equation. The trajectory is shown in Fig. 5.15.

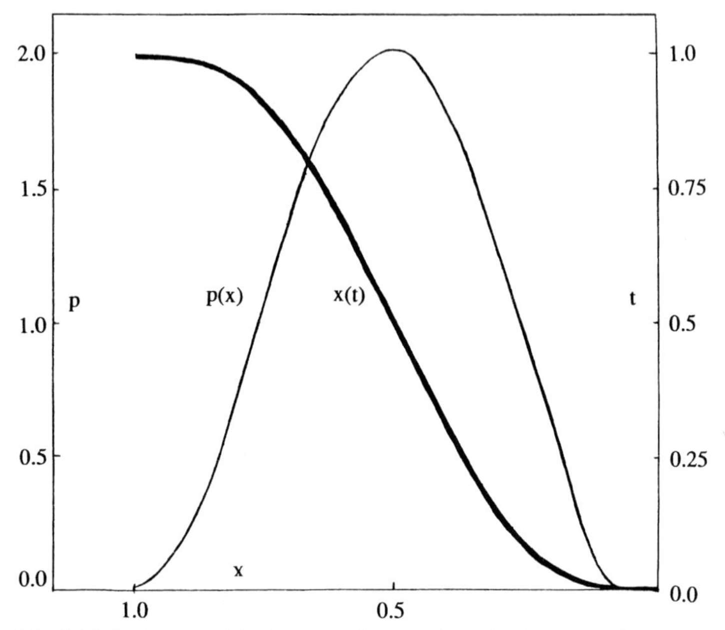

Fig. 5.15. Trajectory (thick line) $x(t)$ for a particle obeying a solution $\psi(x)$ to the Schroedinger wave equation. The thin line shows the histogram of x-values obtained by uniformly sampling the trajectory over time. It checks with the theoretical answer $(2/L)\cos^2(\pi x/L)$ for the problem, to the accuracy defined by the finite bin size $\Delta x = 0.05$ used in the sampling

5.13 Unequal Numbers of Input and Output Variables: "Helper Variables"

The Jacobian problems in Sects. 5.2–5.6 are limited to "square" cases, where the number of input random variables equals the number of outputs. However, there are many problems where they are unequal. We consider some of these next.

5.13.1 Probability Law for a Quotient of Random Variables

Many physical relations are of a simple product form, as examples $F = ma$, $V = IR$, $E = h\nu$, $xx' = -f^2$. Consider in particular the last relation – the Newtonian form of the lens-makers equation. If the variables $-f^2$ and x' are random and independent, and have known probability laws, what is the probability law on x? How may this be found? This is a problem of the following form: Given that

$$x = y/z\ , \tag{5.97}$$

where random variables y and z are independent and obey known laws $p_Y(y)$, $p_Z(z)$, what is $p_X(x)$?

Analysis: There are two input variables y, z but only one output variable x. To allow a 2×2 Jacobian approach (5.18) to be taken, we need a second output variable. Call this a *helper* variable. As with its mate x in (5.97) the helper variable should be some function of the input variables y, z. Any such well-defined function will do, but for practical purposes the function should be as simple as possible and, also, such that we have a one-root case. For example, a variable

$$t = z \tag{5.98}$$

could be chosen. Eqs. (5.97) and (5.98) together now define a 2×2 Jacobian problem. If we invert and solve for the inputs y, z in terms of the outputs we get

$$z = t\,, \quad y = xt\,. \tag{5.99}$$

Note: It is essential that this inversion be carried through, because partial (not *full*) derivatives of these inputs with respect to the outputs are required below. These partial derivatives can only be correctly computed by differentiating *inverse* relations (5.99).

This is indeed a one-root case, and therefore we have made a good choice for the helper variable. In a one-root case (5.18) collapses to just its first term,

$$p_{XT}(x, t) = p_{YZ}(y, z)|J(y, z/x, t)|\,. \tag{5.100}$$

From the inversion (5.99) we see that the Jacobian matrix has elements

$$\begin{bmatrix} \partial y/\partial x = t & \partial y/\partial t = x \\ \partial z/\partial x = 0 & \partial z/\partial t = 1 \end{bmatrix} \tag{5.101}$$

Therefore the absolute Jacobian is $|t \cdot 1 - x \cdot 0| = |t|$. Substituting this into (5.100), using the independence of y and z, the system relations (5.99), and integrating out over the helper variable t, we get the marginal density

$$p_X(x) = \int \mathrm{d}t\, |t|\, p_Y(xt) p_Z(t) \tag{5.102}$$

as required. This is the formal solution. To go further, the individual laws $p_Y(y)$, $p_Z(z)$ must be specified. See individual problems in Exs. 5.2 below.

5.13.2 Probability Law for a Product of Independent Random Variables

Here, instead of (5.97), we have an output random variable x obeying

$$x = yz\,, \tag{5.103}$$

with y and z independent. What is $p_X(x)$? Proceeding as in (5.98)–(5.101) the absolute Jacobian is now $|\mathrm{J}(y, z/x, t)| = |(1/t) \cdot 1 + (x/t^2) \cdot 0)| = 1/|t|$. We again use this in (5.100). The required marginal density $p_X(x)$ is again obtained by integrating out over t,

$$p_X(x) = \int \mathrm{d}t\, |t|^{-1}\, p_Y(x/t) p_Z(t)\,. \tag{5.104}$$

(*Note*: the argument of p_Y is a division x/t and not a conditional event $x|t$.)

These Jacobian problems often allow alternative routes to the same solution. An alternative in the current case is to take the logarithm of both sides of (5.103), which turns the product into a sum $\log x = \log y + \log z$. Renaming these three quantities to new random variables u, v, w, respectively, we now have a problem of finding the probability law $p_U(u)$ for a variable $u = v + w$ with v, w independent. Of course this was already worked out as the convolution answer (4.19). The final step is to use another Jacobian step to transform the (now) known law $p_U(u)$ to the corresponding one $p_X(x)$ via the transformation $u = \log x$. This is a one-root, one-dimensional problem with a solution of the form (5.5).

5.13.3 More Complicated Transformation Problems

Having found how to attack problems of the simple division form (5.97) or product form (5.103), we may now solve more complicated problems such as $x = f(z/y)$ or $x = f(zy)$ where the output variable is a given function f of a division or a product of random variables. This can be solved in two distinct Jacobian steps (*divide and conquer*). For example, consider the problem $x = f(z/y)$. This may be attacked "from the inside out." First find the probability law for the inner combination $z/y \equiv w$. This has the answer (5.102). Next do the outer problem $x = f(w)$ by means of (5.11), giving the required solution $p_X(x)$.

5.14 Use of an Invariance Principle to Find a Probability Law

Sometimes an unknown probability law $p_X(x)$ may be defined by an invariance, or symmetry, property that it obeys [5.13]. The statement of invariance is that, under *a given* transformation $y = f(x)$, the transformed law is to keep the same functional dependence as it had originally,

$$p_Y(y) = p_X(y) \equiv g(y) . \tag{5.105}$$

(Note the argument y of p_X.) The aim is to find the common functional form $g(y)$.

By Jacobian transformation Eq. (5.5) $g(y)$ is to obey a relation

$$g(y) = g[f^{-1}(y)]/|f'(f^{-1}(y))| . \tag{5.106}$$

The remaining problem is to find a $g(y)$ that satisfies this relation. The answer is not always unique. Some examples follow.

(1) Suppose that the law is to be invariant to *linear magnification* of the random variable. This describes a transformation $y = ax$, $a =$ const., $a > 0$. Hence $f(x) = ax, x \equiv f^{-1}(y) = y/a, f'(x) = f'(f^{-1}(y)) = a$, and (5.106) becomes a condition

$$g(y) = g(y/a)/a.$$

A solution to this problem is $g(y) = a/y$. (Try it out!) For normalization purposes this can only hold over a finite range of y, $y_0 \leq y \leq y_1$, where y_0 and y_1 are constants with $0 < y_0 < y_1 < \infty$. Why can't $y_0 = 0$ or $y_1 = \infty$? The answer $g(y)$ to this problem seems to be unique.

Discussion: A probability law $p(x) = a/x$ is called a *reciprocal law*. An interesting case where this kind of invariance arises is as follows. The universal physical constants should obey a histogram that is invariant to choice of units (whether expressed in *mks* units, or *cgs*, *fps*, "*natural*", *etc.*). If the constants didn't have this property they would not obey well-defined statistics. Choice of units is equivalent to multiplication by a constant.

(2) The universal physical constants most often arise *as multipliers in defining equations*, e.g., the way Planck's constant h arises in the equation $E = h\nu$, where E and ν are the energy and frequency of an observed photon. As to whether the constant is a multiplier *or a divisor* in the equation is then completely arbitrary. For example, the equation $E = \nu/g$ could instead have been used to define a different "Planck's" constant g. Given this arbitrariness, the probability law governing the constants should be invariant to *reciprocation* $y = 1/x$ of the random variable. Here $f(x) = 1/x$, $x \equiv f^{-1}(y) = 1/y$, $f'(x) = f'(f^{-1}(y)) = -y^2$, and (5.106) becomes a condition

$$g(y) = g(1/y)/|-y^2| = g(1/y)/y^2 \; .$$

A particular solution to this problem is, as in (1), $g(y) = a/y$, $a =$ const. Again, this should be verified. This serves to independently verify the reciprocal law that the physical constants were previously found to obey.

But, invariance under reciprocation by itself does not lead to a unique probability law. The general solution is a Laurent (reciprocal powers) expansion

$$g(y) = \frac{a}{y} + \sum_{n=0}^{\infty} b_n \left(y^n + \frac{1}{y^{n+2}} \right) . \qquad (5.107)$$

All parameters a, b_n are arbitrary real constants. Each term of (5.107) obeys (5.106), as may be easily verified.

However, there is only one law that obeys *both* invariance principles (1) and (2), and that is the reciprocal law a/y. The latter therefore seems the compelling choice for describing the universal physical constants [5.14]. The known constants in fact agree with the hypothesis of a reciprocal law with high confidence; see Ex. 11.1.9.

(3) *Bertrand's paradox*. This is a problem that can arise in trying to model the probability law of a statistical experiment. An event that is described by *a single* random variable which is selected "randomly" from a finite range of possible values can reasonably be modelled as obeying a flat probability law. An example is the probability (5.32b) on phase angle in laser speckle. However, real events are often defined by the joint behavior of *many* variables. Hence a statement that *the event* is chosen "randomly" in the experiment does not uniquely define *which of its variables* should obey the flat law. Often, constraints are such that if one variable obeys a flat law, others (by Jacobian transformation) can't obey a flat law. Bertrand's paradox is the problem of deciding which random variable obeys the flat law. Examples of Bertrand's paradox are in Jaynes [5.15] and Papoulis [5.16].

Appropriate use of an invariance principle can resolve Bertrand's paradox. An angle viewed from a constant direction remains the same regardless of distance. Hence the probability law on the angle must remain invariant to distance as well. That is, the law remains invariant to linear magnification of the (x, y) space coordinates. Hence, If one of the event variables is this angle, requiring the probability law on the variable to remain invariant under such magnification can give the correct law. This was done by E.T. Jaynes [5.15] to resolve a famous case of Bertrand's paradox.

A class of central limit theorems may be defined by an appropriate invariance principle. See Ex. 9.2.1.

5.15 Probability Law for Transformation of a Discrete Random Variable

The analyses in Sects. 5.2–5.6 described transformations of *continuous* random variables. This is why we could draw continuous curves in Fig. 5.1. What if the RV is instead discrete, as in the following case?

A die has known probabilities $P_i, i = 1 - 6$ of its six face values. A game of chance is devised such that the roller of the die gets a prize of amount $j = i^2$ dollars for any roll outcome i. Given the P_i, what are the resulting probabilities of prize amounts j? Call these probabilities the Q_j. Now, j can only take on the values $j = 1^2, 2^2; \ldots, 6^2$. Also, of course, the event $j = 3^2$ occurs as often as the corresponding event $i = 3$. Thus $Q_9 = P_3$. Or for a general game play, any $Q_j = P_{\sqrt{j}}$.

More generally, with any functional relation $j = f(i)$ or $i = f^{-1}(j)$ connecting the two discrete variables,

$$Q_j = P_i, \ i = f^{-1}(j), \ i = 1, \ldots, N \,. \tag{5.108}$$

Notice in particular *there is no Jacobian factor entering in.* Why not?

The preceding was actually a single-root case. Let us return to a case $j = i^2$, but where the events i are not produced by rolling a die, so that *negative* values of i are now allowed. Then an output event j can be caused by either of two events $i = \pm\sqrt{j}$. These two events are disjoint, so that here $Q_j = P_{-\sqrt{j}} + P_{+\sqrt{j}}$. (With the roll of a die as in the above, only the second term would have meaning.) As in the continuous random variable case (5.11), there is *a sum over roots* contributing to the output probability.

Exercise 5.2

5.2.1 ***Quotient of two Gaussian random variables.*** Show using the formula (5.102) that the probability law governing the quotient of two independent Gaussian random variables is a Cauchy law. Show that, in the special case where the two variances are equal, the output probability law is an absolute entity, independent of the variances. *Note*: This is one of those non-intuitive results that probability theory is famous for – regardless of whether the two Gaussians are both broad or both narrow, the quotient has the same distribution.

5.2.2 ***Quotient of two exponential random variables.*** Using formula (5.102), find the probability law governing the quotient of two independent exponential random variables. Show that when the two exponential variables have the same mean the output probability law is $p_X(x) = (1 + x)^{-2}, x \geq 0$. As in the preceding problem, this law is independent of any free parameters.

5.2.3 ***Quotient of two uniformly random variables.*** Using the formula (5.102) show that the probability law for the quotient of two independent random variables obeying a common uniform law over an interval $(0, b)$ is

$$p_X(x) = 1/2 \text{ for } 0 \leqq x \leqq 1, \text{ or } (2x^2)^{-1} \text{ for } x > 1.$$

The interval length b doesn't matter. Let the student verify that this law obeys normalization. Make a sketch of the law.

5.2.4 ***Quotient of two Rayleigh random variables.*** A Rayleigh probability law has the general form

$$p_Y(y) = \frac{2y}{< y >} \exp(-y^2/ < y >), \quad y \geq 0 . \tag{5.109}$$

Using formula (5.102) show that the probability law for the quotient y/z of two independent Rayleigh variables y, z of respective mean values $\langle y \rangle$, $\langle z \rangle$ obeys a law

$$p_X(x) = \frac{2rx}{(1 + rx^2)^2}, \quad r = < z > / < y >, \quad x \geq 0 .$$

When the two means are equal the output law is once again an absolute, i.e., independent of parameters.

5.2.5 ***Product of two uniformly random variables.*** Show that the product of the two random variables of Ex. 5.2.3 obeys a probability law

$$p_X(x) = \frac{1}{b^2} \ln \left(\frac{b^2}{x} \right), \quad 0 \leqq x \leqq b^2 .$$

Verify that the law obeys normalization. Make a plot of the law, noting regions of high and low probability. Does the curve make sense?

5.2.6 ***Distortion of a probability density due to relative motion.*** Suppose that a probability density has a form $p_{\mathbf{X}}(\mathbf{x})$ in a frame with four-coordinates $\mathbf{x} \equiv (x_0, x, y, z)$ with $x_0 \equiv ct$. A second frame moves in the x-direction at speed u. Its four-coordinates are denoted as primed versions of the first frame's. The transformation law connecting coordinates $\mathbf{x}'$ and $\mathbf{x}$ is $\mathbf{x}' = [L]\mathbf{x}$, with $[L]$ the 4×4 Lorentz matrix (17.83).

(a) Show that the Jacobian $\mathrm{J}(\mathbf{x}'/\mathbf{x}) = 1$.

(b) Show that the probability density in the second frame is given in terms of that of the first as

$$p_{\mathbf{X}'}(\mathbf{x}') = p_{\mathbf{X}}([L]\mathbf{x}) . \tag{5.110}$$

The shape of the density in the second frame is distorted from that of the first according to the distortions of the x- and time coordinates that are caused by the finite speed u.

6. Bernoulli Trials and Limiting Cases

The emphasis until this chapter has been upon continuous random variables. Most discrete RV's may, anyhow, be described as a Dirac limit of corresponding continuous RV's (Sect. 3.12). But, now we want to consider a situation which is *intrinsically discrete* and has many applications to the world around us. This is the situation of repeated trials.

Suppose that a coin is flipped N times, giving rise to N outcomes consisting of heads and tails. Each flip is of course independent, and we assume for generality that the coin is not necessarily fair. The successive heads and tails form a binary sequence of what are called *Bernoulli trial* outcomes. As will be seen, many physical processes may be modelled by Bernoulli trials. Suppose that we want to know the probability of acquiring exactly m heads irrespective of their order in the N outcomes. The answer is a binomial probability law, as is shown next.

6.1 Analysis

Let p be the probability of a head, and $q \equiv 1 - p$ the probability of a tail, at a given trial (flip). Now each trial is independent of its predecessors. Therefore the probability of obtaining m heads on *m particular* trials (say, flip numbers 2, 5, 7 and 10) is by (2.13) simply p^m. This is irrespective of the outcomes for the remaining $N - m$ trials. These could be heads or tails in any order. If, in addition, we demand *all tails* in the remaining $N - m$ trials we come up with a net probability

$$p^m q^{N-m} . \tag{6.1}$$

To recapitulate, this represents the probability of obtaining m heads and $N - m$ tails in a *prescribed order* (e.g., with $N = 10$, heads on flips 2, 5, 7 and 10; tails on flips 1, 3, 4, 6, 8, and 9).

But we want to know the *total* probability of getting m heads and $N - m$ tails. This comprises *all possible* sequences of heads and tails having m heads total. By statistical independence, each such sequence has the same elemental probability, given at (6.1). Also, the sequences are disjoint. Therefore, the desired probability is

$$P_N(m) = W_N(m) \cdot p^m q^{N-m} \tag{6.2}$$

by additivity property (2.2). Factor $W_N(m)$ is the total number of sequences of flips which give rise to m heads. $W_N(m)$ is sometimes called the "degeneracy" for the state m of the experiment.

The degeneracy may be computed as follows. $W_N(m)$ actually describes the compound event "1 head placed in the flip sequence, *and* a second head so placed, ..., *and* an mth head so placed." We consider these placements one at a time, as they fill an initially empty sequence of flip events (Fig. 6.1).

This first head can appear at any of the N flips (i.e., at the first flip, *or* at the second, *or* at the third, ..., or at the Nth). For example, in Fig. 6.1 it appears at the second flip. Therefore, the number of possible ways of placing the first head is simply N.

With one head already placed, there are $N-1$ sequence locations left for placing the second head (Fig. 6.2). Therefore, by the preceding analysis there are $N-1$ ways of placing the second head, and this is independent of where the first head was placed. Since, then, for *each* first-head emplacement there are $N-1$ second-head placements, the number of ways both heads may be placed is simply the product

$$N(N-1)\,.$$

By the same reasoning, the number of ways three heads may be placed is $N(N-1)(N-2)$. Or, for m heads,

$$N(N-1)(N-2)\cdots(N-m+1) = N!/(N-m)! \tag{6.3}$$

Now this number represents the number of ways that a *particular sequence* of m heads can be allocated among the N trials. (We called this sequence the "first" head, the "second" head, etc.) This would be the valid answer $W_N(m)$ if an identifier were attached to each head occurrence, and they were somehow distinguishable. For example, they had different types of shading as in Fig. 6.2. However, they are not in our case – all heads are equivalent.

Therefore (6.3) *must be reduced* by a factor equal to the number of ways m distinguishable heads may occur in a sequence of m heads. But this is simply the answer (6.3) with, now, $m = N$ since *all* sequence positions are now filled with heads. That is, the answer is $m!$

Hence, the net degeneracy for the m *indistinguishable* head placements is

$$W_N(m) = \frac{N!}{(N-m)!\,m!} \equiv \binom{N}{m}\,. \tag{6.4}$$

The latter notation arises because this $W_N(m)$ is precisely the mth coefficient in the expansion of $(a+b)^N$, i.e., $W_N(m)$ is a *binomial coefficient.*

The required probability $P_N(m)$ is finally obtained by combining (6.2) and (6.4),

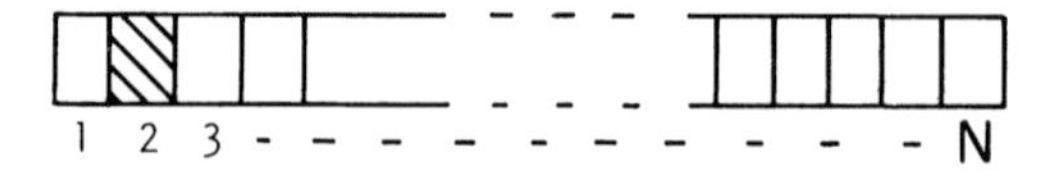

Fig. 6.1. A "head" occurs at the second flip

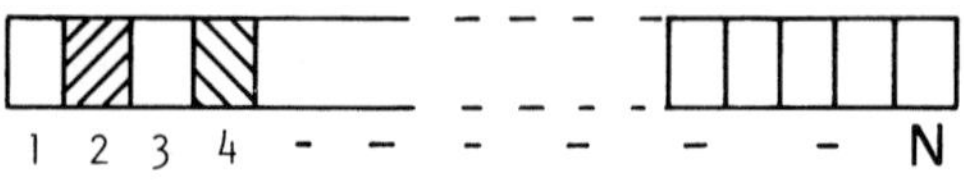

Fig. 6.2. With a "head" already having occurred at flip number 2, the second head can occur at $N-1$ other flips. It occurs at flip number 4

$$P_N(m) = \binom{N}{m} p^m q^{N-m} . \tag{6.5}$$

Because of the binomial nature of the degeneracy factor here, (6.5) is called a "binomial probability distribution." Parameter p is assumed known.[1]

6.2 Illustrative Problems

A fair coin is flipped $N = 4$ times.

(a) What is the probability of obtaining two heads – one on the second flip and one on the fourth?
Since outcomes one and three were not specified, we assume they may be either heads or tails. Therefore, the answer is simply $(1/2)^2 = 1/4$.

(b) What is the probability of obtaining heads on the second flip and fourth flip, as above, but now also tails on the first and third flips?
Now we must use (6.1), so that the answer is $(1/2)^2(1 - 1/2)^{4-2} = 1/16$.

(c) What is the probability of obtaining two heads and two tails?
Since the order of the events was not specified, we must use result (6.5). Hence, the answer is $[4!/(2!2!)](1/16) = 3/8$.

(d) What is the probability of obtaining two or more heads?
Equation (6.5) must now be summed over $m = 2$, 3, and 4. The answer is

$$\sum_{m=2}^{4} \binom{4}{m} (1/2)^m (1/2)^{4-m} = 3/8 + 1/4 + 1/16 = 11/16 .$$

6.2.1 Illustrative Problem: Let's (Try to) Take a Trip: The Last Word

In Sects. 2.8 and 2.13.1 we presented variations on a problem of interest to airline travelers. Here we shall try to include within the problem's framework all realistic assumptions which have a bearing on whether the passenger gets his seat. Accordingly, it is reworded as follows.

Airlines purposely overbook seats on flights, since in practice a significant number of booked passengers do not show up. Suppose a certain airline books 105 people for 100 seats on a given flight. If the probability is p that any given passenger will show up for the flight, what is the probability that a passenger who shows up will get a seat?

Analysis: In Sect. 2.8 we assumed, unreasonably, that every passenger shows up. Here, we allow them to show or not, according to fixed probability p. This affects the analysis, since now the event S (that the person gets a seat) is contingent upon how many other people show up. For example, if $m = 0$ other people show up, the passenger must get his seat, i.e.

[1] A generalization of (6.5) to include *uncertainty* in p is given in Eq. (10.16).

$$P(S|m = 0) = 1 .$$

In fact, more generally

$$P(S|m) = \begin{cases} 1(\text{assurance}) & \text{if } m \leq 99 \\ 100/(m+1) & \text{if } m > 99 \end{cases} .$$

That is, recalling that there are in total 100 seats (fixed), if up to 99 other people show up the given passenger gets his seat; while, if more than 99 others show up, the given passenger has decreasingly less of a chance of getting a seat (see derivation of (2.7)).

By the partition law (2.18) the net $P(S)$ obeys

$$P(S) = \sum_{m=0}^{104} P(S|m)\,P(m) .$$

We already know $P(S|m)$. Probability $P(m)$ is the probability that m other people show up. Now there are 104 "slots" for showing up, each of which is independently filled according to probability p, or not. This defines a Bernoulli trials sequence. Hence

$$P(m) \equiv P_{104}(m) = \binom{104}{m} p^m (1-p)^{104-m} .$$

Combining the last three equations formally solves the problem for $P(S)$. The answer depends upon the free parameter p. If, e.g., $p = 100/105$ (as would be implied by the booking procedure described above), $P(S) = 0.996$. This is a considerable improvement over the answer for the situation where all people show up.

6.2.2 Illustrative Problem: Mental Telepathy as a Communication Link?

Many people claim to have para-normal mental abilities, in particular the ability to pick up messages "through the air" from a remote message sender. We want to show here that if a person with such talents existed, even in the most limited sense of a *mere tendency* to pick up correct messages, that person could be used as a real communication link. Surprisingly enough, the error rate for the link could be made as small as is desired.

Suppose the messages are to be binary, in particular the single symbols 0 or 1 (note that any message can be coded as a sequence of 0's and 1's). The communication link operates as follows: The message sender selects his symbol, such as the 0, and thinks of it for an agreed-upon time, while the message receiver tries to receive an impression of the symbol. Since there are only two possible symbols, if the receiver has *no* telepathic ability, the probability p that he will guess the correct symbol is exactly $1/2$. This is equivalent to the receiver merely guessing randomly at the symbol, for example by flipping a fair coin.

But if the receiver has some telepathic ability, p will be slightly greater than $1/2$. Advantage can be taken of this tendency, and in effect it may be magnified, as follows. *Let the sender transmit the same symbol (say the 0) a large number N of times*. Then, by the law of large numbers (2.8), the number of times this symbol is received will be about Np. Also, the number of times the incorrect symbol (here 1) will be received is about $N(1 - p)$. With N large, Np will be much larger than $N(1 - p)$. That is, he receives the correct message many more times than he receives the incorrect one. The larger N is made, the stronger will be this effect.

Hence, the strategy adopted by the receiver is *simply to decide upon the symbol in question on the basis that it was received the greatest number of times*. This is a simple "majority rule" strategy.

Each new symbol is transmitted in this manner. That is, it is repeated N times by the sender. After the N symbols are received, in some order, by the receiver, he decides upon the one symbol as the symbol he has received more than $N/2$ times.

Let us now analyse the capabilities of this proposed communication link. Suppose $p = 0.6$, indicating a small but significant telepathic ability for the receiver. How large must N be made in order that the probability P (majority rule symbol is correct) $= 0.70$? Notice that we expect a gain, from 0.60 to 0.70, in the ability to correctly identify symbols. We shall show, in fact, how to make the gain arbitrarily large.

Analysis: Consider the N repetitions of one symbol. The received symbols constitute Bernoulli trials with outcomes "correct" or "incorrect." Since p, the probability of a correct outcome, is known, so is $P_N(m)$, the probability of m correct outcomes. Now the receiver's estimate will be correct if in particular $m > N/2$, since he bases his verdict on "majority rule." Hence, his probability of making the correct verdict will be $P(m > N/2)$. Or alternatively, if we equate $P(m > N/2)$ to a value such as 0.70, we are attempting to find the number N of repetitions that will lead to such a probability of making a correct estimate. Hence, we set

$$\sum_{m=(N/2)+1}^{N} \binom{N}{m} 0.6^m 0.4^{N-m} = 0.7\ , \quad (N/2) \equiv \begin{cases} N/2 & \text{for } N \text{ even, or} \\ (N-1)/2 & \text{for } N \text{ odd}\ . \end{cases}$$

By trying different values for N, it is found that a value $N = 7$ suffices. Hence, a moderate amount of repetition is required to increase the receiver reliability to 0.7.

If instead we ask that the probability of a correct decision be 0.9, the above procedure gives $N = 41$. Quite a bit more repetition is required, although it is perhaps surprising that value 0.9 can be attained with *any* value of N. This suggests that any degree of reliability can be attained, if N is merely made big enough.

To test out this hypothesis, assume an exceedingly weak telepathic effect, $p = 0.51$, to be present. At the same time, require the probability of a correct decision to be 0.97, exceedingly high. A value of $N = 9,100$ repetitions is now found to do the trick.

In summary, then, no matter how weak the telepathic effect may be, so long as it exists at all it can be used in a real communication link. The price paid may be a slow rate of data transmission.

In general, simple signal repetition coupled with a "majority rule" decision is a powerful and commonly used method of communication.

6.3 Characteristic Function and Moments

We found in Sect. 4.7 that the characteristic function for the binomial probability law is

$$\varphi(\omega) = (p\,\mathrm{e}^{\mathrm{j}\omega} + q)^N \,. \tag{6.6}$$

It was also found there that the first moments are

$$\langle m \rangle = Np \,, \quad \sigma_M^2 = Npq \,. \tag{6.7}$$

Finally, from these last two, the S/N ratio for a binomial law is

$$\mathrm{S/N} \equiv \langle m \rangle / \sigma_M = [Np/(1-p)]^{1/2} \,. \tag{6.8}$$

This shows an increase with $N^{1/2}$, as is well-known for the Poisson limit (see below).

6.4 Optical Application: Checkerboard Model of Granularity

When viewed close up, a photograph exhibits the individual clumps of grains that form the overall picture. These clumps have an undesirable granular appearance, which ultimately limits picture quality.

Because of such granularity, the energy transmittance at a given point of the photo is a random variable, even if the film was uniformly exposed to light. Various theories have been proposed for accounting for the observed statistics of transmittance, varying in degrees of complexity from the single-layer "checkerboard model" [6.1] described next, to the multiple-layer model of Ex. 6.1.25, to the Monte Carlo theory of Sect. 7.2.

Suppose that a piece of film is uniformly exposed to light. By the checkerboard model, the film is imagined to consist of a rectangular grid of cells, each of which is either (a) developed and opaque, or (b) undeveloped and clear (Fig. 6.3). It is assumed that each cell has *independently* the same probability p of remaining undeveloped and clear.

But then the same checkerboard pattern could equivalently have been produced by repeatedly flipping a biased coin, whose $P(\text{head}) = p$. A head outcome defines a clear cell, a tail outcome an opaque cell. The two problems are statistically equivalent. Therefore, by Sect. 6.1 the checkerboard model is an example of Bernoulli

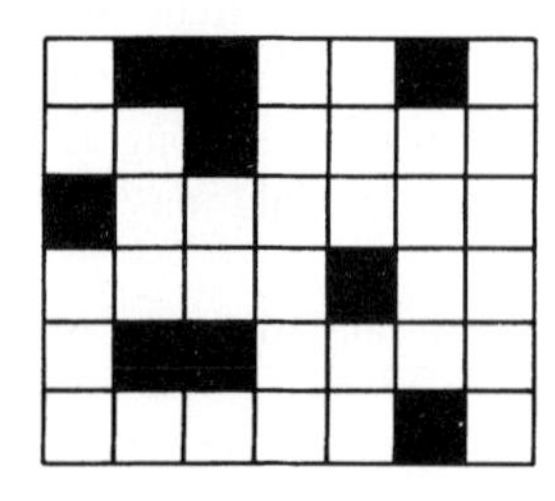

Fig. 6.3. The checkerboard model of granularity

trials. This permits us to analyze the statistics of the checkerboard, and consequently, of granularity.

If m cells remain clear, and there are a total of N in the checkerboard, the net energy transmittance of the entire checkerboard obeys

$$T = m/N \ . \tag{6.9}$$

That is, the net transmittance is the ratio of the clear to total area. T and m are random variables, since if the checkerboard is exposed over and over to the same uniform light, a different number m of cells will remain clear at each exposure. This number is unpredictable. We want to establish the statistics of fluctuation in T over this ensemble of checkerboards.

By the binomial law (6.5), the probability $P_N(m)$ of obtaining m clear cells out of the N total obeys

$$P_N(m) = \binom{N}{m} p^m (1-p)^{N-m} \ . \tag{6.10}$$

Because m is related to T by relation (6.9), Eq. (6.10) defines the statistics for T as well.

Using moment results (6.7) and relation (6.9) we find

$$\langle m \rangle = Np = N \langle T \rangle \ ,$$

so that

$$\begin{aligned} \langle T \rangle &= p \ ; \qquad \text{and} \\ \sigma_M^2 &= Np(1-p) = N^2 \sigma_T^2 \ , \end{aligned} \tag{6.11a}$$

so that

$$\sigma_T^2 = \frac{\langle T \rangle (1 - \langle T \rangle)}{N} \ . \tag{6.11b}$$

If the grain, or cell, size is a and if the checkerboard has area A, then $N = A/a$ and (6.11b) becomes

$$\sigma_T^2 = \frac{a \langle T \rangle (1 - \langle T \rangle)}{A} \ . \tag{6.12}$$

Quantity σ_T^2 actually represents the granularity on the scale defined by area A, since it measures the variability of T from one area A of film to the next. Hence, if one large piece of film were uniformly exposed to light, developed, and then scanned with an aperture of area A, the output transmittance would fluctuate by the amount σ_T^2 given by (6.12). This fluctuation may aptly be called the "granularity" of the film-aperture combination.

That σ_T^2 should vary inversely with A of the aperture is the Selwyn root-area law. Hence, the simple checkerboard model predicts this known experimental law of the photographic emulsion. The model must have a germ of truth to it.

The dependence of σ_T^2 upon the mean transmittance level $\langle T \rangle$ is also close to correct. Equation (6.12) shows that σ_T^2 should be very small when the film is either near saturation black or saturation clear, and conversely that σ_T^2 is close to a maximum

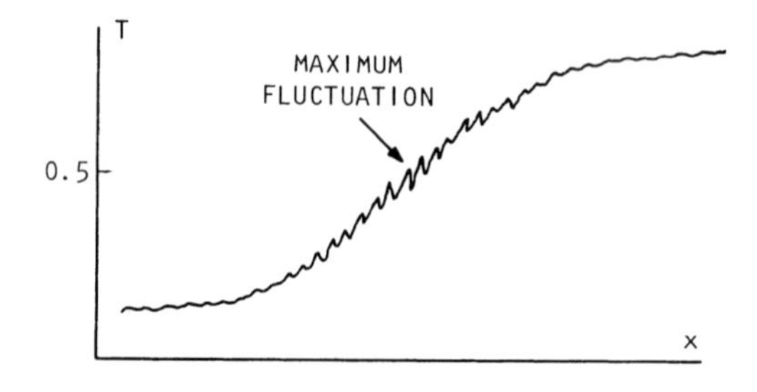

Fig. 6.4. Typical transmittance scan of T vs x across an edge image. Since the signal image has values $0.1 \leqq \langle T \rangle \leqq 0.9$, variance σ_T^2 varies across it, with minimal values at $\langle T \rangle \simeq 0.1$ or 0.9, and maximal values where $\langle T \rangle \simeq 0.5$

when $\langle T \rangle$ is at a mid-range value. These are experimentally observable tendencies as well (Fig. 6.4).

6.5 The Poisson Limit

In most applications of the binomial law (6.5), N is very large while the mean count $\langle m \rangle$ tends to be finite. But by (6.7), $\langle m \rangle = Np$. The implication is then that $p \ll 1$. Let us examine (6.5) under these circumstances, i.e.,

$$N \gg 1 \text{ and } p \ll 1 \text{ such that} \tag{6.13a}$$

$$\langle m \rangle \equiv Np \equiv \lambda \text{ is finite .} \tag{6.13b}$$

6.5.1 Analysis

The binomial law (6.5) is now, by (6.13b)

$$P_N(m) = \binom{N}{m} (\lambda/N)^m (1 - \lambda/N)^{N-m} . \tag{6.14}$$

We want the limiting form for this expression under condition $N \to \infty$. It is convenient to regard $P_N(m)$ as a product

$$P_N(m) = xy , \tag{6.15}$$

where

$$x = \binom{N}{m} (\lambda/N)^m , \quad y = (1 - \lambda/N)^{N-m} .$$

We consider x and y separately, in the limit. First

$$\ln y = (N - m) \ln(1 - \lambda/N) = \frac{\ln(1 - \lambda/N)}{(N - m)^{-1}}$$

and using l'Hopital's rule

$$\begin{aligned}
\lim_{n \to \infty} \ln y &= \lim_{n \to \infty} \frac{\lambda(1 - \lambda/N)^{-1} N^{-2}}{-(N - m)^{-2}} = -\lambda \frac{(N - m)^2}{N^2} \\
&= \lim_{N \to \infty} -\lambda(1 - 2m/N + m^2/N^2) \\
&= -\lambda .
\end{aligned}$$

Hence

$$y \to e^{-\lambda} . \tag{6.16}$$

Next

$$\lim_{N\to\infty} x = \lim_{N\to\infty} \frac{N(N-1)\dots(N-m+1)}{m!N^m}\lambda^m \tag{6.17}$$

by (6.4). But since m is fixed and finite, all factors in (6.17) become

$$\lim_{N\to\infty} (N-k+1) = \lim_{N\to\infty} N , \quad k = 1, 2, 3, \dots, m .$$

Using this in (6.17),

$$\lim_{N\to\infty} x = \lim_{N\to\infty} \frac{N^m}{m!N^m}\lambda^m = \lambda^m/m! , \tag{6.18}$$

independent of N.

Hence, by (6.15),

$$P_N(m) = \frac{\lambda^m}{m!} e^{-\lambda} \tag{6.19}$$

is the asymptotic limit under conditions (6.13a, b). This is the usual origin of the Poisson probability law. However, the law can arise in other circumstances as well (Exercise 7.1.10).

The Poisson law (6.19) has but one free parameter, λ, which equals $\langle m\rangle$. From Ex. 3.1.11, its variance $\sigma_M^2 = \langle m\rangle = \lambda$ as well.

In practice, the requirements (6.13) for the Poisson law to hold are easily met. The approximation is good if the elemental probability

$$p \lesssim 0.1 \tag{6.20}$$

and if $N \gtrsim 10$. An example below shows this.

Since *either* the binomial or the Poisson law suffice to describe many phenomena, the question arises as to which to use. The Poisson is usually preferable, (i) because it is simpler, requiring one parameter while the binomial requires two, and (ii) because of computational advantages described next.

6.5.2 Example of Degree of Approximation

The sensitivity of any grain in a uniformly exposed emulsion containing $N = 10$ grains is $p = 0.1$. What is the probability that at most one grain will be sensitized for development?

Using the precise (6.5)

$$P(m = 0 \text{ or } 1) = P_{10}(0) + P_{10}(1)$$
$$= \binom{10}{0}(0.1)^0(0.9)^{10} + \binom{10}{1}(0.1)^1(0.9)^9 = 0.7361 .$$

Using the Poisson approximation (6.19), we have $\lambda \equiv Np = 10(0.1) = 1.0$. Then

$$P(m = 0 \text{ or } 1) = \frac{1.0^0}{0!}\mathrm{e}^{-1.0} + \frac{1.0}{1!}\mathrm{e}^{-1.0} = 2\mathrm{e}^{-1} = 0.7358\,.$$

Even in this borderline use of the Poisson law, the degree of approximation is very good.

A practical reason for using the Poisson approximation becomes apparent here. The precise Bernoulli law requires evaluation of terms q^m, where q is now close to 1 and m is very large (e.g., the terms 0.9^{10} and 0.9^9 preceding). These are close to the indeterminate form 1^∞, and hence tend to suffer from poor accuracy. The Poisson law avoids this kind of problem.

6.5.3 Normal Limit of Poisson Law

Figure 3.3 showed that the shape of the Poisson law is skewed for low values of the mean, but approaches a symmetric form as the mean gets larger. Here we address the question of the limiting shape of this symmetric form.

As in the derivation of the central limit theorem in Sect. 4.25, we have to cope with the case of an infinite mean. As the trend in Fig. 3.3 indicates, the Poisson probability curve $P(m)$ in that limit will be infinitely flattened and broad. This is inconvenient for purposes of identifying shape. Hence, as in Sect. 4.25 we content ourselves with finding the probability of a variable that is proportional to the required one. The probability law on this variable will have a well-defined shape.

Hence, with m the Poisson RV of Eq. (6.19), define a new RV

$$n = m/\sqrt{\lambda}, \quad \lambda \equiv \langle m \rangle\,.$$

Squaring and averaging the first equation shows that $\sigma_N^2 = \sigma_M^2/\lambda = \lambda/\lambda = 1$. A unit variance does permit the identification of shape, as we required.

We will find $P_N(n)$ through its characteristic function. By definition

$$\begin{aligned}\varphi_N(\omega) &= \langle \exp(\mathrm{j}\omega n)\rangle = \left\langle \exp(\mathrm{j}\omega m/\sqrt{\lambda})\right\rangle = \varphi_M(\omega/\sqrt{\lambda}) \\ &= \exp[-\lambda(1 - \exp(\mathrm{j}\omega/\sqrt{\lambda})].\end{aligned}$$

The second equality is by definition of n, the third recognizes the characteristic function of the old random variable m, and the last is by (4.7). Since $\lambda \to \infty$, we may expand in Taylor series the inner exponential, giving

$$\varphi_N(\omega) = \exp\{-\lambda[1 - 1 - (\mathrm{j}\omega/\sqrt{\lambda}) - \frac{1}{2!}(\mathrm{j}\omega/\sqrt{\lambda})^2 - \frac{1}{3!}(\mathrm{j}\omega/\sqrt{\lambda})^3 - \ldots]\}.$$

Taking the limit $\lambda \to \infty$, all terms beyond the quadratic drop out since they go inversely as powers of λ, and we get the limiting form

$$\varphi_N(\omega) = \exp(\mathrm{j}\sqrt{\lambda}\omega)\exp(-\omega^2/2).$$

By (4.13) this is the characteristic function for a *normal* RV n, with mean $\sqrt{\lambda}$ and unit variance. The latter checks with the value $\sigma_N^2 = 1$ we found before, and of course $\langle n\rangle = \sqrt{\lambda}$ by $n = m/\sqrt{\lambda}$. Thus, n is asymptotically a normal random variable as the mean of the Poisson law approaches infinity.

6.6 Optical Application: The Shot Effect

Suppose that a light source is randomly emitting photons, at an average rate of ν photons/time. The photons are emitted independently, and in *each* time interval $\mathrm{d}t$ either 0 or 1 photons are emitted. See Fig. 6.5. The probability p of emitting a photon during interval $\mathrm{d}t$ is $p = \nu\,\mathrm{d}t$. What is the probability that the source will emit m photons during a total exposure time T?

We recognize that the problem describes a Bernoulli sequence of trials. From the figure, $N = T/\mathrm{d}t$. Then by the binomial law (6.5),

$$P_N(m) = \binom{T/\mathrm{d}t}{m} (\nu\,\mathrm{d}t)^m (1 - \nu\,\mathrm{d}t)^{T/\mathrm{d}t - m} . \tag{6.21}$$

This could be evaluated exactly as we did to derive (6.19), the Poisson law.

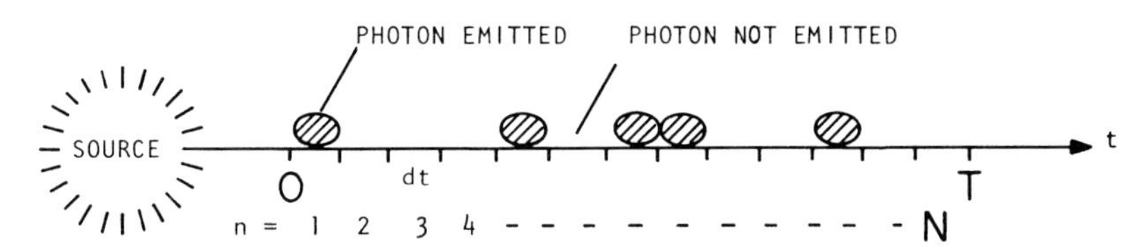

Fig. 6.5. A Bernoulli time sequence of photons. In each slot $\mathrm{d}t$, there is independently one photon or no photon. This model is from the "semiclassical theory" of radiation; see, e.g., [6.2]

Instead, note that the ground rules for use of the law are satisfied:

$$N = T/\mathrm{d}t \text{ is very large, while } Np = (T/\mathrm{d}t)\cdot \nu\,\mathrm{d}t = \nu T \tag{6.22}$$

is finite. Then the Poisson law may be invoked, and

$$P_T(m) = \frac{(\nu T)^m}{m!} \mathrm{e}^{-\nu T} . \tag{6.23}$$

This is the famous probability law for a shot noise effect.

Exposure time T must obey the following constraint. In order for interval $\mathrm{d}t$ to be small enough to contain *at most* one photon, it must be the order of the coherence time (3.76), or one over the spectral frequency range of the light. Hence for $T/\mathrm{d}t \equiv N$ to be large, as was required, exposure time T must be a large multiple of the coherence time.

The Poisson law (6.23) is part of an overall stochastic model of emitted particles, called the "Shot noise" process. See Sect. 8.15.

It is important to note that the Poisson law (6.23) holds for any rate of arrival ν of photons, even high ones. The "rare events" hypothesis (6.13a,b) is already built into the model (6.21) by the results (6.22), and these hold for any finite value of ν.

It is interesting to ask what the statistics would be for *detected* photons, if the detection process *itself* has a finite probability ν' for detection/photon arrival. This is in addition to the Poisson nature of the source, and hence is a kind of "double" Poisson effect. Intuitively, the overall process is still Poisson. Let us see if it is.

Since emission and detection are independent processes, the probability ϱ of *detecting* a photon in time interval $\mathrm{d}t$ is the probability that it was emitted at a particular previous time interval $\mathrm{d}t$ times its probability of being detected *if* emitted, or

$$\rho = p \cdot \nu' = (\nu\nu')\,\mathrm{d}t\ . \tag{6.24}$$

There are still $N = T/\mathrm{d}t$ "slots" available for detection, which is very large, and the mean-value $N\rho$ is now

$$N\rho = (T/\mathrm{d}t)(\nu\nu')\,\mathrm{d}t = \nu\nu' T\ , \ \text{finite}\ .$$

Hence, the Poisson approximation (6.19) applies once more, with $\lambda = \nu\nu' T$:

$$P(m \text{ detected photons}) = \frac{(\nu\nu' T)^m}{m!}\mathrm{e}^{-\nu\nu' T}\ . \tag{6.25}$$

The "double" Poisson effect of emission-detection is itself Poisson. A different kind of double Poisson effect is defined in Ex. 6.1.22.

6.7 Optical Application: Combined Sources

Two independent light sources are emitting photons, at rates ν_1 and ν_2, respectively. What is the probability of their total emitted outputs summing to m, after time T? Intuition again tells us that this is Poisson. Let us see.

After time T, let m_1 photons be emitted by source 1, m_2 by source 2. The total observed is to be m, and so

$$m = m_1 + m_2\ ; \tag{6.26}$$

m_1 and m_2 are each independent and Poisson. Therefore, by Sect. 4.17, so is m. Equation (4.31) shows that the net rate ν obeys

$$\nu = \nu_1 + \nu_2\ . \tag{6.27}$$

6.8 Poisson Joint Count for Two Detectors – Intensity Interferometry

We now consider the shot noise effect at two distinct points $\boldsymbol{r}_1, \boldsymbol{r}_2$ as in Fig. 6.6. Imagine detectors to be sensing incoming photons at the two points. Suppose that after a *short time interval* T, m_1 photons are counted at $\boldsymbol{r}_1$ and m_2 at $\boldsymbol{r}_2$. Will these values m_1, m_2 correlate, over many such time intervals?

We have to further specify how short T is. Let T be the order of the temporal coherence time, or (spectral range)$^{-1}$, for the incoming light. With this short a detection interval, the integrated light flux at a detector is itself a statistical parameter. This turns out to be the basis for the intensity correlation to be described below.

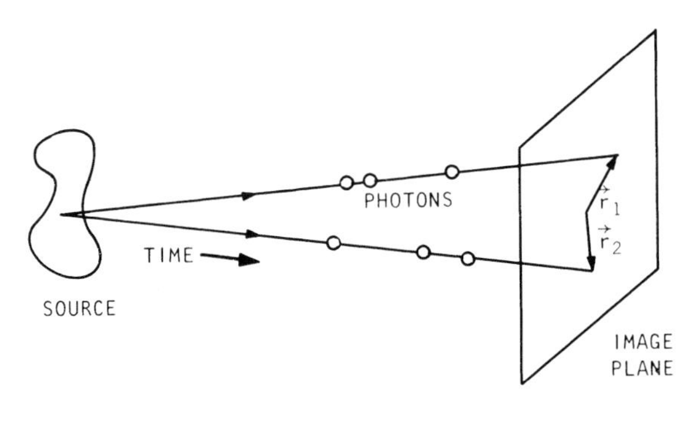

Fig. 6.6. Two simultaneous Poisson effects. These give rise to intensity interferometry

To analyze this situation, we shall use some basic results [6.2] from quantum optics.

We had previously assumed the *average flow rate* ν over one interval $(0, T)$ to be common to all such intervals. This actually hinges on an assumption that T is large. However, with a short T *flow rate* ν *will be random*, as shown next.

Let ν_i be the average flow rate at space point $\boldsymbol{r}_i$, $i = 1, 2$, during some *one* interval $(0, T)$. If $I_i(t)$ is the instantaneous photon arrival rate,

$$\nu_i = (\alpha/T) \int_0^T \mathrm{d}t I_i(t) , \quad i = 1, 2 \tag{6.28}$$

where $\alpha < 1$ measures the sensitivity of the detector (α is identical with ν' in (6.24)). This ν_i is the average rate for *detected* photons.

Now $I_i(t)$ is random with t. Hence if T is not overly large in (6.28), ν_i will also exhibit randomness from one interval $(0, T)$ to the next. This proves the assertion.

Next, consider marginal probabilities. By previous theory, the probability of m_i counts at point $\boldsymbol{r}_i$ follows (6.23), or

$$P_i(m_i) = \frac{(\nu_i T)^{m_i}}{m_i!} \mathrm{e}^{-\nu_i T} . \tag{6.29}$$

Combining this with (6.28)

$$P_i(m_i) = \frac{1}{m_i!} \left(\alpha \int_0^T \mathrm{d}t I_i(t) \right)^{m_i} \exp\left(-\alpha \int_0^T \mathrm{d}t I_i(t) \right) , \quad i = 1, 2 .$$

Note that these P_i represent probabilities of m_i arrivals over any *one* interval $(0, T)$, i.e., due to one particular flow rate ν_i. On the other hand, we want to compute $\langle m_1 m_2 \rangle$, averaging over all pertinent RV's, i.e., over m_1, m_2, *and time*. The latter includes many time intervals $(0, T)$.

Let us first consider $\langle m_1 m_2 \rangle$ over all possible pairs of counts, during *one* time interval $(0, T)$. Flow rates ν_1, ν_2 are here fixed. This leads us to consider $P_{12}(m_1, m_2)$, the joint probability for counts m_1 and m_2 at $\boldsymbol{r}_1$ and $\boldsymbol{r}_2$, respectively. During any *one* time interval there is no temporal correlation, of course, between m_1 and m_2. Hence, by independence (2.13)

$$P_{12}(m_1, m_2) = P_1(m_1) P_2(m_2) .$$

Next, consider the average $\langle m_1 m_2 \rangle$ over all possible counts m_1 and m_2, still with ν_1 and ν_2 fixed. This is simply

$$\begin{aligned}\langle m_1 m_2 \rangle &\equiv \sum_{m_1=0}^{\infty} \sum_{m_2=0}^{\infty} m_1 m_2 P_1(m_1) P_2(m_2) \\ &= \sum_{m_1} m_1 P_1(m_1) \sum_{m_2} m_2 P_2(m_2) = \langle m_1 \rangle \langle m_2 \rangle \\ &= (\nu_1 T)(\nu_2 T) \ . \end{aligned} \tag{6.30}$$

This result holds for any one time interval $(0, T)$, and hence any *one* pair of flow rates ν_1, ν_2. But repeated measurements (m_1, m_2) over many intervals $(0, T)$ means a further averaging process specifically over time. Then the temporal correlation is

$$\langle m_1 m_2 \rangle = T^2 \langle \nu_1 \nu_2 \rangle \ , \tag{6.31}$$

where the right-hand brackets designate the average over many time intervals $(0, T)$. Using (6.28)

$$\langle m_1 m_2 \rangle = \alpha^2 \int_0^T \int_0^T \mathrm{d}t \, \mathrm{d}t' \left\langle I_1(t) I_2(t') \right\rangle \ . \tag{6.32}$$

Further analysis of (6.32) requires detailed knowledge of the joint probability density $p(I_1(t), I_2(t'))$ for photon flow. This is known as follows.

By the scalar theory of light

$$I_i(t) = a_i(t) a_i^*(t) \ , \quad i = 1, 2 \tag{6.33}$$

in terms of a complex light amplitude a_i.

Also, for a "thermal" light source such as a star or a blackbody, amplitudes $a_i(t)$ and $a_j(t')$ obey joint Gaussian statistics. This follows from the *independence* of radiation from each source element of a blackbody. (By this independence, amplitude $a_i = \sum_n \exp(\mathrm{j}k\Phi_n)$ where the RV's Φ_n are independently sampled from one probability law. Hence a_i obeys the central limit theorem, and is Gaussian.)

Now it is easy to show, e.g., by successively differentiating the characteristic function (4.16), that for two jointly Gaussian *real* RV's x, y

$$\left\langle x^2 y^2 \right\rangle = \left\langle x^2 \right\rangle \left\langle y^2 \right\rangle + 2 \langle xy \rangle^2 \ .$$

Likewise, for our two *complex*, jointly Gaussian RV's $a_1(t)$, $a_2(t')$

$$\left\langle |a_1(t)|^2 \left|a_2(t')\right|^2 \right\rangle = \left\langle |a_1(t)|^2 \right\rangle \left\langle \left|a_2(t')\right|^2 \right\rangle + \left| \left\langle a_1^*(t) a_2(t') \right\rangle \right|^2 \ .$$

Then by (6.33)

$$\left\langle I_1(t) I_2(t') \right\rangle = \langle I_1(t) \rangle \left\langle I_2(t') \right\rangle + \left| \left\langle a_1^*(t) a_2(t') \right\rangle \right|^2 \ . \tag{6.34}$$

Now the last term here has further interpretation if the statistics of $a_i(t)$ are "wide-sense stationary" (Sect. 8.3). Then it is only a function of $(t - t')$,

$$\left| \left\langle a_1^*(t) a_2(t') \right\rangle \right|^2 = \left| \Gamma_{12}(t - t') \right|^2 \ . \tag{6.35}$$

This defines the "mutual coherence" function $\Gamma_{12}(t)$, by Ex. 3.1.24.

Substituting (6.34) and (6.35) into (6.32) produces

$$\langle m_1 m_2 \rangle = \alpha \int_0^T \mathrm{d}t \, \langle I_1(t) \rangle \, \alpha \int_0^T \mathrm{d}t' \left\langle I_2(t') \right\rangle + \alpha^2 \int_0^T \int_0^T \mathrm{d}t \, \mathrm{d}t' \left| \Gamma_{12}(t - t') \right|^2 .$$

But by (6.28) the first two integrals are simply the product $\langle \nu_1 T \rangle \langle \nu_2 T \rangle$ or $\langle m_1 \rangle \langle m_2 \rangle$. Hence

$$\langle m_1 m_2 \rangle = \langle m_1 \rangle \langle m_2 \rangle + \alpha^2 \int_0^T \int_0^T \mathrm{d}t \, \mathrm{d}t' \left| \Gamma_{12}(t - t') \right|^2 ,$$

or further reducing the double integral

$$\langle m_1 m_2 \rangle = \langle m_1 \rangle \langle m_2 \rangle + 2\alpha^2 \int_0^T (T - t) \left| \Gamma_{12}(t) \right|^2 \mathrm{d}t . \tag{6.36}$$

This relation describes the "Hanbury-Brown Twiss effect," or "intensity interferometry." Further defining the correlation coefficient ρ_{12} as the correlation in fluctuations from the means,

$$\rho_{12} \equiv \langle (m_1 - \langle m_1 \rangle)(m_2 - \langle m_2 \rangle) \rangle ,$$

we get

$$\rho_{12} = \langle m_1 m_2 \rangle - \langle m_1 \rangle \langle m_2 \rangle .$$

(Note that ρ_{12} is the unnormalized ρ of (3.47).) Then from (6.36)

$$\rho_{12} = 2\alpha^2 \int_0^T (T - t) \left| \Gamma_{12}(t) \right|^2 \mathrm{d}t \tag{6.37}$$

or *the correlation coefficient for intensity is a measure of the mutual coherence between the two points.* This relation between *intensity* and *coherence* is unexpected and interesting. It also allows precise measurement of star geometries (separations, diameters) from ρ_{12} data, since Γ_{12} connects with their intensity profiles as follows.

Let ν denote general light frequency, and $o_\nu(x, y)$ denote the unknown distribution of intensity/frequency interval in the source. For the particular case of a quasi-monochromatic source [6.3] at frequencies near $\bar{\nu}$,

$$o_\nu(x, y) \equiv o(x, y)\delta(\nu - \bar{\nu}) .$$

And by the Van Cittert–Zernike theorem (3.74, 75)

$$\Gamma_{12}(t) = \mathrm{e}^{-2\pi \mathrm{j} \bar{\nu} t} \mathrm{FT}[o(x, y)] ,$$

FT denoting Fourier transform. Consequently,

$$\left| \Gamma_{12}(t) \right|^2 = \left| \mathrm{FT}[o(x, y)] \right|^2 , \tag{6.38}$$

independent of time.

Using this in (6.37), we find that the time integration is now trivial, leaving

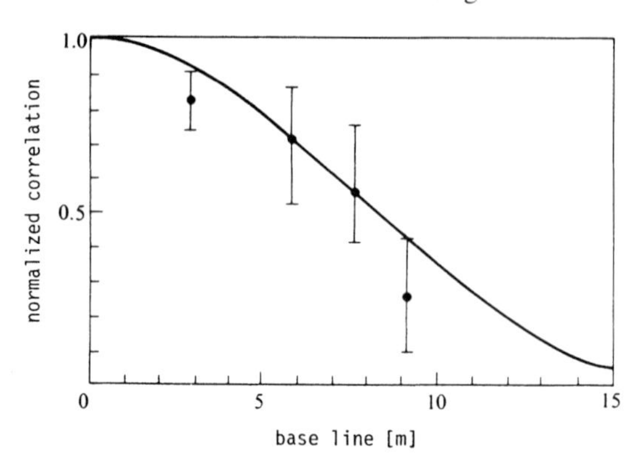

Fig. 6.7. Correlation ρ vs baseline p as obtained for the star Sirius A [6.4]

$$\rho_{12} \equiv \rho(p, q) = \alpha^2 T^2 \left| \iint_{\text{source}} \mathrm{d}x\, \mathrm{d}y\, o(x, y) \exp[-\mathrm{j}\bar{k}(px + qy)] \right|^2 , \qquad (6.39)$$

with $\bar{k} \equiv 2\pi/\bar{\lambda}$, and (p, q) the angular separation as viewed from the object source, of the (x, y) components of data points $\boldsymbol{r}_1, \boldsymbol{r}_2$, respectively.

Hence, by measuring $\rho(p, q)$ due to intensity correlation, information about the source intensity $o(x, y)$ may be obtained. For simple source shapes, such as a disk of unknown radius, measurements $\rho(p, q)$ allow properties of the source (such as diameter) to be accurately known.

For a disk source, one-dimensional measurements $\rho(p)$ (essentially on radius) suffice. Figure 6.7 shows the experimental $\rho(p)$ (dots) obtained for the star Sirius A. The continuous curve is the theoretical result (6.39) for a uniform disk of diameter 0.0069 arc sec [6.4]. This resolution level is a tremendous achievement, when we consider that the ordinary "seeing disk" through turbulence is the order of 1 arc sec (Fig. 8.4).

6.9 The Normal Limit (De Moivre–Laplace Law)

Here we show that the basic Bernoulli problem may be alternatively viewed as a *random walk* phenomenon. This leads to a *normal* approximation to the binomial probability distribution.

6.9.1 Derivation

We return to the basic Bernoulli problem involving N independent flips of a biased coin, and approach it in a different way. We again want to know the probability $P_N(m)$ of obtaining m heads, where at each flip the probability of a head is p.

Let us define a "counter" c_n with the properties

$$c_n = \begin{cases} +1 \text{ if a head occurs on the } n\text{th flip} \\ \;\;0 \text{ if a tail occurs on } n\text{th flip} . \end{cases} \qquad (6.40)$$

Then simply

$$m = \sum_{n=1}^{N} c_n \; , \tag{6.41}$$

which looks like m results from a random walk of N steps. Hence m might be normal. This is clarified next.

The c_n are *independently* determined, and sampled from the *same* probability law $P(\text{head}) = p$, $P(\text{tail}) = 1 - p$. Also, $N \gg 1$. Therefore, by Sect. 4.26 m obeys the central limit theorem, and so its *probability density is approximately normal,*

$$p_N(m) = (2\pi)^{-1/2}\sigma_M^{-1} \exp[-(m - \langle m \rangle)^2/2\sigma_M^2] \; . \tag{6.42}$$

This assumes that m can take on a continuum of values, whereas m can only take on integer values $0, 1, 2, \ldots, N$. Assume, however, that $p_N(m)$ of (6.42) changes very little from one value of m of the next. That is, it is a *smooth* function locally. Then the discrete probability $P_N(m)$ of obtaining m heads obeys, by definition (3.1),

$$P_N(m) \equiv \int_m^{m+1} \mathrm{d}m \, p_N(m) \simeq p_N(m)\Delta m = p_N(m) \cdot 1 \; , \tag{6.43}$$

by the smoothness assumption.

Hence, using (6.42),

$$P_N(m) = (2\pi)^{-1/2}\sigma_M^{-1} \exp[-(m - \langle m \rangle)^2/2\sigma_M^2] \; , \tag{6.44}$$

which is the probability we wanted. This result is called the De Moivre–Laplace theorem, after its co-discoverers. It is unusual in representing a *discrete* RV by a normal law.

6.9.2 Conditions of Use

It remains to find the conditions for which $p_N(m)$ is smooth. Obviously, this will be so if points m are spaced closely compared to σ_M, as illustrated in Fig. 6.8. This is algebraically a requirement $\sigma_M \gg \Delta m$, or since $\Delta m = 1$

$$\sigma_M \gg 1 \; . \tag{6.45}$$

But we know σ_M from the exact Bernoulli results (6.7), which combined with (6.45) form a requirement

$$Np(1 - p) \gg 1 \; . \tag{6.46}$$

Note that this requirement is quite different from (6.13b) for validity of the Poisson approximation. For example, if $Np \propto 1$ the Poisson is valid, but by (6.46) the normal is not.

A final requirement on the accuracy of the approximation (6.44) is that *extreme regions $m \simeq N$ or $m \simeq 0$ be avoided.* A way of expressing these requirements on m is to limit it to values near its mean $\langle m \rangle$,

$$|m - \langle m \rangle| \lesssim \sigma_M \; .$$

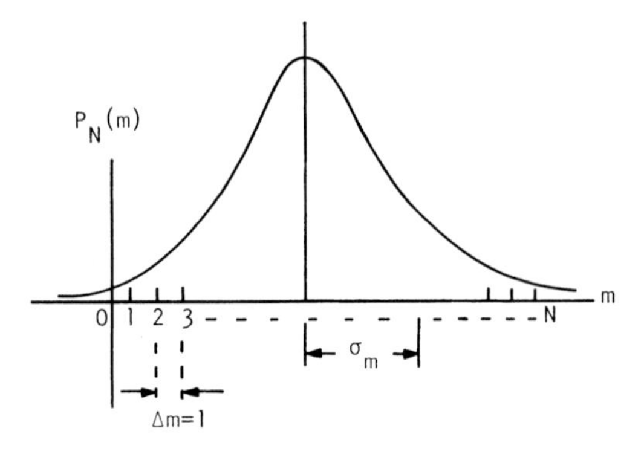

Fig. 6.8. If $\sigma_M \gg \Delta m = 1$, then the De Moivre-Laplace theorem holds. The curve must be smooth, on the scale of changes $\Delta m = 1$

By the Bernoulli results (6.7), this is equivalent to requiring

$$|m - Np| \lesssim [Np(1-p)]^{1/2} . \tag{6.47}$$

Hence, the normal curve best approximates the Bernoulli result within roughly one σ of its mean.

6.9.3 Use of the Error Function

The normal approximation (6.44) is very important because it readily applies to problems where the Bernoulli approach would be very tedious.

For example, a fair coin is flipped 10 000 times. Find the probability that the number of heads lies between 4950 and 5050. We have here $N = 10\,000$, $p = q = 1/2$. The Bernoulli answer would be, by repeated use of (6.5),

$$\begin{aligned} P(4950) \leqq m \leqq 5050 = &\binom{10\,000}{4950}(1/2)^{4950}(1/2)^{5050} \\ &+ \binom{10\,000}{4951}(1/2)^{4951}(1/2)^{5049} + \ldots \\ &+ \binom{10\,000}{5050}(1/2)^{5050}(1/2)^{4950} . \end{aligned}$$

This would be an overwhelming numerical evaluation, if attacked directly. We have here a series of very large numbers ($\sim$ of size 10^{10^4}) multiplied by very small numbers. Clearly, accuracy could not be maintained during these multiplications, even with today's largest computers.

Instead, let us try the normal approach. It applies if requirements (6.46) and (6.47) are met.

We have $Npq = 2500 \gg 1$, so that (6.46) is satisfied. Also, $Np = 5000$ and $4950 \leqq m \leqq 5050$. Then the largest value of $|m - Np|$ is 50, while $(Npq)^{1/2} = 50$, so that (6.47) is satisfied at all m involved.

Hence, the normal law (6.44) applies. Furthermore, by the smoothness property (6.45), this is a probability *density* (6.42) as well, so we may cast the answer as an *integral*

$$P(4950) \leqq m \leqq 5050) = (2\pi)^{-1/2}\sigma_M^{-1} \int_{4950}^{5050} \mathrm{d}m \exp[-(m - \langle m \rangle)^2/2\sigma_M^2] \,.$$

With $\langle m \rangle = Np = 5000$, $\sigma_M = (Npq)^{1/2} = 50$, and the change of variable

$$x = (m - \langle m \rangle)/2^{1/2}\sigma_M \,,$$

we get

$$P(4950 \leqq m \leqq 5050) = \pi^{-1/2} \int_{-\frac{50}{50\sqrt{2}}}^{+\frac{50}{50\sqrt{2}}} \mathrm{d}x\, \mathrm{e}^{-x^2} \,.$$

Now the tabulated function $\mathrm{erf}(z)$ was defined by (4.61). Then

$$P(4950 \leqq m \leqq 5050) = \mathrm{erf}(1/\sqrt{2}) = 0.685 \,.$$

In this way, the normal approximation has cast a numerically overwhelming problem into a mere table lookup!

It is easy to generalize this procedure. If $P(m_1 \leqq m \leqq m_2)$ is required, providing requirements (6.46) and (6.47) are met

$$P(m_1 \leqq m \leqq m_2) = 2^{-1}\left[\mathrm{erf}\left(\frac{m_2 - Np}{(2Npq)^{1/2}}\right) - \mathrm{erf}\left(\frac{m_1 - Np}{(2Npq)^{1/2}}\right)\right] \,. \qquad (6.48)$$

[Compare with the more general result (4.62)].

Finally, it is important to note that *requirement* (6.47) *is not as binding upon the summation result* (6.48) *as it is upon the single probability* (6.44). Suppose that requirement (6.47) is not met for certain terms

$$\binom{N}{m} p^m q^{N-m}$$

in the sum for $P(m_1 \leqq m \leqq m_2)$. *These terms are each much less than those which do obey* (6.47). Hence if at least one value of m satisfies (6.47), those erroneous terms which violate (6.47) will contribute negligibly to $P(m_1 \leqq m \leqq m_2)$. In summary, only requirement (6.46) must be met, if $P(m_1 \leqq m \leqq m_2)$ is to be computed and at least one value of m within (m_1, m_2) does obey requirement (6.47).

Exercise 6.1

The purpose of problems (1) through (4) is to gain practice in recognizing which of the Bernoulli, Poisson, or normal forms is applicable in a given case. First set up the answer (but do not necessarily evaluate it) in the exact Bernoulli form. Then use the appropriate Poisson or normal approximation for evaluation, *if* one of these applies.

6.1.1 A fair coin is flipped six times. What is the probability of obtaining either three or four heads?

6.1.2 A fair coin is flipped 60 times. What is the probability of obtaining from 30 to 35 heads?

6.1.3 A fair die is rolled 120 times. What is the probability that roll outcome 2 will occur between 16 and 24 times?

6.1.4 A pair of dice are rolled 120 times. What is the probability that total roll value 7 will occur between 16 and 24 times?

6.1.5 A major earthquake strikes the U.S. on the average of once every 20 (say) years. Regarding earthquake occurrences, then, what is the probability of

(a) one in a given year.

(b) 3 or more in 10 years.

(c) 3 or more in 20 years.

6.1.6 At a certain university, on the average $1/3$ of entering freshmen drop out after a year. Suppose a class numbering 6000 freshmen enters in a given year. What is the probability that between 1900 and 2100 students will have dropped out after a year?

6.1.7 How could the "telepathic communication link" of Sect. 6.2.2 be used if $p = 0.40$, indicating a higher probability of being incorrect than correct, or "reverse" telepathy? Find the repetition value N required to attain 0.80 as the probability of making a correct decision by your revised strategy.

6.1.8 An emulsion is being scanned with a square aperture of side 10^{-2} cm. Its grains are each 10^{-3} cm on a side. The emulsion had been weakly exposed, with mean transmittance level of 0.01, prior to the scan.

(a) What is the probability that exactly three undeveloped grains will be detected by the aperture at any one position? (Assume that the aperture is linked to a perfect detector.)

(b) Now assume a detection efficiency of 0.50. What is the three-grain probability of detection?

6.1.9 Repeat problem (8), now assuming a mean transmittance level of 0.1.

6.1.10 Photons are streaming toward a perfect detector at the mean rate of $10/\text{s}$. Consider a time exposure of 0.1s.

(a) What is the probability that exactly three photons will be detected in that time period?

(b) Now assume a detection (quantum) efficiency of 0.50. What is the three-photon probability of detection?

6.1.11 Repeat problem (10), now assuming a time exposure of 1.0 s.

6.1.12 A particle is undergoing one-dimensional (1-D) Brownian motion. This is a 1-D random walk situation. After any collision with a molecule of the medium, the particle has the same probability for moving distance $+\Delta x$ as distance $-\Delta x$. Collisions are taking place at the average rate of $10/\text{s}$. After 10 s, what is the probability that the particle will be displaced by a net amount of $40\Delta x$ to $60\Delta x$?

6.1.13 The annual crime rate in Tucson is currently 10%, meaning that the probability p that a resident will be a victim in any given year is $1/10$. An average university student stays 4 years in Tucson.

(a) What is the probability that he will be a crime victim?

(b) After how many years is it *likely* ($p > 1/2$) that he will be a victim? On this basis, should he go on to graduate school?

6.1.14 Redundancies are often built into complex systems to insure a high probability of successful operation. For example, an automobile can be brought to a halt by any one of three different braking systems (front, rear, or emergency). Let a system have a "redundancy of order n," meaning that the system will work if *any one* or more of n components of the system work. Suppose that the probability is p that a given component will fail to work during a time period T. Each component has the same p value. What is P, the probability that the system will work during the entire time period T?

6.1.15 A bicyclist reaches an intersection which he would like to cross. The cross street has traffic which may be modelled as Poisson with an average flow rate of ν cars/s. The bicyclist estimates that he needs at least a time gap of t_0 s between successive cars in order to cross safely. What is the probability that the next car is at least time t_0 away? [*Hint:* He wants the probability of Bernoulli events sequence

$$\underbrace{(\text{no car, no car, no car}, \ldots, \text{no car, car})}_{\longleftarrow \quad n \text{ events} \quad \longrightarrow},$$

where $p(\text{car}) = \nu\,\mathrm{d}t$ and $n = t_0/\mathrm{d}t$.]

Notice that no event *before* the indicated sequence of events enters into consideration. By independence of the events, whether there were cars or not *prior to* the sequence does not influence the probability of any event in the sequence.

6.1.16 The molecules of a gas are undergoing random motion. They randomly collide, with a mean time $\langle t \rangle$ between collisions. What is the probability density for a molecule lasting a time t between successive collisions? (Note the mathematical similarly to the preceding problem.)

6.1.17 Weather forecasting is difficult because of randomly changing wind patterns, etc. TV forecasters often assume that a "heat wave," i.e., a particular mass of high pressure air, has arrived if a large number of cities in the area simultaneously show record highs for the day. Let us test out this hypothesis.

Suppose that, on the average, a city establishes over a year's time 10 days of record high temperature, that is, $p(\text{high}) = 10/365$. Let 20 particular cities (e.g., Chicago, Cincinnati, etc.) represent the "Midwest" group of cities. What is the probability that, completely by chance (and not necessarily due to a high pressure system), 5 or more of these cities will attain record highs on the same day?

6.1.18 Suppose that, on the average, a US citizen experiences 3 days of illness during a given year which are due to "colds." What is the chance that, in a family of five, two or more people will be down with a cold on the same day?

Is this a *significant event*, i.e., one that merits concern? It is commonly assumed in statistics that if an event happens for which $P \leq 0.05$, then the event is not consistent with mere chance, and is "significant."

6.1.19 Repeat problem (18), if instead 200 000 or more people, out of 500 000 (of a city, say), come down with the cold.

6.1.20 When is a grade report significantly bad? Suppose that, for the average college student, a grade worse than C (i.e., D or F) will be received in five out of every 40 grades. It is observed that a certain student obtains grades worse than C in two out of the five courses he has taken during a given semester. Is this a significantly bad term for him, that is, is $P(m \geqq 2) \leqq 0.05$ for the event "grade worse than C?"

6.1.21 ***Cascaded events.*** A "cascaded" or "secondary" event sequence arises as in Fig. 6.9. Each "primary" event i gives rise to, or excites, a number k_i of "secondary" events. All the event numbers m, $\{k_i\}$ are random. Then the total number n of secondary events obeys

$$n = \sum_{i=0}^{m} k_i \,,$$

and n is random. This situation occurs, for example, as one stage of amplification within a photomultiplier tube, where each incident photon causes a random number k_i of secondary electrons to be emitted from a photosensitive surface. The secondary electrons tend to be clustered or "clumped" in time according to the finite sizes of the $\{k_i\}$. If m photons are incident upon the surface, n electrons are emitted. What are the statistics of n?

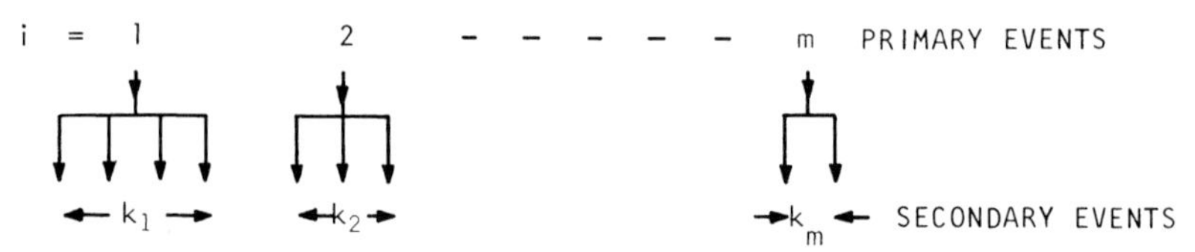

Fig. 6.9. Schematic of a cascaded-event system. Each of the m (random) primary events i gives rise to a random number k_i of secondary events

Let the characteristic function for RV m be known to be $\varphi_M(\omega)$. Let the $\{k_i\}$ be independent samples from one probability law, whose characteristic function is $\varphi_K(\omega)$. Show that:

(a) The characteristic function for n obeys

$$\begin{aligned}\varphi_N(\omega) &= \sum_{m=0}^{\infty} P(m)\varphi_K(\omega)^m \\ &= \varphi_M[-\mathrm{j}\ln\varphi_K(\omega)] \,. \end{aligned} \qquad (6.49)$$

Then, by multiple differentiation of this result, show that

(b) $\langle n\rangle = \langle w\rangle\langle k\rangle$ and

(c) $\sigma_N^2 = \langle k\rangle^2\sigma_M^2 + \langle m\rangle\sigma_K^2$. The latter is called the "Burgess variance theorem" [6.5].

6.1.22 Suppose that the primary events of problem (21) obey Poisson statistics,

$$P(m) = \mathrm{e}^{-\langle m\rangle}\,\langle m\rangle^m/m!\,, \quad m = 0, 1, \ldots$$

(a) From the first Equation of (6.49) show that the number n of secondary events follows a law

$$\varphi_N(\omega) = e^{\langle m\rangle[\varphi_K(\omega)-1]} \,. \tag{6.50}$$

(b) From results (b) and (c) of problem (21), show that now

$$\sigma_N^2 = \langle n\rangle\left(\langle k\rangle + \sigma_K^2/\langle k\rangle\right), \tag{6.51}$$

Hence, the variance in the number n of secondary events is proportional (but not necessarily equal to) its mean. The proportionality constant depends upon the statistics for RV k. See the next problem.

6.1.23 Regarding problem (22):

(a) If RV k is binary, then either $k = 0$ or 1. Then the secondary outputs correspond to a Poisson event sequence where randomly some (for $k = 0$) events are missed. What will the output law $P_N(n)$ be? *Hint:* If k is binary, then

$$P(k = 1) = \langle k\rangle \text{ and } P(k = 0) = 1 - \langle k\rangle$$

for the two alternatives; then

$$\varphi_K(\omega) = 1 + \langle k\rangle\,[\exp(j\omega) - 1]$$

and consequently from (6.50) n is Poisson with mean $\langle m\rangle\langle k\rangle$. The output statistics are still Poisson.

(b) If RV k is Poisson with mean $\langle k\rangle$, then show by (6.50) that n follows a law

$$\varphi_N(\omega) = \exp\{\langle m\rangle\,[e^{\langle k\rangle(e^{j\omega}-1)} - 1]\} \,. \tag{6.52}$$

This is called “Neyman type A.” Show also that the variance of n is now

$$\sigma_N^2 = \langle n\rangle\,(1 + \langle k\rangle) \tag{6.53}$$

by the use of (6.51). Hence, in this cascaded-Poisson case, the variance in the output number exceeds that for a Poisson law of the same mean $\langle n\rangle$. In summary, somewhat surprisingly, the outputs of a doubly-Poisson event sequence are not themselves Poisson, but instead fluctuate more wildly than a Poisson law of the same mean.

6.1.24 ***A measure of reliability.*** The reliability of an electro-optic component can be defined as its mean number of uses before failure occurs. Let the probability of successful use of the component on a given trial be p, and assume that p remains constant from trial to trial.

Show that our measure of reliability is then the first moment of a geometric probability law (3.57)

$$P_n = \frac{a^n}{(1+a)^{n+1}}$$

where $a = p(1-p)^{-1}$. Since $\langle n\rangle = a$ (by Ex. 4.1.11), a is our reliability measure as well.

6.1.25 ***Origin of the Hurter-Driffield (H–D) curve.*** The H–D curve describes the average photographic density $D \equiv -\log_{10} T$ (energy transmittance) in a film that is exposed to a uniform light level E.[2] Specifically, it is a plot of D against $\log_{10} E$. A typical one is shown in Fig. 15.1. It is not widely known that this curve has a probabilistic origin [6.6], specifically arising out of the shot noise nature (6.23) of photons. The derivation follows.

First, a model must be formed for the granular structure of the film. This is imagined to be planar layers of grains, the grains in each layer being placed with uniform randomness. For simplicity, all grains are of the same size and obey the same sensitization property (described below). After the film is developed, a given grain is either opaque (because it was sensitized) or remains transparent.

Let t_i be the resulting transmittance of layer i in the film. Each layer is of total area A. Then if the total cross-sectional area of opaque grains in the layer is A_i, $t_i = (A - A_i)/A$. Suppose, in addition, that the net energy transmittance T obeys a simple product law

$$T = \prod_i t_i \, .$$

Define the corresponding density D by $D = -\log_{10} T$.

(a) Then show that if the grains are sparse, so that $A_i/A \ll 1$, D obeys the *Nutting formula*

$$D = (\log_{10} \mathrm{e}) na/A \, . \tag{6.54}$$

Quantity n is the total number of opaque grains, including all layers, and a is the area of a grain. The dependence of D upon light exposure is through parameter n, as follows.

A simple model for sensitization of a grain is that sensitization occurs when the grain receives a certain threshold of light energy. Assume the light to be monochromatic. Then this amounts to a requirement that at least Q photons strike the grain. Then if E is the average number of photon arrivals over the fixed exposure time, and if the probability $P(m)$ of m photon arrivals at a grain obeys (6.23):

(b) Show that

$$P(\text{sensitization}) \equiv P(m \geq Q) = 1 - \mathrm{e}^{-E} \sum_{m=0}^{Q-1} E^m/m! \tag{6.55}$$

The final result is obtained by combining this effect with the Nutting formula (6.54). The actual number n of opaque grains will obey simply $n = N \cdot P$ (sensitization), where N is the total number of grains in the film. (This assumes already-sensitized grains to not block incident photons from layers beneath.) Then combining this with (6.55) and 6.54),

$$D = (\log_{10} \mathrm{e})(Na/A) \left(1 - \mathrm{e}^{-E} \sum_{m=0}^{Q} E^m/m! \right) \, . \tag{6.56}$$

This is the basic H–D effect. Remarkably, the dependence of D upon E goes simply as 1 minus a cumulative probability law.

[2] The conventional notation for exposure is H. We use instead E since H has already been used to describe entropy.

Plots of (6.56) are shown in Fig. 6.10, for different sensitization thresholds Q. Compare the qualitative features of these curves with the typical H–D curve shown in Fig. 15.1.

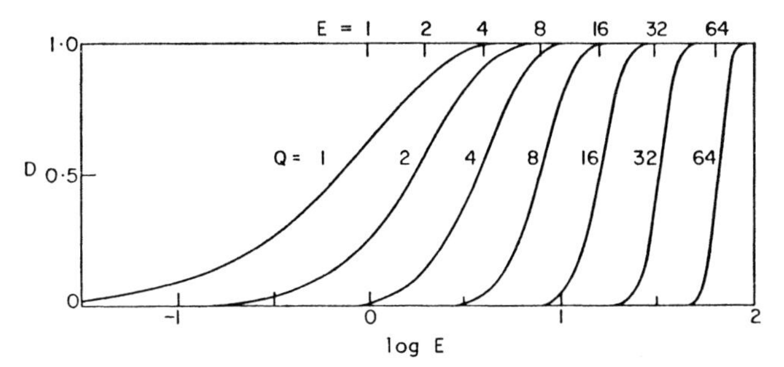

Fig. 6.10. Density D vs $\log_{10} E$ (H–D) curves for different sensitization thresholds Q. Exposure E is the average number of photons per grain. The grains are assumed to be identical [6.6]

6.1.26 Suppose that the human eye has retinal receptors of size Δx equal to $1/2$ the central core width of the retinal point spread function. Approximate the latter by the sinc^2-function (3.16). It has been conjectured that the eye-brain estimates the lateral position of a point source as simply the position of the retinal receptor receiving the largest number of photons. (Of course, this is provided enough occur to sensitize the receptor; this shall be assumed.) Then the probability P_c of correctly identifying source position is the probability that out of N received photons, the maximum number occurs in the central Δx interval of the sinc^2-function.

Note that if $m \geqq N/2$ photons locate within a given Δx, the maximum *must* occur there, although it might also occur there even if $m < N/2$. Hence $P_c \geqq P(m \geqq N/2)$. This enables us to find a lower bound to P_c.

Find $P(m \geqq N/2)$ if the central Δx interval of the sinc^2-function contains $p = 0.77$ of its total area. Evaluate $P(m \geqq N/2)$ if $N = 5$ photons arrive; if 100 photons arrive. (*Hint:* is $Npq \gg 1$?)

6.1.27 ***Probability law for nearest-neighbor distance in a random distribution of particles.*** Suppose that the stars in a given photograph of the night sky are positioned with uniform randomness. Let $p(r)\,\mathrm{d}r$ denote the probability that the nearest neighbor to an arbitrary star A occurs between distance r and $r + \mathrm{d}r$. We want $p(r)$.

Analysis: Of course

$$\int_0^r \mathrm{d}r\, p(r)$$

is the probability that the nearest neighbor to A lies somewhere between distance 0 and r, so that

$$1 - \int_0^r \mathrm{d}r\, p(r)$$

is the probability that the nearest neighbor lies outside distance r. This is equivalent to the probability that no star is closer to A than distance r.

Now $p(r)\,dr$ clearly equals the latter probability times the probability that a star does exist in the ring area between r and $r + dr$. Thus, $p(r)$ satisfies

$$p(r)\,dr = \left(1 - \int_0^r dr\,p(r)\right)(2\pi r\,dr)\nu \tag{6.57}$$

where ν is the average number of stars per unit area.

(a) Show that the solution to the preceding is

$$p(r) = 2\pi\nu r\,e^{-\pi\nu r^2}, \tag{6.58}$$

a Rayleigh probability law. *Hint:* Differentiate (6.57) with respect to r and solve the resulting differential equation.

(b) Show by direct integration that the average distance $\langle r\rangle$ between nearest neighbors obeys

$$\langle r\rangle = 0.5/\sqrt{\nu}\,. \tag{6.59}$$

The three-dimensional version of this problem is solved in [6.7]. The one-dimensional problem has already been solved at Ex. 6.1.16, although in a different way.

6.1.28 The probability law for a RV x is $p_X(x)$. Pairs of events (x_S, x_L) are independently sampled from the law, where x_S is by definition the smaller of the two events and x_L is the larger. What is the probability law $p_S(x)$ for x_S? *Hint*: This is a Bernoulli trials sequence of length $N = 2$, where for any value $x_1 = x$, x_2 is the event $x_2 \geq x$. Thus, in (6.5), $m = 1$, $p = p_X(x)\,dx$ and $q = 1 - F_X(x)$.

6.1.29 Using the answer to Ex. 6.1.28, for the case where both RV's are *uniformly* random over the interval $(0, a)$ show that the PDF on the smaller RV is $p_S(x) = (2/a)(1 - x/a)$, $0 \leq x \leq a$. Also, what is the probability law on the larger RV?

6.1.30 ***Is vitamin C of use in combating the common cold?*** [6.8] A test program is set up whereby 17 people are given vitamin C, and 17 other people a placebo, over a period of time. Each person from the first group is randomly matched with one from the other group. At the end of the time period a talley is made of each matched pair as to which shows better relief from cold symptoms. If the vitamin C person, then this is called an event C; if the placebo person, then this is called an event P. A typical events sequence so obtained would be

$$\text{CPPCPCCPPCPCPCCPP}\,. \tag{6.60}$$

Call the talley of the number of events C, P an event $(m_0, N - m_0)$. Here $N = 17$ and $m_0 = 8$, for a talley of $(8, 9)$.

If vitamin C had no effect, then there would tend to be as many events C as events P in the sequence. Suppose, instead, that vitamin C did have an effect. Then lopsided talleys such as $(13, 4)$ or $(4, 13)$ would result. *Suppose, then, that a talley* $(13, 4)$ *is observed.* (*Note*: This is only a hypothetical example.) Can we state with some degree of confidence that vitamin C had an effect? Call this this the "effect" hypothesis H_1.

It is in fact easier to instead test for the hypothesis H_0 that vitamin C *has no effect*. Call this the "chance" hypothesis. With no effect, each event in the sequence would have occurred as if the outcome of the flip of a fair coin (also see Sect. 11.5). That is, it has probability $p = 1/2$. Also, then the events sequence (6.60) is simply Bernoulli, and we can easily analyze the data talley. In particular, we can compute the probability of the talley $(m_0, N - m_0)$ on the basis of the binomial law (6.5) and the chance probability $p = 1/2$.

According to classical statistics, the chance hypothesis is taken to be rejected if the talley event $(m_0, N - m_0)$ is so lopsided that the probability of a chance talley value m either exceeding m_0 *or* being less than $(N - m_0)$ is very small. Calling this probability α, a value of $\alpha = 0.05$ is considered small enough to reject the hypothesis.

Hence, using

$$\alpha = \left(\sum_{m=m_0}^{N} + \sum_{m=0}^{N-m_0} \right) \binom{N}{m} p^m (1-p)^{N-m} \tag{6.61}$$

show that the chance hypothesis is indeed rejected on the level $\alpha = 0.049$. Vitamin C did have an effect.

The test statistic α given by (6.61) is actually equivalent to the α resulting from a two-tailed chi-squared test of the data. See Sect. 11.4.

6.1.31 A wave of monochromatic photons bounces back and forth between the endplates of a laser cavity. The endplates have an energy transmittance value of t. What is the probability $P(n)$ that a photon makes exactly n bounces before exiting the cavity?

7. The Monte Carlo Calculation

Let us reconsider the problem of determining the output probability law $p_Y(y)$ from a system, when the input law has a known form $p_X(x)$. This problem was originally considered in Chap. 5. We found there that the Jacobian approach called "transformation of a random variable" did the trick. Actually, this analytic approach was somewhat misleading. In a great number of real problems the system is too complicated to take an analytic approach. In these problems, the output RV y is the result of a great many intermediary events, each randomly conditional upon its predecessor. This causes the Jacobian approach to become extremely complicated and unwieldy since even the number of intermediary events that occur under such circumstances is a random variable. Hence, a different approach must be resorted to.

An alternative is called the "Monte Carlo" calculation in recognition of its wheel-of-fortune aspect. Invented by N.C. Metropolis and S. Ulam in 1949, it was used on the first electronic computer – the ENIAC (acronym for "Electronic Numerical Integrator and Computer") – to help develop the hydrogen bomb. The Metropolis algorithm is widely considered to be among the 10 algorithms that have had the greatest influence upon the development and practice of science and engineering in the 20th century [7.0].

The basis of the approach is the law of large numbers (2.8). This says that the probability $P(Y)$ of event Y can be known by observing the number m of times Y occurs, out of a large number N of trials. Hence, if Y is the system output event whose probability $p_Y(y)$ is desired, the latter could in principle be determined *by constructing* typical outputs Y of the system, and noting how many m of these lie in each test interval $(y, y + \mathrm{d}y)$. If enough such outputs are constructed, N will be large enough for the law of large numbers to predict $p_Y(y)$ to a desired degree of accuracy.

But how would such "typical outputs" be constructed? They should be typical of all the probability laws governing the transition from input to final output. More precisely, *they should be random samples from such laws.*

For example, consider the experiment where photons pass through a photographic emulsion, which is later developed to form an image. Any photon meets with many probability laws *en route* through the emulsion: it can be absorbed or scattered by the emulsion surface, then by a grain of the emulsion, then by another, etc., and each such phenomenon obeys its own probability law. Accordingly, the Monte Carlo procedure is to trace each photon through the experiment, and to construct a random *intermediary output*, in turn, at each such encounter, according to its probability law. (We shall show in the next section how to numerically carry this out.) Each such intermediary output state of the photon is made the intermediary *input* to the next, until the photon reaches the *final output* state which, in this case, is absorption somewhere or scattering out of the emulsion. In Sect. 7.2 we shall describe the Monte Carlo approach [7.1] to this emulsion problem. Other applications will also be given.

7.1 Producing Random Numbers That Obey a Prescribed Probability Law

At each step of a Monte Carlo calculation, the user must obtain a random sample from a known probability law. Call the latter $p_X(x)$.

On the other hand, most large computers are capable of generating random numbers of only one variety – *uniformly* random numbers over the interval (0, 1). For example, each call to the *CDC library function* RANF$(\cdot)$ generates a new number chosen *independently and uniformly* over the interval (0, 1).

The question arises, then, as to whether such uniformly random numbers may somehow be transformed by computer operation into numbers characterized by a required probability law, say the normal. In fact, we showed in Sect. 4.29 that the sum of three outputs from RANF is approximately normally distributed. More generally, we now want to know if a procedure can be found for transforming to *any* prescribed probability law, say $p_X(x)$.

Such a procedure may be found, as follows. Suppose that a uniformly RV y is defined by each output of RANF. Then y ranges over (0, 1). We seek a functional relation $y = f(x)$ which for each number y_i will have a root x_i that is characterized by the *prescribed density* $p_X(x)$. Now by the transformation law (5.4)

$$p_Y(y)\,\mathrm{d}y = p_X(x)\,\mathrm{d}x\ . \tag{7.1}$$

Note that we use the unique-root transformation. This is because we *demanded* that $x = f^{-1}(y)$ have a unique output number x_i for each given y_i.

Equation (7.1) simplifies to

$$\mathrm{d}y = p_X(x)\,\mathrm{d}x\ , \quad 0 \leqq y \leqq 1 \tag{7.2}$$

for our uniform case. But (7.2) may be regarded as a differential equation. Its solution is simply

$$y = \int_{-\infty}^{x} \mathrm{d}x'\, p_X(x') \equiv F_X(x)\ , \tag{7.3}$$

the cumulative probability function for RV x.

The required function $f(x)$ we sought for effecting the transformation is then $F_X(x)$. For each input y_i, the root x_i of (7.3) is the required output RV. The situation is illustrated in Fig. 7.1.

The virtues of this approach are that (a) it is exact, i.e., prescribed law $p_X(x)$ is exactly followed by number outputs $\{x_i\}$ (compare with the approximate approach

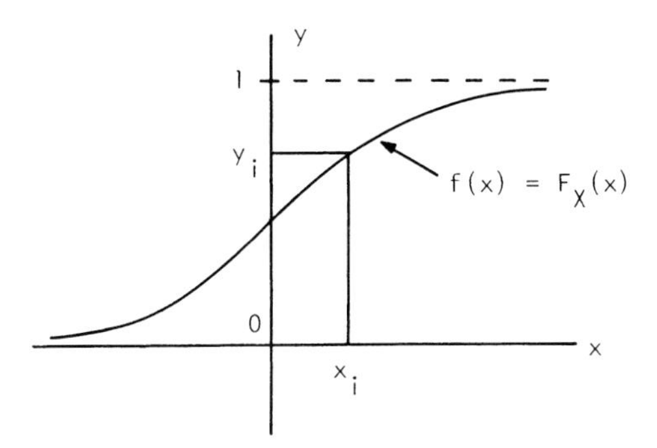

Fig. 7.1. Showing how to transform input numbers y_i from the uniform random number generator to outputs x_i having a prescribed density $p_X(x)$. The inputs are reflected through the cumulative law $F_X(x)$ to create the outputs

in Sect. 4.29); (b) it is easily programmed on the computer, either by direct use of the analytic function $F_X(x)$, if $p_X(x)$ is integrable, or if not, by numerically summing $p_X(x_r)$ over x_r to form a sequence of $F_X(x_r)$ numbers, which are stored in memory and interpolated as needed.

The main drawback is that the function $F_X(x)$ must somehow be inverted for its root x_i at each input y_i value, and this is not usually possible analytically (although see illustrative case below). A numerical procedure must then be found for solving

$$F_X(x_i) - y_i = 0 \tag{7.4}$$

for the root x_i. The well known Newton relaxation method is good for this purpose, because of its fast convergence property.

7.1.1 Illustrative Case

Since the trigonometric functions are easy to integrate and to invert analytically, let us suppose the required probability law to be

$$p_X(x) = \begin{cases} (a/2)\cos ax\ , & |x| \leq on\pi/2a \\ 0\ , & |x| > \pi/2a\ . \end{cases} \tag{7.5}$$

This is a bell-shaped curve [see $s(x)$ in Fig. 7.5]. By direct integration, it has a cumulative probability

$$F_X(x) = (1/2)(1 + \sin ax)\ .$$

According to the recipe (7.3) this expression is set equal to y, and then solved for $x(y)$ to define outputs x that follow the prescribed law. Doing so, we find that

$$x = a^{-1}\sin^{-1}(2y - 1)\ , \quad y \equiv \mathrm{RANF}(\cdot)\ . \tag{7.6}$$

Here, then, we have an *analytic answer* to the problem.

7.1.2 Normal Case

The normal law $p_X(x)$ would not be as easily accommodated. Its cumulative probability function is an error function (4.61) which cannot be analytically inverted for output x. Although the numerical approach suggested previously may be taken, or the approximate approach in Sect. 4.29, an easier method exists. It is both analytic and exact.

Instead of seeking a transformation of a *single* RV y to do the job, as we have been presuming, why not seek a function of two RV's y_1, y_2, each of which is uniformly random over (0, 1)? This, in fact, has an analytic answer.

Consider the transformation from (y_1, y_2) to outputs (x_1, x_2)

$$x_1 = \langle m \rangle + \sigma(-2 \ln y_1)^{1/2} \cos 2\pi y_2 \tag{7.7a}$$

$$x_2 = \langle m \rangle + \sigma(-2 \ln y_1)^{1/2} \sin 2\pi y_2 \, . \tag{7.7b}$$

Use of transformation law (5.18) with $r = 1$ root shows that the joint law $p_{X_1 X_2}(x_1, x_2)$ is a product of normal laws, each of mean $\langle m \rangle$ and standard deviation σ. Hence the marginal $p_{X_1}(x_1)$ or $p_{X_2}(x_2)$ is also a normal law, with mean $\langle m \rangle$ and standard deviation σ. Either of (7.7a, b) may be used to generate the required normal RV x.

7.2 Analysis of the Photographic Emulsion by Monte Carlo Calculation

The optical properties of photographic emulsions are very difficult to predict by analytic methods because of the large number of physical phenomena that characterize latent image formation. These properties, however, were accounted for by *Depalma* and *Gasper* [7.1] on the basis of a classic Monte Carlo calculation. They tested a *simple model* by which incoming photons are either absorbed or scattered in a random direction, as they interact with particles within the emulsion. By this model, the optical properties of a photographic emulsion would result from absorption and scattering of incoming photons by the surface overcoat, by the silver halide grains within the emulsion, and by the undercoat base. This is, of course, contingent upon use of the *correct* probability laws for absorption and scattering.

If the overall model is correct, the Monte Carlo calculation should predict a point spread function and modulation transfer function (MTF) for the emulsion which is *physically observable*. Or, in reverse, observing such an agreement would verify the correctness of the model, by the usual empirical method of physics. However, this would not necessarily mean that the test model is *unique* in producing such an agreement. Other models might do as well. The problem lies in the limited number of physical observables that are at hand. As is the usual case with the empirical method of physics, the more observables there are that can be confirmed, the more precise and unique can be the test model.

In order to carry through the calculation, all possible fates for a photon, as it passes through an emulsion, had to be enumerated. Figure 7.2 shows these possibilities. Furthermore, a probability law had to be assigned to each fate. These are given in [7.1].

A uniformly random number generated by the computer allows a fate to be defined, in turn, by each probability law. The method of Fig. 7.1 was used to transform

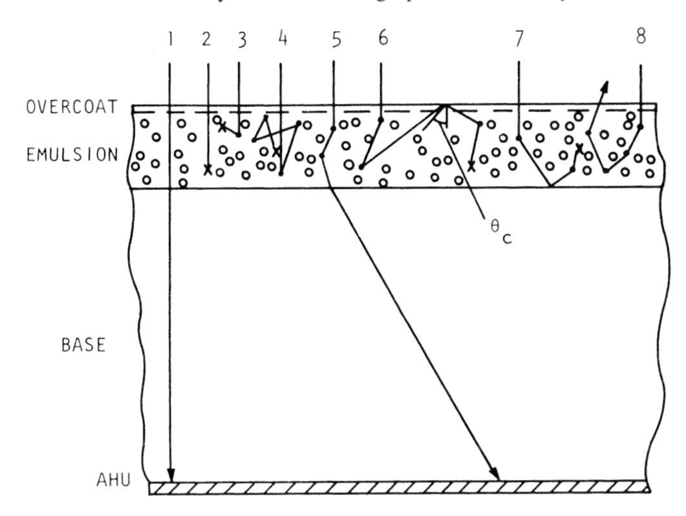

Fig. 7.2. Possible fates 1–8 for a photon. ∘ = silver halide grain; • = scattering event; × = absorption event; θ_c = critical angle [7.1]

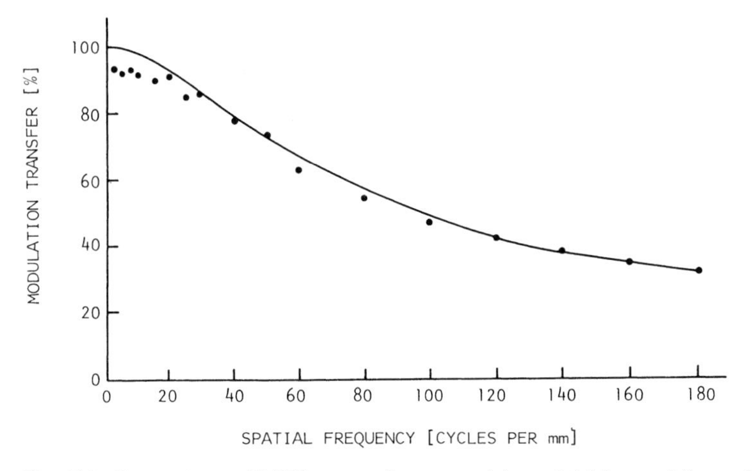

Fig. 7.3. Comparison of MTF curves for an emulsion of thickness 4.5 μm: (— — —) Monte Carlo result; • experimental [7.1]

each uniformly random number into one characterized by the known probability law. A given photon was followed through the emulsion in this manner until it was absorbed somewhere, was scattered out, or (arbitrarily) exceeded 250 events.

The point spread function was computed simply as the number of photon collisions within the volume dV of the emulsion lying beneath a differential surface area dA; i.e., $\mathrm{d}V = t \cdot \mathrm{d}A$, with t the emulsion thickness. The presumption is that each such collision will later develop a grain. MTF was computed as the modulus of the Fourier transform (3.64) of the point spread function. As with all Monte Carlo simulations, *the accuracy in these outputs improves with the number of photons traced through.* Specifically, *the standard deviation in the output goes inversely with the square-root of the photon number;* see (9.21). Acceptable accuracy in these quantities was accomplished by following the fates of a few thousand photons.

A typical MTF curve established in this way is shown by the solid curve in Fig. 7.3. The dotted curve is the *experimentally measured* MTF for the same emulsion. Agreement is seen to be quite good, implying that the interactive model is a good one.

7.3 Application of the Monte Carlo Calculation to Remote Sensing

Monte Carlo simulation has also been extensively used to predict radiative transfer through the ocean and clouds to a remote sensor above the atmosphere (see, e.g., [7.2]). The problem was to construct the curve of *upward directed radiance*, in this problem defined as number of photons/area-steradian, versus *angle of observation* (the "nadir" angle from horizontal). Physical conditions are as follows. The sun is presumed to have a fixed zenith angle θ, the sky is cloudless but contains "typical" water droplets, a constant wind speed fans the ocean, the ocean water follows an idealized "clear ocean model," and the ocean bottom is totally absorbing. *Because of the sheer number* of differing physical phenomena involved here, there is no hope for an analytic solution; *we have a tailor-made situation for a Monte Carlo approach.*

Here the calculation consists in following the fate of a photon as it passes down through the atmosphere, interacts with the ocean surface, interacts with the ocean water, and interacts with the ocean bottom. Each interaction is with a particle of the medium and is *described by a known probability law* governing *absorption or scattering*. The photons that make it back to the sensor form the required radiance curve.

Each radiance curve that was generated in this way followed the fate of about 1.5×10^6 interactions of photons with scattering and absorbing particles of the media involved. The result was curves of radiance versus nadir angle for a variety of solar zenith angles and wind speeds. Typical results are shown in Fig. 7.4. (Note: In these plots, angle ϕ denotes horizontal azimuthal angle from the sun's incident direction.)

A unique benefit of Monte Carlo methods pays off here. Because the user knows the fate of every photon everywhere in the atmosphere and ocean, he can likewise count photons crossing a small area *located anywhere. Therefore, he can compute the radiance at the top of the atmosphere (as shown in Fig. 7.4), or within the atmosphere, just above or just below the ocean surface, etc.* This property was exploited by *Plass* et al. [7.2] to produce radiance curves for a variety of positions within the atmosphere and ocean.

7.4 Monte Carlo Formation of Optical Images

An optical image is, in reality, a random superposition of a finite number of photons, *whose placement in the picture is in accordance with a probability law on position.* This probability law is, by the law of large numbers, the ideal image $i(x)$, i.e., the image which would be formed as the realization of an *infinity* of photon arrivals (Sect. 3.4). Hence $i(x)$ represents both an intensity of events at position x and, simultaneously, the probability of a photon arrival at x, i.e., the "probability of x."

This correspondence between ideal image and probability density arises *physically* out of the quantum nature of light. For example, the diffraction pattern due to a point object (in particular) is well known to also represent the probability density for placement of photons.

Since we have been dealing with the event-by-event simulation of probability laws, it might therefore seem possible to simulate the formation of an optical image by random photon arrivals. We consider this problem next.

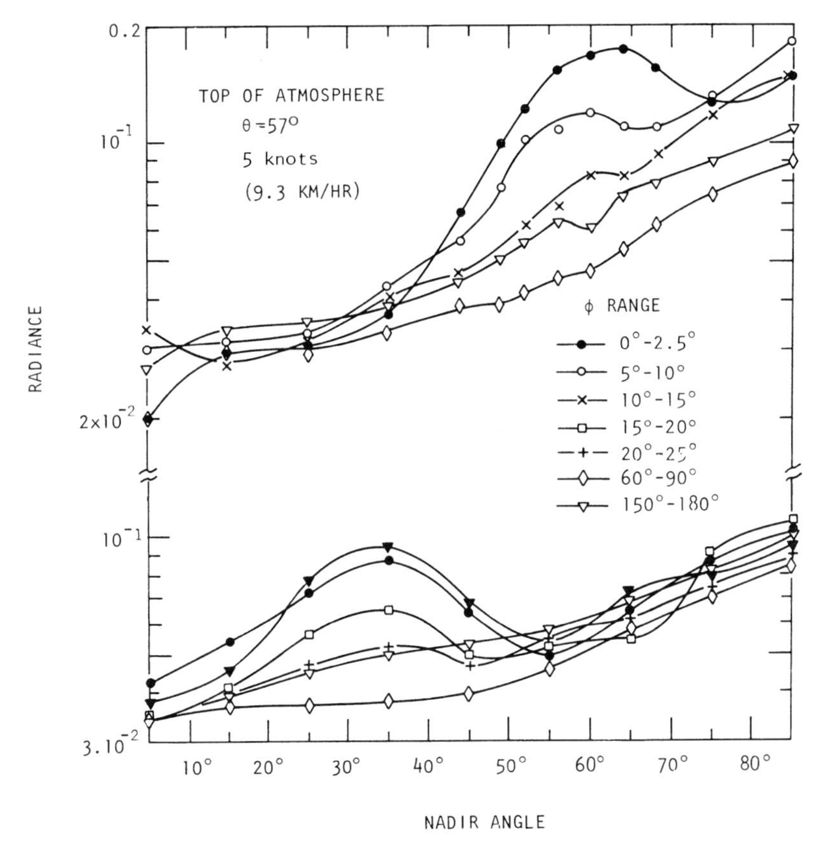

Fig. 7.4. Upward directed radiance at top of atmosphere, for $\theta = 32°$ (*lower curves*) and $\theta = 57°$ (*upper*), and a wind speed of 5 knots [7.2]

The probability density $i(x)$ for the image relates to the object $o(x)$ and point spread function $s(x)$ by the convolution relation (4.22)

$$i(x) = o(x) \otimes s(x) \; . \tag{7.8}$$

Now for the particular case of $o(x) = \delta(x)$, this equation states that $i(x) = s(x)$. Hence, $s(x)$ represents, by the reasoning above, the *probability density* for position in the image due to a point source.

Furthermore, by similar reasoning $o(x)$ represents the probability in x for *radiating* photons (Sect. 3.4).

Hence, the *image-formation equation (7.8) alternatively represents the convolution of two probability densities to form a third.* But then, by (4.17) and (4.19), it also describes the probability of position value x_i in the image, where

$$x_\mathrm{i} \equiv x_\mathrm{o} + x_\mathrm{s} \; , \tag{7.9}$$

x_o a random position in the object, and x_s independently a random position in the spread function.

The latter relation tells us how to simulate random formation of the image due to known object and spread functions (probability laws). *First generate a random sample x_o from probability law $o(x)$, then a sample x_s from law $s(x)$, and add the two samples.* The result is a position coordinate x_i whose frequency of occurrence, and hence whose image, follows the physical effect (7.8) which we wanted to simulate.

The final problem is how to generate the two random samples x_o and x_s. If, as is usual, only the *uniform* random number generator is present, the approach taken in Fig. 7.1 may be used. For example, the cumulative probability law

$$F_X(x) = \int_{-\infty}^{x} \mathrm{d}x \, o(x) \tag{7.10}$$

would be used to generate values x_o. Note the further result that to generate values x_s, a cumulative law

$$F_S(x) = \int_{-\infty}^{x} \mathrm{d}x \, s(x) \; . \tag{7.11}$$

would be used, precisely the *optical edge response function.* This is another interesting correspondence that ultimately arises out of the linear theory that is common to both image formation and probability theory (Chap. 4).

7.4.1 An Example

Suppose we want to Monte Carlo-simulate the photon image due to an object which is a rectangle function, and a spread function which is the cosine bell in (7.5). These are shown along the top of Fig. 7.5. By directly convolving them, the exact or signal image is obtained. This is the raised cosine shown at the right in the figure.

The random photon image is now formed, using the recipe of the preceding section: take a random sample x_o from probability law $o(x)$, independently take a random sample x_s from probability law $s(x)$, add the two to form x_i; and repeat the procedure L times to form the histogram of events x_i. *This histogram is the required photon image due to L photon arrivals.*

With an object length D, and a the free parameter in (7.5) for $s(x)$, values x_o and x_s were formed from outputs of the uniformly random number generator RANF as

$$x_o = D \cdot \mathrm{RANF}(\cdot) \,, \tag{7.12a}$$

and

$$x_s = a^{-1} \sin^{-1}[2\,\mathrm{RANF}(\cdot) - 1] \,, \tag{7.12b}$$

the latter from (7.6). Equation (7.12a) expresses the fact that since RANF($\cdot$) is uniformly random between 0 and 1, $D \cdot$ RANF must be uniformly random between 0 and D, as required for the rectangular object $o(x)$.

Results of this simulation are shown along the bottom of Fig. 7.5, for three different values of L. These noisy images are to be compared with the signal image $i(x)$ in the figure (upper right). We see that for L as small as 100 the random image fluctuates wildly about $i(x)$, while for L of size 2500 or more the fluctuations start becoming insignificant.

As an interesting byproduct of this exercise, we have accomplished the mathematical convolution $o(x) \otimes s(x)$, *without explicitly having to perform the convolution operations* of shift, multiply, and add. Oddly enough, random numbers formed the output instead, just as in nature, through simple additions (7.9).

7.5 Monte Carlo Simulation of Speckle Patterns

We saw in Fig. 5.3 what a speckle pattern looks like. Suppose we now want to *simulate* a typical speckle intensity pattern upon the computer. How may this be

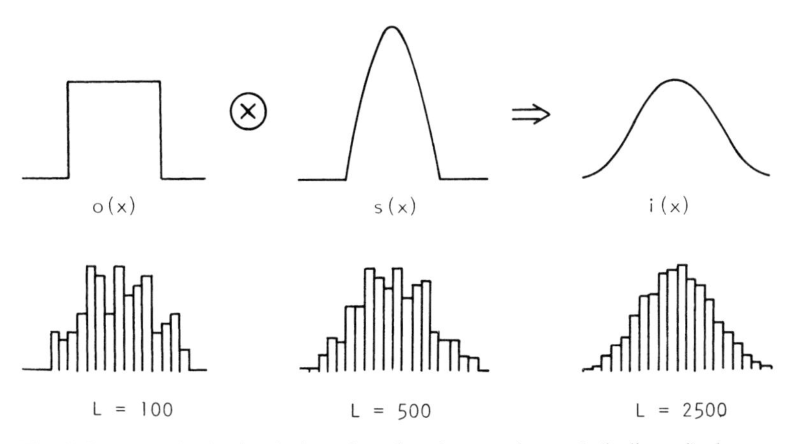

Fig. 7.5. Monte Carlo simulation of random images due to L (indicated) photon arrivals. The top-right image is the ideal image ($L = \infty$)

done? As usual, suppose we have access to the uniformly random numbers between 0 and 1 generated by the library function RANF(·). Let the latter be denoted by values y.

We know the probability law obeyed by speckle intensity to be

$$p_I(v) = \sigma^{-2} \exp(-v/\sigma^2) . \tag{7.13}$$

This was derived at (5.32a). Then we may use the approach of Fig. 7.1 to generate typical values of intensity from RANF numbers y. First the cumulative probability law must be formed. From (7.13),

$$F_I(x) = \sigma^{-2} \int_0^x dv \exp(-v/\sigma^2) = 1 - \exp(-x/\sigma^2) .$$

Therefore by (7.4) we set

$$1 - \exp(-x/\sigma^2) = y .$$

Remarkably, this has an analytic solution

$$x = -\sigma^2 \ln(1 - y) . \tag{7.14}$$

Quantity x is the simulated intensity value in question.

In order to obtain a *spatial* intensity pattern, it is necessary to place the generated intensities x at positions in the image plane. How should this be done? According to a model of *Goodman* [7.3], a speckle image consists of uncorrelated speckles of a characteristic size a. Hence, the simulated intensity values x (which are indeed uncorrelated since RANF values are) should simply by placed, in turn, at positions (x_m, y_n) spaced distance a apart in the image plane. An $N \times N$ point simulation will result.

A speckle pattern also has *phase properties* pointwise across the image. How may these be simulated? It was shown at (5.32b) that the phase Φ at any point is uniformly random between $-\pi$ and $+\pi$. Hence, since the y values are themselves uniformly random, we may simply use numbers x obeying

$$x = \pi(2y - 1) \tag{7.15}$$

to represent the phase.

Computer simulation of speckle patterns has been carried through by *Fujii* et al. [7.4]. They questioned the model of *independent* scattering spots in the diffuser, based on the reasoning that a smooth diffuser, as encountered in high-quality optical systems, ought to have significant correlation across its face. Accordingly, they simulated formation of the speckle image as if the diffuser were Gaussian random at each point but correlated from point to point. The authors note that their resulting speckle images are more in compliance with experimental observation than when *independent* Gaussian randomness is assumed in the simulation.

Exercise 7.1

7.1.1 If y = RANF(·), find a function $g(y) \equiv x$ such that RV x follows an exponential probability law

$$p_X(x) = \begin{cases} a^{-1}\exp(-x/a) & \text{for } x \geqq 0 \\ 0 & \text{for } x < 0\,. \end{cases}$$

7.1.2 Repeat problem (1) if x is instead to follow a *bimodal* (double humped) probability law

$$p_X(x) = (4a)^{-1}\mathrm{Rect}\left(\frac{x-a}{2a}\right) + (4a)^{-1}\mathrm{Rect}\left(\frac{x-4a}{2a}\right)\,.$$

7.1.3 Repeat problem (1) if x is instead to follow a triangle probability law

$$p_X(x) = \begin{cases} a^{-2}x + a^{-1} & \text{for } 0 \geqq x \geqq -a \\ -a^{-2}x + a^{-1} & \text{for } 0 \leqq x \leqq a\,. \end{cases}$$

7.1.4 (Continuation) In most cases, the required form for $p_X(x)$ does not permit a simple function $g(y)$ to represent the output x. This is true of the cases below. For these cases, show what algebraic equation in x and y would have to be solved. Do not attempt to solve them explicitly as in problems (1) through (3) above.

(a) $x = N(0, \sigma^2)$

(b) $p_X(x) = \begin{cases} a^{-2}x\exp(-x/a) & \text{for } x \geqq 0 \\ 0 & \text{for } x < 0\,. \end{cases}$

7.1.5 Write a computer program that generates speckle intensity patterns. Use RANF (or its equivalent on your computer) to generate uniformly random numbers y, and use these in (7.14) to generate the corresponding speckle intensity values. Display these in any convenient manner.

7.1.6 ***Honors problem.*** Carry through the simulation of an optical image $i(x)$ as in Sect. 7.4.1 if, instead of the simple rectangle assumed there for the object, this is replaced with a bimodal object as in problem (2) above. By varying the number L of photons, observe the effect of low light level upon resolution of the two humps in the object.

7.1.7 ***Honors problem.*** Photons are traveling within a certain medium, from generally top to bottom. All photons enter the medium initially heading straight down and colinear. The medium has a random placement of scattering particles within it, and is of finite height. We want to establish, by a Monte Carlo calculation, the probability law on angle of divergence from the initial direction as the photons leave the scattering medium.

We are given the following probability laws modelling the scattering medium. The probability of a photon encountering a particle after travelling distance r (in any direction) obeys

$$p_R(r) = \begin{cases} \bar{r}^{-1} \exp(-r/\bar{r}) & \text{for } r \geq 0 \\ 0 & \text{for } r < 0 \, . \end{cases}$$

The total height of the medium is h. Assume its width is infinite.

When a photon encounters a particle, there is a probability P_A of being absorbed, and a complementary probability P_S of being scattered, where

$$P_A = 0.1 \, , \quad P_S = 0.9 \, .$$

If a photon is absorbed, that ends its history. It is not detected at the output. If a photon is scattered, it can only be scattered within a fixed plane (say, that of the page). At event i, its probability of being scattered by angle θ_i, from an initial direction obeys

$$p_\theta(x) = \begin{cases} (a/2) \cos ax & \text{for } |x| \leq \pi/2a \\ 0 & \text{for } |x| > \pi/2a \, . \end{cases}$$

Write a computer program that builds up, event by event, the histogram of output angles α from the initial direction as the photons leave the medium.

Experiment with different values of parameter a governing the amount of scattering at each encounter. Also, vary the height h of the medium. How should the σ of your histogram vary with a and with h?

7.1.8 ***Histogram equalization*** is a common method of image processing [7.5]. Its aim is to bring out features in an image that are hidden by (usually) an underexposed state.

Let x denote a general intensity value in the image. The given image has a probability law $p_X(x)$ for intensity values x that may be approximated by its histogram of intensities. If the picture is underexposed, the values x will nearly all be bunched close to a small value x_0 of x, causing histogram $p_X(x)$ to be compressed, as in Fig. 7.6a.

On the other hand, had the picture been taken at the proper exposure level, it would have exhibited all intensities x about equally, so that $p_X(x)$ would be nearly flat, as in Fig. 7.6b.

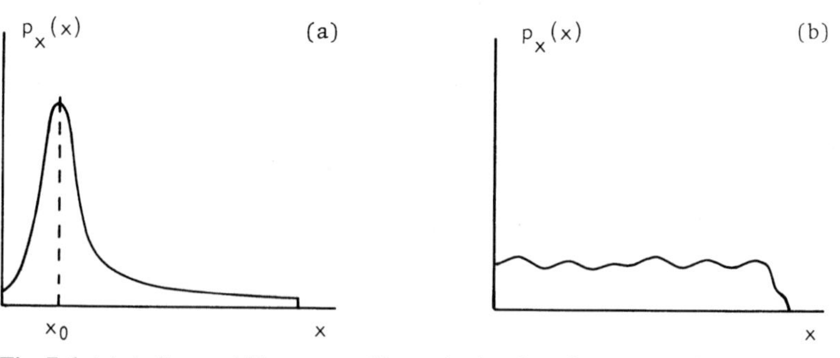

Fig. 7.6. (**a**) A distorted histogram of intensity levels x due to an underexposed condition; (**b**) a nearly ideal histogram of intensities

This leads us to ask whether it is possible to somehow *transform* the underexposed picture into one whose histogram *is* flat. The underexposed features, originally close to intensity x_0, would be transformed into brighter values of x and hence might become visible.

Thus the problem: Denoting an intensity in the transformed picture by y, does a transformation $y = f(x)$ exist such that the new histogram obeys $p_Y(y) = \text{constant}$? Note that the *operation $f(x)$ is a simple one-to-one mapping of each intensity value in the given picture into a new one in the output.* To preserve features from input to output, the new intensity y should be placed at the same position in the scene as the old value x (although this is not required to attain a flat histogram).

Show that the intensity mapping function $f(x)$ is precisely the cumulative histogram $F_X(x)$ in Sect. 7.1.

7.1.9 ***Beam transformations.*** Laser beams are commonly used for the illumination of extended scenes. But if the beam is in a single mode such as TEM_{00} its intensity as a function of radius r is not uniform. Hence, the scene is not uniformly illuminated, and this may be undesirable. A remedy is to *redistribute the laser beam*, ray by ray, such that the redistributed intensity profile is indeed uniform (over a finite area). Hence, a ray at radius r in the laser beam is to be specified to go to radius ϱ in the plane of illumination (Fig. 7.7). *What functional relation $\varrho = \varrho(r)$ should be obeyed to accomplish a uniformly irradiated disk of radius ϱ_0?* The total light flux in the disk will equal that in the laser output.

The general approach of Sect. 7.1 provides the answer. First, we recognize the usual correspondence between light intensity and probability: intensity as a function of r represents the probability of a photon being at radius r. Hence, *the problem of redistributing light energy is in reality one of transforming a probability law.* Let $p_R(r)$ represent the initial (suitably normalized) laser irradiance. Let $p_P(\varrho)$ represent the transformed laser irradiance in the illumination plane. We want the event "a photon is within the ring $(r, r+\mathrm{d}r)$" to occur one to one with the transformed event "that photon is within the ring $(\varrho, \varrho + \mathrm{d}\varrho)$." Hence, as often as the former happens so does the latter, or

$$p_P(\varrho) \cdot 2\pi\varrho\, \mathrm{d}\varrho \equiv p_R(r) \cdot 2\pi r\, \mathrm{d}r \ . \tag{7.16}$$

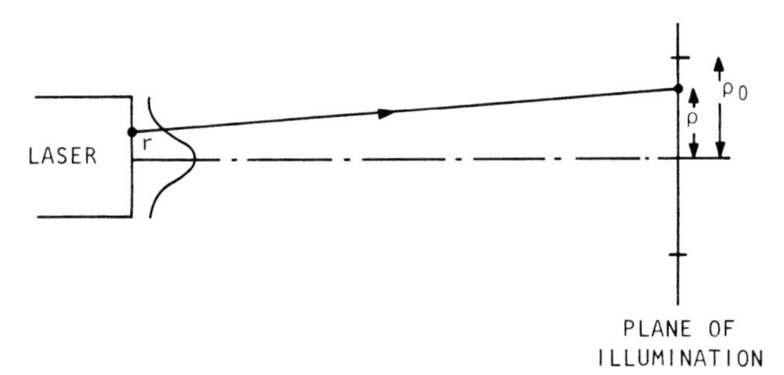

Fig. 7.7. What ray relation $\varrho(r)$ will accomplish uniform irradiance in the image plane within the circle $0 \leqq \varrho \leqq \varrho_0$?

[Compare with (7.1).] But the requirement that

$$p_{\mathrm{P}}(\varrho) = \text{constant for } 0 \le \varrho \le \varrho_0$$

defines

$$p_P(\varrho) = (\pi\varrho_0^2)^{-1}\mathrm{Rect}[(\varrho - \varrho_0/2)/\varrho_0] . \tag{7.17}$$

Using (7.17) in (7.16) and solving the resulting differential equation, as in Sect. 7.1, provides the solution $\varrho(r)$. *Show that this is*

$$\varrho(r) = \pm\varrho_0 \left[\int_0^r \mathrm{d}r' 2\pi r' p_{\mathrm{R}}(r')\right]^{1/2} \tag{7.18}$$

(one sign must be chosen for all r).

Also show that in the particular case

$$p_{\mathrm{R}}(r) = (2\pi\sigma^2)^{-1}\exp(-r^2/2\sigma^2) \tag{7.19}$$

of a Gaussian laser beam, the solution is

$$\varrho(r) = \pm\varrho_0[1 - \exp(-r^2/2\sigma^2)]^{1/2} . \tag{7.20}$$

Methods of accomplishing the ray relations $\varrho(r)$ are given in [7.6].

7.1.10 ***Photons in a random medium: analytic solution.*** Exercise 7.1.7 actually has an analytic solution, which the reader may compare with his Monte Carlo results. Output angle α is $\theta_0 + \cdots + \theta_m$, with m a *random number* of interactions, unfortunately. However, the problem is now essentially "cascaded events," *solved* in Exercise 6.1.21 in terms of a known $P(m)$. How may we find $P(m)$ here?

Let x_i denote a photon's longitudinal distance $r_i \cos\theta_i$ into the medium at interaction i. Event m is defined by the joint occurence of events $0 \leqq x_i \leqq h$ for $i = 0, \ldots, m-1$, but $h \leqq x_m \leqq \infty$, conditional upon no absorption. Then by (2.11) and (3.1), show that

$$P(m) = Q(m)/\sum_{m=0}^{\infty} Q(m) ,$$

$$Q(m) = P_{\mathrm{S}}^m \int_0^h \mathrm{d}x_1 \cdots \int_0^h \mathrm{d}x_{m-1} \int_h^\infty \mathrm{d}x_m\, p(x_1, \ldots, x_m) , \tag{7.21}$$

$$m = 1, 2, \ldots .$$

But each $x_i = x_{i-1} + t_i$, the RV's $\{t_i\}$ i.i.d. from a fixed law $p_X(t)$. Then $\{x_i\}$ constitute a Markov events sequence. Show that then, by identity (2.35), (7.21) becomes

$$Q(m) = P_{\mathrm{S}}^m \int_0^h \mathrm{d}t_1\, p_X(t_1) \ldots$$

$$\int_{-t_1-\cdots-t_{m-2}}^{h-t_1-\cdots-t_{m-2}} \mathrm{d}t_{m-1}\, p_X(t_{m-1}) \int_{h-t_1-\cdots-t_{m-1}}^{\infty} \mathrm{d}t_m\, p_X(t_m) ,$$

$$m = 1, 2, \ldots . \tag{7.22}$$

for general $p_X(t)$.

If $p_X(t)$ is, in particular, exponential with mean distance $\langle x \rangle$, show that $P(m)$ becomes *precisely Poisson*, with mean $hP_S/\langle x \rangle$. Then the required statistics in α obey simply relations (6.50), (6.51).

7.1.11 ***How to form random variables that correlate, (1).*** Many simulations require the construction of pairs of random variables x, y that correlate by a desired amount ϱ. One convenient way is as follows.

Key effect: Suppose that x is sampled normally as $x = N(a, \sigma^2)$, n is sampled normally as $n = N(0, \sigma_N^2)$, and y is generated from them as $y = x + n$. Then if n is constrained to be small (by choosing σ_N^2 small) $y \approx x$. This means that y has close to perfect correlation ($\varrho \approx 1$) with x. Conversely, if σ_N^2 is chosen to be large, the randomly large values of n that result will blind y to corresponding values of x. Thus, x and y will tend not to correlate, or $\varrho \approx 0$.

This suggests that the correlation coefficient can be adjusted at will by adjusting the value of σ_N. In fact, what matter is the ratio σ_N/σ. For a prescribed value of ϱ, the ratio must obey

$$\frac{\sigma_N}{\sigma} = \sqrt{\frac{1-\varrho^2}{\varrho^2}}. \tag{7.23}$$

Show this.

7.1.12 ***How to form random variables that correlate, (2).*** A simple, graphical way to accomplish correlation is as follows. Let RV's x and y initially be independent, obeying marginal laws $p_X(x)$ and $p_Y(y)$. Draw x and y coordinate axes. Draw the parallel lines $y = x - \epsilon$ and $y = x + \epsilon$ with $\epsilon = const$. Randomly choose events x and y according to the given marginal laws. However, only accept as "events" cases (x, y) that lie *in the region between* the two lines $y = x - \epsilon$ and $y = x + \epsilon$. Notice that with ϵ sufficiently small these events obey $y \approx x$. Therefore the events must correlate. Compute the correlation coefficient ϱ when x and y are i.i.d. uniform over an interval $(0, a)$. See also the next problem.

7.1.13 ***How to form a photon-depleted, two-dimensional image.*** We found in Sect. 7.1 how to randomly sample from a one-dimensional probability law. We found in Sect. 2.10 that images are probability laws. Images are, in fact, two-dimensional. How, then, may a random sample from a two-dimensional probability law $p(x, y)$ be formed? (*Note*: In that RV's x and y in general correlate, this problem complements the preceding two in giving a third way to enforce a required degree of correlation.)

A random sample of a one-dimensional law $p_X(x)$ is accomplished by the use of Eqs. (7.3) and (7.4). It would be convenient to produce a two-dimensional photon sample by two such one-dimensional steps. In fact this is possible. By definition (3.5), a two dimensional law $p(x, y)$ is equivalent to a product of laws $p_X(x)p(y|x)$. This suggests first generating the coordinate value x of a photon from the law $p_X(x)$ and then, *for that value of* x, generating a corresponding coordinate y from the law $p(y|x)$. The latter is of course one-dimensional in the RV y.

For a discretized image $\{i_{mn}\}$ this entails summing across each row m to form the marginal i_m, and sampling from this one-dimensional probability law to get a row value $m = m_0$. Next, form the division $\{i_{m_0 n}/i_{m_0}\}$ at each n to form the associated probability law $i(n|m_0)$ on column values n. Then randomly sample a column value of $n = n_0$ from this law. This locates the random photon position (m_0, n_0). The output image is incremented at that pixel position. The procedure is iterated over the N photons that are to form the image.

Write a computer program that constructs an image by randomly selecting photon positions (m_0, n_0) in this way. The use of only a few thousand photons will produce an image that very closely resembles a typical photon-depleted image, such as one sees at low light-levels.

8. Stochastic Processes

Stochastic processes are widespread in optics. Whenever an optical signal exists, if there is some uncertainty in knowing exactly which signal is present it may be regarded as a stochastic process. The signals may exist in space, time, or frequency. Examples are spatial photographic images, photon pulse shapes in time due to multipath propagation through clouds, and the MTF characteristics of emulsions. Even the income distribution of attendees at the *coming* Optical Society meeting is a stochastic process. The uncertainty in choice of signal ultimately lies in the choice of some parameters λ peculiar to the signal, as described next.

8.1 Definition of a Stochastic Process

We are all familiar with the concept of a "family of curves." For example, $f(x) = \lambda/x$ is a family of hyperbolas, as in Fig. 8.1. The individual members of the family are defined by their λ values.

If λ is an RV, then $f(x)$ is denoted $f(x; \lambda)$ and is called a "stochastic process." Note that $f(x)$ does not have to "look" random in x for it to be stochastic. *The randomness lies in which curve $f(x)$ of the family λ was chosen.*

A good example is the case of a radar signal return. This is of the form $f(t; d) = \cos(\omega t - 2kd)$, ω the frequency and $k = 2\pi/\text{wavelength}$. Parameter d is the random distance to the unknown source. Hence, radar returns have random retardation.

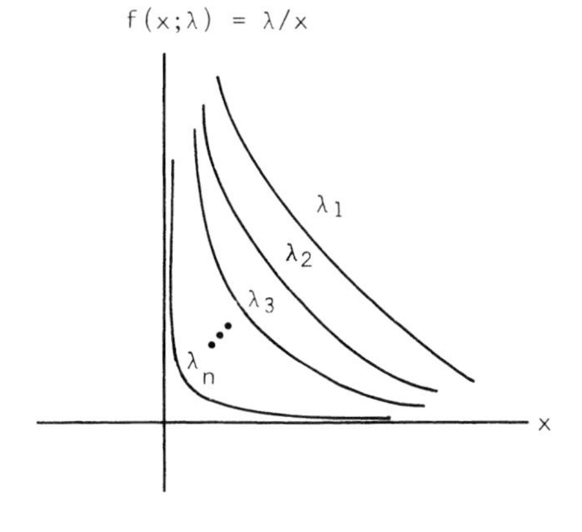

Fig. 8.1. A stochastic family of hyperbolas $f(x; \lambda)$

However, no matter what value d takes, the form of $f(t; d)$ is still a smooth curve, the cosine function.

On the other hand, some stochastic functions do have a random appearance. An example is "additive noise," described below.

In the most general situation, λ is a vector $\boldsymbol{\lambda}$ of random parameters. Then each component λ_n of $\boldsymbol{\lambda}$ is called a "stochastic component" of the process $f(x; \boldsymbol{\lambda})$. An example is the optical transfer function $T(\omega; \boldsymbol{\Phi})$, where ω is a spatial frequency and $\boldsymbol{\Phi}$ is a random phase function across the pupil (see (3.65) with $k\Delta \equiv \boldsymbol{\Phi}$). Each point in the pupil defines a new stochastic component Φ_n.

Consider a stochastic process $f(x; \boldsymbol{\lambda})$. It is also usual to call it an "ensemble" of functions $f(x)$. Each member of this ensemble is defined by a new random vector $\boldsymbol{\lambda}$. All averages, such as $\langle f(x)\rangle$, shall be over the random values of $\boldsymbol{\lambda}$, and hence are called "ensemble averages."

Naturally, all the rules we have developed for handling RV's will apply to RV's $\boldsymbol{\lambda}$ Moreover, no extra methods or rules will be needed for analyzing stochastic processes, except for a logical extension of identity (3.28). Identity (3.28) states that the mean value of a sum equals the sum of the mean values. This may be extended to evaluating the mean value of an *integral*, by merely replacing the word "sum" with "integral" in (3.28). An integral is, of course, merely a limiting form of a sum.

Hence, *the theory of stochastic processes is nothing more than the application of probability theory,* as developed in preceding chapters, *to signals containing parameters that are* RV's.

Note that in this chapter x has the role of a continuous abscissa, as in the hyperbolas above, and is not a random variable. Thus, averaging brackets $\langle\rangle$ denote averages over $\boldsymbol{\lambda}$ and not over x.

Two central concepts of the theory of stochastic processes are the power spectrum and the autocorrelation function. These are defined next.

8.2 Definition of Power Spectrum

The finite Fourier transform of a stochastic process $f(x; \boldsymbol{\lambda})$ is defined as

$$F_L(\omega; \boldsymbol{\lambda}) \equiv \int_{-L}^{L} \mathrm{d}x\, f(x; \boldsymbol{\lambda})\, \mathrm{e}^{-\mathrm{j}\omega x}\,, \quad \mathrm{j} = \sqrt{-1}\,. \tag{8.1a}$$

The power spectrum $S_f(\omega)$ is defined in terms of this quantity as

$$S_f(\omega) \equiv \lim_{L\to\infty} \left\langle |F_L(\omega; \boldsymbol{\lambda})|^2 \right\rangle / 2L\,. \tag{8.1b}$$

Brackets $\langle\rangle$ denote an ensemble average over the RV's $\boldsymbol{\lambda}$. Typically these are random in time, although this depends upon the application. Division by $2L$ will be needed in order to preserve a finite value for $S_f(\omega)$ in the case of a spatially uncorrelated process $f(x; \boldsymbol{\lambda})$ (Sect. 8.5).

Noticing that in the limit $L \to \infty$ the finite Fourier transform (8.1a) becomes the ordinary Fourier transform, we see that (8.1b) actually defines the power spectrum

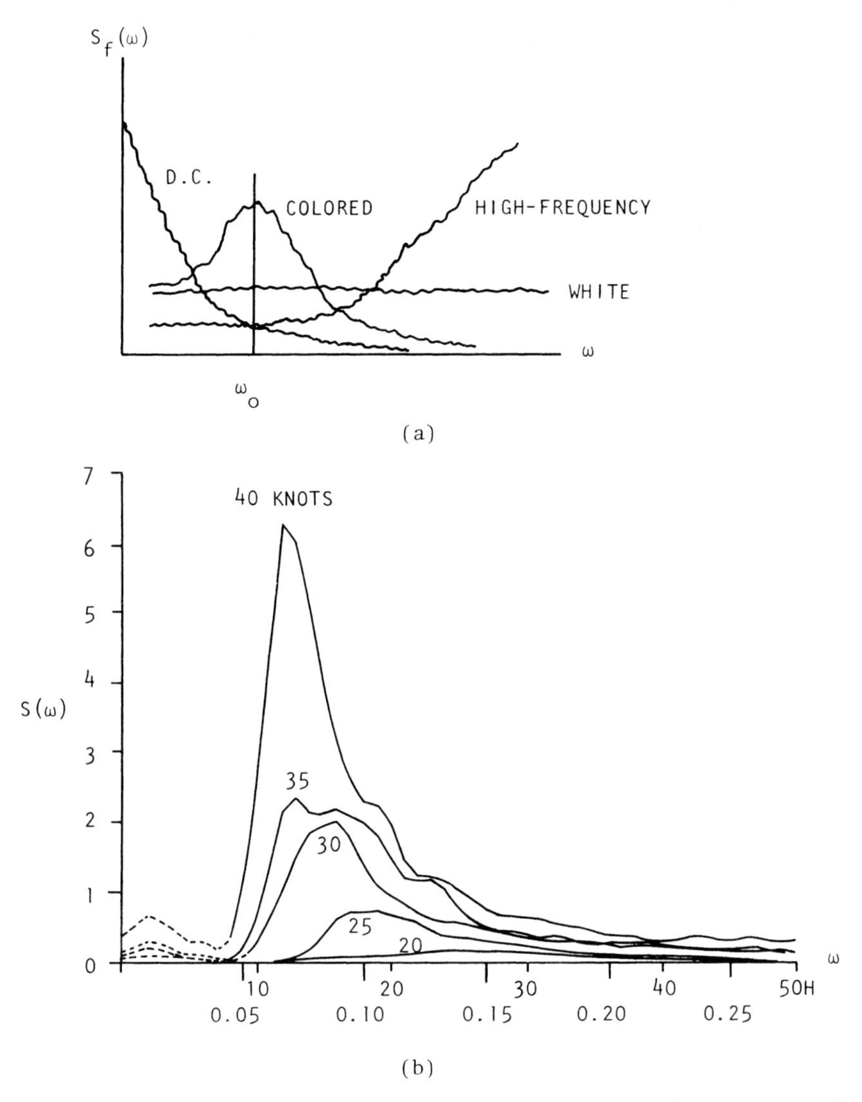

Fig. 8.2. (**a**) Some basic kinds of power spectra. (**b**) Power spectra for ocean waves due to different wind speeds

as *the average modulus-squared of the ordinary spectrum of* $f(x; \boldsymbol{\lambda})$, divided by an infinite length to preserve finiteness. It is usually easier to think of the power spectrum in this way than to visualize (8.1b) directly. Hence,

$$S_f(\omega) \propto \left\langle |F(\omega; \boldsymbol{\lambda}|^2 \right\rangle . \tag{8.2a}$$

A power spectrum $S_f(\omega)$ describes the allocation of average power to the various frequencies comprising $f(x)$. (See Fig. 8.2a.) If S_f is high at high frequencies, this means that $f(x)$ has a lot of high-frequency wiggles. If S_f is higher at lowest frequencies, then $f(x)$ has a lot of constant or D.C. content. If S_f tends to be peaked about some frequency ω_0, we say that it exhibits "color," in analogy to the case of optical temporal signals. Conversely, if S_f tends to be flat with frequency, it is usually called a "white" power spectrum, again in analogy with the temporal case.

8.2.1 Some Examples of Power Spectra

In the theory of optical turbulence, the power spectrum that is widely used is the *Kolmogoroff spectrum*

$$S(\omega) = 8.16\, C_N^2 \omega^{-11/3}\,, \tag{8.2b}$$

where C_N^2 is a constant called the "refractive-index structure constant." Note that (8.2b) has a pole at the origin. This means that its Fourier transform does not exist, and in particular it does not have a well-defined "autocorrelation function" (see subsequent sections). A way around this was found by *Tatarski* [5.9], who multiplies (8.2b) by a factor

$$\exp(-\omega^2/\omega_m^2) \tag{8.2c}$$

where $\omega_m \equiv 5.92/I_0$; parameter I_0 is called the "inner scale of turbulence." The net $S(\omega)$ now has a well-defined Fourier transform.

A second commonly used power spectrum occurs in remote sensing problems, where "sea state" or degree of surface turbulence is desired. Estimates of sea state often use the power spectrum for ocean waves. These are functions of the wind speed directly over the ocean. For low wind speeds, basically all ocean wavelengths occur in equal amounts, and the spectrum is flat. As wind speed increases, the ocean becomes more turbulent and waves of one characteristic length become predominant. This allows for the sport of surfing. These effects are shown in Fig. 8.2b.

8.3 Definition of Autocorrelation Function; Kinds of Stationarity

The autocorrelation $R_f(x; x_0)$ of a stochastic process $f(x; \boldsymbol{\lambda})$ is defined as

$$R_f(x; x_0) \equiv \left\langle f(x_0; \boldsymbol{\lambda}) f^*(x_0 + x; \boldsymbol{\lambda}) \right\rangle\,. \tag{8.3}$$

It is therefore the average, over RV's $\boldsymbol{\lambda}$, of f at some arbitrary point x_0 times f^* at a point a distance x away. The * denotes complex conjugate. This is often redundant notation since most processes f are purely real.

If $R_f(x; x_0)$ equals $R_f(x)$ regardless of position x_0, and if the mean value of $f(x; \lambda)$ is independent of x (i.e., is a constant), the process f is called "wide-sense or weakly stationary." Algebraically,

$$\left\langle f(x; \boldsymbol{\lambda}) f^*(x'; \boldsymbol{\lambda}) \right\rangle = R_f(x - x')\,, \quad \langle f(x; \boldsymbol{\lambda}) \rangle = \text{constant}\,. \tag{8.4}$$

Or, the autocorrelation depends only upon the separation $x - x'$ of the two test points, and not upon their absolute positions in x-space.

A closely related concept to the autocorrelation function R_f is the "autocorrelation coefficient," defined as

$$\rho_f(x; x_0) \equiv \left\langle [f(x_0; \boldsymbol{\lambda}) - \langle f(x_0; \boldsymbol{\lambda})\rangle][f^*(x_0 + x; \boldsymbol{\lambda}) - \left\langle f^*(x_0 + x; \boldsymbol{\lambda})\right\rangle]\right\rangle . \tag{8.5a}$$

This is the correlation in fluctuations of f from its means at the two points. It differs conceptually from ρ of (3.47) only in not being normalized by variance $\sigma_f(x_0)^2$. The autocorrelation coefficient may be expanded out to

$$\rho_f(x; x_0) = R_f(x) - |\langle f(x_0; \boldsymbol{\lambda})\rangle|^2 , \tag{8.5b}$$

provided the process f is wide-sense stationary, so that mean $\langle f(x_0; \boldsymbol{\lambda})\rangle$ equals $\langle f(x_0 + x; \boldsymbol{\lambda})\rangle$.

A stronger form of stationarity than wide-sense is "strict-sense stationarity." Consider the n stochastic processes $f_1(x_1; \boldsymbol{\lambda}), f_2(x_2; \boldsymbol{\lambda}), \ldots, f_n(x_n; \boldsymbol{\lambda})$. In general, the processes are different, and so are their observation points x_i, $i = 1, 2, \ldots, n$. The processes have a joint probability density

$$p(f_1, f_2, \ldots, f_n) .$$

Next, consider the n *shifted* stochastic processes $f_1(x_1 + x; \boldsymbol{\lambda}), f_2(x_2 + x; \boldsymbol{\lambda}), \ldots,$ $f_n(x_n + x; \boldsymbol{\lambda})$. Shift x is arbitrary. They have a probability density

$$p(f_1', f_2', \ldots, f_n') ,$$

where the primes mean evaluation at shifted points $x_i' = x_i + x$. By strict-sense stationarity we mean that

$$p(f_1, f_2, \ldots, f_n) = p(f_1', f_2', \ldots, f_n') \tag{8.6}$$

regardless of the size of x. Hence, *the joint statistics of the n processes are independent of their absolute position in the signal.* Only the mutual separations among the $\{x_i\}$ matter.

This kind of stationarity will be assumed for optical phase errors induced by atmospheric turbulence, for $n = 2$ points (Sect. 8.6.1).

8.4 Fourier Transform Theorem

It is interesting that for a wide-sense stationary process, $R_f(x)$ and $S_f(\omega)$ are a Fourier transform pair. This is shown as follows.

From definitions (8.1a, b)

$$S_f(\omega) = \lim_{L\to\infty} \frac{1}{2L} \int_{-L}^{L} \mathrm{d}x\, \mathrm{e}^{-\mathrm{j}\omega x} \int_{-L}^{L} \mathrm{d}x'\, \mathrm{e}^{\mathrm{j}\omega x'} \left\langle f(x; \boldsymbol{\lambda}) f^*(x'; \boldsymbol{\lambda})\right\rangle . \tag{8.7}$$

Substituting in condition (8.4) and changing variable x' to $y = x - x'$ yields

$$S_f(\omega) = \lim_{L\to\infty} \frac{1}{2L} \int_{-L}^{L} \mathrm{d}x\, \mathrm{e}^{-\mathrm{j}\omega x}\, \mathrm{e}^{+\mathrm{j}\omega x} \int_{x-L}^{x+L} \mathrm{d}y\, \mathrm{e}^{-\mathrm{j}\omega y} R_f(y) .$$

In the limit $L \to \infty$ the $\mathrm{d}y$ integral limits approach $(-\infty, +\infty)$ for any x, and the integrand for $\mathrm{d}x$ is unity, so that

$$S_f(\omega) = \int_{-\infty}^{\infty} \mathrm{d}y \, \mathrm{e}^{-\mathrm{j}\omega y} R_f(y) \,, \tag{8.8}$$

which is what we set out to prove.

8.5 Case of a "White" Power Spectrum

We now show that a white power spectrum arises out of a signal $f(x)$ which is uncorrelated at different points,

$$\langle f(x; \boldsymbol{\lambda}) f^*(x'; \boldsymbol{\lambda}) \rangle = 0 \quad \text{for} \quad x \neq x' \,.$$

At the same time we want to permit $f(x)$ to correlate "perfectly" with itself at the same point x. These two properties may be combined as

$$\langle f(x; \boldsymbol{\lambda}) f^*(x'; \boldsymbol{\lambda}) \rangle = m_2(x)\delta(x - x') \tag{8.9}$$

in terms of the Dirac delta function $\delta(x)$ (see Sect. 3.11 for properties). The quantity

$$m_2(x) \equiv \langle |f(x; \boldsymbol{\lambda})|^2 \rangle \tag{8.10}$$

is the second moment of f at the point x. We note in passing that (8.10) allows for generally shift-*variant* (i.e., *not* stationary) statistics, since m_2 is generally a function of position x. The proof will therefore be general in this regard.

Substituting (8.9) into (8.7), and using the sifting property (3.30d) of $\delta(x)$,

$$S_f(\omega) = \lim_{L \to \infty} (1/2L) \int_{-L}^{L} \mathrm{d}x \, m_2(x) \, \mathrm{e}^{-\mathrm{j}\omega x + \mathrm{j}\omega x} \,,$$

which simplifies to

$$S_f(\omega) = \lim_{L \to \infty} (1/2L) \int_{-L}^{L} \mathrm{d}x \, m_2(x) = \text{constant} \,. \tag{8.11}$$

This constant is the average second moment m_2 of $f(x)$ across x. Note that, in the particular case of a constant second moment $m_2(x) = m_2$, (8.11) becomes

$$S_f(\omega) = m_2 \,. \tag{8.12}$$

The power spectrum is precisely the second moment of $f(x)$!

But in either case (8.11) or (8.12), the power spectrum $S_f(\omega)$ is constant, or white. This is what we set out to prove.

In summary, *an uncorrelated stochastic process* $f(x; \boldsymbol{\lambda})$ *has a white power spectrum.* This is independent of the probability law obeyed by f at any x, and is independent of any assumption of stationarity for f. An example of an uncorrelated stochastic process is the speckle pattern of Fig. 5.3. It provides a visual picture of "white noise" when viewed at a sufficiently large distance that speckle correlation lengths are no longer discernible.

8.6 Application: Average Transfer Function Through Atmospheric Turbulence

Given these preliminaries, we may tackle a basic problem of atmospheric seeing, evaluation of the average transfer function of a turbulent medium. This analysis will make particular use of the concepts of the characteristic function and strict-sense stationarity. It also will provide a good example of forming a "statistical model" for a stochastic process.

Consider a lens system, as in Fig. 8.3, which is attempting to bring to focus light from a distant point source. The light is quasi-monochromatic, of wavelength λ, and has been perturbed by atmospheric turbulence before entering the lens. For simplicity, assume the lens to be one-dimensional (cylindrical), and of focal length F. The derivation may easily be generalized to two dimensions.

At any instant of time, the net phase Δ at point β of the pupil is composed of a deterministic contribution Δ_0 plus *a random part* Δ_R (as sketched),

$$\Delta(\beta) = \Delta_0(\beta) + \Delta_R(\beta) \ .$$

Component $\Delta_0(\beta)$ is constant in time, due to fixed system aberrations, while $\Delta_R(\beta)$ is random due to the turbulence. Thus, function $\Delta_R(\beta)$ takes on a randomly different form at different times. Coordinate β is proportional to Cartesian distance x,

$$\beta = kx/F \ ,$$

where $k = 2\pi/\lambda$.

The transfer function in the image plane at distance F away is given by (3.65),

$$T(\omega) = K \int_{\omega-\beta_0}^{\beta_0} d\beta \exp\{jk[\Delta(\beta) - \Delta(\beta - \omega)]\} \ , \tag{8.13}$$

the autocorrelation of the pupil function $\exp(jk\Delta)$.

The time average transfer function is then

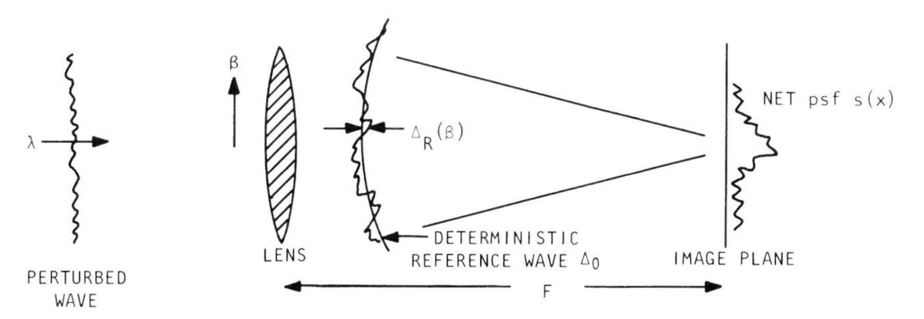

Fig. 8.3. The optical phase $\Delta_R(\beta)$ across a pupil, at any time instant, due to atmospheric turbulence. The source of light is a point at infinity to the left. The resulting spread function $s(x)$ is then called a "turbulence speckle pattern" (see Fig. 8.5)

$$\langle T(\omega)\rangle = K \int_{\omega-\beta_0}^{\beta_0} \mathrm{d}\beta \exp\{\mathrm{j}k[\Delta_0(\beta) - \Delta_0(\beta-\omega)]\}$$
$$\times \langle \exp\{\mathrm{j}k[\Delta_\mathrm{R}(\beta) - \Delta_\mathrm{R}(\beta-\omega)]\}\rangle \tag{8.14}$$

by (8.13). Note that the $\langle\rangle$ brackets have been brought *within* the integral to operate solely upon the randomly varying part. The expectation of a sum (or integral) is the sum of the expectations. To proceed further we have to define the statistics for Δ_R at the two points β and $\beta - \omega$.

8.6.1 Statistical Model for Phase Fluctuations

As in all analyses of stochastic effects, the key step is in selecting a *proper model* for the fluctuating quantities. Here, the ingenuity and imagination of the analyst come into full use. For a good model is *both* (a) descriptive of the phenomenon, and (b) convenient to analyze. These are often conflicting aims. To permit (b), often the *simplest* model that can satisfy (a) with any degree of confidence is chosen. Hence, the choice is sometimes controversial, and only really verifiable when a proper laboratory experiment can demonstrate its correctness. The following model of atmospheric turbulence (indicated by dots •) is commonly assumed. It is not controversial, at least for *weak* turbulence conditions.

- Assume that Δ_R obeys the central limit theorem, and is therefore normal at each point β. This can be justified on the grounds that Δ_R in the pupil is the sum total perturbation due to many *independent* phase perturbations on the way from source to pupil (Sect. 4.28.3).
- Further, it is reasonable to suppose that Δ_R has zero mean, since phase fluctuations can be expected to go negative as often as positive. Then Δ_R is a Gaussian random variable.
- Finally, assume that the phase is strict-sense stationary (8.6), for $n = 2$ points in the pupil. Thus, denoting

$$\Delta_\mathrm{R} \equiv \Delta_\mathrm{R}(\beta)\,, \quad \Delta'_\mathrm{R} \equiv \Delta_\mathrm{R}(\beta-\omega)\,, \tag{8.15}$$

  we conclude that

$$p(\Delta_\mathrm{R}, \Delta'_\mathrm{R}) \text{ is independent of } \beta\,. \tag{8.16}$$

  That is, the joint statistics for phase at two points of separation ω is independent of their absolute position in the pupil. Other properties of $p(\Delta_\mathrm{R}, \Delta'_\mathrm{R})$ are found next.

By integrating out the RV Δ'_R, it follows that the marginal law

$$p(\Delta_\mathrm{R}) \text{ is independent of } \beta \tag{8.17}$$

as well. Analogously for $p(\Delta'_\mathrm{R})$.

Let σ and σ' denote the standard deviations in Δ_R, and Δ'_R respectively. Then by property (8.17),

$$\sigma' = \sigma\,. \tag{8.18}$$

Or, every point in the pupil suffers the same turbulence effects.

A probability law that has the preceding properties is the Gaussian bivariate (3.44). Hence, we shall assume that *the joint probability density for fluctuations* Δ_R *at the two pupil points* β, $(\beta - \omega)$ *is Gaussian bivariate.* Furthermore, by property (8.16) its correlation coefficient should obey

$$\rho = \rho(\omega) , \tag{8.19}$$

that is, be only a function of the distance between the two points. This is reasonable.

8.6.2 A Transfer Function for Turbulence

We now apply the model. Consider the $\langle\rangle$ part of (8.14), where β is fixed. Designate this as $T_R(\beta;\omega)$. Then by notation (8.15)

$$T_R(\beta;\omega) \equiv \langle \exp[jk(\Delta_R - \Delta'_R)] \rangle . \tag{8.20}$$

The right-hand side of (8.20) has two important properties. First, by property (8.16) it may be taken outside the integration $d\beta$ of (8.14). Thus, $T_R(\beta;\omega) = T_R(\omega)$ alone, and

$$\langle T(\omega) \rangle = T_0(\omega) T_R(\omega) , \tag{8.21}$$

where

$$T_0(\omega) \equiv K \int_{\omega-\beta_0}^{\beta_0} d\beta \exp\{jk[\Delta_0(\beta) - \Delta_0(\beta-\omega)]\}$$

is the *fixed transfer function* due to system aberrations and diffraction. Equation (8.21) states that the average transfer function may be expressed as the product of two distinct transfer functions, one due to fixed or intrinsic lens properties, and the other due to turbulence alone.

The second interesting property of the right-hand side of (8.20) is that by definition (4.14) it is precisely the characteristic function $\varphi(\omega_1, \omega_2)$ for the joint behavior of Δ_R and Δ'_R, evaluated at $\omega_1 = k$, $\omega_2 = -k$ in particular. That is,

$$T_R(\omega) = \varphi_{GB}(\omega_1, \omega_2) ,$$

with

$$\omega_1 = k , \quad \omega_2 = -k . \tag{8.22}$$

Subcript GB denotes Gaussian bivariate, our model for the joint behavior of Δ_R and Δ'_R.

But, by (4.16), we know the characteristic function for GB behavior,

$$\varphi_{GB}(\omega_1, \omega_2) = \exp[-2^{-1}(\sigma^2\omega_1^2 + \sigma'^2\omega_2^2 + 2\rho\sigma\sigma'\omega_1\omega_2)] . \tag{8.23}$$

We may now combine results and form the answer. By (8.18, 19, 22 and 23)

$$T_R(\omega) = \exp\{-k^2\sigma^2[1 - \rho(\omega)]\} . \tag{8.24}$$

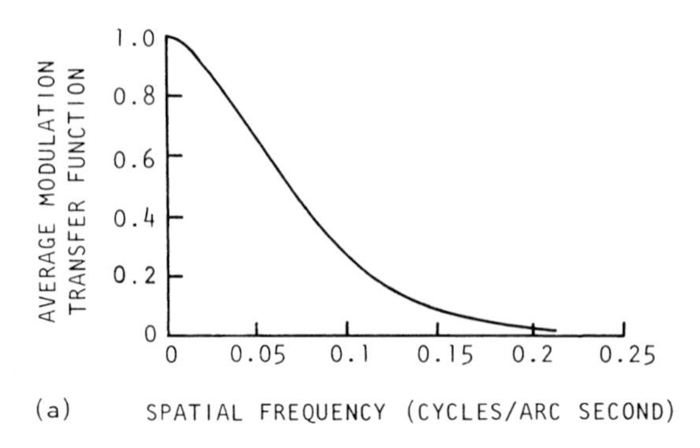

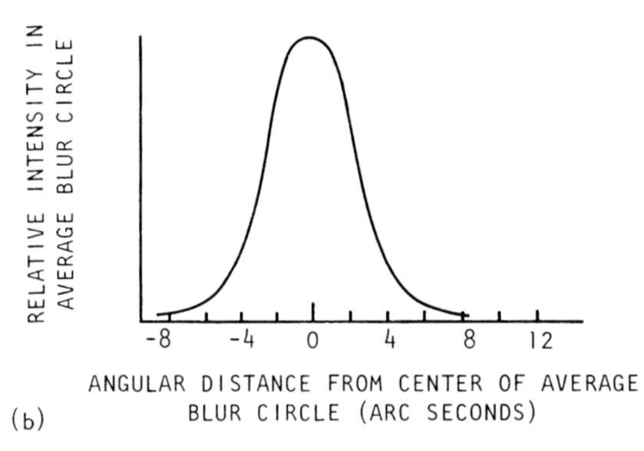

Fig. 8.4. Long-term average transfer function (**a**), and its corresponding spread function (**b**), for zenith viewing through the whole earth's atmosphere [8.1]

This result shows that atmospheric turbulence affects seeing through three factors: the wavelength λ of light, the absolute level of fluctuation σ in phase at any point in the pupil, and the correlation in such phase fluctuations at two points separated by ω in the pupil. The last effect is particularly interesting. It states that the greater is the correlation distance for phase, the higher is the cutoff frequency in $T_{\mathrm{R}}(\omega)$.

The multiplier $k^2\sigma^2$ in the exponent of (8.24) exerts a particularly bad effect upon $T_{\mathrm{R}}(\omega)$. If, for example, $\sigma = \lambda/4$, which by most standards represents rather modest fluctuation, $k^2\sigma^2 = 2.5$ rad, so that $T_{\mathrm{R}}(\omega) = 0.08^{1-\rho(\omega)}$. Unless the correlation ρ is particularly high, then, $T_{\mathrm{R}}(\omega)$ must be very low. At frequencies for which $\rho(\omega) = 0$, $T_{\mathrm{R}}(\omega) = 0.08$, which is severe attenuation.

Various authors have used different forms for $\rho(\omega)$. For one such form [8.1], Fig. 8.4 shows the resulting $T_{\mathrm{R}}(\omega)$ and its Fourier transform, the effective point spread function due to turbulence.

8.7 Transfer Theorems for Power Spectra

An optical image $i(x)$ is related to its object $o(x)$ and point spread function $s(x)$ through (4.22)

$$i(y) = \int_{-\infty}^{\infty} \mathrm{d}x\, o(x)s(y-x)\ . \tag{8.25}$$

We ask what the power spectrum $S_\mathrm{i}(\omega)$ of the image is, under two differing circumstances:

(i) $o(x)$ is a stochastic process; or
(ii) $s(x)$ is a stochastic process.

Optical examples of these complementary problems follow.

8.7.1 Determining the MTF Using Random Objects

Situation (i) arises when one optical system, specified by a fixed $s(x)$, is used to image many randomly chosen objects $o(x)$ from an "ensemble" of objects. Then $o(x) = o(x; \boldsymbol{\lambda})$, where $\boldsymbol{\lambda}$ is random and designates the choice, and $o(x)$ is a stochastic process. For example, the object ensemble might consist of letters of the alphabet which are to be relayed by an optical relay lens.

Since $i(y)$ is functionally related to $o(x; \boldsymbol{\lambda})$ by Eq. (8.25), i(y) must also be a stochastic process. Hence we denote it as $i(y, \boldsymbol{\lambda})$.

We use definition (8.1a,b) to compute $S_\mathrm{i}(\omega)$. By (8.1a), we must form

$$I_L(\omega; \boldsymbol{\lambda}) = \int_{-L}^{L} \mathrm{d}y\, i(y; \boldsymbol{\lambda})\, \mathrm{e}^{-\mathrm{j}\omega y}\ .$$

Taking this finite Fourier transform of (8.25) produces

$$I_L(\omega; \boldsymbol{\lambda}) = \int_{-\infty}^{\infty} \mathrm{d}x\, o(x; \boldsymbol{\lambda}) \int_{-L}^{L} \mathrm{d}y\, s(y-x)\, \mathrm{e}^{-\mathrm{j}\omega y}\ .$$

Now use an integration variable $t = y - x$. Then

$$I_L(\omega; \boldsymbol{\lambda}) = \int_{-\infty}^{\infty} \mathrm{d}x\, o(x; \boldsymbol{\lambda})\, \mathrm{e}^{-\mathrm{j}\omega x} \int_{-L-x}^{L-x} \mathrm{d}t\, s(t)\, \mathrm{e}^{-\mathrm{j}\omega t}\ .$$

Anticipating that the limit $L \to \infty$ will be taken, this becomes

$$I_L(\omega; \boldsymbol{\lambda}) = \int_{-L}^{L} \mathrm{d}x\, o(x; \boldsymbol{\lambda})\, \mathrm{e}^{-\mathrm{j}\omega x} \int_{-\infty}^{\infty} \mathrm{d}t\, s(t)\, \mathrm{e}^{-\mathrm{j}\omega t}\ . \tag{8.26}$$

Now by (3.64)

$$T(\omega) = \int_{-\infty}^{\infty} \mathrm{d}t\, s(t)\, \mathrm{e}^{-\mathrm{j}\omega t} \tag{8.27}$$

defines the optical transfer function $T(\omega)$ of the lens system. Its modulus is called the "modulation transfer function" and denoted MTF(ω). Also, by definition (8.1a) the first integral in (8.26) is $O_L(\omega; \boldsymbol{\lambda})$. Then taking the modulus-squared of (8.26) yields

$$|I_L(\omega; \boldsymbol{\lambda})|^2 = |O_L(\omega; \boldsymbol{\lambda})|^2 \,\mathrm{MTF}(\omega)^2 \,. \tag{8.28}$$

Finally, taking the average over $\boldsymbol{\lambda}$ of both sides, dividing by $2L$, and using definition (8.1b) yields

$$S_\mathrm{i}(\omega) = S_\mathrm{o}(\omega)\mathrm{MTF}(\omega)^2 \,. \tag{8.29}$$

Or, *the power spectrum for the image is the power spectrum for the object modulated by the* MTF*-squared of the lens system.*

Equation (8.29) may be used to experimentally determine MTF(ω) for an optical film [8.2]. As a test object, a laser speckle pattern is used, whose power spectrum $S_\mathrm{o}(\omega)$ is white (Sect. 8.5). Then the experimentally observable quantity $S_\mathrm{i}(\omega)$ relates to MTF(ω) through a simple proportionality

$$S_\mathrm{i}(\omega) = K \cdot \mathrm{MTF}(\omega)^2 \,. \tag{8.30}$$

Use of a coherent processor, as described in [8.2], enables $S_\mathrm{i}(\omega)$ to be observed. Then the square root of $S_\mathrm{i}(\omega)$ yields the required MTF(ω) curve.

8.7.2 Speckle Interferometry of Labeyrie

An example of a stochastic process of type (ii), where now $s(x)$ is stochastic, occurs commonly in astronomy. If a stellar object is bright enough, it can be imaged by telescope during a time exposure of 0.01 s or less. This causes atmospheric turbulence, which changes at a slower pace, to be effectively "frozen." The result is a phase function across the telescope pupil that is temporally constant (during the exposure) but spatially random; see Fig. 8.3. Such a pupil function causes, via (3.62) and (3.63), a point spread function $s(x)$ that is said to suffer from "short-term" turbulence. Typically, such a spread function consists of randomly scattered blobs or "speckles," as in Fig. 8.5.

Suppose that many (the order of 100) such short-term exposures of the object are taken. Since each causes randomly a new pupil phase function to be formed, $s(x)$ must be considered a stochastic process.

Since all the while one object is being observed, $o(x)$ is fixed. We now have exactly the reverse situation of Sect. 8.7.1, where $o(x)$ was instead stochastic. Going through similar steps, it is easy to show that the power spectrum $S_\mathrm{i}(\omega)$ in the image now obeys

$$S_\mathrm{i}(\omega) = |O(\omega)|^2 \, S_\mathrm{s}(\omega) \,, \tag{8.31}$$

where $S_\mathrm{s}(\omega)$ is the power spectrum of $s(x)$.

Now both $S_\mathrm{i}(\omega)$ and $S_\mathrm{s}(\omega)$ are easily formed from the short-exposure images, the former by averaging over the individual spectra $I(\omega)$ of the images, the latter by averaging over the transfer functions $T(\omega)$ of an isolated star (supposing one to exist in the field of view) in the images. Hence, (8.31) allows $|O(\omega)|^2$ to be found. The modulus $|O(\omega)|$ of the unknown object is then known.

Fig. 8.5. Typical speckle pattern due to a short-exposure image of a single star [8.3]

Although the phase $\Phi(\omega)$ of the spectrum $O(\omega)$ is still unknown, knowledge of the modulus $|O(\omega)|$ by itself often allows the object $o(x)$ to be known. This is in cases where either (a) the phase $\Phi(\omega)$ is known to be zero because of symmetry in the object, or (b) the object may be parameterized. As an example of the latter, consider the case where the object is a double star of equal intensities B and separation a. Then parameters a and B are all that are needed to specify the object. Moreover, since for this object $o(x) = B\delta(x-a/2) + B\delta(x+a/2)$, we have $|O(\omega)|^2 = 4B^2\cos^2(a\omega/2)$. Or, $|O(\omega)|^2$ *will be a set of fringes* (origin of the term "speckle interferometry"). Then a can be estimated from the frequency of the fringes, while B is known from their contrast. Typical fringe images for simulated double stars are shown in Fig. 8.6.

How far may this procedure be pushed? That is, how small a separation a could be estimated in this manner? We may note from (8.31) that power spectrum $S_s(\omega)$ acts as a transfer function in relaying object information $|O(\omega)|^2$ into the image data $S_i(\omega)$. Hence, the cutoff frequency for $S_s(\omega)$ acts to limit object frequency-space information in data $S_i(\omega)$. Amazingly, $S_s(\omega)$ actually cuts off at the diffraction limit of the telescope doing the viewing, independent of the turbulence! This was shown by *Korff* [8.5a, 8.5b]. Of these two derivations, the latter is simpler and we show it next.

8.7.3 Resolution Limits of Speckle Interferometry

This derivation requires a two-dimensional approach. Vector $\boldsymbol{\omega}$ denotes (ω_1, ω_2). By (8.31), the transfer function $S_s(\omega)$ for speckle interferometry is the power spectrum

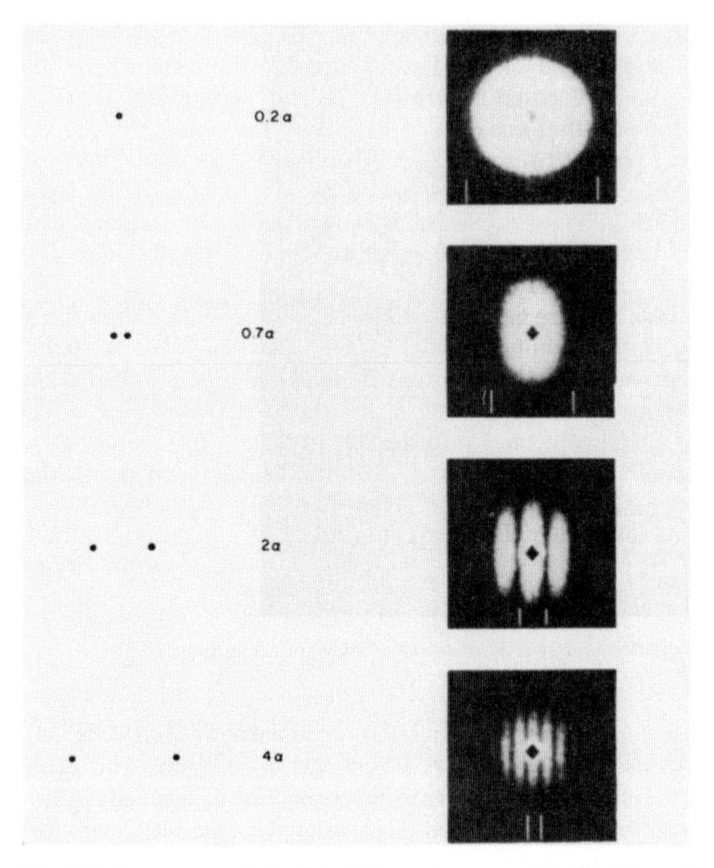

Fig. 8.6. A sequence of simulated binary-star cases. On the left are the binary stars, with separations indicated to their right. On the far right are the corresponding images $S_i(\omega)$. Tick marks locate the fringe widths [8.4]

of spread function s. By (8.2a),

$$S_s(\boldsymbol{\omega}) = \left\langle |T(\boldsymbol{\omega})|^2 \right\rangle , \tag{8.32}$$

where the brackets denote an average over the many short-exposure images at hand. (We have ignored an unimportant constant of proportionality here.) What are the resolution properties of $\langle |T(\boldsymbol{\omega})|^2 \rangle$?

By (3.66)

$$\left\langle |T(\boldsymbol{\omega})|^2 \right\rangle = K^2 \iint_{\substack{\text{overlap}\\ \text{regions}}} \mathrm{d}\boldsymbol{\beta}\, \mathrm{d}\boldsymbol{\beta}' \Big\langle \exp\{\mathrm{j}k[\Delta_\mathrm{R}(\boldsymbol{\beta}) - \Delta_\mathrm{R}(\boldsymbol{\beta} - \boldsymbol{\omega}) + \Delta_\mathrm{R}(\boldsymbol{\beta}' - \boldsymbol{\omega}) - \Delta_\mathrm{R}(\boldsymbol{\beta}')]\} \Big\rangle . \tag{8.33}$$

We have assumed the pupil to be diffraction-limited, i.e. $\Delta_0(\boldsymbol{\beta}) = 0$. Let us denote the deterministic transfer function $T_0(\boldsymbol{\omega})$ for this case as $\tau_0(\omega)$, the diffraction-limited transfer function.

We may note from the form of Eq. (8.33) that the integrand is evaluated at points $\boldsymbol{\beta}$ and $\boldsymbol{\beta}'$ which lie in the overlap region between two circles (assuming the pupil to be circular) separated by distance $\boldsymbol{\omega}$. Let us suppose now, *as a model for the turbulence*, that RV Δ_R is perfectly correlated over distances less than or equal to a value ρ_0, but is uncorrelated beyond this distance. This means that the pupil consists of uncorrelated "blobs" of phase, each of a characteristic size ρ_0. In this case, the integrals in (8.33) may be replaced by sums over the blobs. Each blob has area $\pi\rho_0^2/4$. Then, how many such blobs will lie within the overlap region?

Denote this number by N. Let the pupil have diameter D. Assume that $\rho_0 < D$, there is more than just one blob. Then N is just the number of circles of area $\pi\rho_0^2/4$ which will fit into the overlap region. But by (3.66) $\tau_0(\omega) =$ overlap area/pupil area, where the numerator is $N \cdot \pi\rho_0^2/4$ and the denominator is $\pi D^2/4$. Combining these, we get

$$N \simeq (D/\rho_0)^2 \tau_0(\omega) \ . \tag{8.34}$$

As discussed, we replace (8.33) *by a sum* over the uncorrelated blobs,

$$\begin{aligned} \left\langle |T(\boldsymbol{\omega})|^2 \right\rangle &\equiv (\pi\rho_0^2/4)^2 K^2 \sum_{m=1}^{N} \sum_{n=1}^{N} \\ &\cdot \langle \exp\{jk[\Delta_R(\boldsymbol{\beta}_m) - \Delta_R(\boldsymbol{\beta}_m - \boldsymbol{\omega})]\} \cdot \exp\{jk[\Delta_R(\boldsymbol{\beta}_n) - \Delta_R(\boldsymbol{\beta}_n - \boldsymbol{\omega})]\}\rangle \ . \end{aligned} \tag{8.35}$$

This may be evaluated at terms for which $m = n$, and then for $m \neq n$. The first give a sum

$$(\pi\rho_0^2/4)^2 K^2 \sum_{n=1}^{N} \langle 1 \rangle = (\pi\rho_0^2/4)^2 K^2 (D/\rho_0)^2 \tau_0(\omega) \tag{8.36}$$

by (8.34). This is linear in the diffraction-limited transfer function! Hence it preserves all object frequencies in (8.31) out to optical cutoff.

The terms for which $m \neq n$ cause a contribution

$$\begin{aligned} &(\pi\rho_0^2/4)^2 K^2 \sum_{m=1}^{N} \langle \exp\{jk[\Delta_R(\boldsymbol{\beta}_m) - \Delta_R(\boldsymbol{\beta}_m - \boldsymbol{\omega})]\}\rangle \\ &\cdot \sum_{\substack{n=1 \\ n\neq m}}^{N} \langle \exp\{-jk[\Delta_R(\boldsymbol{\beta}_n) - \Delta_R(\boldsymbol{\beta}_n - \boldsymbol{\omega})]\}\rangle \ . \end{aligned} \tag{8.37}$$

This has broken into a product of averages because, by the model, different blobs do not correlate. But each average in (8.37) is the *long exposure average* $T_R(\omega)$ of (8.20) since consecutive short-term exposures are being added. We found in Sect. 8.6.2 that this is close to zero at high frequencies $\boldsymbol{\omega}$. Hence, at high frequencies only contributions (8.36) remain,

$$\left\langle |T(\boldsymbol{\omega})|^2 \right\rangle \simeq (\pi\rho_0^2/4)^2 K^2 (D/\rho_0)^2 \tau_0(\boldsymbol{\omega}) \ . \tag{8.38}$$

Finally, by normalization $T(0) \equiv 1$, (8.13) shows that

$$K = (\pi D^2/4)^{-1} . \tag{8.39}$$

This gives us the remarkably simple and elegant answer

$$\langle |T(\boldsymbol{\omega})|^2 \rangle \simeq (\rho_0/D)^2 \tau_0(\omega) , \tag{8.40}$$

for ω high.

Thus, speckle interferometry works because *the data contain spatial frequencies out to cutoff* in the diffraction-limited transfer function $\tau_0(\omega)$. The fact that the transfer function for the technique, a squared quantity, is *linear* in $\tau_0(\omega)$ is a serendipitous effect which further aids the high-frequency content.

The proportionality to $(\rho_0/D)^2$ shows that the effect becomes weaker as the correlation (blob) size becomes smaller. This makes sense since the smaller the blobs are the more uncorrelated in phase is the pupil.

Finally, regarding statistical modeling per se, we have found that two different models may be used to describe the same phenomenon! The reader may have noticed that in Sect. 8.6.1 we used a Gaussian bivariate (GB) phase model, whereas here we used a "phase blob" model. Obviously, both cannot be correct. In fact, *Korff* in [8.5a] used the GB model *as well* to prove the results of this section. (But since it was a much more difficult derivation than by his alternative blob approach [8.5b], we presented the latter here.) Overall, the GB model may be regarded as the correct choice. However, the blob model often gives good insights into the turbulence problem, despite its oversimplicity.

Exercise 8.1

In problems (1) through (5) following, it will be convenient to use an *unnormalized power spectrum* $\tilde{S}_f(\omega)$ defined by simply

$$\tilde{S}_f(\omega) \equiv \langle |F(\omega; \boldsymbol{\lambda})|^2 \rangle . \tag{8.41a}$$

As discussed at (8.2a), this quantity is simply $S_f(\omega)$ without the constant multiplier $1/2L$. The latter is irrelevant here. Spectrum $F(\omega; \boldsymbol{\lambda})$ is defined as

$$F(\omega; \boldsymbol{\lambda}) \equiv \int_{-\infty}^{\infty} \mathrm{d}x\, f(x; \boldsymbol{\lambda})\, \mathrm{e}^{-\mathrm{j}\omega x} . \tag{8.41b}$$

8.1.1 A stochastic process $s(x; \theta)$ is of the form

$$s(x; \theta) = \mathrm{e}^{-x^2} \cos(ax + \theta)$$

where a is fixed, and θ is uniformly random from 0 to 2π. Thus, each randomly selected θ defines a randomly shifted sinusoid from the ensemble of signals. Find

(a) The Fourier spectrum $S(\omega; \theta)$, defined at (8.41b);

(b) The power spectrum $\tilde{S}_s(\omega)$ defined at (8.41a). Note that the averaging operation is with respect to RV θ, and in the usual sense (3.12).

8.1.2 A function $f(x)$ is bandlimited, with cutoff frequency Ω rad/length. It may therefore be represented by its sampled values $f(n\pi/\Omega) \equiv f_n$, as

$$f(x) = \sum_{n=-N}^{N} f_n \operatorname{sinc}(\Omega x - n\pi) .$$

This formula is to be used to compute $f(x)$ from measured values $\{f_n\}$. The latter are in error by amounts e_n, with a triangular correlation function

$$\langle e_m e_n \rangle = \sigma^2(\delta_{mn} + 2^{-1}\delta_{m,n+1} + 2^{-1}\delta_{m,n-1}) .$$

δ_{mn} is the Kronecker delta function.

Thus the erroneous part of $f(x)$, call it $e(x)$, is inadvertently formed by the user from the same sampling relation

$$e(x) = \sum_{n=-N}^{N} e_n \operatorname{sinc}(\Omega x - n\pi) .$$

Note that $e(x)$ is a stochastic process $e(x; \mathbf{e})$.

(a) What is Fourier spectrum $E(\omega)$ in terms of RV's $\{e_n\}$?

(b) From this show that $\langle |E(\omega)|^2 \rangle \equiv \tilde{S}_e(\omega)$, the un-normalized power spectrum, obeys

$$\tilde{S}_e(\omega) = 4\pi^2 N(\sigma/\Omega)^2 \cos^2(\pi\omega/2\Omega) + (\pi\sigma/\Omega)^2 .$$

8.1.3 An image $i(x)$ is repeatedly formed from unknown and randomly selected object intensities $o_1, o_2, \ldots, o_N$ located, respectively, at fixed points $x_1, \ldots, x_N$. The points have a common spacing Δx. The image obeys

$$i(x) = \sum_{n=1}^{N} o_n s(x - x_n) ,$$

where $s(x)$ is a fixed spread function. Thus $i(x; \mathbf{o})$ is a stochastic process. Find $\tilde{S}_\mathrm{i}(\omega)$ if

(a) $\langle o_m o_n \rangle = \sigma^2 \delta_{mn}$; the object has zero correlation length.

(b) $\langle o_m o_n \rangle = \sigma^2(\delta_{mn} + 2^{-1}\delta_{m,n+1} + 2^{-1}\delta_{m,n-1})$; triangular correlation.

(c) $\langle o_m o_n \rangle = \sigma^2(\delta_{mn} + \delta_{m,n+1} + \delta_{m,n-1})$; boxcar correlation.

The procession from (a) to (b) to (c) has stronger and stronger correlation. What is the trend in the power spectrum $S_\mathrm{i}(\omega)$?

8.1.4 An image $i(x)$ is formed as

$$i(x) = \sum_{n=1}^{N} o_n s(x - x_n)$$

where the o_n are *fixed* intensities, obeying the constraints

$$\sum_{n=1}^{N} o_n = P , \quad \sum_{n=1}^{N} o_n^2 = Q^2 .$$

The *positions* $\{x_n\}$ are random variables, selected independently from the law

$$p(x) = a^{-1}\operatorname{Rect}\left(\frac{x - a/2}{a}\right) .$$

This model fits a uniformly random scattering of stars in an image field, where each star gives rise to the same point spread function $s(x)$. The transfer function is $T(\omega)$. Show that

$$\tilde{S}_i(\omega) = |T(\omega)|^2 \left[Q^2 + (P^2 - Q^2)\,\mathrm{sinc}^2(a\omega/2)\right] .$$

8.1.5 The derivation (of Sect. 8.4) that $S_f(\omega)$ and $R_f(y)$ are Fourier transform mates does not hold for $\tilde{S}_f(\omega)$ and $R_f(y)$. Use definition (8.41a, b) to show what happens when it is attempted to relate the un-normalized $\tilde{S}_f(\omega)$ and $R_f(y)$.

8.1.6 Images are *two*-dimensional signals $f(x, y)$ which may be regarded as stochastic processes. The two-dimensional power spectrum $S_f(\omega)$ is defined as

$$S_f(\boldsymbol{\omega}) = \lim_{L\to\infty} \left\langle |F_L(\boldsymbol{\omega})|^2 \right\rangle / (2L)^2 ,$$

where

$$F_L(\boldsymbol{\omega}) = \iint_{-L}^{L} \mathrm{d}\boldsymbol{r}\, f(\boldsymbol{r})\,\mathrm{e}^{-\mathrm{j}\boldsymbol{\omega}\cdot\boldsymbol{r}} .$$

Vector notation is used for convenience, with $\boldsymbol{r} = (x, y)$, $\boldsymbol{\omega} = (\omega_1, \omega_2)$. Suppose also that the two-dimensional autocorrelation function defined as

$$R_f(\boldsymbol{r}; \boldsymbol{r}_0) \equiv \left\langle f(\boldsymbol{r}_0) f^*(\boldsymbol{r}_0 + \boldsymbol{r}) \right\rangle$$

obeys *isotropy* (the two-dimensional analog of wide-sense stationarity)

$$R_f(\boldsymbol{r}; \boldsymbol{r}_0) = R_f(r) .$$

(a) Then show that $S_f(\boldsymbol{\omega})$ is related to $R_f(r)$ via a J_0-Hankel transform. (*Hint:* Stick with vector notation until final integration over angle.)

(b) It has been conjectured that natural scenes have about the same power spectrum irrespective of the degree of magnification used to view them. Hence, the Earth looks statistically the same from any distance. What form does this imply for the autocorrelation function $R_f(r)$ for natural scenes? (*Hint:* Let $r' = Mr$, M a variable magnification, use $\mathrm{J}_0(x) \simeq x^{1/2}$ for x large.)

8.8 Transfer Theorem for Autocorrelation: The Knox–Thompson Method

As in the preceding section, suppose that a telescope is repeatedly imaging one object $o(x)$ through short-term turbulence. Many images $i(x; \boldsymbol{\lambda})$ are formed, and these are stochastic because of the stochastic nature of the spread function $s(x; \boldsymbol{\lambda})$. But contrary to the preceding section, *we now want to form an average over the image data that will preserve some phase information about the object.* What kind of an average would accomplish this? *Knox* and *Thompson* [8.6] found that one such average is

$$\left\langle I(\omega) I^*(\omega + \Delta\omega) \right\rangle ,$$

the autocorrelation of the image spectrum. This may be found approximately, by summing $I(\omega)I^*(\omega + \Delta\omega)$ over an ensemble of images, typically numbering 50 to 100. The imaging equation (8.25) shows that the autocorrelation in question obeys

$$\left\langle I(\omega)I^*(\omega + \Delta\omega)\right\rangle = \left\langle T(\omega)T^*(\omega + \Delta\omega)\right\rangle O(\omega)O^*(\omega + \Delta\omega) \,, \tag{8.42a}$$

in terms of the autocorrelation of the turbulence transfer function $T(\omega)$ and the one object spectrum $O(\omega)$. This is a *transfer theorem*. It may be used to reconstruct $O(\omega)$, as follows.

The displacement frequency $\Delta\omega$ is actually two-dimensional, as are x and ω here because of the two-dimensional nature of the images. Suppose that vector $\Delta\omega$ takes on two coordinate displacements, $(0, \Delta\omega)$ and $(\Delta\omega, 0)$, as well as no displacement $(0, 0)$. The latter produces speckle interferometry data once again, permitting $|O(\omega)|$ to be known (Sect. 8.7.2).

With $|O(\omega)|$ known, it only remains to find the object phase function $\Phi(\omega)$ in order to reconstruct its entire complex spectrum $O(\omega)$. To see how this may be done, we note that equality (8.42a) holds for both the modulus and phase separately. In particular, the phase part obeys

$$\begin{aligned}\text{Phase}\left\langle I(\omega)I^*(\omega + \Delta\omega)\right\rangle = {} & \Phi(\omega) - \Phi(\omega + \Delta\omega) \\ & + \text{Phase}\left\langle T(\omega)T^*(\omega + \Delta\omega)\right\rangle \end{aligned} \tag{8.42b}$$

(Note that Phase $\langle\rangle$ means the phase of the function inside $\langle\rangle$.) Now the left-hand phase in (8.42b) may be computed from the data. Also, the right-hand phase of $\langle TT^*\rangle$ can be known from the image of an isolated star. This allows the unknown object

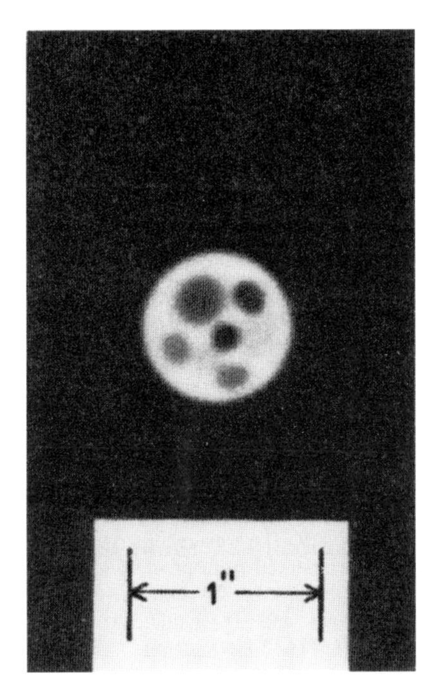

Fig. 8.7. The test object [8.6]

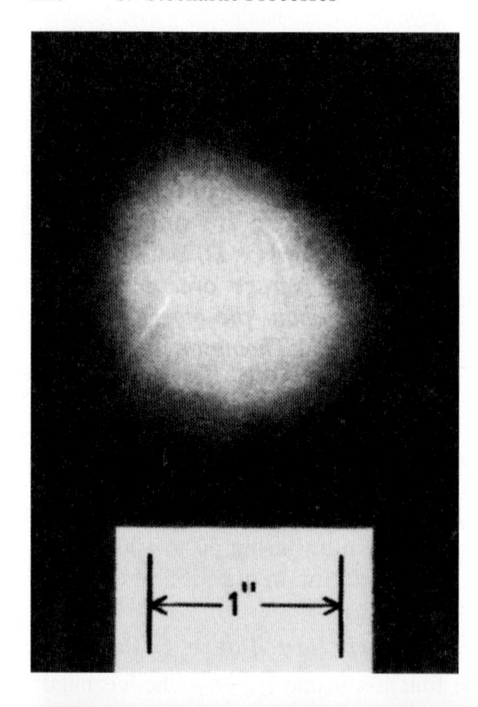

Fig. 8.8. One of 40 simulated short-exposure images of the object [8.6]

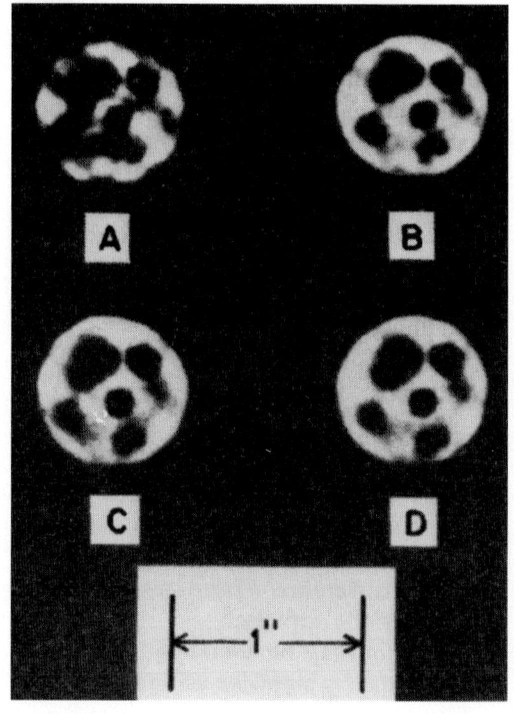

Fig. 8.9. The four reconstructions that resulted from processing (a) 10 images; (b) 20 images; (c) 30 images; and (d) 40 images [8.6]

phase difference $\Phi(\omega) - \Phi(\omega + \Delta\omega)$ to be found in (8.42b), and is the touchstone of the approach.

The phase Φ at any particular vector ω may be found by noting that $\Phi(0, 0) = 0$, since $o(x)$ is a real function, and by stepwise using Eq. (8.42b) to find $\Phi(\Delta\omega)$, then $\Phi(2\Delta\omega)$, etc., until ω is reached. In fact, being a two-dimensional problem, there are any number of paths in frequency space that can be taken to produce Φ at the required ω. Each path should give rise to the correct $\Phi(\omega)$ value. Therefore, in the case of noisy data, the different path values may be averaged to effect some gain in accuracy.

A computer simulation of the approach is shown in Figs. 8.7–9. The object in Fig. 8.7 was meant to model a moon of Jupiter. Figure 8.8 shows one, typical short-exposure image of the object. Figures 8.9a–d show four reconstructions based on algorithm (8.42b), using different numbers of images to form the ensemble averages. Certainly the results Fig. 8.9b–d are much better than the typical, unprocessed image Fig. 8.8. Work on this approach is continuing.

8.9 Additive Noise

What is "noise"? In Sect. 2.15 we defined a "noisy message" as an *informationless message*. This is a message (e.g., consisting of photon image positions) that contains zero information about a set of events (corresponding object space positions for the photons). The concept of noise is always defined in a specific context. As a consequence, what is considered noise in one case may be considered "signal" in another. One man's weed is another man's wildflower. For example, in Sect. 8.7.1 a *random, uncorrelated* object is to be used to determine the MTF of a lens system. Hence, any systematic or correlated tendency in the test object would there be unwanted, and rightfully regarded as "noise." This is of course contrary to our usual notion of noise.

Whatever its application, the concept of noise always refers to the departure of an observed or otherwise *used signal* from its ideal or "true" value. Therefore, let noise in a data value d_m be defined as

$$n_m \equiv d_m - s_m \ , \tag{8.43a}$$

where s_m, is its ideal value, called the "signal." From this, one normally expresses an "output" or data value d_m in terms of its "input" or signal value s_m as

$$d_m = s_m + n_m \ . \tag{8.43b}$$

The preceding has been a discrete (in m) presentation. Quantities s, d, and n may instead be functions of a continuous coordinate x, as in

$$d(x) = s(x) + n(x) \ . \tag{8.44}$$

For example, x might be a space coordinate, with d, s, and n intensity values.

Let us regard s and n as stochastic processes $s(x; \boldsymbol{\lambda})$ and $n(x; \boldsymbol{\lambda}')$, where each of s and n obey arbitrary probability laws $p_S(s)$ and $p_N(n)$ at x. Then by (8.44), d is also stochastic,

$$d(x; \boldsymbol{\lambda}, \boldsymbol{\lambda}') = s(x; \boldsymbol{\lambda}) + n(x; \boldsymbol{\lambda}') \,. \tag{8.45}$$

Now if all possible sets of $\boldsymbol{\lambda}$ and $\boldsymbol{\lambda}'$ specifying signal s and noise n are realized, in general it will be observed that RV's $s(x)$ and $n(x')$ correlate at certain coordinates x, x'. On the other hand, if they do not correlate at any x and x' (including even the common point $x = x'$), then n is statistically independent of s and is called "additive noise." In this situation,

$$\left\langle s(x; \boldsymbol{\lambda}) n(x'; \boldsymbol{\lambda}') \right\rangle = \langle s(x; \boldsymbol{\lambda}) \rangle \left\langle n(x'; \boldsymbol{\lambda}') \right\rangle \tag{8.46a}$$

by identity (3.27).

If, in addition, the noise n has mean of zero at each x', this equation becomes

$$\left\langle s(x; \boldsymbol{\lambda}) n(x'; \boldsymbol{\lambda}') \right\rangle = 0 \,, \tag{8.46b}$$

which is the starkest way of saying that s and n do not correlate.

A final distinctive feature of additive noise arises when we attempt to find the joint probability density $p_{DS}(d, s)$ between input s and output d of (8.44). Coordinate x is arbitrary and fixed. A logical starting point is identity (3.5),

$$p_{DS}(d, s) = p(d|s) p_S(s) \,. \tag{8.47a}$$

The probability law $p_S(s)$ for input s may be assumed to be known, but what is $p(d|s)$?

From the form of (8.44) we see that if s is fixed, the fluctuations in d will follow those in n; in other words,

$$p(d|s) = p_N(n|s) \tag{8.47b}$$

where $n = d - s$. But since noise n is additive, it is independent of s. Consequently, by identity (3.6)

$$p_N(n|s) = p_N(n) \,. \tag{8.47c}$$

Combining the last two relations,

$$p(d|s) = p_N(d - s) \,. \tag{8.47d}$$

This is a widely used property of additive noise. Combining it with (8.47a) gives the sought-after relation,

$$p_{DS}(d, s) = p_N(d - s) p_S(s) \,. \tag{8.47e}$$

8.10 Random Noise

So far we have treated signal s and noise n symmetrically. The theory in the preceding section allows s and n to be interchanged, depending upon the user's viewpoint. In real life, however, there is always a strong distinction between what constitutes signal and what constitutes noise. Usually the point of departure is the degree of correlation in x obeyed by the process. A signal s usually has strong correlation over some finite range of x while noise n is uncorrelated in x, obeying

$$\langle n(x;\boldsymbol{\lambda})n(x';\boldsymbol{\lambda})\rangle = m_2(x)\delta(x-x')\ , \tag{8.48a}$$

by (8.9). (Note: $n^* = n$ since in all practical cases we shall consider, n will be a real quantity.) A noise process obeying (8.48a) will be called "random" noise. Note that this term is used irrespective of the probability law $p_{\mathrm{N}}(n)$ obeyed by the noise at x. It may be Gaussian, Poisson, etc.

Quite often, the noise has a mean of zero,

$$\langle n(x;\boldsymbol{\lambda})\rangle = 0\ . \tag{8.48b}$$

In this case, by identity (3.15) $m_2(x) = \sigma_n^2(x)$, the variance at x, and (8.48a) becomes

$$\langle n(x;\boldsymbol{\lambda})n(x';\boldsymbol{\lambda})\rangle = \sigma_n^2(x)\delta(x-x')\ . \tag{8.48c}$$

Because of the very useful sifting property of $\delta(x)$, *this expression vastly simplifies the analysis of noise in many cases.*

If furthermore the noise process is wide-sense stationary, then the whole right-hand side of (8.48c) must solely be a function of the shift $(x-x')$. But then quantity σ_{n}^2 could no longer be a function of absolute position x. Hence, it must be a constant, σ_{n}^2. Then (8.48c) becomes simpler,

$$\langle n(x;\boldsymbol{\lambda})n(x';\boldsymbol{\lambda})\rangle = \sigma_{\mathrm{n}}^2\delta(x-x')\ . \tag{8.48d}$$

Is additive noise ever random noise? Yes, it is often random noise, since its independence from the signal *at least permits* the noise to lack spatial correlation. On the other hand, the noise might still have its own, distinctive correlation effect, independent of the signal. Then it would *not* be random.

We showed in Sect. 8.5 that a process obeying (8.48a) has a white power spectrum. Hence, random or uncorrelated noise has a flat power spectrum, exhibiting all frequencies equally. This is such an important property that it is frequently considered the *definition* of a random noise process. That is, white processes are called "random noise," often quite irrespective of what they *physically* represent.

8.11 Ergodic Property

As in Fig. 8.1, let subscript k of $\boldsymbol{\lambda}_k$ denote the kth particular set of $\boldsymbol{\lambda}$ values. Consider an ensemble of stochastic, random noise processes $n(x;\boldsymbol{\lambda}_k)$, $k = 1, 2, \ldots, N$ as in Fig. 8.10. At any fixed $x = x_0$, there is a probability law $p_{X_0}(n)$ for random values of noise n. This would be formed as the histogram of occurrences of values n at the indicated points x_0 down the figure.

We showed in Sect. 8.10 that random noise is sometimes *additive or signal-independent.* Suppose that this is now true. Then the values *across* any one noise

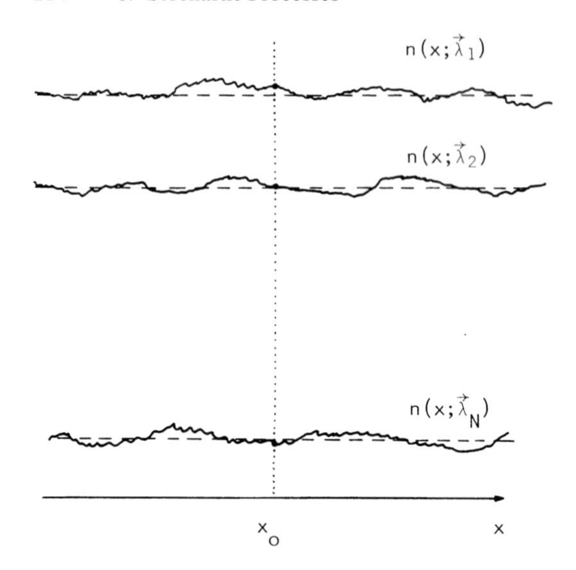

Fig. 8.10. N ergodic random noise processes. The statistics at $x = x_0$ down all the curves equal those *along any one* curve

curve $n(x;\,\boldsymbol{\lambda}_k)$ must be independent samples from the same law $p_{x_0}(n)$ that describes noise from curve to curve *at* x_0. Therefore, a histogram built up of noise occurrences across one curve $n(x;\,\boldsymbol{\lambda}_k)$ should be identical to the probability law $p_{x_0}(n)$, with x_0 fixed, as formed from curve to curve. Finally, it becomes apparent that the law $p_{x_0}(n)$ must be independent of position x_0, and so may simply be specified as $p(n)$. Of course, $p(n)$ remains an arbitrary law.

This is called the "ergodic" theorem or property [8.7]. In practice, it permits the statistics of the noise to be known from observations along one noise profile (provided enough samples of the profile are used to build up the histogram accurately, of course).

As a corollary, any ensemble average, such as $\langle n^2 \rangle$ at a fixed $x = x_0$, also equals the corresponding average across any member of the ensemble. For example, the mean is now

$$\langle n \rangle = \int_{-\infty}^{\infty} \mathrm{d}x\, n(x;\,\boldsymbol{\lambda}_k)\ , \tag{8.49}$$

the second moment is

$$\langle n^2 \rangle = \int_{-\infty}^{\infty} \mathrm{d}x\, n^2(x;\,\boldsymbol{\lambda}_k)\ , \tag{8.50}$$

and so on. These hold for any index value k.

The ergodic theorem plays an invaluable role in practical signal processing problems. For example, in the image restoration problem ordinarily but one image is at hand, from which it is generally impossible to form noise statistics such as $\langle n^2 \rangle$, $R_{\mathrm{n}}(x)$, etc. These require an ensemble average for their calculation. However, it is often possible to find a part of the picture where it is known from context that the signal part of the image must be constant, e.g. where the sky is present, so that all observed fluctuations must be due to noise. Then if the noise obeys the ergodic property, a suitable average *across* it, as in (8.49) or (8.50), will produce the required statistic.

Exercise 8.2

8.2.1 The operation of linear filtering of data has widespread application. Here the data are convolved with a processing kernel $f(x)$ to effect the output. Suppose that the data suffer from additive, zero-mean, random noise $n(x)$, with variance $\sigma_{\mathrm{n}}^2(x)$. Then by (8.44) the error $e(x)$ in the output is itself simply the convolution of $f(x)$ with $n(x)$,

$$e(x) = \int_{-\infty}^{\infty} \mathrm{d}y\, n(y) f(x - y)\ .$$

Since the noise $n(y)$ is a stochastic process, so is the output error $e(x)$. We want to establish all pertinent statistical properties of $e(x)$.

Show that

(a) $\langle e(x) \rangle = 0$

(b) $\sigma_{\mathrm{e}}^2(x) = \int_{-\infty}^{\infty} \mathrm{d}y\, \sigma_{\mathrm{n}}^2(y) f^2(x - y)$

(c) $R_{\mathrm{e}}(x; x_0) = \int_{-\infty}^{\infty} \mathrm{d}y\, \sigma_{\mathrm{n}}^2(y) f(x_0 - y) f(x_0 + x - y)$

(d) $S_{\mathrm{e}}(\omega) = |F(\omega)|^2 \lim_{L\to\infty} \frac{1}{2L} \int_{-L}^{L} \mathrm{d}y\, \sigma_{\mathrm{n}}^2(y)\ .$

Also, if $n(y)$ obeys wide-sense stationarity, show that

(e) $\sigma_{\mathrm{e}}^2(x) = \sigma_{\mathrm{e}}^2 = \sigma_{\mathrm{n}}^2 \int_{-\infty}^{\infty} \mathrm{d}x\, f^2(x)$

(f) $R_{\mathrm{e}}(x) = \sigma_{\mathrm{n}}^2 \int_{-\infty}^{\infty} \mathrm{d}y\, f(y) f(y + x)$

(g) $S_{\mathrm{e}}(\omega) = |F(\omega)|^2\, \sigma_{\mathrm{n}}^2\ .$

8.2.2 It is often important to know whether the real and imaginary parts of spectral processes correlate. For example, let $f(x)$, $g(x)$ be two stochastic processes, with corresponding spectra

$$F(\omega) = F_{\mathrm{re}}(\omega) + \mathrm{j}F_{\mathrm{im}}(\omega)\ ,$$
$$G(\omega) = G_{\mathrm{re}}(\omega) + \mathrm{j}G_{\mathrm{im}}(\omega)\ .$$

We now ask, what are $\langle F_{\mathrm{re}}(\omega) G_{\mathrm{re}}(\omega) \rangle$, $\langle F_{\mathrm{re}}(\omega) G_{\mathrm{im}}(\omega) \rangle$, $\langle F_{\mathrm{im}}(\omega) G_{\mathrm{re}}(\omega) \rangle$, and $\langle F_{\mathrm{im}}(\omega) G_{\mathrm{im}}(\omega) \rangle$? These represent cross correlations at the same frequency ω.

To aid in the calculation, we shall assume that the processes $f(x)$ and $g(x)$ are *strict-sense stationary*, meaning that probability density $p(f, g)$ at any one x equals $p(f, g)$ at any other (shifted) x. We also assume that f and g do not correlate at different points. [Certainly the latter is a reasonable assumption if f and g represent different, and hence independent, physical processes. Even if they are the same physical process, if they are random noise they again do not correlate at different points; see (8.48a).]

Combining the two assumptions, it must be that

$$\langle f(x)g(x')\rangle = \langle fg\rangle\, \delta(x - x') \,.$$

Quantity $\langle fg\rangle$ is the necessarily constant (because of stationarity) cross correlation between f and g *at the same* x.

Using the latter identity in conjunction with definition (8.41b) of the spectra, *show that correlations* $\langle F_{\mathrm{re}} G_{\mathrm{re}}\rangle$ *(ω suppressed) and* $\langle F_{\mathrm{im}} G_{\mathrm{im}}\rangle$ *much exceed* $\langle F_{\mathrm{re}} G_{\mathrm{im}}\rangle$ *and* $\langle F_{\mathrm{im}} G_{\mathrm{re}}\rangle$. This is in the sense that the former two correlations exhibit proportionality to integrals (∞ limits) $\int \mathrm{d}x \cos^2 \omega x$ and $\int \mathrm{d}x \sin^2 \omega x$, respectively, while the latter correlations are proportional to integral $\int \mathrm{d}x \cos \omega x \sin \omega x$. The latter integral is much less than the former two, since the two only add in x, while the latter integral periodically goes to zero in x (by orthogonality).

8.2.3 A stochastic process such as $n(x; \boldsymbol{\lambda})$ may be expanded, over a finite interval, as a series of orthogonal functions in x multiplied by random coefficients $\boldsymbol{\lambda}$. For example, suppose

$$n(x; \boldsymbol{\lambda}) = \sum_{n=1}^{\infty} \lambda_n \cos(n\pi x/L) \text{ for } \langle x\rangle \leqq L \,,$$

where the $\{\lambda_n\}$ are random variables defining the stochastic process. In fact, the $\{\lambda_n\}$ are more specifically here a measure of the "power" allocated to each frequency component of the process.

Suppose that the process is given to be *ergodic*. Then one set of data $n(x; \boldsymbol{\lambda})$ over the interval $|x| \leqq L$ suffices to find the statistics of $n(x; \boldsymbol{\lambda})$. Ensemble averages $\langle\rangle$ are replaced by integrals over x, as in (8.49) and (8.50).

Based on ergodicity, show that

(a) $\langle n(x)\rangle = 0$

(b) $\langle n(x)^2\rangle = L \sum_{n=1}^{\infty} \lambda_n^2$

(c) $R_{\mathrm{n}}(x; x_0) = L \sum_{n=1}^{\infty} \lambda_n^2 \cos(n\pi x/L)$

(d) $S_{\mathrm{n}}(\omega_m) = L^2 \lambda_m^2$, where frequency $\omega_m \equiv m\pi/L$. By Sect. 4.21, $\{\omega_m\}$ are the sampling frequencies corresponding to a situation of "limited support" in $n(x)$, $n(x) = 0$ for $|x| > L$.

8.2.4 ***The Karhunen–Loeve expansion.*** Suppose that the process $n(x)$ has a known autocorrelation function $R_{\mathrm{n}}(x; y - x)$. Let $\{\psi_n(x)\}$ be a complete, orthogonal set of functions defined on interval (a, b), which satisfy the "eigenvalue equation" for R_{n},

$$\int_a^b \mathrm{d}x R_{\mathrm{n}}(y - x; x)\psi_n(x) = \lambda_n \psi_n(y) \,, \quad n = 0, 1, 2, \ldots \,.$$

Functions $\{\psi_n(x)\}$ are called "eigenfunctions," coefficients $\{\lambda_n\}$ "eigenvalues."

Let the process $n(x)$ be represented by an orthogonal series *in terms of eigenfunctions* $\{\psi_n(x)\}$,

$$n(x) = \sum_{n=0}^{\infty} c_n \psi_n(x) \,.$$

This particular choice of series for $n(x)$ is called a "Karhunen-Loeve expansion."

Then coefficients $\{c_n\}$ are RV's that in particular *do not correlate*, obeying

$$\langle c_m c_n \rangle = \lambda_n \delta_{mn} .$$

Prove this.

8.12 Optimum Restoring Filter

Suppose that an image $i(y)$ suffers from blur according to a deterministic spread function $s(x)$ and also suffers from additive noise $n(y)$. It is desired to estimate the object $o(x)$ which gave rise to the image [8.8].

Since the object and noise are unknowns, we regard them as stochastic processes $o(x; \boldsymbol{\lambda})$, $n(y; \boldsymbol{\lambda}')$, with corresponding power spectra $S_o(\omega)$ and $S_n(\omega)$. The imaging equation (4.22) is now

$$i(y) = \int_{-\infty}^{\infty} \mathrm{d}x\, o(x; \boldsymbol{\lambda}) s(y - x) + n(y; \boldsymbol{\lambda}') . \tag{8.51}$$

Hence, the image $i(y)$ itself is a stochastic process.

It is convenient to Fourier transform both sides of (8.51), producing

$$I(\omega) = T(\omega) \cdot O(\omega) + N(\omega) \tag{8.52}$$

in terms of corresponding spectral quantities (8.41b). The notation $\boldsymbol{\lambda}$, $\boldsymbol{\lambda}'$ has been suppressed for brevity. Spectra O, N, and I are all stochastic processes. I is known from the data; O and N are to be estimated. As we shall see, the answer will depend upon knowledge of their power spectra $S_o(\omega)$ and $S_n(\omega)$.

8.12.1 Definition of Restoring Filter

One way to form an estimate $\hat{O}(\omega)$ of $O(\omega)$ is to form it as

$$\hat{O}(\omega) \equiv Y(\omega) I(\omega) , \tag{8.53}$$

where $Y(\omega)$ is a *restoring filter function* that is to be found. $Y(\omega)$ is defined as follows.

Let the estimate $\hat{o}(x)$ corresponding to (8.53) depart minimally from the true $o(x)$ in the mean-square sense,

$$e \equiv \left\langle \int_{-\infty}^{\infty} \mathrm{d}x \left| o(x) - \hat{o}(x) \right|^2 \right\rangle \equiv \min . \tag{8.54}$$

The minimum is with respect to the choice of $Y(\omega)$. This is called a minimum mean-square error (mmse) criterion. It may be transformed into frequency space, enabling power spectra to be introduced, as follows.

Parseval's theorem is the mathematical identity

$$\int_{-\infty}^{\infty} \mathrm{d}x\, |f(x)|^2 = \int_{-\infty}^{\infty} \mathrm{d}\omega\, |F(\omega)|^2 . \tag{8.55}$$

(This may easily be proven by substituting for $F(\omega)$, on the right-hand side, its definition (8.41b), integrating through $d\omega$ to obtain a $\delta(x - x')$, and then using the sifting property (3.30d) of $\delta(x)$ to produce the left-hand side.)

Applying Parseval's theorem (8.55) to (8.54) yields

$$e = \left\langle \int_{-\infty}^{\infty} d\omega \left| O(\omega) - \hat{O}(\omega) \right|^2 \right\rangle = \min . \tag{8.56a}$$

Using the theorem (3.28) that the mean of a sum (or integral) equals the sum (or integral) of the means, (8.56a) becomes

$$e = \int_{-\infty}^{\infty} d\omega \left\langle \left| O(\omega) - \hat{O}(\omega) \right|^2 \right\rangle . \tag{8.56b}$$

Squaring out the modulus yields

$$e = \int_{-\infty}^{\infty} d\omega \left[\langle |O|^2 \rangle + \left\langle \left|\hat{O}\right|^2 \right\rangle - \left\langle O\hat{O}^* \right\rangle - \left\langle O^*\hat{O} \right\rangle \right] . \tag{8.56c}$$

Arguments ω have been suppressed for brevity. Substitution of (8.52) and (8.53) yields

$$\begin{aligned} e = \int_{-\infty}^{\infty} & d\omega [\langle |O|^2 \rangle + |Y|^2 \left(|T|^2 \langle |O|^2 \rangle + T^* \langle O^* N \rangle + T \langle O N^* \rangle + \langle |N|^2 \rangle \right) \\ & - Y^* T^* \langle |O|^2 \rangle - Y^* \langle O N^* \rangle - Y T \langle |O|^2 \rangle - Y \langle O^* N \rangle] \end{aligned} \tag{8.56d}$$

8.12.2 Model

To proceed further we have to assume something about the physical nature of the two processes o and n. This comprises a model. We realistically assume o and n to be physically different processes, so that they do not correlate at any points x and x' whether $x \neq x'$ or $x = x'$, and that n has zero mean at each x. Effectively then, *n is additive noise.*

Then we find by direct use of definition (8.41b) that

$$\langle O^* N \rangle = \int_{-\infty}^{\infty} dx\, e^{-j\omega x} \int_{-\infty}^{\infty} dx'\, e^{-j\omega x'} \langle o(x) n(x') \rangle = 0 , \tag{8.57}$$

since $\langle o(x) n(x') \rangle = 0$ by the model.

Accordingly, (8.56d) simplifies to

$$\begin{aligned} e = \int_{-\infty}^{\infty} & d\omega \{ \tilde{S}_o(\omega) + YY^* [|T|^2 \tilde{S}_o(\omega) + \tilde{S}_n(\omega)] - Y^* T^* \tilde{S}_o(\omega) - Y T \tilde{S}_o(\omega) \} \\ & = \text{minimum} \end{aligned} \tag{8.58}$$

where we temporarily used form (8.41a) of the power spectrum.

8.12.3 Solution

Recalling that the processing filter $Y(\omega)$ sought is complex, it then has two degrees of freedom, its real and imaginary parts, which must be determined. It turns out to be simpler yet to instead regard Y and Y^* as the two, distinct degrees of freedom. Then we may merely solve (8.58) by using ordinary Euler–Lagrange theory for variable Y^*. By Eq. (G.14) of Appendix G, the Lagrange solution for Y^* is

$$\frac{\mathrm{d}}{\mathrm{d}\omega}\left(\frac{\partial \mathcal{L}}{\partial \dot{Y}^*}\right) = \frac{\partial \mathcal{L}}{\partial Y^*}\,, \tag{8.59}$$

where $\mathcal{L}$ is the integrand of (8.58) and $\dot{Y}^* \equiv \mathrm{d}\dot{Y}^*/\mathrm{d}\omega$. We note from (8.58) that $\mathcal{L}$ is *not* a function of $\dot{Y}^*$, so that $\partial\mathcal{L}/\partial\dot{Y}^* = 0$. Then (8.59) becomes

$$\partial\mathcal{L}/\partial Y^* = 0\,. \tag{8.60}$$

Applying this principle to (8.58) yields an immediate answer

$$Y(\omega) = \frac{T^*(\omega)\tilde{S}_\mathrm{o}(\omega)}{|T(\omega)|^2\,\tilde{S}_\mathrm{o}(\omega) + \tilde{S}_\mathrm{n}(\omega)} \tag{8.61}$$

in terms of the associated power spectra $\tilde{S}_\mathrm{o}$, $\tilde{S}_\mathrm{n}$. The answer may alternatively be put in terms of the ordinary power spectra S_o, S_n by merely dividing numerator and denominator by length $2L$ of the image, and letting $L \to \infty$. The result is

$$Y(\omega) = \frac{T^*(\omega)S_\mathrm{o}(\omega)}{|T(\omega)|^2\,S_\mathrm{o}(\omega) + S_\mathrm{n}(\omega)}\,. \tag{8.62}$$

This is the processing filter we sought on the basis of minimum mean-square error. It is called a *Wiener-Helstrom* filter, in honor of its discoverers.

By substituting the solution (8.61) back into (8.58), we find what the minimized error $e_{\min}$ is,

$$e_{\min} = \int_{-\infty}^{\infty} \mathrm{d}\omega \frac{\tilde{S}_\mathrm{n}(\omega)\tilde{S}_\mathrm{o}(\omega)}{|T(\omega)|^2\,\tilde{S}_\mathrm{o}(\omega) + \tilde{S}_\mathrm{n}(\omega)}\,. \tag{8.63}$$

Equations (8.61) and (8.63) together show what the optimum filter does in its quest to minimize error e, and what the effect is upon $e_{\min}$. We see from (8.61) that at frequencies ω for which $\tilde{S}_\mathrm{n} \gg \tilde{S}_\mathrm{o}$, i.e., high noise regions, $Y(\omega) \simeq 0$. By processing operation (8.53) this has the effect of blanking out these data frequencies from the output $\hat{o}(x)$. The data are too untrustworthy since the noise swamps the signal. Blanking-out also occurs for frequencies beyond cutoff Ω in $T(\omega)$, since that data $I(\omega)$ is *purely noise* $N(\omega)$.

Conversely, where there is low noise ($\tilde{S}_\mathrm{o} \gg \tilde{S}_\mathrm{n}$), (8.61) becomes $Y(\omega) \simeq 1/T(\omega)$ or an "inverse filter." By (8.52, 53), at these ω unattenuated, and true, object values $O(\omega)$ are permitted to be output into $\hat{o}(x)$. The data can be trusted at these ω.

Regarding error e, (8.63) shows that where the noise is high the whole object power spectrum $\tilde{S}_\mathrm{o}$ contributes to the error, whereas where the noise is low, as long as $|T(\omega)|$ is finite, essentially zero times the object power spectrum contributes to the error.

In summary, the optimal, Wiener-Helstrom filter $Y(\omega)$ does its job by rejecting high-noise frequency components and accepting low-noise components. Although this sounds like the proper thing for a filter to do, it turns out to produce outputs $\hat{o}(x)$ that are, to the eye, often rather blurred. Even when the noise is modest, the method sacrifices too much high-frequency detail. Caveat emptor!

Figure 8.11 shows some typical restorations by this approach, along with corresponding restorations using pure inverse filtering $Y = 1/T(\omega)$. The image that contains a significant amount of noise (at top left) is poorly restored by either filter.

The unsatisfactory level of blur permitted by this approach may be reduced somewhat by means of a "sharpness constraint" tacked onto the mmse principle (8.54). See Ex. 8.3.2 below.

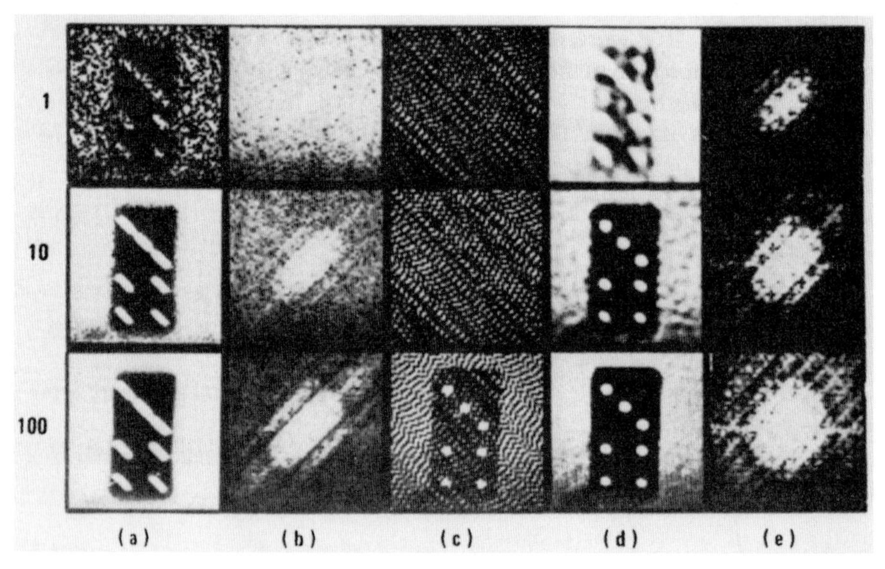

Fig. 8.11. Comparison of restorations by Wiener-Helstrom, and inverse, filtering, (**a**) Images suffering motion blur, with three states of noise, (**b**) Their Fourier transforms, (**c**) The images restored by inverse filtering, (**d**) The images restored by Wiener-Helstrom filtering, (**e**) Fourier transforms of restorations [8.9]

Exercise 8.3

8.3.1 Suppose that the object process is close to being "random noise," with $\tilde{S}_o(\omega) = S_o \operatorname{Rect}(\omega/2\Omega)$, Ω large, and that the noise is "white," $\tilde{S}_n(\omega) = S_n =$ constant for all ω. Also, the transfer function $T(\omega)$ is a triangle function $T(\omega) = (1 - |\omega|/\Omega) \operatorname{Rect}(\omega/2\Omega)$.

(a) Determine the optimum Wiener filter $Y(\omega)$ and its resulting error $e_{\min}$ under these conditions.

(b) Suppose that now the object power spectrum obeys $\tilde{S}_o(\omega) = S_o \operatorname{Rect}(\omega/2\Omega')$, where $\Omega' > \Omega$. What are the new $Y(\omega)$ and $e_{\min}$?

8.3.2 ***Sharpness constrained-Wiener filter.*** The "sharpness" $\mathcal{L}$ in a function $\hat{o}(x)$ is defined as

$$\mathcal{L} \equiv \int_{-\infty}^{\infty} \mathrm{d}x \, [\hat{o}'(x)]^2 \tag{8.64}$$

where $\hat{o}'(x)$ is the derivative $\mathrm{d}\hat{o}(x)/\mathrm{d}x$. We note that if there is an edge-gradient somewhere in the scene $\hat{o}(x)$, the steeper that gradient is the more it contributes to $\mathcal{L}$. Hence, $\mathcal{L}$ measures the amount of "edge content" there is within the scene. If $\mathcal{L}$ were permitted to be an input variable to a restoring method, the user could at liberty "tune" the output to whatever degree of edge content he wished.

By Parseval's theorem (8.55) and the definition (8.41b) of a spectrum, (8.64) becomes

$$\mathcal{L} = \int_{-\infty}^{\infty} \mathrm{d}\omega \, \omega^2 \left| \hat{O}(\omega) \right|^2 . \tag{8.65}$$

It is desired to supplement the mmse criterion (8.54) with a constraint on average sharpness,

$$\langle \mathcal{L} \rangle = \mathcal{L}_0 ,$$

$\mathcal{L}_0$ fixed. This constraint may be accomplished by use of the Lagrange multiplier technique. The new problem is, by Eq. (G.18) of Appendix G,

$$\left\langle \int_{-\infty}^{\infty} \mathrm{d}\omega \left| O(\omega) - \hat{O}(\omega) \right|^2 \right\rangle + \lambda \left[\left\langle \int_{-\infty}^{\infty} \mathrm{d}\omega \, \omega^2 \left| \hat{O}(\omega) \right|^2 \right\rangle - \mathcal{L}_0 \right] = \text{minimum} . \tag{8.66a}$$

Parameter λ is a Lagrange multiplier. Assume (8.52) and (8.53) to still hold.

Show that the $Y(\omega)$ achieving the minimum in problem (8.66a) is a *serial* processor

$$Y_s(\omega) = Y(\omega) \left(\frac{1}{1 + \lambda\omega^2} \right) , \tag{8.66b}$$

where $Y(\omega)$ is the Wiener-Helstrom filter (8.62). The quantity in parentheses is called a "sharpness" filter. By adjusting parameter λ, particularly its sign, the user can selectively enhance or blur high-frequency details. For example, a negative λ will make the sharpness filter much exceed 1 for frequencies near $\sqrt{-1/\lambda}$. If λ is small enough, this corresponds to the high-frequency range.

8.13 Information Content in the Optical Image

The information $I(X, Y)$ in an optical image $i(x)$, or equivalently in its spectrum $I(\omega)$, is a measure of its quality [8.10]. The information in spectrum $I(\omega)$ might be expected to depend upon power spectra, such as $S_o(\omega)$ and $S_n(\omega)$. These measure the randomness of the input to the image channel, and its noise properties. We shall see that this dependence is indeed present.

But correspondingly, we found in Sect. 8.12 that the Wiener-Helstrom filter $Y(\omega)$ and its attendant error also depend upon quantities $S_o(\omega)$ and $S_n(\omega)$. It might be possible, then, to combine results and *directly express $Y(\omega)$ and e in terms of the available image information.* This would be an expression of the ability to retrieve or restore object structure, as a function of the amount of image information at hand. Certainly, we could expect the concept of "information" to make such a prediction. These steps will be carried out in Sect. 8.14. The results show an interesting and simple tie-in between the concepts of information and restorability.

The imaging equation (8.52) at any one frequency ω may be expressed as

$$D = S + N\,, \quad S = T \cdot O\,. \tag{8.67}$$

The notation is meant to express that information theory regards I as primarily an output datum D, and $T \cdot O$ as primarily an input signal S. We now seek the trans-information $I(S, D)$ between input signal and output datum. According to (3.25) this obeys

$$I(S, D) = H(D) - H(D|S) \tag{8.68}$$

at any one frequency ω.

The general plan of analysis is to determine $H(D)$ and $H(D|S)$, form their difference (8.68), which represents the information at one frequency ω, and then sum over all frequencies to yield the total information. Only those frequencies ω within the "optical passband" $(-\Omega, \Omega)$ will be seen to contribute.

8.13.1 Statistical Model

• We assume that each frequency acts statistically independent of the others, i.e., the joint probability separates,

$$p_{D_1, D_2, \ldots, D_n}(d_1, d_2, \ldots, d_n) = p_{D_1}(d_1) p_{D_2}(d_2) \ldots p_{D_n}(d_n)\,. \tag{8.69}$$

The subscripts denote frequency number. Similar products hold for the joint behavior, over frequencies, for S and for N.

• Note further that the spectral quantities D, S, and N are complex numbers, having real and imaginary parts. We shall assume, as derived in Ex. 8.2.2, that real and imaginary parts of D, S, and N *do not correlate* at each frequency.

• Recalling that the probability of a complex number is *defined* as the joint probability of its real and imaginary parts (Sect. 2.24), each of the quantities D in (8.69) represents a number couplet $(D_{\text{re}}, D_{\text{im}})$; similarly for S and N.

• For simplicity, it shall be assumed that *real and imaginary parts* of any spectral quantity at a given frequency *are identically distributed.* Hence, if RV $D_{2\text{re}}$ is normal with mean m and variance σ^2, so is RV $D_{2\text{im}}$.

• Finally, consider the noise N. This shall be assumed to be additive. It may be recalled that this was assumed as well in the model (Sect. 8.12.2) for deriving the optimum restoring filter $Y(\omega)$. Ultimately, use of the same model here will allow us to combine the results of this section, on information, with the optimum filter $Y(\omega)$ given by Eq. (8.61). An additional property for noise N is specified in Sect. 8.13.4.

8.13.2 Analysis

Quantity $H(D|S)$ is defined in Sect. 3.7 as

$$H(D|S) \equiv -\iint_{-\infty}^{\infty} \mathrm{d}u\,\mathrm{d}v\, p_{\mathrm{SD}}(u,v) \ln p(v|u)\ . \tag{8.70}$$

(Again note that all quantities u, v, S, D have implicitly real and imaginary parts, so that $\mathrm{d}u = \mathrm{d}u_{\mathrm{re}}\,\mathrm{d}u_{\mathrm{im}}$, etc.)

In order to evaluate this integral we have to further specify probability $p(v|u)$. Since N is additive noise, we may use the result (8.47d),

$$p(v|u) = p_{\mathrm{N}}(v-u)\ . \tag{8.71}$$

Also, by identity (8.47e),

$$p_{\mathrm{SD}}(u,v) = p_{\mathrm{N}}(v-u)\,p_{\mathrm{S}}(u)\ . \tag{8.72}$$

Substitution of (8.71) and (8.72) into (8.70) produces

$$H(D|S) = -\int_{-\infty}^{\infty} \mathrm{d}u\, p_{\mathrm{S}}(u) \int_{-\infty}^{\infty} \mathrm{d}v\, p_{\mathrm{N}}(v-u) \ln p_{\mathrm{N}}(v-u)\ .$$

Changing variable $v - u \equiv n$, we get a separation of integrals,

$$H(D|S) = -\int_{-\infty}^{\infty} \mathrm{d}u\, p_{\mathrm{S}}(u) \int_{-\infty}^{\infty} \mathrm{d}n\, p_{\mathrm{N}}(n) \ln p_{\mathrm{N}}(n)\ .$$

Then by normalization of RV u,

$$H(D|S) = H(N)\ , \tag{8.73a}$$

where

$$H(N) \equiv -\int_{-\infty}^{\infty} \mathrm{d}n\, p_{\mathrm{N}}(n) \ln p_{\mathrm{N}}(n)\ , \tag{8.73b}$$

the entropy of the noise.

This result is very important, because it states that $H(D|S)$ does not depend upon signal S and is solely a function of the noise statistics. In particular, information equation (8.68) is now

$$I(S,D) = H(D) - H(N)\ . \tag{8.74}$$

The trans-information is the difference between two entropies, that of the data D and that of the noise N. These are separately fixed by probability laws $p_{\mathrm{D}}(d)$ and $p_N(n)$, once the latter are specified.

8.13.3 Noise Entropy

The noise has so far been specified as additive, and with independent real and imaginary parts. Let it also be zero mean and Gaussian. Then the joint (since complex) probability $p_{\mathrm{N}}(n)$ separates into a product of Gaussians

$$p_N(n) = \frac{1}{\sqrt{2\pi \langle n_{re}^2 \rangle}} \exp\left(-n_{re}^2/2\langle n_{re}^2 \rangle\right) \cdot \frac{1}{\sqrt{2\pi \langle n_{im}^2 \rangle}} \exp\left(-n_{im}^2/2\langle n_{im}^2 \rangle\right) . \tag{8.75}$$

But the associated power spectrum $\tilde{S}_n$ is defined at (8.41a) as

$$\tilde{S}_n = \langle n_{re}^2 \rangle + \langle n_{im}^2 \rangle .$$

Also, we assumed that real and imaginary parts of all quantities were to be identically distributed. Then

$$\langle n_{re}^2 \rangle = \langle n_{im}^2 \rangle = \tilde{S}_n/2$$

and (8.75) becomes simply

$$p_N(n) = \frac{1}{\pi \tilde{S}_n} \exp(-|n|^2/\tilde{S}_n) . \tag{8.76}$$

By substituting this result into (8.73b) and integrating out the real and imaginary parts, it is found that

$$H(N) = 1 + \ln(\pi \tilde{S}_n) . \tag{8.77}$$

8.13.4 Data Entropy

The preceding equation shows that if the power spectrum $\tilde{S}_n$ of the noise is known, the entropy $H(N)$ is known. Let us *suppose* correspondingly *that the power spectrum* $\tilde{S}_d$ *of the data is known,* from observing and averaging over many data sets. What will be the data entropy $H(D)$?

Of course, many different possible laws $p_D(d)$ can give rise to the given power spectrum, and each law will have a generally different entropy $H(D)$. Which, then, should be chosen?

We see from (8.74) that, with $H(N)$ fixed, the $p_D(d)$ that produces a maximum $H(D)$ will also maximize the trans-information. What is the significance of attaining a maximum trans-information? Obviously, *being maximized with respect to data statistic* $p_D(d)$, the resulting $I(S, D)$ *reflects the ability of the information channel per se to transmit information.* That is, it measures the capacity of the image-forming medium for transmitting information, regardless of its output. We shall take this to be of fundamental importance in our investigation. The maximized information $I(S, D)$ is called the "channel capacity," and is a quality measure of the imaging channel.

Hence, we seek the function $p_D(d)$ whose entropy $H(D)$ is a maximum, when the power spectrum $\tilde{S}_d$ is specified. Again under the assumption that real and imaginary parts are identically distributed, the answer is (Sect. 10.4.9)

$$p_D(d) = \frac{1}{\pi \tilde{S}_d} \exp(-|d - \langle d \rangle|^2/\tilde{S}_d) . \tag{8.78}$$

That is, the solution is the normal distribution. The resulting entropy $H(D)$ turns out to be independent of mean $\langle d \rangle$ and to hence have the same form as (8.77),

$$H(D) = 1 + \ln(\pi \tilde{S}_d) . \tag{8.79}$$

8.13.5 The Answer

Combining results (8.77, 79) with (8.74), we have

$$I(S, D)_{\max} \equiv C = \ln(\tilde{S}_{\rm d}/\tilde{S}_{\rm n}) \ . \tag{8.80}$$

Hence, the channel capacity is the logarithm of the ratio of the power spectrum of the image data to that of the noise. We note that because $\tilde{S}_{\rm d}$ and $\tilde{S}_{\rm n}$ are ratioed in (8.80), by definitions (8.1b) and (8.41a) the maximized information may also be expressed in terms of ordinary power spectra, as

$$I(S, D)_{\max} \equiv C = \ln(S_{\rm d}/S_{\rm n}) \ . \tag{8.81}$$

It is mathematically more convenient, however, to continue using the form (8.80).

This is the information at any one frequency. According to plan, we merely sum over frequency to get the total information,

$$I(S, D)_{\rm total} = \int_{-\infty}^{\infty} {\rm d}\omega \ln[\tilde{S}_{\rm d}(\omega)/\tilde{S}_{\rm n}(\omega)] \ . \tag{8.82}$$

In this manner, the functional forms of the power spectra enter in. This result was first derived by *Fellgett* and *Linfoot* [8.10].

By squaring both sides of (8.67), we find that

$$\tilde{S}_{\rm d}(\omega) = \tilde{S}_{\rm s}(\omega) + \tilde{S}_{\rm n}(\omega) + \langle S^* N\rangle + \langle SN^*\rangle \tag{8.83}$$

after taking the expectation. If we assume that signal image and noise arise from different physical causes, then as in Sect. 8.12.2, $\langle S^* N\rangle = \langle SN^*\rangle = 0$. Accordingly, (8.83) becomes

$$\tilde{S}_{\rm d}(\omega) = |T(\omega)|^2\, \tilde{S}_{\rm o}(\omega) + \tilde{S}_{\rm n}(\omega) \ . \tag{8.84}$$

Identity (8.67) was also used. Substituting this into (8.82) yields

$$I(S, D)_{\rm total} = \int_{-\infty}^{\infty} {\rm d}\omega \ln\left[1 + \frac{|T(\omega)|^2\, \tilde{S}_{\rm o}(\omega)}{\tilde{S}_{\rm n}(\omega)}\right] \ . \tag{8.85}$$

This is the final answer. Ordinary power spectra may replace those in (8.85), as at (8.81).

8.13.6 Interpretation

Equation (8.85) is worth discussing for the insight it provides regarding image formation. First we see that all frequencies contribute in like manner. Hence, insofar as information is concerned, high and low frequencies convey potentially the same amount of information. Or, *information is not a measure of pre-existing resolution.* This goes against intuition. However, we shall show in Sect. 8.14 that the amount of information present has a bearing on the ability to enhance resolution by a *later* restoring step. The higher the existing information the better the *ultimate* resolution can be made. Hence, information does measure ultimate resolution, but only in a restored version of the given image.

Next, frequencies outside the optical passband $(-\Omega, \Omega)$ obey $T(\omega) = 0$, and hence do not contribute to the information. This is what one would intuitively expect, since such image frequencies contain only noise. Hence, the infinite limits in result (8.85) may be replaced by $\pm\Omega$.

The system optics enter expression (8.85) through $|T|^2$, which depends in no way upon the phase of the transfer function. This is at first surprising, since it says that an image suffering from a lot of phase shift error, including contrast reversal and spurious oscillation, can still have a great deal of information. However, as in the resolution dilemma above, the situation is remedied by allowing the image to be later restored. With high information present, phase errors can be corrected out.

Finally, we note the interesting ratio $|T|^2/\tilde{S}_n$ in (8.85). This states that if the noise power $\tilde{S}_n$ is low enough, even poor-quality optics, with a low $|T|$, can pass high information into the image. So *ultimately, it is noise, and not system aberrations, that degrades the information content.* This will become clearer in the next section, where it will be shown that noise is also the ultimate limitor of the ability to restore an image.

To illustrate this point, we ask the reader to judge which of the two photos in Fig. 8.12 contains more information. The left-hand one is strongly blurred, with low $|T(\omega)|$, but contains no noise (ideally). The right-hand one has noise, but no blur (ideally). According to (8.85), the left-hand image is superior. In fact, it contains an infinite amount of information! (Since it has no noise, it can be perfectly restored.)

In summary, information is a valid and intuitive figure of merit for an image that is later to be restored. In effect, the restoring step removes most of the systematic or deterministic kinds of degradation in the image, leaving only the random ones behind. Information is in the main sensitive to the random kinds of degradation.

8.14 Data Information and Its Ability to be Restored

In Sect. 8.12 we derived the optimum mmse filter $Y(\omega)$ and its ensuing mean-square error e, under the *sole condition* that signal and noise arise from different physical

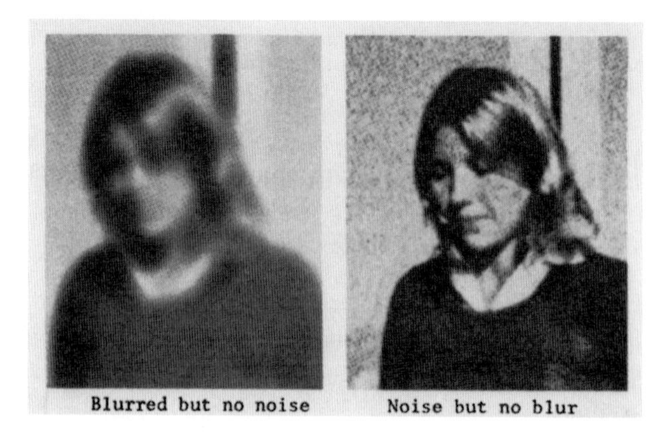

Fig. 8.12. Which picture contains more information?

causes and hence are independent. In Sect. 8.13 we determined the maximum information in an image, under the same condition of independence, and also assumptions on the independence of real and imaginary parts of spectra. This is then a special case of the model used in derivation of $Y(\omega)$. Hence, the results may be combined, giving some insight into how the availability of information limits the ability to restore [8.11].

We let an information *density* $C(\omega)$ be defined as

$$C(\omega) \equiv \ln\left[1 + \frac{|T(\omega)|^2\, \tilde{S}_o(\omega)}{\tilde{S}_n(\omega)}\right] . \tag{8.86}$$

Then (8.85) becomes simply

$$I(S, D)_{\text{total}} = \int_{-\infty}^{\infty} d\omega\, C(\omega) . \tag{8.87}$$

We now solve (8.86) for $\tilde{S}_n(\omega)$ and substitute this into (8.61, 63) for $Y(\omega)$ and $e_{\min}$. The results are the surprisingly simple expressions

$$Y(\omega) = [1/T(\omega)][1 - e^{-C(\omega)}] \tag{8.88a}$$

and

$$e_{\min} = \int_{-\infty}^{\infty} d\omega\, \tilde{S}_o(\omega)\, e^{-C(\omega)} . \tag{8.88b}$$

These relate the optimum filter and its resulting error to the information density $C(\omega)$ *per se.*

Equations (8.88a) give important insight into the manner by which available information limits the ability to restore. From (8.88a), at any frequency ω for which the information density $C(\omega)$ is high, the optimum filter $Y(\omega)$ is virtually a pure inverse filter $1/T(\omega)$. From (8.88b), in this case zero error is contributed to $e_{\min}$ as well. These effects make sense, since by (8.86) $C(\omega)$ can only be very high if the noise $\tilde{S}_n(\omega)$ is very low, and we found below (8.63) that in this case the *true* object value $O(\omega)$ is restored.

Conversely, where information density $C(\omega)$ is low, (8.88a) shows that $Y(\omega)$ is about equal to $C(\omega)/T(\omega)$, indicating near rejection of these frequency components. Also the penalty paid in error is shown in (8.88b) to be the largest contribution possible, $\tilde{S}_o$.

8.15 Superposition Processes; the Shot Noise Process

In many physical processes, data are formed as the random superposition over space, or over time, of one function. The latter is called a "disturbance" function $h(x)$ and has a known form. For example, we found that short-term atmospheric turbulence causes the image of a point (star) to be a random scatter of Airy-like functions, as in Fig. 8.5. Each Airy-like function has about the same form $h(x)$, and these are centered upon random points $\{x_n\}$. For simplicity, we use one-dimensional notation. (Note that some of the $\{x_n\}$ in the figure are so close together that their $h(x)$ functions overlap, forming wormlike features.) Let there be randomly m disturbances in all. Accordingly, we postulate the superposition process to obey the following statistical model:

- A known disturbance function $h(x)$ exists.
- It is centered upon m random positions x_n, $n = 1, \ldots, m$, where m is random according to a known law $P(m)$.
- The disturbances add, so that the total disturbance at any point x obeys

$$i(x) = \sum_{n=1}^{m} h(x - x_n) . \tag{8.89}$$

  The situation is illustrated in Fig. 8.13.
- The $\{x_n\}$ are random variables that are independent samples from one, known probability law $p_X(x)$.

From the last model assumption and (8.89), $i(x)$ is a RV at each x. In fact,

$$i(x) = i(x; \{x_n\}) \tag{8.90}$$

so that $i(x)$ is a stochastic process. *We are interested in establishing the statistics of i at any fixed point $x = x_0$.*

Problems of the form (8.89) abound in optics. The following three are typical, and shall be considered further on.

By one model of the photographic emulsion, the individual grains are randomly scattered over the emulsion, and each grain has the same density profile $d(\boldsymbol{r})$. We want to know the statistics of the net density at any one point.

Another problem is the "swimming pool" effect. We have all seen the shimmering light pattern on the bottom of a pool due to the random refraction of sunlight at the water surface. If i is the net intensity at a point on the pool bottom, what is $p(i)$? This problem can be analyzed by assuming that the net intensity at the given point is due to the simultaneous presence of one intensity profile $h(x)$ centered at random positions $\{x_n\}$ across the pool bottom. Each originated at an individual wave $n = 1, 2, \ldots$ of the pool surface. The random positions x_n are related to the corresponding surface wave positions and random tilts by simple application of Snell's law. A relation like (8.89) is once again obeyed.

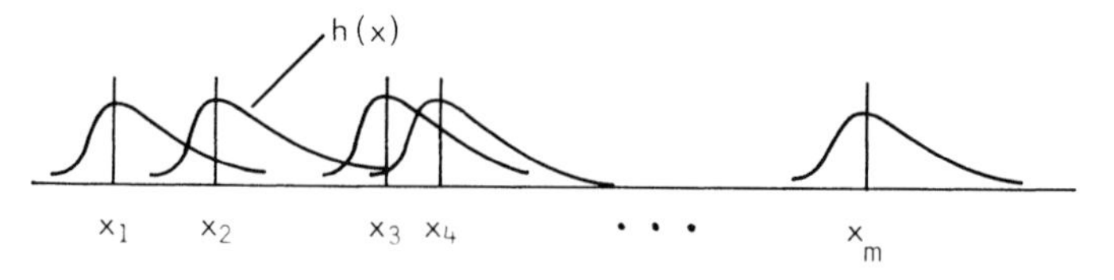

Fig. 8.13. A *superposition process* $i(x; x_1, \ldots, x_m)$ is the superposition of a disturbance function $h(x)$ displaced randomly by values $x_1, \ldots, x_m$

Finally, consider the "shot noise" process. Here photons arrive at a detector at random times $\{t_n\}$, where $p_{T_n} \equiv p_T(t)$ since each t_n obeys the same law. Probability $p_T(t)$ is a uniform law, and m a Poisson RV. Each arrival causes a transient current of the form $h(t)$ in the detector circuit, so that the total current $i(t)$ obeys the form (8.89) with x replaced by t.

8.15.1 Probability Law for *i*

We first want to know the probability density for $i(x_0)$, x_0 an arbitrary but fixed value of x. Denote this probability as $p_I(i)$.

Examining (8.89) at $x = x_0$ fixed, we see that i is a sum of independent RV's $y_n \equiv h(x_0 - x_n)$, where h has the new role of a transformation function (5.1a) from old RV's $(x_0 - x_n)$ to new RV's y_n. The $\{y_n\}$ are independent because the $\{x_n\}$ are given as independent. Thus

$$i(x_0) = \sum_{n=1}^{m} y_n \,, \tag{8.91}$$

and with $m \gtrsim 3$, intensity $i(x_0)$ obeys all the requirements for satisfying the central limit theorem (Sect. 4.27). Hence, $i(x_0)$ is normally distributed. It remains to find the mean $\langle i \rangle$ and second moment $\langle i^2 \rangle$.

8.15.2 Some Important Averages

It will facilitate all the analysis to follow by taking the Fourier transform of (8.89), yielding

$$I(\omega) = H(\omega) \sum_{n=1}^{m} \mathrm{e}^{-\mathrm{j}\omega x_n} \,. \tag{8.92}$$

This separates out the deterministic part $H(\omega)$ from the random part $\exp(-\mathrm{j}\omega x_n)$, and easily allows averages over RV's x_n to be taken. The following identities will be needed.

Since the $\{x_n\}$ all follow the one probability law $p_X(x)$, it follows that

$$\left\langle \mathrm{e}^{-\mathrm{j}\omega x_n} \right\rangle \equiv \int_{-\infty}^{\infty} \mathrm{d}x\, p_X(x)\, \mathrm{e}^{-\mathrm{j}\omega x} \tag{8.93a}$$

$$= \varphi_X(-\omega) \,, \tag{8.93b}$$

which is the characteristic function (4.1). Also, since the x_n are idependent

$$\left\langle \mathrm{e}^{-\mathrm{j}(\omega x_n + \omega' x_k)} \right\rangle \equiv \iint \mathrm{d}x_n\, \mathrm{d}x_k\, p_X(x_n) p_X(x_k)\, \mathrm{e}^{-\mathrm{j}\omega x_n}\, \mathrm{e}^{-\mathrm{j}\omega' x_k}$$

$$= \varphi_X(-\omega)\varphi_X(-\omega') \,, \quad n \neq k \tag{8.94a}$$

while

$$\left\langle \mathrm{e}^{-\mathrm{j}(\omega x_n + \omega' x_n)} \right\rangle \equiv \int \mathrm{d}x_n\, p_X(x_n)\, \mathrm{e}^{-\mathrm{j}x_n(\omega + \omega')} = \varphi_X(-\omega - \omega') \,. \tag{8.94b}$$

(Note: Unless otherwise indicated, all integrals have infinite limits.) Some important averages over the x_n may now be found.

From (8.92, 93b),

$$\langle I(\omega)\rangle = H(\omega)\cdot\sum_{n=1}^{m}\left\langle \mathrm{e}^{-\mathrm{j}\omega x_n}\right\rangle = H(\omega)\cdot m\varphi_X(-\omega) \tag{8.95}$$

since the result (8.93b) holds *for each* of the m RV's x_n.

Also, from (8.92) and (8.94a, b)

$$\left\langle I(\omega)I(\omega')\right\rangle = H(\omega)H(\omega')[(m^2-m)\varphi_X(-\omega)\varphi_X(-\omega') + m\varphi_X(-\omega-\omega')] \,. \tag{8.96}$$

Factors $(m^2 - m)$ and m arise because product $I(\omega)I(\omega')$ is proportional to

$$\sum_{n=1}^{m}\mathrm{e}^{-\mathrm{j}\omega x_n}\sum_{k=1}^{m}\mathrm{e}^{-\mathrm{j}\omega' x_k} \,,$$

which has m^2 terms total, of which m have $n = k$ and consequently $(m^2 - m)$ have $n \neq k$.

We want to emphasize at this point that the probability law $p_{X_n}(x) \equiv p_X(x)$ for the random positions $\{x_n\}$ is left completely arbitrary in the derivations that follow. This is to permit the broadest possible application of the results. In most presentations, only the shot noise case of a uniform $p_X(x)$ is considered. We take the shot noise case as a particular example.

Those advantaged readers who already have seen the standard shot noise derivations will also note that performing the averaging processes (8.95) and (8.96) in frequency space leads to some noteworthy simplifications.

8.15.3 Mean Value $\langle\, i(x_0)\,\rangle$

Of course $i(x_0)$ may be represented in terms of its spectrum $I(\omega)$,

$$i(x_0) = (2\pi)^{-1}\int \mathrm{d}\omega\, \mathrm{e}^{\mathrm{j}\omega x_0} I(\omega) \,. \tag{8.97}$$

Then

$$\langle i(x_0)\rangle = (2\pi)^{-1}\int \mathrm{d}\omega\, \mathrm{e}^{\mathrm{j}\omega x_0}\langle I(\omega)\rangle \,, \tag{8.98}$$

and by (8.95)

$$\langle i(x_0)\rangle = (2\pi)^{-1} m\int \mathrm{d}\omega\, H(\omega)\varphi_X(-\omega)\,\mathrm{e}^{\mathrm{j}\omega x_0} \,. \tag{8.99a}$$

The answer simplifies further yet if spectral quantities $H(\omega)$ and $\varphi_X(\omega)$ are replaced by their transforms, and the $\mathrm{d}\omega$ integration is carried through (Ex. 8.4.2). The result is a simple convolution

$$\langle i(x_0)\rangle = m\int \mathrm{d}x\, h(x)p_X(x_0 - x) \,. \tag{8.99b}$$

The latter average holds for any one value of m. Since m is itself a RV, an additional average of (8.99b) over m must be taken, giving directly

$$\langle i(x_0)\rangle = \langle m\rangle \int \mathrm{d}x\, h(x)\, p_X(x_0 - x)\ . \tag{8.99c}$$

8.15.4 Shot Noise Case

In the case of shot noise, m is a Poisson random variable (6.23)

$$P(m) = \frac{\mathrm{e}^{-\nu T}(\nu T)^m}{m!}\ . \tag{8.100}$$

Also, $p_T(t)$ is uniform over an interval $0 \leqq t \leqq T$,

$$p_T(t) = T^{-1}\mathrm{Rect}[(t - T/2)/T]\ , \tag{8.101}$$

and by Sect. 6.6

$$\nu T = \langle m\rangle\ , \tag{8.102}$$

the average number of photon arrivals.

The total current in a detector circuit due to random arrival times $\{t_n\}$ of photons, each of which contributes a disturbance current of the form $h(t)$, is

$$i(t) = \sum_{n=1}^{m} h(t - t_n)\ . \tag{8.103}$$

This has the basic form (8.89). Therefore we may use result (8.99c), which by (8.102) becomes

$$\langle i(t_0)\rangle = \nu T \int \mathrm{d}t\, h(t)\, p_T(t_0 - t)\ . \tag{8.104}$$

If now we use (8.101) for $p_T(t)$ and further stipulate that t_0 is not too close to either value $t = 0$ or $t = T$, we get simply

$$\langle i(t_0)\rangle = \langle i\rangle = \nu \int \mathrm{d}t\, h(t)\ . \tag{8.105}$$

That is, the mean net current is the photon arrival rate times the total integrated disturbance. The dependence upon time t_0 has dropped out, as intuition tells us it should.

8.15.5 Second Moment $\langle\, i^2(x_0)\, \rangle$

Starting from (8.97) once again, we have

$$\left\langle i^2(x_0)\right\rangle = (2\pi)^{-2} \int \mathrm{d}\omega\, \mathrm{e}^{\mathrm{j}\omega x_0} \int \mathrm{d}\omega'\, \mathrm{e}^{\mathrm{j}\omega' x_0} \left\langle I(\omega) I(\omega')\right\rangle\ , \tag{8.106}$$

with the last average given by identity (8.96). The answer simplifies a great deal if spectral quantities $H(\omega)$ and $\varphi_X(\omega)$ are replaced by their transforms, and the $\mathrm{d}\omega$ and $\mathrm{d}\omega'$ integrations are carried through. The result is (Ex. 8.4.3)

$$\langle i^2(x_0)\rangle = (m^2 - m)\left[\int \mathrm{d}x\, h(x)p_X(x_0 - x)\right]^2 + m\int \mathrm{d}x\, h^2(x)p_X(x_0 - x)\ . \tag{8.107a}$$

Next, averaging over the RV m gives directly

$$\langle i^2(x_0)\rangle = (\langle m^2\rangle - \langle m\rangle)\left[\int \mathrm{d}x\, h(x)p_X(x_0 - x)\right]^2 + \langle m\rangle\int \mathrm{d}x\, h^2(x)p_X(x_0 - x)\ . \tag{8.107b}$$

8.15.6 Variance $\sigma^2(x_0)$

Using identity (3.15), (8.99b) and (8.107a) give

$$\sigma^2(x_0) = m\int \mathrm{d}x\, h^2(x)p_X(x_0 - x) - m\left[\int \mathrm{d}x\, h(x)p_X(x_0 - x)\right]^2\ . \tag{8.108}$$

Again, this follows from just averaging over the x_n. Note the interesting convolution form for each of the terms. Extending the averaging to now include the RV m gives, by the use of Eqs. (3.15), (8.99c) and (8.107b), the result

$$\sigma^2(x_0) = (\sigma_M^2 - \langle m\rangle)\left[\int \mathrm{d}x\, h(x)p_X(x_0 - x)\right]^2 + \langle m\rangle\int \mathrm{d}x\, h^2(x)p_X(x_0 - x)\ , \tag{8.109}$$

where

$$\sigma_M^2 = \langle m^2\rangle - \langle m\rangle^2\ . \tag{8.110}$$

It is interesting that (8.108, 109) for the variance are not of the same form, as *were* (8.99b, c) for the mean.

8.15.7 Shot Noise Case

We note in (8.109) that if $\sigma_M^2 = \langle m\rangle$, the entire first term vanishes. This is, in fact, the case when m is Poisson (Sect. 3.15.1). Also using identity (8.102), we have

$$\sigma^2(t_0) = \nu T\int \mathrm{d}t\, h^2(t)p_T(t_0 - t)\ . \tag{8.111}$$

But in particular $p_T(t)$ is of the form (8.101). Using this in (8.111), along with the requirement that t_0 not be too close to either $t = 0$ or $t = T$, we get simply

$$\sigma^2(t_0) = \sigma^2 = \nu\int \mathrm{d}t\, h^2(t)\ . \tag{8.112}$$

Or, the variance in the net current is the photon arrival rate times the total square-integrated disturbance. Again the dependence upon time t_0 has dropped out.

8.15.8 Signal-to-Noise (S/N) Ratio

The act of measuring something usually leads to a gain in knowledge. There are many ways to measure the *quality* of the knowledge so gained. One of them is the concept of "relative error," N/S which is called the "coefficient of variation" in statistics. The inverse of N/S is the signal-to-noise ratio S/N. Relative error N/S is not as widely used as S/N simply because it seems more natural to associate high quality with a high number rather than a low one.

Quantitatively, given an ensemble of measurements i at the particular position x_0, the signal-to-noise ratio S/N is defined as simply

$$\mathrm{S/N} = \langle i(x_0)\rangle / \sigma(x_0) \,. \tag{8.113}$$

In general, results (8.99b) or (8.99c) may be used for the numerator, and (8.108) or (8.109) for the denominator. In this way, S/N may be expressed in terms of the known m or $\langle m\rangle$, and σ_M^2, $h(x)$, and $p_X(x)$.

In the particular case of a shot noise process, we may use results (8.105) and (8.112), giving

$$\mathrm{S/N} = \sqrt{\nu}\left(\int \mathrm{d}t\, h(t)\right) \Big/ \left(\int \mathrm{d}t\, h^2(t)\right)^{1/2} . \tag{8.114}$$

This is the origin of the familiar maxim that S/N *goes as the square root of the arrival rate* ν. We see that this is true if the process is shot noise. But see also Ex. 8.4.4.

We may note that S/N also depends upon disturbance function $h(t)$, but exactly in what manner is hard to know from the form of (8.114). To clarify matters, suppose $h(t)$ to be of Gaussian form,

$$h(t) = (2\pi)^{-1/2}\sigma_h^{-1} \exp(-t^2/2\sigma_h^2) \,. \tag{8.115a}$$

Then of course

$$\int \mathrm{d}t\, h(t) = 1 \,, \tag{8.115b}$$

and also

$$\int \mathrm{d}t\, h^2(t) = (2\sqrt{\pi}\sigma_h)^{-1} \,. \tag{8.115c}$$

Using these in (8.114) gives

$$\mathrm{S/N} = (2\sqrt{\pi})^{1/2}(\sigma_h \nu)^{1/2} \cong 1.88(\sigma_h \nu)^{1/2} \,. \tag{8.115d}$$

Hence, the S/N depends upon the disturbance function width parameter σ_h exactly as it depends upon the arrival rate ν! It increases with each as their square root. Intuitively, the finite width σ_h of each randomly arriving disturbance must tend to average out the signal fluctuations.

Exercise 8.4

8.4.1 We previously analyzed the shot noise process when the disturbance $h(t)$ is Gaussian. For a shot noise process, find the S/N when instead

(a) $h(t) = w^{-1}\text{Rect}(t/w)$, w the width of a rectangle disturbance.

(b) $h(t) = t_0^{-1}\exp(-t/t_0)$, the case of a transient current h in a photomultiplier tube circuit that is detecting incoming photons.

(c) $h(r) = (2\pi\sigma^2)^{-1}\exp(-r^2/2\sigma^2)$, a two-dimensional spatial Gaussian disturbance function. The process is two-dimensional spatial shot noise. What is ν defined as here?

8.4.2 Prove the "Fourier transform theorem," that

$$(2\pi)^{-1}\int d\omega F(\omega)G(\omega)\,e^{j\omega x_0} = \int dx\, f(x)g(x_0 - x)\ , \tag{8.116}$$

where capitalized functions are Fourier transforms (8.41b) of corresponding lower cased functions. *Hint:* Directly substitute the Fourier integrals for the left-hand functions, switch orders of integration so that the integral $d\omega$ is on the extreme right, and use identity (3.30g),

$$\int d\omega e^{j\omega y} = 2\pi\delta(y)\ . \tag{8.117}$$

8.4.3 Prove (8.107a), based on (8.96, 106), and the Fourier transform theorem (8.116).

8.4.4 We found that for a shot noise process, S/N $\propto \sqrt{\langle m\rangle}$. Examining equation (8.109), find at least two other probability laws $P(m)$, and hence two other types of processes, that obey S/N $\propto \sqrt{\langle m\rangle}$.

8.4.5 We found that for a shot noise process for which $p_X(x)$ is uniform, neither $\langle i(x_0)\rangle$ nor $\sigma(x_0)$ depend on x_0. But, in fact, this would be true as well for other processes: Law $p_X(x)$ does not have to be uniform. Notice that x_0 enters into these moments only through convolution of $p_X(x)$ with either $h(x)$ or $h^2(x)$. Thus, *for a general function* $p_X(x)$ establish a condition on the relative supports of $h(x)$ and $p_X(x)$ that would still cause $\langle i(x_0)\rangle$ and $\sigma(x_0)$ to be independent of x_0.

8.15.9 Autocorrelation Function

Because the disturbance function $h(x)$ has finite extension, we would suspect there to be correlation in $i(x)$ over some finite distance. This, despite the individual disturbances being randomly placed. We want to find, then,

$$R_i(x; x_0) \equiv \langle i(x_0)i^*(x_0 - x)\rangle\ , \tag{8.118}$$

by definition (8.3). The average is first over all positions $\{x_n\}$, and then over all placements m as well. In all cases of interest $i(x)$ will be real. We shall also specialize to the shot noise case to illustrate each result. The derivation follows the same steps as in Sect. 8.15.5.

Representing $i(x_0)$ and $i(x_0 + x)$ by their Fourier transforms, we have

$$\langle i(x_0)i(x_0 + x)\rangle = (2\pi)^{-2} \int d\omega\, e^{j\omega x_0} \int d\omega'\, e^{j\omega(x_0+x)} \cdot \left\langle I(\omega)I(\omega')\right\rangle\ . \quad (8.119)$$

Substituting identity (8.96) for the last average, replacing all spectral quantities by their transforms in x, integrating through $d\omega$ and $d\omega'$ to obtain delta functions via identity (8.117), and then using the sifting property (3.30d) of the delta function, we find that

$$\begin{aligned} R_i(x; x_0) = &\ (m^2 - m) \int dx' h(x') p_X(x_0 - x') \int dx' h(x') p_X(x_0 + x - x') \\ &+m \int dx' h(x') h(x' + x) p_X(x_0 - x')\ . \end{aligned} \quad (8.120)$$

The final expression for R_i is additionally averaged over m, with $\langle m\rangle$ replacing m and $\langle m^2\rangle$ replacing m^2 in (8.120).

8.15.10 Shot Noise Case

Once again $p_X(x)$ is to be replaced by a uniform law (8.101) for $p_T(t)$. Then provided fixed point t_0 is not too close to either $t = 0$ or to $t = T$, in all integrals dx' of (8.120) the entire $h(x')$ will be integrated over. Then (8.120) becomes

$$R_i(t; t_0) = \left(\left\langle m^2\right\rangle - \langle m\rangle\right)T^{-2}\left[\int dt' h(t')\right]^2 + \langle m\rangle\, T^{-1} \int dt' h(t') h(t' + t)\ ,$$

which is independent of absolute position t_0. This result and (8.105) show that a shot noise process obeys wide-sense stationarity (8.4). It also allows us to replace the general notation $R_i(t; t_0)$ with $R_i(t)$.

Finally the Poisson result

$$\sigma_M^2 = \langle m\rangle = \nu T$$

of Sect. 3.15.1 gives

$$R_i(t) = \nu^2\left[\int dt' h(t')\right]^2 + \nu \int dt' h(t') h(t' + t)\ . \quad (8.121)$$

This famous result shows that, aside from an additive constant (the first term), the autocorrelation of a shot noise process goes simply as the autocorrelation function of the disturbance function. Also the absolute strength of autocorrelation R_i is proportional to the arrival rate ν of photons. This was also true of the mean and variance.

The correlation coefficient $\rho_i(t; t_0)$ in fluctuations *from the mean* was defined at (8.5b). Using (8.5b, 105, 121) yields

$$\rho_i(t; t_0) = \rho_i(t) = R_i(t) - \langle t\rangle^{2} = \nu \int dt' h(t') h(t' + t)\ , \quad (8.122)$$

showing that the coefficient ρ_i goes directly as the autocorrelation of the disturbance function.

8.15.11 Application: An Overlapping Circular Grain Model for the Emulsion

We investigated in Sect. 6.4 a very simple model of granular structure for the photographic emulsion. This is the checkerboard model. Although it agrees with some experimental results, such as the Selwyn-root law (6.12), it suffers from being overly simplistic. First of all, emulsions are not as orderly as checkerboards in their structure. Also, any real emulsion has a finite thickness, and hence has overlapping grains, something not permitted by the checkerboard model. Overlapping grains also implies the possibility of a wide range of density levels at any point, which is impossible by the binary checkerboard model.

A better approximation to the real photographic emulsion is one that permits overlapping, circular grains, as illustrated in Fig. 8.14. Each grain has the same radius a and a constant density d_0. This permits multiple density levels to exist. Regarding the distribution of these grains over the emulsion, we shall model this as uniformly random in lateral positions $\boldsymbol{r}_n$.

Hence we assume that the grains are round (sometimes referred to as "poker chips"), occur at any position $\boldsymbol{r}_n$ within the emulsion with uniform randomness, and with an average density of ν grains/area. Given their round shape and constant d_0, the grains have a common density profile

$$h(\boldsymbol{r}) = d_0 \mathrm{Rect}\left(\frac{r - a/2}{a}\right) . \tag{8.123}$$

This is the disturbance function for the process.

The total density d at a point $\boldsymbol{r}$ is a superposition

$$d(\boldsymbol{r}) = \sum_{n=1}^{m} h(\boldsymbol{r} - \boldsymbol{r}_n) . \tag{8.124}$$

This follows because density d is related to transmittance t by

$$d \equiv -\log_{10} t ,$$

and the net transmittance t at a point $\boldsymbol{r}$ obeys a product

$$t(\boldsymbol{r}) = \prod_{n=1}^{m} t(\boldsymbol{r} - \boldsymbol{r}_n)$$

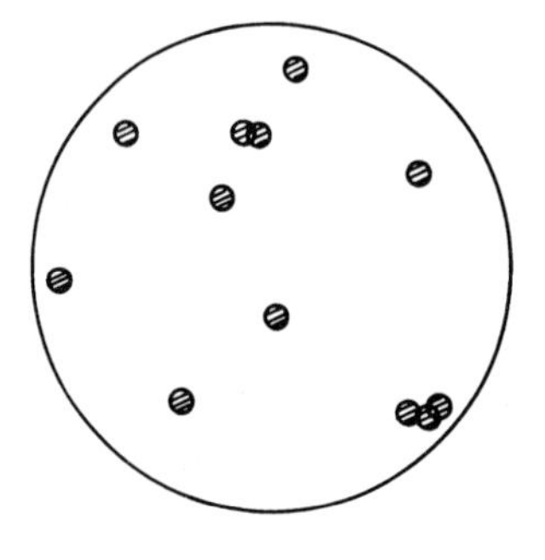

Fig. 8.14. An overlapping circular grain model of the emulsion. All grains have the same radius and photographic density

over transmittance disturbance functions. Taking the logarithm of the right-hand product gives the sum in (8.124).

Hence, total density $d(r)$ is a shot noise process in two spatial dimensions. This enables us to use the key results previously derived, suitably generalized to two dimensions.

By results (8.105, 112, 114, and 122),

$$\langle d(\boldsymbol{r}_0)\rangle = \langle d\rangle = \nu \int_0^a \mathrm{d}r 2\pi r h(r) = \pi\nu\, d_0 a^2 \,, \tag{8.125a}$$

$$\sigma^2(\boldsymbol{r}_0) = \sigma^2 = \nu \int_0^a \mathrm{d}r 2\pi r h^2(r) = \pi\nu\, d_0^2 a^2 \tag{8.125b}$$

$$\mathrm{S/N} = \sqrt{\pi\nu}\, a \tag{8.125c}$$

and

$$\rho_{\mathrm{d}}(r) = \begin{cases} 2\nu\, d_0^2 a^2\{\cos^{-1}(r/2a) - (r/2a)[1-(r/2a)^2]^{1/2}\} & r \leqq 2a \\ 0 & r > 2a \end{cases} . \tag{8.125d}$$

The latter is of the same mathematical form as the transfer function $T(\omega)$ for an incoherent, diffraction limited optical system [8.12]. Hence, the correlation coefficient falls off as nearly a triangle function in distance r, and there is zero correlation beyond the diameter of a grain (Fig. 8.15). This of course makes sense.

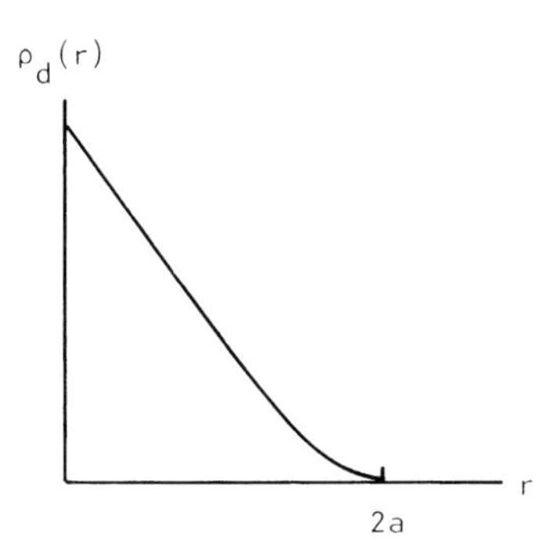

Fig. 8.15. Correlation coefficient as a function of separation across the model emulsion. a is the common grain radius

8.15.12 Application: Light Fluctuations due to Randomly Tilted Waves, the "Swimming Pool" Effect

This application will show that the superposition process may sometimes be used as a simplifying model for an otherwise very complicated physical situation.

We have all seen the shimmering light distribution at the bottom of a sunlit swimming pool. At any fixed point x_0 on the pool bottom, there will be random fluctuations of light intensity. These are due to surface waves which are tilted at just the proper angle to refract incident light waves from the sun toward x_0. The tilts obey the Cox-Munk law (3.56) and are of course random.

Suppose we want to establish the statistics of light fluctuation on the pool (or ocean) bottom. At first, the problem seems extremely complicated. If we imagine ourselves to be at point x_0 on the pool bottom, we have to somehow add or integrate all light flux contributions reaching us from all elements of the pool surface and intervening pool volume. What sort of theory would permit this to be done?

The situation might seem, at first, to resemble that of Fig. 5.4 for the analysis of laser speckle. The diffuser surface is replaced by the pool surface. However, the sunlight that illuminates the pool is incoherent, and hence cannot obey Huygens' law (5.20) assumed in the speckle analysis. Also, in the intervening water between surface and observer each element of water has its own flux contribution to the total intensity at x_0. We then have a volume sum replacing the area sum (5.20), and hence an extra degree of complexity.

A way out is provided by an experimental look at the problem. By putting on swimming goggles and actually watching the pattern of dancing light on a pool bottom, an observer will see that the light pattern consists of randomly oscillating filaments of light that occasionally intersect to form extra-bright regions. Most important, *the filaments have pretty much the same brightness pattern;* call it $h(x)$. (For simplicity, we use one-dimensional notation.)

Then assuming incoherence to the light, the total light intensity $i(x_0)$ is simply a superposition

$$i(x_0) = \sum_{n=1}^{m} h(x_0 - x'_n) \ . \tag{8.126}$$

The situation is illustrated in Fig. 8.16. Each disturbance profile $h(x)$ arises from a corresponding wave on the surface. Specifically, a wave centered on point y_n of the surface refracts its incident sunlight (coming straight down for simplicity) via Snell's law toward a coordinate x'_n on the pool bottom. The resulting disturbance $h(x)$ is centered on x'_n. For simplicity, the surface waves are all assumed to be of the same size, so that the $\{y_n\}$ are at a constant spacing Δy. The number of such waves is $2m$.

The problem (8.126) now nearly obeys the model (see beginning of Sect. 8.15) for a superposition process. We say "nearly" because here m will be fixed, not random, and also the last model assumption in Sect. 8.15 is violated. That requires the RV's $\{x'_n\}$ to all be samples from one probability law. Instead, using simple geometry and Snell's law Fig. 8.16 shows that each

$$x'_n = y_n + Rs_n \ , \quad R \equiv R_0(\eta - 1)/\eta \ , \tag{8.127}$$

with η the refractive index of the pool water. Quantity s_n is the random slope of the water wave centered upon point y_n. It is assumed that every s_n, $n = 1, \ldots, m$, is small and obeys the same probability law (3.56). Then, why don't all the x'_n obey the same probability law? Because

$$y_n = n\Delta y \ , \quad n = 1, 2, \ldots, m \ , \tag{8.128}$$

the sum $y_n + Rs_n$ is systematically shifted to higher values as n increases. Therefore x'_n gets larger with n, and hence each x'_n cannot obey one probability law.

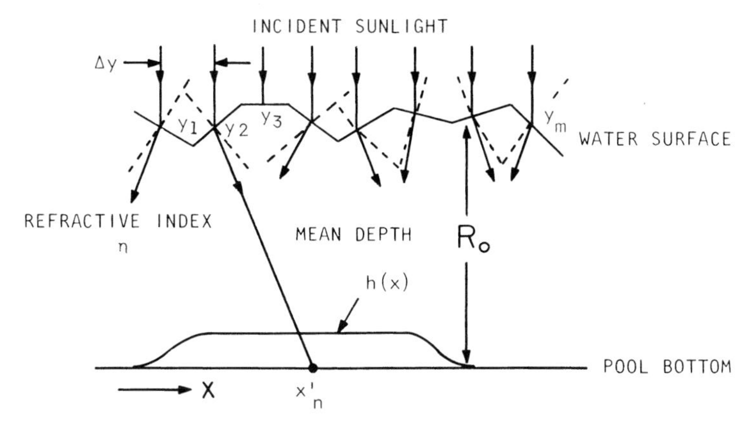

Fig. 8.16. The swimming pool effect. Each randomly refracted surface ray forms a randomly placed disturbance function $h(x)$ on the pool bottom. All $h(x)$ are of the same form

Our process is then slightly different from an ordinary superposition process. Substituting (8.127) into (8.126),

$$i(x_0) = \sum_{n=1}^{m} h(x_0 - y_n - x_n) \, , \quad x_n \equiv R s_n \, . \tag{8.129}$$

This new superposition problem has the $h(x)$ functions *systematically* shifted by y_n and, on top of that also *randomly* shifted by x_n. The systematic shifts are the new twist to the problem. However, they are easy to handle, as discussed next.

It is not too hard to show that the effect of the systematic shifts $n\Delta y$ in $h(x)$ is to alter the general results (8.99b) and (8.108) on mean and variance in a simple way. We need only replace the multiplier m in those expressions by a sum over n from $-m$ to $+m$, and replace the convolutions evaluated at x_0 by convolutions evaluated at successively shifted points $x_0 - n\Delta y$. Thus

$$\langle i(x_0) \rangle = \sum_{n=-m}^{m} [h(x) \otimes p_X(x)]_{x=x_0-n\Delta y} \tag{8.130a}$$

and

$$\sigma(x_0)^2 = \sum_{n=-m}^{m} \{h^2(x) \otimes p_X(x) - [h(x) \otimes p_X(x)]^2\}_{x=x_0-n\Delta y} \, . \tag{8.130b}$$

These results, although reasonable, are not obvious. They may be derived by using the same overall approach as in Sect. 8.15.2. That is, averages are taken in frequency space, and these are Fourier transformed back to x space for the final answer.

To proceed further with the analysis would require knowledge of the relative widths of $h(x)$ and $p_X(x)$, and the size of Δy. If, e.g., $p_X(x)$ were much broader than $h(x)$, and also smooth, essentially shot noise results like (8.105) and (8.112) would result.

Experimentally, $p_X(x)$ is known, since $x = Rs$ and the Cox-Munk law (3.56) gives $p_S(s)$. Quantity $h(x)$ is then the only remaining unknown. This quantity is analogous to an impulse response for the medium. That is, for a single ray input it represents the resulting intensity profile on the pool bottom due to all reflections, refractions, and scattering events en route. We are not aware of any experimental determination of $h(x)$. This is a good example of theory leading experiment into a novel area. Anyway, swimming pools are pleasant places in which to conduct research.

Exercise 8.5

8.5.1 We saw that in order to model a physical situation as a superposition process, four properties must be obeyed: (i) the existence of a unique disturbance function $h(x)$, (ii) additivity of the $h(x)$ to form the total output disturbance $i(x_0)$, (iii) randomness in the positions x_n of the disturbances, and (iv) randomness in the number m of disturbances.

(a) Star field images at least superficially seem to obey these properties. Discuss pro and con aspects of this model for a star field.

(b) When a mirror is polished, an abrasive rouge is used. Then surface pits remain which have a size that is characteristic of the rouge particle diameter. Would the net deformation at a point due to such pits be a superposition process?

(c) When a surface is spray painted, under what conditions would the total paint thickness at any point obey a superposition process?

8.5.2 The phase function $\Phi(y)$ across an optical pupil causes a point amplitude response function $a(x)$ obeying (Ex. 3.1.23)

$$a(x) = \int_{-y_0}^{y_0} \mathrm{d}y\, \mathrm{e}^{\mathrm{j}(\Phi(y) - kxy/f)} \tag{8.131}$$

where $k = 2\pi/\lambda$, λ is the wavelength of light, and f is the distance from pupil to image plane. The point spread function $s(x)$ relates to $a(x)$ through

$$s(x) = a(x) \cdot a^*(x) \,. \tag{8.132}$$

It is possible to model optical turbulence by a superposition process

$$\Phi(y) = \sum_{n=1}^{N} h(y - y_n)\,, \quad N \text{ fixed}\,, \tag{8.133}$$

where the $\{y_n\}$ are uniformly random across the pupil and function $h(y)$ represents a phase distribution function. Then $a(x)$ is a stochastic process.

Assuming that the total disturbance in phase $\Phi(y)$ is small:

(a) Find $\langle a(x_0)\rangle$ and $\langle a(x_0) \cdot a^*(x_0)\rangle \equiv \langle s(x_0)\rangle$, the average spread function due to the turbulence.

(b) Specialize to the case where $h(y) = (2\alpha_0)^{-1}\mathrm{Rect}(y/2\alpha_0)$. This represents phase blobs of size $2\alpha_0$.

8.5.3 Evaluate the "swimming pool" results (8.130), when all of the following occur:

(a) $h(x)$ is Gaussian with variance σ_h^2.

(b) $p_X(x)$ is Gaussian with variance $\sigma_X^2 \equiv R^2\sigma_S^2$, the latter variance that of the Cox-Munk law.

(c) Wave size $\Delta y \ll \sigma_h$ and $\Delta y \ll \sigma_X$ so that sums may be replaced by integrals.

(d) The number $2m$ of waves sending images $h(x)$ to point x_0 is so large that $m\Delta y \to \infty$.

Show that $\langle i(x_0)\rangle \simeq 1/\Delta y$ and

$$\sigma(x_0)^2 \cong \frac{1}{2\sqrt{\pi}\Delta y}\left(\frac{1}{\sigma_h} - \frac{1}{(\sigma_h^2 + R^2\sigma_S^2)^{1/2}}\right) . \tag{8.134}$$

8.5.4 ***Fractal process.*** This is a stochastic process that obeys a power-law autocorrelation function

$$R_f(x; x_0) = Ax^\alpha, \quad A = const., \ \alpha < 0 . \tag{8.135}$$

Fig. 8.17. Simulated random growth of a crystal from a single seed point. 3600 particles are randomly emplaced to form the crystal pattern. [8.14]

The power α is called the "Hausdorff dimension" [8.13] of the process. Typical values of the power are $\alpha = -1/3$ for two- and $\alpha = -2/3$ for three-dimensional fractal processes. The spatial structures of lightning, crystals, thin films and atmospheric turbulence are three dimensional fractal processes.

A two-dimensional fractal process can be represented as an image. The image in Fig. 8.17 simulates crystal growth from a "seed" point as follows. A rectangular image lattice of constant spacing is defined. The first particle of the process is a seed point that is placed in the center of the image lattice. Then each new particle is randomly emplaced within the lattice at a distant point from the seed, and made to walk randomly until it visits a site adjacent to any particle within the image. The walking particle then becomes part of the image. A dimension value $\alpha = -0.34$ results.

Under a linear magnification $y = bx$ of the coordinate x, show that the autocorrelation function $R_f(y; y_0)$ preserves a power law dependence (8.135). This is called a property of "self-similarity", in that the process preserves similarity of statistics at ever-finer scales. Nature obeys self-similarity over a wide class of phenomena.

9. Introduction to Statistical Methods: Estimating the Mean, Median, Variance, S/N, and Simple Probability

So far, we have been dealing with problems of predicting properties of data (e.g., their mean) from *known probability laws*. These are problems of the discipline called "probability theory".

On the other hand, there is a wide class of problems that is complementary and reverse to these. This is where a particular set of data is given, and from these we want to somehow estimate the probability law, or some property of the law such as the mean, that formed the data. This type of problem falls within the domain of "statistics", the topic covered by the rest of this book.

Hence, problems encountered in statistics are inverse to those in probability theory. It is usual for inverse problems, in general, to be more difficult than their corresponding direct ones, and statistics is no exception to the rule. The data at hand about the unknown probability law or parameters is almost always insufficient, either in accuracy or in scope, to *uniquely define* an answer. Hence, the answers that are provided by statistics always have a degree of arbitrariness about them. This regards the norm or criterion that is used to pick out a unique answer, and the manner by which the arbitrariness of the answer is phrased. Thus, should a maximum likelihood norm, or a minimum mean-squared error criterion, be used? Is it acceptable to find the variance of error in an estimate, or should "confidence limits" be found? Should data be considered "significant" at a 5% level of confidence, or at a stricter 1% level? For reasons such as these, answers to statistical problems can always be questioned, on the grounds that the wrong criterion was used to form the answer or to measure its uncertainty.

The progression of this chapter will be from simple to complex, i.e., estimation of a mean, a single probability, a variance, a S/N ratio, and finally, the median. Attention will be paid to the errors in the estimates. A summary of these results and other related ones may be found in Appendix E.

Later chapters will attack other problems of statistics, including hypothesis testing of given data. These results have been summarized in Appendix F. Finally, Chap. 17 gives an analytical approach for deriving the statistical laws of *physics*. Background material for this approach is found in Appendix G.

9.1 Estimating a Mean from a Finite Sample

Suppose that the speed of light c is known to be approximately 3.0×10^8 m/s, from the result of one experiment. The standard deviation in c due to any one experiment is judged to be

$$\sigma_c = 0.1 \times 10^8 \text{ m/s}$$

In an attempt to obtain a more accurate value for c, someone suggests repeating the same experiment a number N of times, and estimating the true c by taking an arithmetic average

$$\bar{c} \equiv N^{-1} \sum_{n=1}^{N} c_n \tag{9.1}$$

of the data $\{c_n\}$. In statistics, such data are called "a sample". The arithmetic average (9.1) when taken over a sample is often called the "sample mean". This, and other averages over samples will usually be denoted by bars, as in $\bar{c}$.

Is this a sound approach? Does the sample mean improve indefinitely with the number N of data in the sample? If so, how large must N be for the expected error in $\bar{c}$ to be less than 1%? By comparison, the error of $\pm 0.1 \times 10^8$ amounts to a 3% relative error for any *one* experiment.

9.2 Statistical Model

In order to solve this problem, we shall need a statistical model. The following are simple and plausible assumptions comprising such a model:

• Each experiment is performed by the identical procedure, so that each output c_n is sampled from *the same* probability law.

• The experimental procedure is deterministically sound, in that the mean of the probability law is exactly the true value, c. Then for any n,

$$\langle c_n \rangle = c. \tag{9.2}$$

Note: This is not the sample mean in (9.1). It is the *true mean* (3.13a) of the one probability law describing all the outputs. What (9.2) states is that the experimental procedure is *unbiased*. It has no systematic error that would lead to an average bias away from the true value c. This is a somewhat restrictive assumption, as most experimental procedures do suffer at least some systematic error or bias.

• Finally, we may reasonably assume the experiments to be truly *independent*, so that there is zero correlation between any two outputs c_m and c_n, $m \neq n$,

$$\langle (c_m - c)(c_n - c) \rangle = \delta_{mn} \sigma_c^2 \tag{9.3a}$$

where δ_{mn} is the Kronecker delta

$$\delta_{mn} = \begin{cases} 1 & \text{for } m = n \\ 0 & \text{for } m \neq n. \end{cases} \tag{9.3b}$$

These model assumptions define an RV which is *identically and independently distributed*, called an *i.i.d. random variable* in statistics.

9.3 Analysis

We want to know what error to expect by use of the sample mean formula (9.1). A good measure of error is the mean-square error ε^2 of the sample mean $\bar{c}$ from the true value c. This is easily computed on the basis of the statistical model above.

By definition

$$\varepsilon_{\bar{c}}^2 \equiv \langle (c - \bar{c})^2 \rangle. \tag{9.4}$$

But by (9.1) and (9.2), $\langle \bar{c} \rangle = c$. Therefore $\varepsilon = \sigma_{\bar{c}}$, the ordinary standard deviation. We replace ε by $\sigma_{\bar{c}}$. Using $c \equiv N^{-1} \sum_n c$ and (9.1),

$$\sigma_{\bar{c}}^2 = N^{-2} \left\langle \left[\sum_n (c - c_n) \right]^2 \right\rangle.$$

Squaring out the sum,

$$\sigma_{\bar{c}}^2 = N^{-2} \sum_n \sum_m \langle (c - c_m)(c - c_n) \rangle.$$

By lack of correlation (9.3a)

$$\sigma_{\bar{c}}^2 = N^{-2} \sum_{n=1}^{N} \sigma_c^2,$$

so that

$$\sigma_{\bar{c}} = \sigma_c / \sqrt{N}. \tag{9.5}$$

This famous result states that, yes, the error in the estimate $\bar{c}$ does indeed decrease indefinitely with number of experiments N. However, the effect is a rather slow one, because of the square-root dependence on N.

Equation (9.5) may be recast as a requirement on N, given a required relative error $e \equiv \sigma_{\bar{c}}/c$ in the sample mean. By (9.5),

$$N = (\sigma_c / ec)^2. \tag{9.6}$$

This relation may be used to solve the problem posed at the outset, of how large N must be for accomplishing error $e = 1\%$.

$$\begin{aligned} N &= (0.1 \times 10^8 / 0.01 \times 3.0 \times 10^8)^2 \\ &= 11 \text{ experiments.} \end{aligned}$$

9.4 Discussion

This result can, of course, be generalized to the case where N independent and unbiased determinations of *any* experimental quantity c are known, and from these N numbers a better estimate of c is sought. Also, note that the derivation assumed nothing about the *form* of the probability law obeyed by outputs c_n: whether normal, exponential, or whatever. (However, its variance must be finite – see Exs. 9.1.15, 16.) Hence, the result (9.5) is *distribution free*, and has widespread applicability.

The sample mean is also optimum in a certain sense (Ex. 9.1.4).

We may note further from (9.1) that $\bar{c}$ is the sum of N statistically independent RV's and hence, by the central limit theorem (4.48) $\bar{c}$ is a normal RV. We also had $\langle \bar{c} \rangle = c$, so that

$$\bar{c} = N(c, \sigma_{\bar{c}}^2). \tag{9.7}$$

This knowledge is an important supplement to mere knowledge of $\sigma_{\bar{c}}$. For example, now it is possible to know what fraction of determinations $\bar{c}$ will lie within the intervall $c \pm \sigma_{\bar{c}}$: value 0.68, by Sect. 4.30.

A further generalization of the model is noted. We assumed that each experiment was operationally identical, leading to the i.i.d. model for RV's $\{c_n\}$. However, this assumption is not really necessary for this model to be true. That is, each output c_n, from a *different* operational experiment might still be a sample from the same probability law, or nearly so. Much use is made of this generalization. The universal constants (such as c) given in physical tables are usually sample means (9.1) formed from many different types of experiments. In this case, a *weighted* sample mean often is taken, large weights delegated to more accurate experiments (Sect. 9.5).

Probably the weakest assumption made in the model is (9.2), that the theoretical mean of the experiment is *exactly* the true value c. In reality, all experimental procedures suffer *some* systematic bias away from the true value. The existence of such bias will cause result (9.5) to break down at sufficiently high N. Even with very high N the error $\sigma_{\bar{c}}$ will remain finite.

Sometimes, instead of estimating the mean from a finite sample, it is desired to estimate the variance. This will be taken up in Sect. 9.10.

9.5 Error in a Discrete, Linear Processor: Why Linear Methods Often Fail

Rather than perform a simple average (9.1), in image processing it is common to form an output $\bar{c}$ from N finite image values $\{c_n\}$ *as the weighted superposition*

$$\bar{c} \equiv \sum_{n=1}^{N} w_n c_n \Big/ \sum_{n=1}^{N} w_n. \tag{9.8}$$

Here the $\{c_n\}$ are image intensity values across the given picture. The form (9.8) is linear in the $\{c_n\}$, and hence is often termed a "linear processor" of the image data $\{c_n\}$. These lie within a finite "window" $\{w_n\}$ of fixed weights. This window is first placed upon image values $(c_1, \ldots, c_N)$, then upon $(c_2, \ldots, c_{N+1})$, etc. The operation (9.8) is performed at each such placement, to form respective outputs $\bar{c}_1$, $\bar{c}_2$, etc.

The weights $\{w_n\}$ are numbers selected so as to either smooth the data $\{c_n\}$, in which case they are all positive, or to enhance the resolution in $\{c_n\}$, in which case some are negative (see [9.1]). The form (9.8) includes division by $\{w_n\}$ so as to enforce the requirement that $\langle \bar{c} \rangle = c$ (Ex. 9.1.3). In general, an estimate which on the average equals its true value is called "unbiased". Hence, $\bar{c}$ is an unbiased estimate of c.

Note that the form (9.8) includes the particular case (9.1) of pure data averaging, when all $w_n = 1$. This permits a simple check on the results below.

We are interested in the situation where data values $\{c_n\}$ consist of true or signal values $\{C_n\}$ plus additive noise values $\{e_n\}$,

$$c_n = C_n + e_n. \tag{9.9}$$

In addition, the noise is zero-mean and independent,

$$\langle e_n \rangle = 0, \quad \langle e_m \, e_n \rangle = \sigma_c^2 \delta_{mn}. \tag{9.10}$$

We further suppose that the weights $\{w_n\}$ have been chosen so that when they operate upon noiseless data the result is the ideal value c (which might represent a restored ideal object value, e.g., or a smoothed image signal value):

$$\sum_{n=1}^{N} w_n C_n \Big/ \sum_{n=1}^{N} w_n = c. \tag{9.11}$$

Since $\langle \bar{c} \rangle = c$, the mean-squared error at any point m over many sets of data $\{c_n\}$ is again the variance

$$\sigma_{\bar{c}}^2 \equiv \langle (c - \bar{c})^2 \rangle.$$

This error is then

$$\sigma_{\bar{c}}^2 = \left\langle \left(c - \sum w_n (C_n + e_n) \Big/ \sum w_n \right)^2 \right\rangle$$

by (9.8) and (9.9), or

$$\sigma_{\bar{c}}^2 = \left\langle \left(\sum w_n e_n \right)^2 \right\rangle \Big/ \left(\sum w_n \right)^2$$

by (9.11). Then squaring out the numerator and using independence (9.10) yields

$$\sigma_{\bar{c}}^2 = \sigma_c^2 \frac{\sum_{n=1}^{N} w_n^2}{\left(\sum_{n=1}^{N} w_n \right)^2} \equiv \sigma_c^2 \cdot r. \tag{9.12}$$

Parameter r measures the gain in noise as propagated from the input data to the output. It only depends upon the weights w_n. As a check on (9.12), for the case of unit weights $w_n = 1$ we once again get expression (9.5) for the simple average window.

To see the effect of the weights on the output error $\sigma_{\bar{c}}^2$, consider first the case of *data smoothing*, where all $w_n \geqq 0$. This corresponds to pure "apodization" in the case of image processing. The noise-gain parameter r in (9.12) now obeys

$$r = \frac{\left(\sum w_n\right)^2 - \sum_{m \neq n} w_m w_n}{\left(\sum w_n\right)^2}.$$

Since all $w_n \geqq 0$ here, we see that

$$r \leq 1. \tag{9.13}$$

Hence, by (9.12) the *output error is less than the data error,*

$$\sigma_{\bar{c}}^2 \leqslant \sigma_c^2. \tag{9.14}$$

This is a good state of affairs, but does not imply getting something for nothing, since relation (9.11) has been presumed to hold; i.e., that the weights correctly process the signal. The set of weights satisfying (9.11) will not always turn out to be positive, as assumed here. For example, consider the following case.

If the image $\{c_n\}$ is a *blurred version* of the object (of ideal value c at the output point considered), some of the weights now have to be *negative* [9.1] in order to enforce (9.11). In this case, r will usually exceed 1, and if any significant increase in resolution (such as a factor of two or more) is demanded, will *greatly exceed* one. Then the result (9.12) predicts a *noise-boost phenomenon*. In this case, *the weights* $\{w_n\}$ *are primarily intended to boost resolution, via (9.11), rather than to decrease noise via (9.12)*. Evidently, this tradeoff will be acceptable only if σ_c in (9.12) is small enough. Linear methods of restoring images always suffer such a tradeoff, directly because of the noise propagation effect (9.12). (For demonstrations, see Exercises 9.1.2 and 5.1.21, 22.)

9.6 Estimating a Probability: Derivation of the Law of Large Numbers

One general aim of statistics is to infer a parameter of a probability law, or the probability law itself, from given data. We saw in Sects. 9.1–4 how to estimate a mean, and how to estimate the error in that estimate. Here we similarly attack the problem of estimating the probability of an event, from observation of the number of times the event occurs.

We assumed in Sect. 2.9 that if an event occurs m times out of N total trials, then a reasonable estimate $\hat{p}$ of the probability p of the event obeys $\hat{p} = m/N$, if N is very large. Now we are in a position to derive this result from a basic premise, *even if N is finite*.

First of all, note that the events referred to by the law (2.8) of large numbers are actually binomial events. That is, at each repetition of the experiment either the event in question happens, or it doesn't. Also, the experiments are presumed independent.

Then the probability $P_N(m)$ of m occurrences of the event obeys (6.5), the Bernoulli law. Suppose that a value of m is observed. Given this number, the value of the elementary probability p could be anything in the range $(0, 1)$. What value from this range should be chosen as the estimate $\hat{p}$ of p?

Regard the unknown p as a parameter to be estimated, and assume that the probability law $P(m|p)$ on the datum m in the presence of the parameter p is known. This probability law is called a "likelihood law". The assumption we make is that, since the data value m occurred it must have been *maximum probable to occur.* Hence, the estimate $\hat{p}$ is the value of the parameter p that would make the observed data value m maximum probable, obeying $P(m|p) = \max$. Since the likelihood law is thereby maximized, this is called the *principle of maximum likelihood*, or, ML. See also Chapt. 17.

Applying the ML principle to our problem, we demand that

$$P(m|p) = P_N(m) = \binom{N}{m} p^m (1-p)^{N-m} = \max. \tag{9.15}$$

through choice of p. As in most ML problems, it is convenient to first take the natural logarithm,

$$\ln P_N = \ln \binom{N}{m} + m \ln p + (N-m)\ln(1-p) = \max. \tag{9.16}$$

since $\ln P_N$ is monotonic with P_N. Assume that the extremum is an interior point to the range $(0, 1)$ of p. Then we can differentiate (9.16) to obtain the maximum,

$$\frac{\partial}{\partial p}(\ln P_N) = 0 + \frac{m}{p} - \frac{N-m}{1-p} \equiv 0. \tag{9.17}$$

The solution is directly

$$p \equiv \hat{p} = \frac{m}{N}. \tag{9.18}$$

This is of the same form as (2.8), the law of large numbers!

What does this surprisingly simple result mean? It states that if a probability is unknown, then even for a finite number N of trials the simple ratio m/N serves to estimate it, in a reasonable way. Of course, the use of (9.18) will incur an error, since it is only an estimate with no pretense of being perfect. We find this error next.

9.7 Variance of Error

Suppose p is fixed but unknown, and the N trials are repeated many times over. Each will give rise to a different value of m, and hence a different estimate $\hat{p}$ by (9.18). We first ask what the resulting average value $\langle \hat{p} \rangle$ will be.

Taking the average of both sides of (9.18) and using first-moment result (6.7), we find that

$$\langle \hat{p} \rangle = \langle m \rangle / N = p \tag{9.19}$$

precisely. Hence, although any one $\hat{p}$ will be in error, their average will not. The estimate $\hat{p}$ is unbiased.

The more interesting question to ask is *by how much*, on average, any one value $\hat{p}$ will be in error. A measure of this is the mean-squared error

$$\varepsilon^2 \equiv \langle (p - \hat{p})^2 \rangle, \tag{9.20}$$

where the brackets again denote the average over the many determinations of $\hat{p}$. Using identity $p = Np/N$ and relation (9.18),

$$\varepsilon^2 = N^{-2}\langle (Np - m)^2 \rangle.$$

Then using the first moment (6.7),

$$\varepsilon^2 = N^{-2}\langle (\langle m \rangle - m)^2 \rangle \equiv N^{-2}\sigma_m^2,$$

so that by the second moment (6.7)

$$\varepsilon^2 = N^{-1}p(1 - p). \tag{9.21}$$

Of particular interest is that $\varepsilon \to 0$ as $N \to \infty$. *This, then, proves the law of large numbers* (2.8), *which only pretends to hold true for* $N \to \infty$.

9.8 Illustrative Uses of the Error Expression

9.8.1 Estimating Probabilities from Empirical Rates

Error expression (9.21) is widely used to answer questions such as the following. If $N = 10$ students take an exam, and three fail it, with what *confidence* can we say that the *probability* that a student will fail the exam is 0.3?

By relation (9.21), $\varepsilon^2 = 10^{-1}0.3(1 - 0.3) = 0.021$, or $\varepsilon = 0.15$. Note that we simply used $\hat{p}$ for p in (9.21). Since p is by definition unknown, this is the best we can do. Also, by self-consistency, if the ensuing ε turns out to be small, so ought to be the error in the estimated error caused by this substitution.

This estimated ε of 0.15 should be compared with the estimated p of 0.30. The relative error to be expected, then, in the value $\hat{p} = 0.30$ is $0.15/0.30 = 50\%$! Hence, with very little confidence can we infer that the probability is 0.3. The evidence, based as it is upon only $N = 10$ observations, is too meagre to support that claim.

Often, the experimenter has control over the number N of observations that can be made. He then asks, how small can N be to guarantee a relative error $e \equiv \varepsilon/p$ in p? Recasting expression (9.21) this way produces a requirement

$$N = \frac{1 - p}{pe^2}\,. \tag{9.22}$$

For example, if in the preceding problem $p = 0.30$ truly, how large does N have to be to guarantee 1% error or less in estimating p? By (9.22), the required minimum is $N = 23,333$ students!

In general, (9.22) shows that N must be very large if either (a) the event in question is rare (p is small) or (b) the required accuracy in p is stringent (e is small).

9.8.2 Aperture Size for Required Accuracy in Transmittance Readings

Suppose an aperture area A covers N grain sites of a uniformly exposed emulsion (Fig. 9.1). If m clear grain sites occur, the inferred T is given by (6.9). This frequency-of-occurrence ratio m/N is an approximation to the mean transmittance t, which it would approach as $N \to \infty$ (aperture grows beyond bounds). How accurate an estimate of t is T, when N (area A) is finite? This question arises when a finite aperture is used, in a scanning densitometer, to pick up film density values.

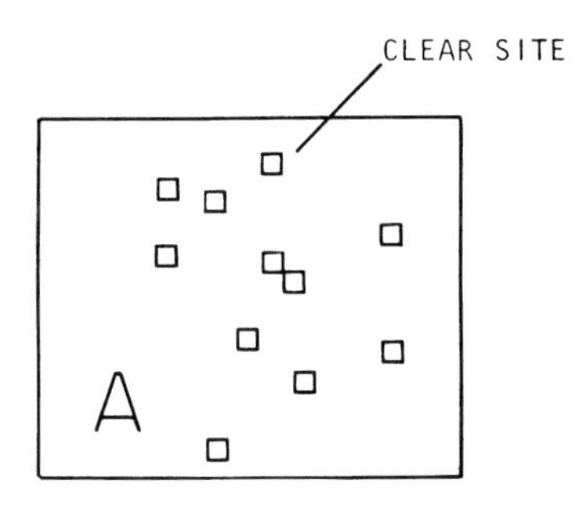

Fig. 9.1. An aperture of area A encloses m clear grain sites out of N total. How reliable an estimate of energy transmittance is m/N?

First we note from (6.11a) that

$$t = p,$$

the probability of a clear grain-site. Hence, T in (6.9) is actually an estimate $\hat{p}$ of a probability $p = t$. We have already found the error involved in estimating a probability based on a finite sample. That was the result (9.22), here

$$N = (1 - t)/(t\,\mathrm{e}^2). \tag{9.23}$$

But

$$N = A/a,$$

where a is the grain area. Then from (9.23),

$$A = a(1 - t)/(t\,\mathrm{e}^2) \tag{9.24}$$

is the requirement on aperture size.

This shows that the larger grain size a is, the larger aperture A must be. Also, for $t \simeq 1$ (clear emulsion) $A \simeq 0$. This describes the situation where there are so many clear grain sites present per unit area, that only a small area A is needed to encompass enough clear sites to achieve high accuracy in m/N. The opposite effect holds for $t \simeq 0$ (black emulsion).

This analysis acts as if the only consideration regarding aperture area A is statistical fluctuation of its outputs. Of course, we are neglecting the resolution aspect. If the film density is changing, as in an ordinary picture, then a large A which causes small output fluctuation would also cause an undesirable averaging, and hence blurring, in the output picture. Unfortunately, then, the user must trade off output noise against resolution when deciding upon A.

9.9 Probability Law for the Estimated Probability; Confidence Limits

Problems such as in Sect. 9.8 are limited to questions regarding the variance of error in the estimate $\hat{p}$ of p. But a wider scope of problems asks questions such as, how many trials N must be observed so as to guarantee that an error in $\hat{p}$ of prescribed value ε_0 or less will occur with high probability α? This fixes *confidence limits* of α on the acceptable error ε_0. In $\alpha\%$ of the estimates $\hat{p}$ of a fixed p, an error of amount ε_0 or less must occur.

The preceding is a requirement that $P(m/N|p)$ be narrow in m/N about p. What is $P(m/N|p)$?

We observe that P is a discrete probability, obeying

$$P(m/N|p) = P_N(m|p),$$

since with fixed N each occurrence of m/N corresponds to a unique value of m. Note further that

$$P_N(m|p) = P_N(m), \tag{9.25}$$

the binomial probability law, since Bernoulli trials characterize the problem. Also, the requirement of error limits $\pm\varepsilon_0$ in m/N corresponds to size limits $\pm N\,\varepsilon_0$ in m. The upshot is a requirement

$$P(Np - N\varepsilon_0 \leqq m \leqq Np + N\varepsilon_0) = \alpha \tag{9.26}$$

on N.

Combining the last relation with identity (9.25) requires that

$$\sum_{m=Np-N\varepsilon_0}^{Np+N\varepsilon_0} P_N(m) = \alpha. \tag{9.27}$$

Let us seek to use the De Moivre-Laplace theorem and replace this cumbersome summation with error function integrals. Accordingly, note that requirement (6.46) is that $\sigma^2 = Np(1-p) \gg 1$, which is satisfied if

$$Np \gg 1. \tag{9.28}$$

Also, requirement (6.47) translates into

$$N\varepsilon_0 \lesssim \sqrt{Np(1-p)}. \tag{9.29}$$

However, this requirement may be ignored since the term $m = Np$ occurs in the sum (9.27); see the discussion following (6.48). The first requirement (9.28) will be checked out once N is found.

Accordingly, we use (6.44) for the probability density whose integral replaces the sum (9.27). The latter becomes

$$\sum_{m=Np-N\varepsilon_0}^{Np+N\varepsilon_0} P_N(m) = \sum_{m=0}^{Np-N\varepsilon_0} P_N(m) - \sum_{m=0}^{Np-N\varepsilon_0} P_N(m)$$

$$\to (2\pi)^{-\frac{1}{2}}\sigma^{-1}\left(\int_0^{Np+N\varepsilon_0} - \int_0^{Np-N\varepsilon_0}\right) e^{-(x-Np)^2/2\sigma^2}\,dx$$

$$= \mathrm{erf}\left(\frac{N\varepsilon_0}{\sqrt{2}\,\sigma}\right) \tag{9.30}$$

after transformation of the variable of integration. Hence the requirement on N is

$$\mathrm{erf}\left(\frac{N\varepsilon_0}{\sqrt{2}\,\sigma}\right) = \alpha, \tag{9.31}$$

where $\sigma^2 = Np(1-p)$. Since p is, of course, the unknown in these problems, we replace it by its estimate $\hat{p} = m/N$. This is an approximation which improves the smaller ε_0 is made. The net requirement on N is then

$$\mathrm{erf}\left(\sqrt{\frac{N}{2}}\frac{\varepsilon_0}{\sqrt{\hat{p}(1-\hat{p})}}\right) = \alpha. \tag{9.32}$$

As an example, consider the following problem. How many trials N must be made to have at least 99% probability of correctness for the inference that $0.38 \leqq p \leqq 0.42$, if the estimate $\hat{p} \equiv m/N = 0.4$?

Here $\alpha = 0.99$, $\varepsilon_0 = 0.02$ and $\hat{p} = 0.4$. According to Appendix A, the parenthesis in (9.32) must have value 1.82. Then $N = 4,050$ trials are required.

We now check whether De Moivre-Laplace assumption (9.28) was satisfied. Now $Np \simeq N\hat{p} = 1620 \gg 1$, which satisfies (9.28).

This calculation was carried through using "classical" statistical methods. The same problem may be attacked using "Bayesian" statistical methods, and in this case the answers are the same (Chap. 16).

9.10 Calculation of the Sample Variance

In Sect. 9.1 we found how to estimate a mean from a finite number of samples $c_n, n = 1, \ldots, N$. Here we take on the corresponding problem of estimating the variance σ_c^2. We shall first derive the well-known formula for an estimate S^2 of σ_c^2, and then find the expected error in the estimate. The latter involves use of the chi-square probability law. The statistical model of Sect. 9.2 shall be assumed for quantities $\{c_n\}$. That is, they are i.i.d. random variables.

9.10.1 Unbiased Estimate of the Variance

Let us form a *sample variance* S^2 from data $\{c_n\}$ in the form

$$S^2 \equiv A \sum_{n=1}^{N} (c_n - \bar{c})^2. \tag{9.33}$$

Coefficient A has to be found. Let A be such that S^2 is an unbiased estimate of σ_c^2. Thus we demand that

$$\langle S^2 \rangle \equiv \sigma_c^2 \tag{9.34}$$

over many estimates of σ_c^2.

Taking the average of (9.33), we get

$$\langle S^2 \rangle = A \sum_{n=1}^{N} (\langle c_n^2 \rangle - 2\langle c_n \bar{c} \rangle + \langle \bar{c}^2 \rangle) \tag{9.35}$$

after squaring out and averaging termwise.

To evaluate each of the right-hand means we have to know $\langle c_m c_n \rangle$ for all m, n. This is known, from the independence requirement (9.3a), to obey

$$\langle c_m c_n \rangle = c^2 + \delta_{mn}(\langle c^2 \rangle - c^2). \tag{9.36}$$

Then the means in (9.35) obey, respectively,

$$\langle c_n^2 \rangle = \langle c^2 \rangle, \tag{9.37}$$

$$\langle c_n \bar{c} \rangle = N^{-1} \sum_{m=1}^{N} \langle c_m c_n \rangle$$
$$= N^{-1}[(N-1)c^2 + 1\langle c^2 \rangle], \tag{9.38}$$

and

$$\langle \bar{c}^2 \rangle = N^{-2} \sum_{m,n} \langle c_m c_n \rangle$$
$$= N^{-2}[(N^2 - N)c^2 + N\langle c^2 \rangle]. \tag{9.39}$$

The latter used the fact that the double sum in m and n has N^2 terms total, of which N have $m = n$ and $(N^2 - N)$ have $m \neq n$. The i.i.d. model assumption of Sect. 9.2 was repeatedly used, to remove unnecessary subscripts from moments of c_n.

Substituting results (9.37) through (9.39) into (9.35) yields

$$\langle S^2 \rangle = A[(N-1)\langle c^2 \rangle - (N-1)c^2]$$
$$= A(N-1)\sigma_c^2. \tag{9.40}$$

Combining this result with requirement (9.34) of an unbiased estimate yields

$$A = (N-1)^{-1}$$

as the requirement on A. Hence, in summary the sample variance (9.33) obeys

$$S^2 = (N-1)^{-1} \sum_{n=1}^{N} (c_n - \bar{c})^2. \tag{9.41}$$

This is a good example of where intuition fails us. It is, after all, intuitive to instead have $A = N^{-1}$ in (9.41), on the basis that N terms were summed over[1]. May we state that intuition and reality are "at variance" here?

[1] However, only $N - 1$ of them are actual degrees of freedom, as shown in the next section. This accounts for the strange factor $(N-1)^{-1}$.

9.10.2 Expected Error in the Sample Variance

Because N is finite, the sample variance calculated by (9.41) will be in error. We want to predict the expected error, defined as the root mean-square deviation of S^2 from the true value σ_c^2. But since by requirement (9.34) the latter equals $\langle S^2 \rangle$, the expected error is simply the standard deviation of RV S^2. Denote this as σ_{S^2}.

The latter could be computed, if we knew the probability law obeyed by S^2. From the form of (9.41) as the sum of squares of RV's $(c_n - \bar{c})$, it appears that S^2 might obey a chi-square probability law.

By condition (5.54) this would be so if RV's $(c_n - \bar{c})$ were Gaussian and independent. In fact, we shall see that RV's $(c_n - \bar{c})$ may be made Gaussian, but since $\bar{c}$ depends upon all the $\{c_n\}$ the $(c_n - \bar{c})$ are not all independent. The result is that S^2 is indeed chi-square, but with $(N-1)$ degrees of freedom instead of N. This is shown next.

We may expand (9.41) as

$$
\begin{aligned}
(N-1)S^2 &= \sum_{n=1}^{N} [(c_n - c) - (\bar{c} - c)]^2 \\
&= \sum_{n=1}^{N} (c_n - c)^2 - [\sqrt{N}(\bar{c} - c)]^2
\end{aligned}
\tag{9.42}
$$

after use of (9.1).

Assume that the RV's c_n are i.i.d. *normal,*

$$c_n = N(c, \sigma^2). \tag{9.43}$$

This somewhat restricts the scope of the derivation. Next, denote the characteristic functions for $(N-1)S^2$, the right-hand sum in (9.42) and the last term in (9.42) as, respectively, φ_1, φ_2, and φ_3. The RV $\bar{c}$ is statistically independent of S^2; see Ex. 9.2.5. Therefore the last term in (9.42) is independent of $(N-1)S^2$, so that by (4.18)

$$\varphi_2(\omega) = \varphi_1(\omega)\varphi_3(\omega) \tag{9.44a}$$

or equivalently

$$\varphi_1(\omega) = \varphi_2(\omega)/\varphi_3(\omega). \tag{9.44b}$$

By (9.43), the RV's $(c_n - c)$ are Gaussian, and of variance σ^2. Therefore the sum in (9.42) is the sum of squares of N independent Gaussian RV's. Therefore, by condition (5.54) the sum is a chi-square RV, with characteristic function

$$\varphi_2(\omega) = (1 - 2\mathrm{j}\omega\sigma_c^2)^{-N/2} \tag{9.45}$$

by (5.55b).

Characteristic function $\varphi_3(\omega)$ is found as follows. By result (9.7)

$$(\bar{c} - c) = N(0, \sigma_c^2), \tag{9.46}$$

so that

$$\sqrt{N}(\bar{c} - c) = N(0, \sigma_c^2) \tag{9.47}$$

by (9.5). Hence, $[\sqrt{N}(\bar{c} - c)]^2$ is itself a chi-square RV, of 1 degree of freedom according to (5.54). By (5.55b) its characteristic function is

$$\varphi_3(\omega) = (1 - 2\mathrm{j}\omega\sigma_c^2)^{-1/2}. \tag{9.48}$$

Combining results (9.44b), (9.45) and (9.48) we find that

$$\varphi_1(\omega) = (1 - 2\mathrm{j}\omega\sigma_c^2)^{(N-1)/2}. \tag{9.49}$$

Thus, the characteristic function for RV $(N - 1)S^2$ is chi-square, with $(N - 1)$ degrees of freedom.

The shift theorem (4.5) then yields

$$\varphi_{S^2}(\omega) = \left(1 - 2\mathrm{j}\omega\frac{\sigma_c^2}{N-1}\right)^{-(N-1)/2}. \tag{9.50}$$

The moments of S^2 may be found in the usual way (4.3) by differentiating (9.50). In particular,

$$\sigma_{S^2}(\omega) = \left(\frac{2}{N-1}\right)^{1/2} \sigma_c^2. \tag{9.51}$$

Since the ideal value of S^2 is σ_c^2, we see that the relative error $e \equiv \sigma_{S^2}/\sigma_c^2$ obeys simply

$$e = \left(\frac{2}{N-1}\right)^{1/2}. \tag{9.52}$$

Remarkably, there is no dependence upon either experimental quantities c or σ_c. In particular, for $N = 3$, $e = 1$ (rms error of 100%); while for $N = 20,001$, $e = 0.01$.

Finally, we should note that the result (9.51) is not as general as the corresponding result (9.5) for the error in the sample mean. We had to assume here that each RV c_n obeys a particular probability law, the normal one (9.43).

9.10.3 Illustrative Problems

Results (9.51) and (9.52) may be used to attack problems of the following type.

An optical mirror is being tested interferometrically for surface roughness. Its roughness, as measured by variance in surface thickness, is known to be about $(\lambda/4)^2$. We want to *estimate* the variance by the sample variance formula (9.41).

(a) How many individual thicknesses must be measured to ensure that the sample variance, or roughness estimate, has an rms error of $(\lambda/10)^2$?

(b) How many must be measured to ensure an error in S^2 of $(\lambda/10)^2$ or less, with probability 0.80 (80% confidence limits)?

It is given that $\sigma_c \cong \lambda/4$. To answer (a), we set $\sigma_{S^2} = (\lambda/10)^2$ and use (9.51). The answer is $N = 29$.

Question (b) requires the value of N satisfying

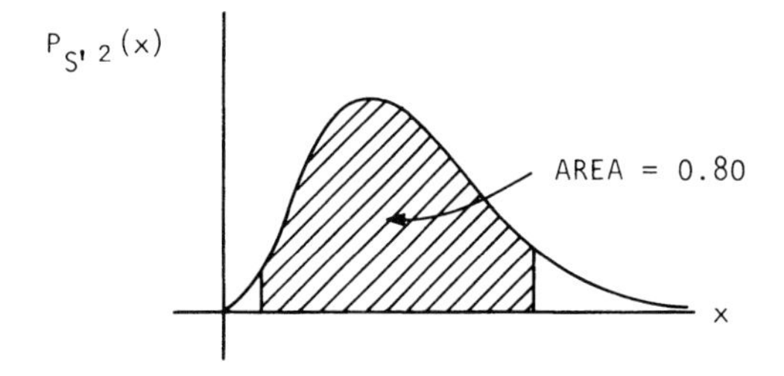

Fig. 9.2. Establishing confidence limits for a chi-square RV

$$P(\sigma_c^2 - (\lambda/10)^2 \le S^2 \le \sigma_c^2 + (\lambda/10)^2) = 0.80. \tag{9.53}$$

The left-hand probability requires integration of the probability density for S^2 over the indicated interval. It would be nice if this were reduced to a table lookup, and this is possible by the following considerations.

Comparison of (5.55b) and (9.50) shows that S^2 is a chi-square RV whose degrees of freedom $\{x_n\}$ each have variance $\sigma_c^2(N-1)^{-1}$. However, tables of chi-square assume unit variance for each x_n. Then note that a new RV $S'^2 \equiv (N-1)S^2/\sigma_c^2$ would have unit variance as required and still be chi-square. (Apply the shift theorem (4.5) to (9.50).)

The problem may be recast in terms of this new RV as

$$P\left[(N-1)\left(1-\frac{(\lambda/10)^2}{\sigma_c^2}\right) \leqq S'^2 \leqq (N-1)\left(1+\frac{(\lambda/10)^2}{\sigma_c^2}\right)\right] = 0.80\,. \tag{9.54}$$

Using the knowledge that $\sigma_c^2 = (\lambda/4)^2$ yields

$$P[(N-1)\,21/25 \leqq S'^2 \leqq (N-1)\,29/25] = 0.80 \tag{9.55}$$

as the requirement on N.

It is equivalent to find a value N satisfying both

$$P[(N-1)\,21/25 \leqq S'^2] = 0.90 \tag{9.56a}$$

and

$$P[(N-1)\,29/25 \leqq S'^2] = 0.10\,. \tag{9.56b}$$

This is because the probability of events S'^2 on the interval in (9.55) is the shaded area of the curve in Fig. 9.2, which may be represented as the area past the lower limit minus the area past the upper. If the former area is required to be 0.90 and the latter is required to be 0.10, then (9.55) will be satisfied.

It may be questioned whether the N that satisfies (9.56a) will necessarily satisfy (9.56b) as well. In fact, surprisingly enough, the same N will satisfy both equations under rather broad conditions [9.2].

To solve the problem (9.56a) for N, we use the chi-square tables of Appendix B, with $\nu = N-1$. Look down the column $\alpha = 0.90$ and for each ν observe the ratio of the tabulated value [our $(N-1)\,21/25$] to ν. When this is $21/25$, the indicated ν is the solution $N-1$. This occurs for approximately $N = 101$, the answer.

The reader may check that the same N accomplishes condition (9.56b) as well.

Comparing the answers to the problems (a) and (b), we see that the requirement of 80% confidence limits on error $(\lambda/10)^2$ was a more stringent condition on N than was the rms error requirement of $(\lambda/10)^2$. It is harder *to bracket* an error with high confidence than to merely specify its σ.

9.11 Estimating the Signal-to-Noise Ratio; Student's Probability Law

Consider a RV c_n obeying the i.i.d. model of Sect. 9.2. Its S/N ratio is defined as

$$\mathrm{S/N} = c/\sigma_c. \tag{9.57}$$

Thus, S/N measures the ratio of the mean of c_n to its standard deviation.

Let c_n be a random *intensity*[2] *value* at position n in an image. Also, let the overall intensity at any position be the sum of a constant background c and zero-mean noise. Then S/N will measure the ratio of the background level to noise fluctuation in the image. This is often regarded as a useful measure of quality for an image.

What we consider here is *how ratio* S/N *may be estimated from real data* $\{c_n\}$ across the image. An intuitive measure of S/N is the sample signal to noise ratio SNR. This is defined as the ratio of the sample mean $\bar{c}$ to the sample standard deviation $\sqrt{S^2}$,

$$\mathrm{SNR} \equiv \bar{c}/\sqrt{S^2}. \tag{9.58}$$

Obviously, in the limit as the number N of samples which define $\bar{c}$ and S^2 approaches infinity, SNR will approach S/N. However, for finite N we shall find that SNR is actually a *biased estimate* of S/N (Sect. 9.11.2).

Since $\bar{c}$ and S^2 are random variables, so then is SNR. So, in order to determine the accuracy in SNR, we must find its probability law.

9.11.1 Probability Law for SNR

From (9.7) and (9.50), SNR of (9.58) is the ratio of a normal RV to the square-root of a chi-square RV. We show next that this defines a RV that obeys a parabolic cylinder function probability law; and that, as $c \to 0$ it goes over into a "Student" probability law.

To simplify the notation, let

$$x = \bar{c} \text{ and } z = S^2, \quad \text{with} \quad t = \mathrm{SNR}. \tag{9.59}$$

We want $p_T(t)$. In order to find this, we shall (a) form $p_{XZ}(x, z)$; regard (9.58) as the transformation of RV x to RV t, and so form $p_{TZ}(t, z)$; and finally, (c) integrate through over z to find $p_T(t)$.

[2] or irradiance

We previously noted that RV's $\bar{c}$ and S^2 are statistically independent. Therefore, by (9.5, 9.7, 9.50) and (5.55a,b)

$$p_{XZ}(x, z) = p_X(x) p_Z(z) \tag{9.60}$$
$$= \left(\frac{N}{2\pi\sigma_c^2}\right)^{1/2} \exp\left[-\frac{N(x-c)^2}{2\sigma_c^2}\right] \frac{z^{(N-3)/2} \exp\left[-\frac{(N-1)z}{2\sigma_c}\right]}{\left(\frac{2\sigma_c^2}{N-1}\right)^{(N-1)/2} \Gamma\left(\frac{N-1}{2}\right)}.$$

Next, regard (9.58) as a transformation

$$t = x/\sqrt{z} \tag{9.61}$$

from old RV x to new one t, for given z. Then by principle (5.4)

$$p_{TZ}(t, z)\,\mathrm{d}t = p_{XZ}(x, z)\,\mathrm{d}x$$
$$= p_{XZ}(\sqrt{z}\,t, z)\sqrt{z}\,\mathrm{d}t\,. \tag{9.62a}$$

Cancelling common factor $\mathrm{d}t$ on both sides and substituting in the form (9.60) yields

$$p_{TZ}(t, z) = A\sigma_c^{-N} z^{(N-2)/2} \exp\left[-\frac{N(\sqrt{z}\,t - c)^2 + (N-1)z}{2\sigma_c^2}\right], \tag{9.62b}$$

where

$$A \equiv \left(\frac{N}{2\pi}\right)^{1/2} \left(\frac{N-1}{2}\right)^{(N-1)/2} \frac{1}{\Gamma\left(\frac{N-1}{2}\right)}.$$

The marginal law $p_T(t)$ is found by integrating this through over z from 0 to ∞. After a change of integration variable $z = y^2$, the result[3] is [9.3]

$$p_T(t) = B(N-1+Nt^2)^{-N/2} \exp\left[-\frac{N}{2}\left(\frac{c}{\sigma_c}\right)^2 \frac{2(N-1)+Nt^2}{2(N-2)+2Nt^2}\right]$$
$$\times D_{-N}\left(-\frac{Nt(c/\sigma_c)}{\sqrt{N-1+Nt^2}}\right) \tag{9.63}$$

where

$$B = 2A\Gamma(N).$$

Function D_n is the parabolic cylinder function of order n. Unfortunately, the integration of (9.63) to find its variance would be awkward to say the least. Luckily, there is a more elementary route.

9.11.2 Moments of SNR

Observe that (9.61), defining t, has x independent of z (as previously assumed). Therefore, by principle (3.27)

[3] due to student S. Ovadia

$$\langle t^n \rangle = \langle x^n \rangle \langle z^{-n/2} \rangle. \tag{9.64}$$

The nth moment of SNR is the product of the nth moment of x with the $(-n/2)$th moment of z. These moments are easy to calculate since x is known to be normal and z chi-square.

For example, with $n = 1$ we have

$$\langle t \rangle = \langle x \rangle \langle z^{-1/2} \rangle,$$

with

$$\langle x \rangle = c$$

and

$$\langle z^{-1/2} \rangle = \left(\frac{N-1}{2\sigma_c^2} \right)^{(N-1)/2} \frac{1}{\Gamma\left(\frac{N-1}{2}\right)} \int_0^\infty \mathrm{d}z\, z^{-1/2} z^{(N-3)/2} \exp\left[-\frac{(N-1)z}{2\sigma_c^2} \right]$$

by (9.7) and (9.60). This is easily evaluated as

$$\langle z^{-1/2} \rangle = \sigma_c^{-1} \sqrt{(N-1)/2}\, \Gamma\left(\frac{N-2}{2} \right) \Big/ \Gamma\left(\frac{N-1}{2} \right).$$

Combining, we get

$$\langle t \rangle = (c/\sigma_c) \sqrt{(N-1)/2}\, \Gamma\left(\frac{N-2}{2} \right) \Big/ \Gamma\left(\frac{N-1}{2} \right) \equiv \langle \mathrm{SNR} \rangle \tag{9.65}$$

by correspondances (9.59).

Recall from (9.57) that the ideal value for SNR is c/σ_c. An unbiased estimator would have $\langle SNR \rangle = c/\sigma_c$, even for finite N. Hence, the estimator (9.58) is biased. It can be shown (Ex. 9.1.12) that the bias factor in (9.65) approaches unity as $N \to \infty$.

The result (9.65) may be used to form an *unbiased* estimator. It is simply SNR of (9.58) divided by the bias factor.

The second moment $\langle t^2 \rangle$ is found in the same way. Then, using identity (3.15) we find the variance to obey

$$\sigma_T^2 = (c/\sigma_c)^2 \mu + N^{-1}(N-3)^{-1}(N-1),$$
$$\mu \equiv \left(\frac{N-1}{N-3} \right) - \left(\frac{N-1}{2} \right) \Gamma^2 \left(\frac{N-2}{2} \right) \Big/ \Gamma^2 \left(\frac{N-1}{2} \right). \tag{9.66}$$

The last term in μ is the square of the bias factor in (9.65), and approaches unity as $N \to \infty$. Then $\mu \to 1 - 1 = 0$ as $N \to \infty$, so that $\sigma_T^2 \to 0$. This states that as the number N of samples approaches infinity, the error in SNR approaches zero, as expected.

Equation (9.66) shows that the variance in the sample signal-to-noise ratio SNR increases as the square of the true signal-to-noise S/N. This does not comply with intuition, at first. On the other hand, the squared *relative error* $\sigma_T^2 \div (c/\sigma_c)^2$ goes approximately as μ for c/σ_c large. Since $\mu \to 0$ as $N \to \infty$ this *is* a reasonable dependence.

9.11.3 Limit $c \to 0$; A Student Probability Law

With $c \to 0$ in (9.62b), the problem vastly simplifies. The z-dependence goes simply as $\exp(-z)$ times z to a power, and this integrates out to a gamma function, so that

$$p_T(t) = \frac{1}{\sqrt{\pi}} \left(\frac{N}{N-1}\right)^{1/2} \frac{\Gamma(N/2)}{\Gamma\left(\frac{N-1}{2}\right)} \left(1 + \frac{N}{N-1}t^2\right)^{-N/2}. \tag{9.67}$$

Oddly enough, there is no longer a dependence upon σ_c [compare with general result (9.63)]. Then the variance σ_T^2 as well will not depend upon σ_c In fact, from (9.66) with $c = 0$.

$$\sigma_T^2 = N^{-1}(N-3)^{-1}(N-1), \tag{9.68}$$

a pure number depending solely upon the number N of data. This unusual circumstance is taken advantage of in Chap. 12, to estimate the error in a sample mean when the standard deviation σ_c in the data is unknown.

Probability law (9.67) is closely related to "Student's" probability law; compare with (12.16).

9.12 Properties of a Median Window

One way to reduce noise in a noisy image is to convolve it with a rectangle function of length N (Fig. 9.3). This is equivalent to forming each output image point as the sample mean over N local data image points.

From the figure, however, it is apparent that while the method beats down the noise, it also reduces the resolution of edges and spike features in the original image. This is of course the drawback to all methods of processing by convolution (Sects. 5.10 and 9.5). To progress further, *nonlinear* methods must be resorted to.

We might ask whether a processing method exists which can both reduce noise in an image and maintain the resolution of (say) edge features in the image. In fact, the answer is yes, and the processing method is the use of a sample "median window", described next.

The theoretical median value x_0 for a continuous RV x is the value of x such that as many events x are less than x_0 as exceed x_0. Simply put, the area under $p_X(x)$ to the left of x_0 equals that to the right, and both equal $1/2$. For example, the exponential law (9.71) has a median $x_0 = \bar{I}\ln 2$. For *discrete* data the related quantity is the "sample median", defined next.

If N numbers are given, the sample median is the number that has as many above it as below it. Hence, if the numbers are ordered in size, the median (for brevity, we often drop the modifier "sample") will be the $(N + 1)/2$th largest. The median is always a member of the given set of numbers. For example, if the numbers 5, 8, 4, 6 and 12 are given, then $N = 5$ and the median is the third largest number, or the 6. Note that the median is not necessarily the same as the mean, the latter 7 in the case cited. However, it tends to be close enough to the mean to have close to the same

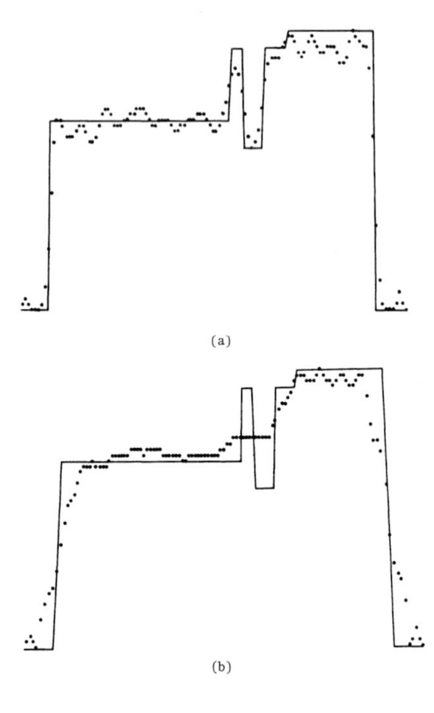

Fig. 9.3. (**a**) Object (*solid*) and image (*dots*). The image contains noise oscillations that we desire to process out.
(**b**) Object (*solid*) and image (*dots*) processed by convolution. The noise has been reduced, but at the expense of reduced gradients

statistical properties, such as a variance of fluctuation going roughly as N^{-1}. This will be shown later.

At this point we shall establish the important non-statistical properties of the median. These are (a) the obliteration of rapid oscillations in a signal, but (b) the preservation of monotonic regions of the signal, such as edge gradients. Consider the top line of numbers in Fig. 9.4, to be considered image data. The data are illustrative of a noisy edge-gradient region, that is, a fluctuating low plateau at the left, followed by the edge region (numbers 3, 5, 7) and then a fluctuating high plateau at the right. Suppose that the median of the leftmost $N = 7$ numbers is taken first, and output at the center position of the window; then the median of number positions 2 through 8 in the same way; etc. In this way a "median window" operation of length $N = 7$ is carried through across the image. The output image, the numbers on the second line of Fig. 9.4, shows some rather unexpected properties. First, the leftmost plateau region is now nearly constant at value 2, as is the rightmost plateau at value 7 or 8. But (craziest of all) the edge-gradient region 3, 5, 7 is *passed unaltered*. The median window operation gives us the best of both worlds.

Experimentation of this kind with median windows of various lengths N shows the following strange, but useful, effects: (a) if an image feature consists of a "bump" (e.g., a point spread function) against constant background, the bump will be obliterated if it is of extension $(N + 1)/2$ *or less*; (b) but if a feature is monotonic over extension N or more, it will be passed unaltered. Property (a) has been found useful

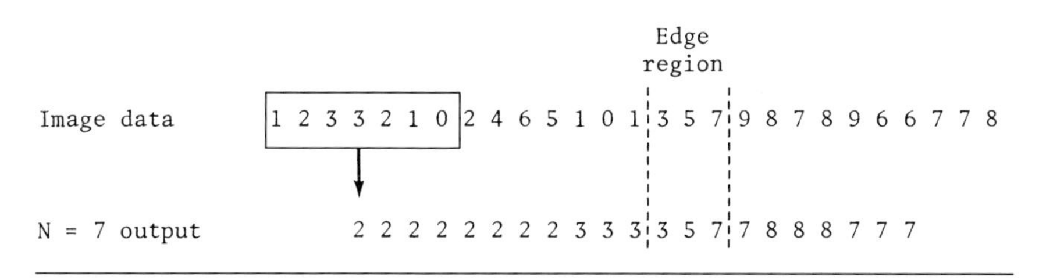

Fig. 9.4. Median window outputs for a line of input data. The first window position is the box on the left; its output is the 2 indicated by the arrow

in separating star images from a background cloud of interstellar dust in certain galactic pictures.

To see these tendencies graphically, Fig. 9.5 shows what happens when the image data of Fig. 9.3a are now processed using a median window of length $N = 7$. Approximately the same amount of noise suppression has been accomplished as in Fig. 9.3b, but now without any loss of edge gradients. This is certainly a very useful, and nonlinear, effect.

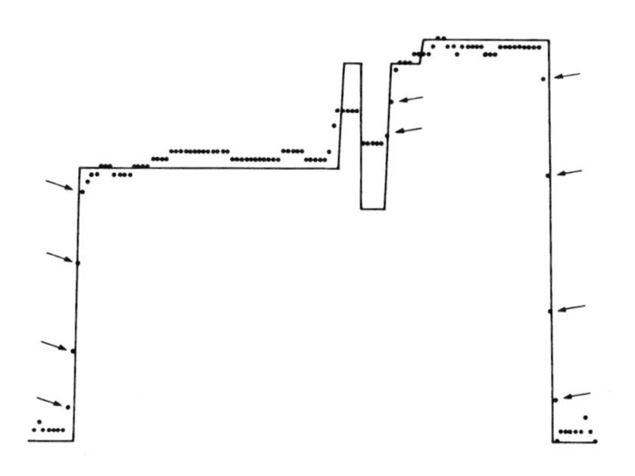

Fig. 9.5. Object (*solid*) and the output (*dots*) of processing the image in Fig. 9.3a by median window. A window length $N = 7$ was used. Noise has been strongly reduced, while gradients remain unchanged. Points denoted by arrows are *the same* as in the image data in Fig. 9.3a

9.13 Statistics of the Median

Any edge detail in an image consists of a low plateau followed by an edge gradient and a high plateau. Having established that the median operation does not reduce edge gradients, we now want to show that it does reduce random fluctuation over plateau (constant) regions. This would make it a very effective restorer of edge-gradient details in an image.

The sample median of a set of N numbers is often close to the sample mean. We might then expect the standard deviation of the median to obey a law like (9.5), i.e., go as $N^{-1/2}$. This will be shown to be approximated below. First we must establish the probability law for the median.

9.13.1 Probability Law for the Median

This turns out to be a remarkably straightforward application of Bernoulli trials (Sects. 6.1 and 10.3). The window is placed upon N (odd) intensities $\{c_n\}$, giving rise to a median window (MW) value x. We want the probability density $p_{\text{MW}}(x)$ due to repeatedly finding the median over independent sets $\{c_n\}$. The N numbers $\{c_n\}$ are i.i.d. samples from a known probability law $p_C(x)$. Denote its cumulative probability by $F_C(x)$.

In any one placement of the window, x will be the median if (a) some one value c_n lies in interval $(x, x + \mathrm{d}x)$, and of the remaining $N - 1$ intensities (b) $(N - 1)/2$ exceed value x and (c) $(N - 1)/2$ are less than value x. (Note: only an infinitesimal number will equal the value x.) Therefore, the probability of the simultaneous occurrence of events (a), (b), and (c) equals the required $p_{\text{MW}}(x)\,\mathrm{d}x$.

Now for an arbitrary x, any one value c_n either lies in interval $(x, x+\mathrm{d}x)$, exceeds x, or is less than x. That is, each trial c_n has three possible outcomes. Also, each c_n is statistically independent of the others. This defines a Bernoulli trials sequence of order three. Moreover, the probability of each of the three outcomes is known:

$$
\begin{aligned}
&P(x \leqq c_n \leqq x + \mathrm{d}x) = p_C(x)\,\mathrm{d}x,\\
&P(c_n < x) = F_C(x) \quad \text{and} \quad P(c_n > x) = 1 - F_C(x).
\end{aligned} \tag{9.69}
$$

The probability of the simultaneous occurrence of events (a), (b) and (c) is then a simple application of (10.10) with $M = 3$,

$$
\begin{aligned}
p_{\text{MW}}(x)\,\mathrm{d}x = &\frac{N!}{[(N-1)/2]![(N-1)/2]!}\\
&\cdot F_C(x)^{(N-1)/2}[1 - F_C(x)]^{(N-1)/2} p_C(x)\,\mathrm{d}x.
\end{aligned} \tag{9.70}
$$

This result holds for any probability law $p_C(x)$ on the individual intensities. The specific form taken by $p_{\text{MW}}(x)$ will of course depend upon the form of $p_C(x)$.

The median window value is an example of a *rank-order* statistic. Imagine the N (odd) numbers $\{c_n\}$ to be sorted according to decreasing size, with a general position in the list denoted as r. Then position $r = 1$ defines the the maximum number, $r = (N + 1)/2$ defines the median number and $r = N$ defines the minimum number in the list. Eq. (9.70) defines the probability of the median, i.e., that of the order $(N + 1)/2$ statistic. The equation may be generalized to the probability of any order r statistic; see (9.97).

9.13.2 Laser Speckle Case: Exponential Probability Law

We found in Sect. 5.9 that intensity follows an exponential law

$$
p_C(x) = \bar{I}^{-1}\mathrm{e}^{-x/\bar{I}}, \quad \bar{I} \equiv \langle I \rangle = 2\sigma^2 \tag{9.71}
$$

in the case of laser speckle. (Note that $\bar{I}$ is here the theoretical - not sample - mean.) The intensity has an intrinsic S/N of unity, as shown at (5.36). We found that one way of augmenting S/N is to use aperture averaging (Sect. 5.10). With N points within the aperture, S/N= $N^{1/2}$ results, an improvement over no averaging. However, the convolution aspect of this operation upon an image results in lost high-frequency details, notably edges. What would a median window operation do instead?

We found in Sect. 9.12 that edge-details would come through unaltered. And in Sect. 9.13.1 we found the probability law that will be obeyed at flat or background parts of the image. We now want to find the resulting S/N, particularly how it measures up to the result $N^{1/2}$ of ordinary averaging.

Use of the particular law (9.71) in (9.70) results directly in

$$p_{\mathrm{MW}}(x) = K(1-\mathrm{e}^{-x/\bar{I}})^{(N-1)/2}(\mathrm{e}^{-x/\bar{I}})^{(N-1)/2}\mathrm{e}^{-x/\bar{I}} \tag{9.72}$$

as the probability law for the median. K is a normalization constant. However, as our ultimate aim is S/N, it is more convenient to proceed from (9.72) to the characteristic function $\varphi_{\mathrm{MW}}(\omega)$. By the change of integration variable

$$y = \mathrm{e}^{-x/\bar{I}}$$

it is found directly [9.3, p. 284] that

$$\varphi_{\mathrm{MW}}(\omega) = \frac{B[(N+1)/2, (N+1/2 - \mathrm{j}\omega\bar{I}]}{B[(N+1)/2, (N+1)/2]} \tag{9.73}$$

where B is the Beta function defined as

$$B(m,n) \equiv \int_0^1 \mathrm{d}x(1-x)^{n-1}x^{m-1} . \tag{9.74}$$

By differentiation of (9.73) we establish the first two moments of MW. The first, the mean value of the median, is

$$\langle \mathrm{MW} \rangle = 2\bar{I} \sum_{n=0}^{(N-1)/2} \frac{1}{N+2n+1} . \tag{9.75a}$$

Note that this contains but a finite number of terms. It is well approximated by

$$\langle \mathrm{MW} \rangle \simeq \bar{I}\left[\ln\left(\frac{2N}{N-1}\right) - \frac{N+1}{2N(N-1)}\right] \tag{9.75b}$$

for $N \gtrsim 5$. As $N \to \infty$, it is evident that $\langle \mathrm{MW} \rangle \to \bar{I}\ln(2)$, which was also the median point for the exponential law (9.71). Since MW is actually a *sample* median (extracted from N data values), this simply states that MW is an unbiased estimate of the true median as $N \to \infty$.

The variance is found to obey

$$\sigma^2_{\mathrm{MW}} = 4\bar{I}^2 \sum_{n=0}^{(N-1)/2} \frac{1}{(N+2n+1)^2} . \tag{9.76a}$$

Again, this has a finite number of terms. For $N \gtrsim 5$, this is well-approximated by

$$\sigma^2_{MW} \cong \frac{1}{2}\bar{I}^2\left[\ln\left(\frac{N-1}{N-3}\right) - \frac{3N-1}{(N+1)(N-1)(N-3)}\right]. \tag{9.76b}$$

For N very large, the latter is dominated by a first-order expansion of the logarithmic term,

$$\sigma^2_{MW} \to \frac{\bar{I}^2}{N-3}. \tag{9.76c}$$

Hence, the variance of the median obeys basically an N^{-1} law, as did the variance of the sample mean. *The median output should therefore fluctuate about as little as the sample mean.* Hence, the plateau regions of speckle images will be processed about as effectively by the median window as by the sample mean. However, they will tend to be depressed by a constant factor of $ln(2) \cong 0.69$ below the mean, as was shown.

Finally, the S/N ratio,

$$\text{S/N} \equiv \langle \text{MW} \rangle / \sigma_{\text{MW}} \tag{9.77}$$

may be computed from (9.75a) and (9.76a). This is plotted as a function of the number of samples N, in Fig. 9.6. It is evident from (9.75b) and (9.76c) that

$$\lim_{N\to\infty} \text{S/N} = N^{1/2}\ln(2)\,. \tag{9.78}$$

Hence, S/N for the median approaches S/N for the sample mean times $\ln(2) \simeq 0.69$. Surprisingly, *there is about a 30% penalty in S/N by use of the median*, somewhat offsetting its benefits at the gradient regions. To see how fast with N the S/N approaches this limit, the curve of $N^{1/2}\ln(2)$ is also plotted in Fig. 9.6.

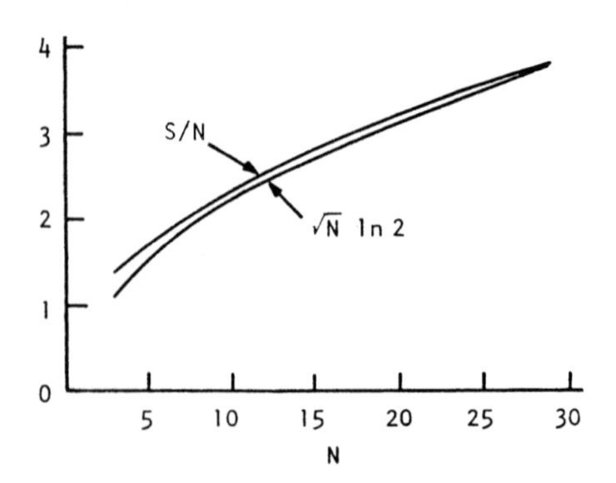

Fig. 9.6. Signal-to-noise ratio S/N in median outputs vs window length N

Exercise 9.1

9.1.1 A thin film is being deposited upon a substrate. During deposition a number N of readings on thickness t of the film are made. There is an uncertainty in each such reading amounting to a σ of 0.125λ. How many readings are needed in order to infer the mean thickness $\langle t \rangle$ with a σ of 0.0125λ?

9.1.2 Linear processors $\{w_n\}$ in (9.8) have a wide variety of uses. The Hanning filter $\{w_n\} = (1/2, 1, 1/2)$ is used to *smooth out* a particular band of ripples in a signal $\{c_n\}$. Another filter F1 has weights [9.4]

$$\{w_n\} = (0.086, -0.288, 1.0, -0.288, 0.086)$$

which accomplish a resolution *enhancement* of factor 1.6 over that of a diffraction-limited image $\{c_n\}$ which is to be processed. Going a bit more extreme, the filter F2 has weights [9.4]

$$\{w_n\} = (-0.142, 0.330, -0.514, 0.676, -0.819, 1.0, \\ -0.819, 0.676, -0.514, 0.330, -0.142)$$

which accomplish a resolution enhancement of factor 2.6 over that of diffraction-limited image $\{c_n\}$.

Use (9.12) to predict the boost (or attenuation) r in noise for the output, for each of the preceding three filters. What kind of tradeoff is being made?

9.1.3 Show that $\bar{c}$ formed according to (9.8) is unbiased, obeying $\langle \bar{c} \rangle = c$. *Hint:* Use the model for outputs $\{c_n\}$ in Sect. 9.2.

9.1.4 Regard the weights $\{w_n\}$ in (9.12) as unknowns. Show that the set of weights that minimizes noise propagation factor r is the uniform set.

9.1.5 A manufacturer of plastic lenses notices that about three out of every 100 lenses have to be discarded because of inferior quality (air bubbles, scratches, etc.). How many lenses must he so observe in order to infer the true discard rate with 10% accuracy? How many must he observe in order to guarantee an error of less than 0.01 in $\hat{p}$ with probability 0.95?

9.1.6 A scanning densitometer is to be used upon a picture whose transmittance values range from 0.1 to 0.9. By what factor must the chosen scanning aperture exceed the film grain size (assuming only one grain size present), in order to achieve an accuracy of 5% or better in each reading?

9.1.7 In the long run, a binary communication link should be "on" as often as it is "off", so that $p(\text{on}) = 1/2$. It is suspected that a certain link is occasionally "on" even when it receives an "off" message, so that its $p(\text{on})$ is biased away from $1/2$. It is decided to test the link by simply observing the number m of times the link is "on" during N trials, and to replace the link if $p(\text{on})$ differs from $1/2$ with a relative error of 0.01. (The link is rather expensive so that a low e is called for.) How large must N be to accomplish such reliability?

9.1.8 A laser beam that propagates through the turbulent atmosphere will suffer random deformations of its wavefront. A scheme for precorrecting the wavefront is suggested. The effectiveness of the scheme depends upon accurately knowing the variance σ^2 of optical path across the observed wavefront $\{W_n\}$. W_n is the nth optical path deformation. The sample variance

$$S^2 = N^{-1} \sum_{n=1}^{N} W_n^2$$

is used as the estimated value of σ^2. Assume that the $\{W_n\}$ are statistically independent samples from one probability law $N(0, \sigma^2)$.

(a) What is the probability law obeyed by S^2? Is S^2 an unbiased estimate? How many degrees of freedom does S^2 have?

(b) How large does N have to be to ensure a relative error $e \equiv \sigma_{S^2}/\sigma^2$ of 1%?

(c) How large must N be to ensure an error in S^2 of $0.1\sigma^2$ or less, with probability 0.75?

9.1.9 N independently moving quantum oscillators are observed in position. Let x_n denote the nth observed position, which for our purposes incurs negligible error (Heisenberg notwithstanding). Each is in its ground state, for which probability of position x is Gaussian,

$$x = N(0, \hbar/2\sqrt{mK}).$$

K is the elasticity constant, m the oscillator mass and $\hbar$ Planck's constant divided by 2π. An experimental quantity

$$S^2 \equiv N^{-1} \sum_{n=1}^{N} x_n^2$$

is formed, in order to verify the theoretical spread $\hbar/2\sqrt{mK}$ in oscillator positions. How many oscillators must be so observed to guarantee 1% relative error e in S^2?

9.1.10 A target of diameter D is aimed at by a laser beam of much smaller diameter. However, atmospheric turbulence in the intervening space causes each photon in the beam to depart from its initial trajectory, as in Fig. 9.7.

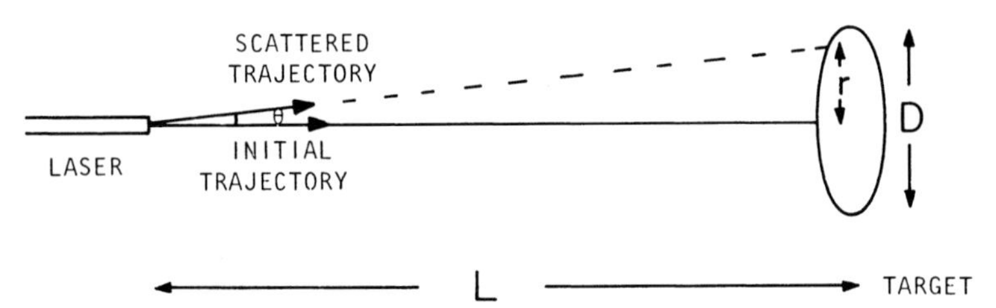

Fig. 9.7. A problem in photon scattering

Let scattering angle θ be Gaussian,

$$\theta = N(0, \sigma_\theta^2)$$

and assume a worst-possible scattering situation – all scattering by angles θ occurs just as photons leave the laser, at distance L from the target. Assume D/L small, so that all angles θ under consideration are small.

(a) What is $p_R(r)$, the probability density for radius r on the target for a given photon? What is the average radius r for photons at the target?

Quantity

$$S^2 \equiv N^{-1} \sum_{n=1}^{N} r_n^2$$

is formed from observation of individual photon radii r_n on the target. S^2 represents the "concentration" of photons on the target, since $\sqrt{S^2}$ is the rms departure of photons from the center of the target.

(b) What is the probability density for S^2?

(c) What is the average concentration?

(d) To avoid overheating the target, no more than a fluctuation/s of σ_S^2 in concentration can be tolerated. What photon flux N/s is needed?

9.1.11 The signal-to-noise ratio is to be estimated for a portion of an image which is at background. The true S/N value is thought to be about 10:1. The estimated S/N is the ratio of sample mean to sample standard deviation in the background region.

(a) What number N of image samples must be taken to ensure a relative error $\sigma_T \div (c/\sigma_c)$ of 5%?

(b) Repeat this calculation if the true S/N value is estimated to be about 100:1.

9.1.12 Show that the bias factor

$$\left(\frac{N-1}{2}\right)^{1/2} \Gamma\left(\frac{N-2}{2}\right) \Big/ \Gamma\left(\frac{N-1}{2}\right)$$

in (9.65) approaches unity as $N \to \infty$. (*Hint:* Use Stirling's approximation

$$\Gamma(m) \cong \sqrt{2\pi}\, \mathrm{e}^{-m} m^{m-1/2}.$$

Take the natural logarithm of the bias, then use l'Hôpital's rule to evaluate this in the limit.)

9.1.13 An image is processed using a median window of length N. The data are independent samples from RANF, i.e., uniformly random over interval (0, 1). Use (9.70) to establish the probability law for the outputs from the window. Find its mean and variance as a function of N. (*Hint:* The required integrals are in [9.3, p. 284].)

9.1.14 A portion of a speckle image has a mean background level of $\langle I \rangle \equiv \bar{I} = 1.0$. It is processed using a median window of length $N = 5$. Find the mean, variance and S/N for the median outputs. Compare with the mean, variance and S/N if a sample mean window is used instead.

9.1.15 The result (9.5) requires that σ_c be finite. A law that violates this premise is the Cauchy law (3.59). Suppose that N Cauchy RV's are averaged, where each is i.i.d. from a law (3.59) with parameter a. Show that the average is still Cauchy, with parameter a. Hence, there is no advantage in precision to be gained by averaging Cauchy RV's. *Hint:* See Exercise 4.1.12b.

9.14 Dominance of the Cauchy Law in Diffraction

In Ex. 9.1.15 we found that the average of N Cauchy RV's remains Cauchy no matter how large N is. That is, *Cauchy RV's do not obey the standard central limit theorem.* This is basically because the variance of each Cauchy datum is infinite, thus violating a premise of the derivation of the central limit theorem (Sect. 4.26). We now show that the Cauchy law arises under much more general conditions. In particular, the average of a large number N of RV's sampled from a *sinc*^{2} or slit diffraction law is *also Cauchy*. Equivalently, N convolutions of the diffraction pattern with itself approach a Cauchy result. This might be considered an *optical* central limit theorem, where the limit is not normal but Cauchy. Consider the following problem of estimation.

9.14.1 Estimating an Optical Slit Position: An Optical Central Limit Theorem

A slit is located an unknown distance b off axis in the object space of an optical system of unit magnification. The slit is illuminated by quasi-monochromatic light of mean wavelength λ. In image space, with a general coordinate y, *the intensity pattern is a sinc*^{2} *diffraction pattern* (3.16), centered upon position b,

$$p_Y(y|b) = p(x) = \frac{\beta_0}{\pi}\mathrm{sinc}^2(\beta_0 x), \quad x = y - b.$$

See (8.47d). Here $\beta_0 = \pi w/\lambda F$ where w is the slit width and F is the image conjugate distance. The RV x defines a random fluctuation in photon position *from* the true position b.

In an attempt to estimate b, the image positions $y_i = b + x_i$, $i = 1, ..., N$ of N photons are independently measured within the *sinc*^{2} diffraction pattern. Each x_i obeys the above law $p(x)$. Hence the x_i are i.i.d. RV's. The fluctuations x_i ought to occur about equally on each side of the mean position b. Hence, the arithmetic average (9.1)

$$\bar{y} \equiv \frac{1}{N}\sum_{i=1}^{N} y_i = b + \bar{x}, \quad \bar{x} \equiv \frac{s}{N}, \quad s = \sum_{i=1}^{N} x_i, \tag{9.79}$$

is taken as *an estimate* of b. Notice that the RV $\bar{x}$ is just a scaled version of the sum s. *What probability law does $\bar{x}$ now obey in the presence of diffraction*?

9.14.2 Analysis by Characteristic Function

For maximum generality we do the analysis for *any* incoherent optical point spread function $p(x)$ as the underlying single-data probability law. (Later we return to the sinc^2 case.) In general *this can suffer any amount of diffraction and aberration*. Let $p(x)$ have a generally nonzero characteristic function $\varphi_X(\omega)$ over a finite band of frequencies $-2\beta_0 \le \omega \le 2\beta_0$, as is the case in optical diffraction. (See Ex. 3.1.23 for an introduction to diffraction.)

Since by (9.79) s is a sum of the x_i, by (4.26) the characteristic function for s is *a product* of N characteristic functions, or $\varphi_S(\omega) = [\varphi_X(\omega)]^N$. Eqs. (9.79) also show that $\bar{x}$ is a scaled version of s. Then by the shift theorem (4.5) the characteristic function

$$\varphi_{\bar{X}}(\omega) = \varphi_S(\omega/N) = [\varphi_X(\omega/N)]^N, \quad -N(2\beta_0) \le \omega \le N(2\beta_0) \tag{9.80}$$

Also, the range of ω is extended as indicated since cutoff in $\varphi_X(\omega/N)$ manifestly occurs at these higher frequencies.

It is to be noted that $\varphi_X(0) = 1$, as compared with $|\varphi_X(\omega)| < 1$ for ω *appreciably away* from value $\omega = 0$. Then a large power of φ_X will exaggerate this effect, tending toward zero *except at frequencies obeying ω small*. Hence, in the limit

$$\lim_{N\to\infty} [\varphi_X(\omega)]^N \to 0 \tag{9.81}$$

for all ω except near the origin $\omega \approx 0$. We may therefore expand φ_X in power series about this point, so that

$$\lim_{N\to\infty} \varphi_X(\omega/N) = 1 + (\omega/N)\varphi'_X(0) + (\omega/\sqrt{2}N)^2\varphi''_X(0) + \ldots \tag{9.82}$$

where we used $\varphi_X(0) = 1$.

In the given problem, the characteristic function $\varphi_X(\omega) = T^*(\omega)$, the complex conjugate of the optical transfer function at the same frequency (compare (3.64) with (4.1)). Then a problem arises in that *$T(\omega)$ has a cusp at the origin*, as in the case of a triangle function [see (3.20)], so that the derivatives $\varphi'_X(0)$, $\varphi''_X(0)$ are undefined if they are to be *central* derivatives. Hence we must resort to one-sided derivatives. For example, a right-handed first derivative $\varphi'(0) \equiv [\varphi(+\epsilon) - \varphi(0)]/\epsilon$, $\epsilon > 0$, in the limit $\epsilon \to 0$. All right-handed derivatives in (9.82) are well-defined and finite. See Ex. 9.2.2. With right-handed derivatives used in (9.82), it becomes a formula for extrapolating to any frequency ω/N *to the right* of the origin. These are positive frequencies ω. We temporarily only consider such frequencies.

Taking the ln of (9.80) and using (9.82) gives

$$\lim_{N\to\infty} \ln \varphi_{\bar{X}}(\omega) = \lim_{N\to\infty} N \ln[1 + (\omega/N)\varphi'_X(0) + (\omega/\sqrt{2}N)^2\varphi''_X(0) + \ldots]. \tag{9.83}$$

In the limit $N \to \infty$ all terms beyond the 1 become sufficiently small to use $\ln(1+\epsilon) = \epsilon$. After multiplying through by N, taking the indicated limit and then exponentiating, we find that

$$\lim_{N\to\infty} \varphi_{\bar{X}}(\omega) = e^{\omega\varphi'_X(0)} \tag{9.84}$$

for $0 \le \omega \le \infty$. Remarkably, all functional dependence upon N dropped out! Also, because $N \to \infty$ the range of positive frequencies in (9.80) is now extended to $+\infty$.

Negative frequencies are handled analogously. Now all indicated derivatives in (9.82) are left-handed instead. See Ex. 9.2.2. The result (9.84) likewise follows for the negative range of frequencies $-\infty \le \omega \le 0$. Hence, *(9.84) now holds for the full range of positive and negative frequencies.*

Let the optical pupil be generally aberrated, with a pupil function $U(\beta) = \exp[j\Phi(\beta)]$ where $\Phi(\beta)$ is the *aberration function*. Then it is shown in Ex. 9.2.2 that

$$\varphi'_X(0) = (j\Delta\Phi - 1)/2\beta_0 \text{ or } (j\Delta\Phi + 1)/2\beta_0, \tag{9.85}$$

respectively, according to whether the derivative $\varphi'_X(0)$ is right- or left-handed. In (9.85)

$$\Delta\Phi \equiv \Phi(-\beta_0) - \Phi(\beta_0) \tag{9.86}$$

is the difference in the phase of the pupil at its margins $\beta = \pm\beta_0$. Using (9.85) in (9.84) gives

$$\lim_{N\to\infty} \varphi_{\bar{X}}(\omega) = \mathrm{e}^{j(\Delta\Phi/2\beta_0)\,\omega}\mathrm{e}^{-\,|\omega|/2\beta_0} \text{ for } -\infty \le \omega \le \infty. \tag{9.87}$$

As a verification, this result *can alternatively be derived* by the use of an invariance principle (Sect. 5.14). This is to demand that $[\varphi_X(\omega/N)]^N$ approach a form invariant to N in the lim $N \to \infty$. See Ex. 9.2.1.

9.14.3 Cauchy Limit, Showing Independence to Aberrations

Comparing (9.87) with (4.29), and by the shift theorem (4.5), we see that $\bar{x}$ obeys a shifted Cauchy density function,

$$p_{\bar{X}}(x) = \frac{a/\pi}{(x - \langle\bar{x}\rangle)^2 + a^2}, \quad a = (2\beta_0)^{-1}, \ \langle\bar{x}\rangle = \Delta\Phi/2\beta_0. \tag{9.88}$$

Remarkably, the probability density on the sample mean of a huge number of photon position values is the same as that on *a single* Cauchy variable. There is nothing to gain by taking the sample mean!

It is also of interest that *the form of the probability law (9.88) does not depend upon the particular aberrations in the pupil, either as to type or as to size.* It is always Cauchy. Only the amount $\langle\bar{x}\rangle$ by which the law is shifted depends upon the aberrations. This is through the difference of the pupil phase values *at the margins.* Interior values of the phase do not matter.

The usual central limit theorem of statistics, taken up in Chap. 4, can approach its limiting normal form after about $N \ge 4$ convolutions (see Sect. 4.27). How many are needed to attain the Cauchy form (9.87) or (9.88)? The answer is generally much more than 4. For example, with pure third-order aberration present of size $\lambda/2$ at the margins, it takes N values of 100 or more to start seriously approaching the Cauchy limit. This is shown in Fig. 9.8, where the characteristic function $\varphi_{\bar{X}}(\omega)$ is shown (exclusive of phase shift) for various values of N.

We started the problem by taking the sample mean $\bar{y}$ as an estimate of the true position b of the source. We can now judge how good an estimate this is. By (9.79), $\bar{y}$ is likewise shifted Cauchy, with a mean $\langle\bar{y}\rangle = \Delta\Phi/2\beta_0 + b$. Again this is shifted from the true position b by something proportional to the difference of the pupil phases at the margins. This implies that, in the case of a diffraction-limited pupil, where the phase is identically zero, *the mean value of* $\bar{y}$ *equals the true position value* b. However, one must not conclude from this that $\bar{y}$ is *a good* estimate of b: Since the density function for $\bar{y}$ is Cauchy, *its variance is infinite*. In reality, $\bar{y}$ is a terrible estimate of b.

It should be noted that the effect (9.80) also describes the *net optical transfer function for N serial relays of a point-source image* (Sect. 4.28.1). Each transfer function is assumed to suffer the same aberrations. The division by N in (9.79) corresponds to linearly demagnifying the output image by this factor. Hence, the output point spread function for such a relay system will be Cauchy. However, as an approximation, at small coordinate values x this approaches a normal law (see Ex. 9.2.2).

The limiting exponential form (9.87) was confirmed by computer simulation [9.6]. The phase function $\Phi(\beta)$ was made a third-order polynomial $\Phi(\beta) = a_1\beta + a_2\beta^2 + a_3\beta^3$ whose coefficients a_1, a_2, a_3 are three randomly selected aberrations. As might be expected, the larger are the sizes of the aberrations the higher are the values of N that are needed to well-approximate the Cauchy limit . As examples, to achieve 3% accuracy: in the diffraction limited case $a_1 = a_2 = a_3 = 0$ a value $N = 8$ is needed; whereas with *quarter-wave* aberration at the margins a value of about $N = 50$ is needed. Half-wave aberration requires about $N = 100$ terms, according to the curves in Fig. 9.8.

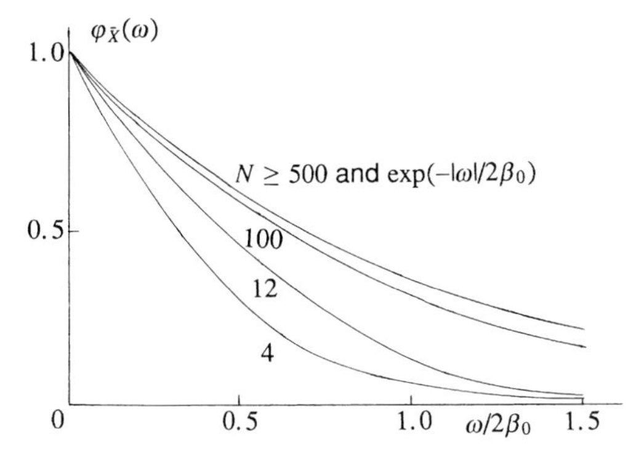

Fig. 9.8. Characteristic function $\varphi_{\bar{X}}(\omega)$ for the average of N position values x in a diffraction pattern. Pure 3rd-order aberration of size $\lambda/2$ at the margins is present. This also represents the optical transfer function for N self-convolutions of an aberrated point spread function. Once N exceeds value 500 the result is indistinguishable from an exponential law, corresponding to a Cauchy random variable.

9.14.4 Widening the Scope of the Optical Central Limit Theorem

The derivation of the Cauchy limit (9.88) presumed each RV x_i to obey the same underlying law $p(x)$. This restriction can now be lifted. This also corresponds to the use of different lenses in the relay system of Sect. 4.28.1. Suppose that each RV type x_i is repeated N_i times, with a corresponding sub sum s_i, $\sum_i s_i = s$, $\sum_i N_i = N$. If each N_i is suitably large, each sub sum s_i obeys, by the above derivation, a Cauchy law. Then, since the sub sums are independent, the characteristic function for the total sum s is a product of exponentials, again an exponential (Ex. 4.1.12). Hence s obeys a Cauchy law and, consequently, so does $\bar{x}$.

Exercise 9.2

9.2.1 ***Use of an invariance principle to derive central limit theorems.*** The problem of Sect. 9.14.1 is mathematically identical to that of multiple incoherent relays of a point spread function image (Sect. 4.28.1). In turn, this is analogous to what goes on within a laser resonator. There, a coherent image is relayed back and forth between the end mirrors of the resonator until an equilibrium image is reached. This image may be found by demanding equilibrium in the form of *a proportionality between its* N*th and* $(N+1)$*st images* in the limit $N \to \infty$. The result is an eigenvalue equation (4.52) whose solution is the amplitude mode functions of the resonator [9.7].

We may analogously attack the problem of finding $\varphi_{\bar{X}}(\omega)$. Regard ω as arbitrary but fixed, throughout the derivation. According to (9.80), $[\varphi_X(\omega/N)]^N$ should equal $[\varphi_X(\omega/(N+1))]^{N+1}$in the limit $N \to \infty$. Another way of saying this is to require $[\varphi_X(\omega/N)]^N = g(\omega)$ to be a function that is *independent of* N. In order to attack a few problems at once, we demand the more general condition, that

$$\varphi_{\bar{X}}(\omega) \equiv [\varphi_X(\omega/N^\alpha)]^N = g(\omega)\,, \tag{9.89}$$

where α is some prescribed constant. The problem of Sect. 9.14 uses an $\alpha = 1$. An analogous problem, that of deriving the central limit theorem of statistics, has $\alpha = 1/2$ (See the definition of the transformed RV s in Sect. 4.26).

The requirement that (9.89) be independent of N is equivalent to demanding $dg/dN = 0$. Actually, it is easier to enforce $d(\ln g)/dN = 0$. *Show that the solution to this problem is a* φ_X *obeying*

$$\varphi_X(\omega) = \exp(B\omega^{1/\alpha}), \quad |\omega| \text{ small}, \tag{9.90a}$$

and a $\varphi_{\bar{X}}$ obeying

$$\varphi_{\bar{X}}(\omega) = \exp(B\omega^{1/\alpha}), \quad \text{any } \omega. \tag{9.90b}$$

The two characteristic functions differ only in their indicated domains of ω. The constant B is generally complex since characteristic functions are complex. By back substitution, show that the form (9.90a) explicitly satisfies requirement (9.89) of independence to N.

In our case $\alpha = 1$, (9.90a) becomes $\varphi_X(\omega) = \exp(B\omega)$, and (9.90b) gives

$$\varphi_{\bar{X}}(\omega) = \exp(B\omega). \tag{9.91}$$

This checks with the result (9.87) for the choice $B = (-1 + \mathrm{j}\Delta\Phi)/2\beta_0$ for a positive ω, or $B = (1 + \mathrm{j}\Delta\Phi)/2\beta_0$ for a negative ω.

The case $\alpha = 1/2$ corresponds to the transformation from RV x to s in Sect. 4.26. The result there was the central limit theorem, i.e., that s obeys a normal law. Our answer (9.90b) is directly

$$\varphi_{\bar{X}}(\omega) = \exp(B\omega^2). \tag{9.92}$$

This corresponds to a Gaussian law with $B = -\sigma^2/2$, which is the conventional result (4.47). Why must B be chosen as purely real in (9.92)?

There is a caveat to the general solution (9.90b). Notice that (9.90a) *is a requirement on the asymptotic form of the underlying statistic in x*. It is not guaranteed that (9.90a) will be obeyed in each application. In cases where it isn't, (9.90b) does not provide the solution $\varphi_{\bar{X}}(\omega)$. Otherwise it does. Examples follow. In the case $\alpha = 1$, show that (9.90a), which only need hold for $|\omega|$ small, is indeed satisfied by our diffraction transfer functions (3.65) and (9.94) in that frequency limit. *Hint*: Use equations (9.85).

The premise (9.90a) is satisfied by each underlying random variable $x \equiv y_i$ in Sect. 4.26. Show this.

9.2.2 Some one-sided derivatives of the optical transfer function. Starting with the pupil function representation (3.65) of the optical transfer function, where $k\Delta(\beta) \equiv \Phi(\beta)$, the pupil phase function, and where $\omega \geq 0$, show by direct differentiation $\partial/\partial\omega$ that the right-handed derivative (indicated by the subscript +) obeys

$$T'_{+}(0) = -(2\beta_0)^{-1}(\mathrm{j}\Delta\Phi + 1). \tag{9.93}$$

This is also $\varphi'_X(0)$ in the problem of Sect. 9.14.2.

At *negative* frequencies $\omega \leq 0$, show that the OTF takes the form

$$T(\omega) = \frac{1}{2\beta_0}\int_{-\beta_0}^{\beta_0+\omega} \mathrm{d}\beta\; U(\beta)U^*(\beta-\omega),\;\; U \equiv \exp(\mathrm{j}\Phi)\,. \tag{9.94}$$

By differentiating this $\partial/\partial\omega$ show that the left-handed derivative (indicated by the subscript -) obeys

$$T'_{-}(0) = (2\beta_0)^{-1}(-\mathrm{j}\Delta\Phi + 1)\,. \tag{9.95}$$

Hint: Note *Leibnitz's Rule*,

$$\frac{\partial}{\partial\omega}\int_{f(\omega)}^{g(\omega)} \mathrm{d}y\, h(\omega, y) = \int_{f(\omega)}^{g(\omega)} \mathrm{d}y\;\partial h/\partial\omega + h(\omega, g(\omega))g'(\omega) - h(\omega, f(\omega))f\,'(\omega).$$

Because $T^*(\omega) = \varphi(\omega)$, the results (9.93), (9.95) correspond to (9.85).

Second- and higher-order derivatives $T''_{+}(0),\; T''_{-}(0)$,... may be computed in like manner.

9.2.3 *A Cauchy law approaches a Gaussian law near the origin.* A Cauchy law (9.88) obeys

$$p(x) \equiv \frac{a/\pi}{a^2+x^2} = \frac{1}{\pi a}\frac{1}{1+(x/a)^2} \approx \frac{1}{\pi a}[1-(x/a)^2] \approx \frac{1}{\pi a}\mathrm{e}^{-(x/a)^2} \tag{9.96}$$

which has the form of a Gaussian law. However, if this is integrated over all x the result is not unity. Why not? Find the needed factor for normalizing the law.

9.2.4 *Special case of diffraction limited imagery.* The transition (9.87) to a Cauchy density function was, historically, first accomplished in the case of zero aberrations. Here the pupil function $U(\beta) = 1$, so that now (9.80) becomes $\varphi_{\bar{X}}(\omega) = [\varphi_X(\omega/N)]^N = [1 - |\omega|/(2N\beta_0)]^N$. By taking the logarithm, show that in the limit $N \to \infty$ the result (9.87) again follows, now with $\Delta\Phi = 0$.

9.2.5 ***Independence of the sample mean and sample variance.*** This independence is true in the presence of any number N of data, as can be proven with some difficulty [9.5]. However, the case $N = 2$ is easy to deal with. Let RV's x_1, x_2 be i.i.d. $N(a, \sigma^2)$. Form new RV's $u = (x_1 + x_2)/2$ and $v = (x_1 - u)^2 + (x_2 - u)^2$. RV's u and v are the sample mean and variance for this low-dimension case.

(a) Show that $\langle u \rangle = a$ and $\langle v \rangle = \sigma^2$. (This is easy.)

(b) Show that $\langle uv \rangle = a\sigma^2$.

Hint: Directly evaluate the expectation, ultimately using $\langle x_1^3 \rangle = \langle x_2^3 \rangle = 3a\sigma^2 + a^3$ for this normal case.

(c) Use definition (3.47) to finish the proof.

9.2.6 ***Probability density for a general rank-order statistic.*** An rth-order statistic has $(r - 1)$ numbers exceeding it and $(N - r)$ numbers less than it. (Why don't we have to count numbers that *equal* it?) Using these facts, generalize the probability law (9.70) for the median to that of the rth-order statistic,

$$p_R(x) = \frac{N!}{(N-r)!(r-1)!} F_C(x)^{N-r}[1 - F_C(x)]^{r-1} p_C(x). \tag{9.97}$$

As a check, notice that the particular rank $r = (N + 1)/2$ defining the median gives back (9.70).

9.2.7 ***The median universal constant is unity.*** On the basis of both theory (Sect. 5.14) and empirical data (Ex. 11.1.9), the *logarithms* of the universal physical constants are distributed uniformly (11.24). The corresponding law for a physical constant value y is

$$p_Y(y) = \frac{1}{2y \ln b}, \quad 10^{-b} \le y \le 10^{b}. \tag{9.98}$$

Show that the median value of this law is 1. Thus, the median value of the universal constants has a definite value, of unity, and (by Sect. 5.14) this is independent of the choice of units! Empirically, the *sample median* of the numbers in the table of Ex. 11.1.9 is 2.73×10^2. Considering the relatively small sample of numbers in the table and, also, that the numbers range over 51 orders of magnitude, this agrees rather well with the prediction of unity. We predict that as more fundamental physical constants are discovered, and added to the list, their sample median will tend toward the theoretical value of unity.

10. Introduction to Estimating Probability Laws

In statistics, one has data and from this tries to infer something about its source. Estimating the *probability law* that gave rise to the data is one of the chief problems of this kind. Once known, its variance, confidence limits, and all other parameters describing fluctuation may be determined. Two major schools of thought on forming the estimates are the *classical* and the *Bayesian* approaches. These are compared in Chap. 16. A third approach is the use of an invariance principle (Sect. 5.14, Ex. 9.2.1 and Chap. 17). A fourth is based upon the use of *Fisher information* (Chap. 17). A major difference among these four approaches is their aims: the Bayesian approach is content to form a "reasonable estimate" of the unknown law, while the others seek *the exact answer*.

The Fisher information-based approach is treated in depth in Chap. 17. That approach aims to find the true probability law describing a physical phenomenon by deriving a wave equation that defines the law. An example is the Schroedinger wave equation, which defines an amplitude function whose modulus-square is the sought probability law.

The unknown law $\{p_j\}$ or $p(x)$ will be allowed to have a general form. Hence, it cannot be specified by estimating merely a couple of parameters, as in the normal case, for example. The latter would be a problem of *parametric estimation*, as it is commonly called. However, our problem is one of estimating the probability values p themselves. If these are regarded as the parameters of the problem, then depending upon whether the unknown law is discrete $\{p_j\}$ or continuous $p(x)$ there is either a large number or an infinite number of unknown parameters present. In the continuous case the problem is obviously outside the realm of parameter estimation, so that it is often called one of *nonparametric estimation*.

Two methods of probability law estimation are presented in this chapter – orthogonal expansions and maximum probability (MP). The first may be considered as classical, and has the virtues of being simple to understand and easy to carry through. The second takes a Bayesian viewpoint, and therefore allows the user's subjective and objective biases to be injected into the estimate. This provides major advantages, it will be seen, although the actual calculation is more difficult to implement than by orthogonal expansion.

The MP estimate also becomes at times a *maximum entropy* estimate [10.1], depending upon conditions. The maximum entropy estimate, usually denoted as M.E. or MaxEnt, is quite popular nowadays. The approach is practical to use and overall gives good results although it is not a panacea (see Sect. 10.4.12 and Ex. 17.3.5). One of our major aims is to indicate its achievements and drawbacks, and to find when it arises out of the MP approach and when *other estimates arise instead.*

But, "entropy" is even more than a technical concept. It has a certain mystique that occasionally transforms even hard-nosed engineers into poets. The following poem was presented to Dr. E.T. Jaynes, the inventor of the MaxEnt approach, by one of his students in a short course on the subject. (We thank Dr. Jaynes for allowing its publication, and thank as well the unknown and inspired student.) Although written with tongue in cheek, it shows some of the insights, as well as misconceptions, that people have had about the concept:

MAXIMUM ENTROPY

There's a great new branch of science which is coming to the fore,
And if you learn its principles you'll need to learn no more.
For though others may be doubtful (which can make them fuss and fret),
You'll be certain your uncertainty's as large as it can get.

This is no toy of theory, to invite the critic's jeering,
But a powerful new method used in modern engineering.
For uncertainty, which often in the past just gave us fidgets,
Has proved to be a vital tool in mass-producing widgets.

The procedure's very simple (it is due to Dr. Jaynes),
You just maximize the entropy, which doesn't take much brains.
Then you form a certain function which we designate by Z,
Differentiate its log by every lambda that you see.

AND LO! – we see before us (there is nothing more to do)
All the laws of thermal physics, and decision theory too.
Possibilities are endless, no frontier is yet in sight,
And regardless of your ignorance, you'll always know you're right!

So, are you faced with problems you can barely understand?
Do you have to make decisions, though the facts are not at hand?
Perhaps you'd like to win a game you don't know how to play.
Just apply your lack of knowledge in a systematic way.

Maximum entropy estimation is taken up in Sect. 10.4.2 and thereafter. We shall first consider estimation by orthogonal expansion, and in particular the Karhunen–Loeve expansion.

10.1 Estimating Probability Densities Using Orthogonal Expansions [10.2]

Let a random variable x have an unknown probability density $p(x)$. It is known that $a \leqq x \leqq b$; for example if x is an *energy*, $a = 0$ and $b = \infty$, expressing the fact that energies must be positive.

We wish to estimate $p(x)$ from N data $\{x_n\}$, which are independent samples x from $p(x)$. A direct approach would be the "brute force" method of building up the histogram of values x, using a suitable "bin" or interval size Δx, counting up how many times m_i each interval $(x_i, x_i + \Delta x)$ occurs among the N occurrences. Then the estimate (9.18) yields an answer $m_i/N \equiv \hat{p}(x_i)$ for each probability.

The trouble with this approach is that, because the number of occurrences is finite, the estimate $\hat{p}(x_i)$ will be in error by an amount proportional to $m_i^{-1/2}$; see (9.22). This will make $\hat{p}(x_i)$ a "lumpy" and erratic estimate, especially in the tails of $p(x)$ where m_i is bound to be small. For example, see the histogram (solid curve) in Fig. 10.1. The true $p(x)$ is the dotted curve, which is of the cosine form (7.5).

A way out of this dilemma is to *force* a smooth estimate by representing $p(x)$ as a series

$$p(x) = \sum_{m=0}^{M} c_m f_m(x) \tag{10.1a}$$

of smooth functions $\{f_m(x)\}$, which obey orthonormality

$$\int_a^b \mathrm{d}x\, f_m(x) f_n(x) = \delta_{mn} , \tag{10.1b}$$

and completeness.

If this sounds familiar, it is because we have already encountered use of such an approach – the Gram–Charlier series of Sect. 3.15.8. There the $\{f_m(x)\}$ were, in particular, Gaussian-weighted Hermite polynomials, which were appropriate since it is known a priori that $p(x)$ is close to Gaussian. But here we want to remain "context-free" and so leave the $\{f_m(x)\}$ general.

With the $\{f_m(x)\}$ chosen, the problem is to find coefficients $\{c_m\}$. By (10.1a), the latter define the solution $\hat{p}(x)$. One way to estimate the $\{c_m\}$ would be to least-squares fit the series (10.1a) to the histogram.

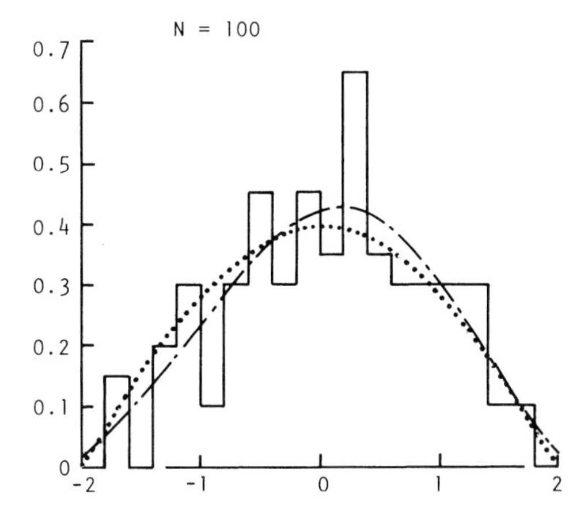

Fig. 10.1. Estimating a probability law by using an orthogonal expansion. The solid curve is the histogram of data. The dotted curve is the true law, and its estimate is the dash-dot curve

But an ingenious and much simpler answer to this problem was recently discussed [10.2]. First of all, observe from (10.1a) that for the true $p(x)$

$$c_m = \int_a^b \mathrm{d}x\, p(x) f_m(x) \; . \tag{10.1c}$$

This derives from property (10.1b). But since $p(x)$ is a probability density, by definition (3.12)

$$c_m \equiv \langle f_m(x) \rangle \; , \tag{10.2a}$$

i.e., c_m is the mean-value of function f_m over random arguments x. But, in fact, we have a way to estimate $\langle f_m(x) \rangle$. Since the data $\{x_n\}$ are given, we may form an estimate using them,

$$\widehat{\langle f_m(x) \rangle} = N^{-1} \sum_{n=1}^{N} f_m(x_n) \tag{10.2b}$$

by the sample-mean formula (9.1). Finally, combining (10.2a) and (10.2b),

$$\hat{c}_m = N^{-1} \sum_{n=1}^{N} f_m(x_n) \; . \tag{10.2c}$$

Hence, *the estimated coefficient is simply the sample mean of its corresponding orthogonal function* f_m.

An example of this approach is in Fig. 10.1. A sample of $N = 100$ data $\{x_n\}$ were generated independently from the theoretical law (7.5) for $p(x)$ (dotted in the figure). Hermite-Gaussian functions

$$(m!2\pi)^{-1/2} H_m(x)\, \mathrm{e}^{-x^2/4} \equiv f_m(x) \tag{10.3}$$

were chosen. Coefficients $\{c_m\}$ were generated via (10.2c). The resulting estimate $\hat{p}(x)$ by the use of series (10.1a) is illustrated by the dash-dot curve in Fig. 10.1. The maximum polynomial order $M = 10$ was used, although beyond $m = 5$ there was small contribution. The fit to the theoretical curve is rather remarkable. The main error consists of an overall shift to the right, which was probably caused by the large number of events that form the large spike in the histogram.

How much error will the resulting estimate $\hat{p}(x)$ suffer? Since each coefficient $\hat{c}_m$ is a sample mean (10.2c), where the $\{x_n\}$ are statistically independent samples from the same probability law $p(x)$, we may use result (9.5) for the error in the mean,

$$\sigma_{\hat{c}_m} = \sigma_{f_m} / \sqrt{N} \; . \tag{10.4}$$

Here $\sigma^2_{f_m} = \langle f_m(x_n)^2 \rangle - \langle f_m(x_n) \rangle^2$, *any* x_n, since the $\{x_n\}$ are identically distributed. Note, then, that for all m the error goes as $N^{-1/2}$. From the form of (10.1a), so then must go the error in the desired quantity $\hat{p}(x)$. Recall, by contrast, that by the direct histogram method the error at point x_i would go as $-m_i^{-1/2}$. Since $m_i \ll N$, we see that an advantage in accuracy has been gained.

Finally, the maximum order M to use in expansion (10.1a) must be defined. A value $M = \infty$ cannot be used since then the estimated $p(x)$ would merely reconstruct the histogram of data $\{x_n\}$! There would be no smoothing effect. One criterion is the value of m for which $\hat{c}_m$ is so small that it equals its fluctuation, $\overline{\sigma}^2_{\hat{c}} = \hat{c}^2_m$. The left-hand quantity may be computed using the sample (barred) version of (10.4), i.e., where all expectations $\langle\rangle$ are replaced by sample means. This choice of cutoff M minimizes the integral of error $[p(x) - \hat{p}(x)]^2$.

There is, however, a major drawback to this method of estimation. It does not permit *prior knowledge* about $p(x)$ to be used. For example, it is known a priori that any $p(x)$ must be positive, whereas the estimate (10.1a) may go negative at certain x values. Also, the observer may know that $p(x)$ should be close to a uniform law. There is no way to force such a bias into (10.1a). The maximum-probability (MP) approach of Sect. 10.4.1 will permit the injection of such prior knowledge into the estimate.

10.2 Karhunen–Loeve Expansion

So far $\{f_m(x)\}$ have been *any set* of complete, orthogonal functions. Indeed, they are arbitrary, except under special circumstances such as the following.

Sometimes the observer knows that the unknown probability law $p(x)$ is one of a class of laws, where the class is defined by a known autocorrelation function

$$R_{\mathrm{p}}(x;\, y - x) \equiv \langle p(x)p(y)\rangle \ . \tag{10.5}$$

Because R_{p} is real, positive, and symmetric in x and y, its eigenvalue equation

$$\int_a^b \mathrm{d}y\, R_{\mathrm{p}}(x;\, y - x)\psi_n(y) = \lambda_n \psi_n(x) \tag{10.6}$$

admits of real eigenfunctions $\{\psi_n(x)\}$ and eigenvalues $\{\lambda_n\}$, $n = 0, 1 \ldots$.

Suppose that the $\{f_n(x)\}$ in orthogonal expansion (10.1a) are chosen to be functions $\{\psi_n(x)\}$. For this choice, (10.1a) is called a "Karhunen–Loeve" expansion of $p(x)$. What property will the expansion coefficients $\{c_n\}$ have?

By property (10.1c)

$$c_m = \int_a^b \mathrm{d}x\, p(x)\psi_m(x) \ .$$

Then

$$\langle c_m c_n\rangle = \int_a^b \mathrm{d}x\, \psi_m(x) \int_a^b \mathrm{d}y\, \psi_n(y)\, \langle p(x)p(y)\rangle$$

after rearranging integration orders; and by definition (8.3) of the autocorrelation

$$\begin{aligned}\langle c_m c_n\rangle &= \int_a^b \mathrm{d}x\, \psi_m(x) \int_a^b \mathrm{d}y\, \psi_n(y) R_{\mathrm{p}}(x;\, y - x) \\ &= \lambda_n \int_a^b \mathrm{d}x\, \psi_m(x)\psi_n(x)\end{aligned}$$

due to the eigenvalue property (10.6). Finally by the orthonormality property (10.1b)

$$\langle c_m c_n \rangle = \lambda_n \delta_{mn} , \tag{10.7}$$

i.e., the random coefficients are *statistically independent*.

Hence, a Karhunen–Loeve expansion of a probability law contains only independent coefficients. This is an important and useful situation, since the individual terms in the expansion (10.1a) are now independent as well. Hence, the number of such terms required to well-approximate $p(x)$ tends to be minimal.

10.3 The Multinomial Probability Law

We found in Sect. 9.6 how to estimate the probability of a single event. This required knowledge of the statistics obeyed by a binary or Bernoulli trials experiment.

We now want to find a method for estimating a probability law for M *distinct events*. Ergo, we have to know the statistics of an *M'ary* experiment. This is an experiment with M possible outcomes. An example is the tossing of a die, which has $M = 6$ possible outcomes.

10.3.1 Derivation

Regarding the die experiment, let this be repeated N times. Let p_1 denote the probability of event 1 (die outcome 1), etc. for $p_2, \ldots, p_M$. Let m_1 describe the number of times die value 1 appears, etc., for $m_2, \ldots, m_M$. We want to know the joint probability $P_N(m_1, \ldots, m_M)$.

The analysis is completely analogous to the binary case of Sect. 6.1, so that we can be brief here.

Any one sequence of trials giving rise to the required m_1 events 1, m_2 events 2, $\ldots$, m_M events M *will have the same probability* of occurrence

$$p_1^{m_1} p_2^{m_2} \cdots p_M^{m_M} . \tag{10.8}$$

The product form is the result of independence from trial to trial, and independence among the events.

Therefore, we have only to enumerate the number of ways $W(m_1, \ldots, m_M)$ the events in question can occur. Multiplying this by (10.8) forms the answer.

$W(m_1, \ldots, m_M)$ may be thought of as the number of ways m_1 indistinguishable red balls, and m_2 green balls, etc., may be drawn out of a bag containing a total of N balls. This is known to describe the multinomial coefficient

$$W_N(m_1, \ldots, m_M) \equiv \frac{N!}{m_1! \ldots m_M!} \tag{10.9}$$

(see Ex. 10.1.18 for an interesting derivation).

Combining (10.8) and (10.9),

$$p_N(m_1, \ldots, m_M) = \frac{N!}{m_1! \ldots m_M!} p_1^{m_1} \cdots p_M^{m_M} \tag{10.10}$$

where $N = m_1 + \cdots + m_M$. This is the multinomial probability law.

10.3.2 Illustrative Example

At a party for 27 people, prizes of the following kind were to be distributed one to a person: 4 noisemakers, 10 paper hats, 8 balloons, and 5 water pistols. All the noisemakers are identical, etc. for the other prizes. In how many ways can the prizes be distributed among the party goers?

This is the number of ways of forming $M = 4$ sets consisting, respectively, of $m_1 = 4$ (indistinguishable) elements, $m_2 = 10$ elements, $m_3 = 8$ elements and $m_4 = 5$ elements. The answer is then the multinomial coefficient (10.9)

$$\frac{(4+8+10+5)!}{4!\,8!\,10!\,5!}$$

or about 8 quadrillion.

10.4 Estimating an Empirical Occurrence Law as the Maximum Probable Answer

An empirical, or, occurrence rate (9.18) is the frequency with which *an event actually happens* in the data. The occurrence rate of an event is to be distinguished from its corresponding theoretical probability, which may not be known from a theoretical model. As a first step toward forming such a model, it is often useful to know the occurrence rate of the event. A simple example is the occurrence rates of the six possible outcomes of the roll of an unknown die. One of keener interest is the occurrence rates of water droplet sizes in a cloud under specific weather conditions.

Of course an occurrence rate equals its corresponding probability law if the number of occurrences is large enough. Also, in some applications ergodicity (Sect. 8.11) must be obeyed. However, these conditions are not generally met.

Sparsity of data is one of the chief problems we encounter when attempting to estimate occurrence rates. For example, the detectors in use might only provide data on the first few moments of the occurrence rate law. Obviously, knowing only the mean of the law does not uniquely determine it. If you are told only that the mean roll of a die is 3.5, what can you infer to be the relative occurrence rates of the 1, the 2, ..., the 6? Certainly *equal* occurrence rates *for all the possible outcomes* would give rise to the 3.5, but so would equal occurrence rates of $1/2$ *for the outcomes 3 and 4*, with zero rate for all other outcomes. Other possibilities exist as well. Which solution should be chosen?

One general procedure for lifting redundancy of the solution is to seek an extreme one. The maximum probable solution is an example. This method was proposed by *Macqueen* and *Marschak* [10.3] in 1975, and is probably the most accurate in existence for estimating an occurrence law. Optical applications of the approach are in [10.12]. Besides using the data, the method permits the *observer's biases* in the form of *prior knowledge* to be built into the estimate. These biases include physical constraints, such as positivity, and preferred tendencies toward known numbers. This is discussed next.

10.4.1 Principle of Maximum Probability (MP)

Suppose an experiment (say, rolling a die having *unknown fairness*) is repeated N times. Each repetition has possible outcomes $i = 1, 2, 3, \ldots, M$, and m_i represents the number of times outcome i occurs over the N experiments. For example, die outcomes

$$2\ 5\ 3\ 4\ 4\ 1\ 6\ 2\ 2\ 4\ 3\ 1\ 1\ 5\ 3\ 4\ 2 \tag{10.11}$$

occur. Note that here $n = 17$, $M = 6$, and $m_i = (3, 4, 3, 4, 2, 1)$ describes the occurrence of outcome 1, outcome 2, ..., outcome 6, respectively. Associated with these absolute occurrences are occurrence rates $m_i/N \equiv p_i$, $i = 1, \ldots, M$. Here, the $p_i = (3/17, 4/17, 3/17, 4/17, 2/17, 1/17)$. *The problem we consider is how to estimate the* $\{p_i\}$ *without direct observation of the outcome events* (10.11). Instead, the data consist of quantities F_k, $k = 1, \ldots, K$ relating indirectly to the $\{p_i\}$,

$$\sum_{i=1}^{M} f_k(i) p_i = F_k\ , \quad k = 1, 2, \ldots, K\ , \tag{10.12}$$

with

$$K < M - 1\ .$$

For example, in the particular case $f_k(i) = i^k$, $\{F_k\}$ are moments of the unknown relative occurrences $\{p_i\}$. In addition to the data, the $\{p_i\}$ must also of course obey normalization

$$\sum_{i=1}^{M} p_i = 1\ . \tag{10.13}$$

Another form of information about the unknown $\{p_i\}$ is called *prior knowledge*. Each event i has an *assumed probability* q_i *prior to* knowing the data $\{F_k\}$. For example, the die in question might be assumed to be fair a priori, so that all $q_i = 1/M$. The $\{q_i\}$ are called "prior probabilities," or simply "priors."

More generally, prior knowledge takes the form of an *assumed probability law*

$$p(q_1, \ldots, q_M) \tag{10.14}$$

for the priors $\{q_i\}$. This concept may seem strange at first, but it serves to allow the observer's state of uncertainty about the $\{q_i\}$ to be expressed. Prior knowledge is then the observer's judgment as to the probable values of the $\{q_i\}$, *and* his assessment of the strength of this judgment. If he has any preconceived biases about $\{q_i\}$, these come through as local maxima in $p(q_1, \ldots, q_M)$. The narrower are these maxima, the more conviction is expressed that the $\{p_i\}$ should equal them. These points will be clarified below; see especially Sect. 10.4.3. In physics, the state of prior knowledge is asserted by a physical model; see, e.g., (10.56). The probability law $p(q_1, \ldots, q_M)$ is called the "prior knowledge" law.

Note that if $K = M - 1$, the (now) M equations (10.12, 13) in $\{p_i\}$ would be sufficient to uniquely solve for the unknown $\{p_i\}$. However, it is more usual to have insufficient data, i.e., a situation where there are more unknowns $\{p_i\}$ than equations. In this case, we have an underconstrained problem at hand. This means *that there are an infinity of sets* $\{p_i\}$ that could give rise to the observed $\{F_k\}$.

To get a unique answer, we ask, which of these sets is *maximum probable* (MP) to have given rise to the F_k? An answer is provided by the result (10.10). This states that *if* $\{q_i\}$ are the assumed or *prior* probabilities, the probability of any set $\{m_i\}$ of occurrences will obey

$$P(m_1, \ldots, m_M | q_1, \ldots, q_M) = \frac{N!}{m_1! \ldots m_M!} q_1^{m_1} \ldots q_M^{m_M} . \tag{10.15}$$

However, we noted before that a *unique set* of prior probabilities $\{q_i\}$ is not usually known. Instead, the user must assume that all possible $\{q_i\}$ may be present, weighted by their probability (10.14) of occurring. Then by the partition law (3.7)

$$P(m_1, \ldots, m_M) = \frac{N!}{m_1! \ldots m_M!} \int_{\sum q_i = 1} \cdots \int \mathrm{d}q_1 \ldots \mathrm{d}q_M \times q_1^{m_1} \ldots q_M^{m_M} p(q_1, \ldots, q_M) . \tag{10.16}$$

This brings prior knowledge $p(q_1, \ldots, q_M)$ into the picture.

The *maximum probable* set of $\{p_i\}$ will then obey a principle[1]

$$\frac{N!}{m_1! \ldots m_M!} \int_{\sum q_i = 1} \cdots \int \mathrm{d}q_1 \ldots \mathrm{d}q_M q_1^{m_1} \ldots q_M^{m_M} p(q_1, \ldots, q_M) = \text{maximum} , \quad m_1 \equiv N p_i \tag{10.17}$$

This is *the general estimation principle* we sought. Particular cases depend upon the form of prior knowledge $p(q_1, \ldots, q_M)$ at hand. We examine in detail one particular case next. (Other cases of prior knowledge are taken up in the problems, Exercise 10.1.10–16.)

It is of interest that this maximum-probable solution is not a MAP solution (Sect. 17.2.3) in the usual sense. See Ex. 10.1.20.

10.4.2 Maximum Entropy Estimate

The form $p(q_1, \ldots, q_M)$ depends entirely upon the particular problem. For example, take the die problem considered at the outset. The vast majority of dice are fair, i.e. have equal probability for outcomes $1, 2, \ldots, 6$. Hence, even without knowing the moment data $\{F_k\}$ the observer might expect equality for the unknown $\{p_i\}$, or equivalently that $q_i = 1/M$, $i = 1, \ldots, M$. This may be called the "equal weights" hypothesis. The prior knowledge that expresses this bias with the strongest possible conviction is

$$p(q_1, \ldots, q_M) = \delta(q_1 - 1/M)\delta(q_2 - 1/M) \ldots \delta(q_M - 1/M) . \tag{10.18}$$

[1] This will be augmented by data in any real problem; see (10.24).

The peaks centered on points $q_i = 1/M$ are infinitely narrow.

Using this in general principle (10.17), we get

$$\frac{N!}{m_1! \dots m_M!}(1/M)^{m_1+\dots+m_M} = \text{maximum} .$$

But since $\sum m_i = N = \text{constant}$, the solution $\{m_i\}$ obeys simply

$$\frac{1}{m_1! \dots m_M!} = \text{maximum} .$$

Taking the natural logarithm of both sides and using Stirling's approximation

$$\ln(k!) \simeq k \ln k , \tag{10.19}$$

the principle becomes

$$-\sum_i m_i \ln m_i = \text{maximum} .$$

Finally, since $m_i = Np_i$ we obtain

$$-\sum_{i=1}^{M} p_i \ln p_i \equiv H = \text{maximum} \tag{10.20}$$

as the estimation principle (after ignoring irrelevant multiplicative and additive constants). This is a principle of *maximum entropy* H (compare with (3.23)).

10.4.3 The Search for "Maximum Prior Ignorance"

The ideal form of prior knowledge is one that biases the estimate toward *the true values* $\{p_i\}$ with maximum conviction,

$$p(q_1, \dots, q_M) = \delta(q_1 - p_1) \dots \delta(q_M - p_M) .$$

However, as $\{p_i\}$ are the very unknowns of the problem, it is impossible to use such a form. Short of this ideal, the choice $p(q_1, \dots, q_M)$ should at least not bias the estimate toward incorrect values. Let $\{Q_i\}$ be the chosen biases. Then, $p(q_1, \dots, q_M)$ should exhibit as broad a spread in values $\{q_i\}$ about bias values $\{Q_i\}$ as is consistent with the level of conviction that the $\{Q_i\}$ *are* the true values of $\{p_i\}$. The less conviction, the broader the spread should be made. In this way, the resulting estimate of $\{p_i\}$ will be optimally "fair," i.e., unbiased toward spurious values.

This brings us to the concept of statistical "ignorance." The broader the spread is made, the more ignorance the observer is admitting about the hypothesis that the biases $\{Q_i\}$ and the $\{p_i\}$ are equal. Hence, in order to obtain a "fair" estimate of the $\{p_i\}$ the observer should permit as large a state of ignorance to exist in $p(q_1, \dots, q_M)$ as is consistent with his problem.[2] This state of "maximum prior ignorance" obviously depends upon the particular problem. However, in an attempt

[2] This does not necessarily mean that the resulting estimates $\{p_i\}$ will all be overly smooth or blurred. The data, of course, force the $\{p_i\}$ in other directions.

to derive an estimation principle from (10.17) that would be *universally applicable*, some authors have tried to define a *universal state of maximum prior ignorance*. This choice of $p(q_1, \ldots, q_M)$ would be context-free, or independent of the confines of any given problem. For example, *Jaynes* [10.1] in effect uses equal weights $Q_i = 1/M$, Eq. (10.18), as his definition of maximum prior ignorance. This yields principle (10.20) of maximum entropy as the resulting "universal estimator."[3]

By contrast, *Macqueen* and *Marschak* [10.3] argued that an assertion that all $q_i = Q_i = 1/M$, with no uncertainty, is a quite definitive statement, and hence defines a much higher state of prior knowledge than *maximum* ignorance. Hence, their definition of maximum prior ignorance allows a spread in each q_i value, in fact the maximum spread possible over (0, 1),

$$p(q_1, \ldots, q_M) = (M-1)!\mathrm{Rect}(q_1 - 1/2)\ldots\mathrm{Rect}((q_M - 1/2)\ , \qquad (10.21)$$
$$\sum_i q_i = 1\ .$$

As might be expected with this vague an assertion on the probable $\{p_i\}$, when substituted into general principle (10.17) all dependence upon the $\{p_i\}$ drops out. All $\{p_i\}$ are therefore equally probable, so that no estimation principle at all results (Ex. 10.1.10)! The search goes on.

10.4.4 Other Types of Estimates (Summary)

Another common state of prior knowledge is one of "negative prior knowledge." Here, the user knows only that the die in use (returning to that problem) is definitely *not* described by one set of biases $\{q_i\} = \{Q_i\}$. Or, if the data are that of a picture, the unknown picture is known *not to be* the Mona Lisa. In this case, the estimation principle becomes one of *minimum cross-entropy* (Ex. 10.1.11). This is the exact opposite of maximum entropy (10.20).

Another case of interest is that of the "empirical prior." Here the user has observed a particular set of occurrences $\{m_i\} = \{M_i\}$ in a sample of the same class as the unknowns. The estimate of $\{p_i\}$ should then be biased toward $\{M_i/N\}$. In fact, it obeys a principle (Ex. 10.1.15)

$$-\sum_i m_i \ln m_i + \sum_i (M_i + m_i)\ln(M_i + m_i) = \text{maximum}\ , \qquad (10.22)$$
$$m_i \equiv Np_i\ .$$

Again, this is different from maximum entropy.

Finally, we consider the problem of estimating optical objects. The events are now *photon occurrences* from resolution cells. Then the degeneracy factor in the

[3] It should be noted that *Jaynes* did not derive principle (10.20) in this manner. Instead of seeking an MP estimate, he maximized the multinomial coefficient W_N *by itself*, believing this to have some universal meaning. We disagree since W_N is but one constituent of the total MP principle. Many claims of the "universality" of maximum entropy estimates follow ultimately from this unfounded assumption. Maximum entropy has its place, but is no panacea.

general principle (10.17) is no longer the multinomial coefficient, but the product of the Bose–Einstein degeneracy factors for each of the cells. The estimation principle now becomes [see (10.39)]

$$-\sum_i m_i \ln m_i + \sum_i (m_i + z - 1)\ln(m_i + z - 1) = \text{maximum}\,, \qquad (10.23)$$

$$m_i \equiv Np_i\,,$$

where z is the number of quantum degrees of freedom for a photon in a cell. We have assumed here a condition of maximum ignorance (10.18). Only when the object is of low intensity, obeying $m_i \ll z$, does the principle (10.23) become maximum entropy (Sect. 10.4.13).

In summary, then, the maximum entropy principle (10.20) is not universally valid. It results only from one circumstance of prior knowledge, namely (10.18) coupled with degeneracy $W(m_i, \ldots, m_M)$ described by the multinomial coefficient. For optical objects in particular, this would require a situation where the object is known to be nearly uniform and of such low intensity that the photons behave like classical particles.

It is often thought that the maximum entropy principle (10.20) arises out of a quest for a maximum-likelihood (ML) estimate, as in the problem of a die with *unknown* biases $q_1, \ldots, q_6$. (These are what we called before the "prior probabilities". In this unknown-die problem, the biases define the die and, so, are to be estimated.) But to the contrary, it has been found [10.7] that the general ML solution for these biases *does not* obey maximum entropy.

10.4.5 Return to Maximum Entropy Estimation, Discrete Case

Possibly the most common situation of prior ignorance is (10.18), even though it may not express *maximum* prior ignorance. Dice are usually fair, as are coins, the occurrence of binary symbols in coded messages, etc. Furthermore, the estimation principle that it leads to – maximum entropy – is distinguished by its lack of free parameters, such as z for the photon object case. This makes it preferable on the basis of Occam's Razor, or simplicity. Also, there is the allure of using a principle that has a physical counterpart in the Second Law, which states that the *thermodynamic entropy* of the universe is always increasing and therefore tending toward a maximum. Finally, the "equal weights" hypothesis (10.18) is often obeyed by physical phenomena.

For these reasons we shall proceed with the development of the maximum entropy principle in particular. It should be kept in mind, however, that the other estimation principles (10.22), (10.23), etc., may be developed in about the same way.

The $\{p_i\}$ obeying maximum entropy must also obey the constraints (10.12,13) due to the data. Standard use of Lagrange multipliers μ, $\{\lambda_k\}$ as in Eq. (G.18) of Appendix G permits us to define a constrained solution:

$$-\sum_i p_i \ln p_i + \mu\Big(\sum_i p_i - 1\Big) + \sum_{k=1}^{K} \lambda_k \Big[\sum_i f_k(i)p_i - F_k\Big] = \text{maximum}\,. \quad (10.24)$$

Rather than solve (10.24) for this discrete case, we shall proceed to the case of a continuous RV. Both the discrete and continuous versions are easily solved, as will be seen.

10.4.6 Transition to a Continuous Random Variable

The situation is easily modified to the continuous limit. Here, each p_i is represented as an integrated segment of a continuous $p(x)$ curve,

$$p_i \equiv \int_{x_i}^{x_i+\Delta x} p(x)\,\mathrm{d}x \ .$$

Then $p(x)$ is the new occurrence law to be estimated. We choose Δx small enough that the approximation

$$p_i = p(x_i)\Delta x$$

is good to whatever degree of accuracy is wanted. Substituting this into principle (10.24),

$$-\sum_i \Delta x p(x_i) \ln p(x_i) - \sum_i \Delta x p(x_i) \ln \Delta x + \mu\left[\sum_i p(x_i)\Delta x - 1\right]$$
$$+\sum_{k=1}^{K} \lambda_k \left[\sum_i f_k(x_i) p(x_i)\Delta x - F_k\right] = \text{maximum} \ .$$

This is to be evaluated in the continuous limit $\Delta x \to 0$.

The second sum here may be evaluated as the additive constant (albeit large)

$$-\ln \Delta x \ ,$$

which does not affect the maximization. Hence it may be ignored. (Purists may note that the left-hand side of (10.24) can be added to by any constant, without affecting the maximization. Choose the constant $+\ln \Delta x$. Then it is exactly cancelled here, *before* taking the lim $\Delta x \to 0$.)

The remaining sums converge toward corresponding integrals, and we get

$$-\int \mathrm{d}x p(x) \ln p(x) + \mu[\int \mathrm{d}x p(x) - 1]$$
$$+\sum_{k=1}^{K} \lambda_k [\int \mathrm{d}x f_k(x) p(x) - F_k] = \text{maximum} \ . \quad (10.25)$$

This is the estimation principle for the continuous case. It is basically a principle of maximum entropy

$$H \equiv -\int \mathrm{d}x p(x) \ln p(x) = \text{maximum} \quad (10.26)$$

for the *continuous* probability density $p(x)$ to be estimated.

It is noted that if the system described by the RV x obeys *ergodicity* (Sect. 8.11), the occurrence law $p(x)$ becomes as well the *probability density function* for x. Hence in the following we will sometimes regard $p(x)$ as a probability density.

10.4.7 Solution

Problem (10.25) is very easy to solve. The Lagrangian (or integrand) $\mathcal{L}$ for dx is

$$\mathcal{L}[x, p(x)] = -p(x) \ln p(x) + \mu p(x) + \sum_k \lambda_k f_k(x) p(x) .$$

Hence, $\mathcal{L}$ does not have terms $dp(x)/dx$, and the well known solution is the Euler-Lagrange equation (G.14) of Appendix G,

$$\partial \mathcal{L} / \partial p(x) = 0 . \tag{10.27}$$

Taking this derivative,

$$\mu - 1 - \ln \hat{p}(x) + \sum_k \lambda_k f_k(x) = 0 ,$$

with the solution

$$\hat{p}(x) = \exp\left[-1 + \mu + \sum_k \lambda_k f_k(x)\right] . \tag{10.28}$$

The coefficients μ, $\{\lambda_k\}$ are found by substituting form (10.28) into the $K + 1$ constraint conditions (10.12, 13). The $\{\lambda_k\}$ and μ are iteratively varied until the data values $\{F_k\}$ are attained. This usually requires a numerical relaxation technique such as the Newton–Raphson method, in a rather sophisticated computer program. The iterative search for a solution also usually requires much more computer time and cost than simpler (and less ambitious) approaches such as orthogonal expansions (10.1a).

10.4.8 Maximized *H*

The maximum in H attained by the solution (10.28) in (10.26) is

$$H_{\max} = 1 - \mu - \sum_k \lambda_k F_k . \tag{10.29}$$

It may be observed that the size of $H_{\max}$ is a measure of the *observer's ignorance* of the physical processes causing $\hat{p}(x)$. This arises from the fact that $H_{\max}$ grows smaller with the addition of each new information constraint F_k to the principle (10.25). Any maximum grows smaller with each added constraint.

10.4.9 Illustrative Example: Significance of the Normal Law

Suppose nothing is known about a random variable x except that it has a mean $\langle x \rangle$, a variance σ_X^2, and the state (10.18) of prior knowledge. What is the MP estimate of $p(x)$ in this case? Because (10.18) is obeyed, the MP estimate obeys a principle (10.25) of maximum entropy.

Here

$$f_1(x) = x\,, \quad f_2(x) = (x - \langle x \rangle)^2\,,$$
$$F_1 = \langle x \rangle \quad \text{and} \quad F_2 = \sigma_X^2\,.$$

Then by (10.28)

$$\hat{p}(x) = \exp[-1 + \mu + \lambda_1 x + \lambda_2 (x - \langle x \rangle)^2]\,.$$

Parameters $\lambda_1, \lambda_2, \mu$ are found by substituting this into the constraint equations

$$\int \mathrm{d}x\, \hat{p}(x) = 1$$
$$\int \mathrm{d}x\, f_k(x)\, \hat{p}(x) = F_k\,, \quad k = 1, 2\,.$$

It is not hard to show that μ, λ_1 and λ_2 satisfying these constraints yield a law

$$\hat{p}(x) = (2\pi)^{-1/2} \sigma_X^{-1} \exp[-(x - \langle x \rangle)^2 / 2\sigma_X^2]\,.$$

That is, x is a *normal* random variable! *The normal curve is then the most probable law consistent with knowledge of its first two moments and a state* (10.18) *of prior knowledge.*

Notice that the normal curve shows bias toward only one value, the mean. That is, there is only one maximum. Does this imply that maximum entropy solutions (10.28) are generally smooth, in some sense?

10.4.10 The Smoothness Property; Least Biased Aspect

Solution (10.28) has an interesting smoothness property. If a function $p_1(x)$ has an entropy H_1, and a new function $p_2(x)$ is formed from $p_1(x)$ by perturbing $p_1(x)$ toward a smoother curve, then *the new entropy H_2 must exceed H_1*. (See Fig. 10.2 and Ex. 10.1.4.) Hence, maximum entropy means *maximally smooth*, in a certain sense, as well as maximum probable.

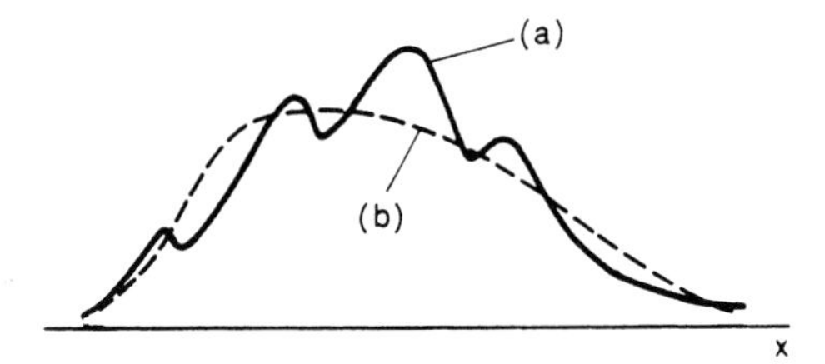

Fig. 10.2. Curves (a) and (b) are two probability laws obeying a constraint F_k on area. The smoother curve (b) has the higher entropy H

In probabilistic terms this means that $\hat{p}(x)$ is least biased toward any particular value of x. That is, $\hat{p}(x)$ has a minimum of spikes about particular events x_n. This is a useful tendency, since with insufficient data $\{F_k\}$ to uniquely determine $p(x)$, *we would want* an estimate of $p(x)$ to be as equinanimous or impartial in x as the data would allow.

Why $\hat{p}(x)$ should be so smooth is fairly easy to see. It derives from prior knowledge (10.18), which expresses *flatness* for the $\{q_i\}$ or $p(x)$ a priori.

However, entropy is not an extremely strong measure of smoothness. If, for example, any two or more segments of a curve are interchanged in any way the resulting entropy value stays the same while the new curve becomes discontinuous and "bumpy". This follows because entropy is a *global measure* of smoothness: it has the same value irrespective of any reordering of the points on the given curve. We shall have more to say about this in Chap. 17. See Ex. 17.3.5.

One normally regards the data $\{F_k\}$ for a problem as forces for constraining the estimated curve $\hat{p}(x)$ *away from maximum* smoothness (see Sect. 10.4.8). However, since a curve $\hat{p}(x)$ has to obey the data $\{F_k\}$ this would not generally permit an interchange of portions of it as described above. Hence, oddly enough, it is the combination of maximum entropy *and data constraints* that preserve a smooth solution.

We note in passing that the Gram-Charlier estimate (3.52) of a probability law cannot coincide with a maximum entropy solution (10.28), except in the purely Gaussian case. The maximum entropy estimate must always be a pure exponential, whereas the Gram-Charlier is an exponential Gaussian *times a polynomial*. Similarly for the more general orthogonal expansion (10.1a).

10.4.11 A Well Known Distribution Derived

We will show next that the maximum entropy approach gives correct answers in certain physical problems.

Suppose N particles with average energy $\langle E\rangle = kT$ are in Brownian motion. What is the MP occurrence law $p(E)$ for describing the energy E of a particle?

The data we have at hand about the random variable E is evidently the first moment alone,

$$\int_0^\infty \mathrm{d}E\, E\, \hat{p}(E) = kT\ .$$

Of course, there is always the normalization constraint in addition,

$$\int_0^\infty \mathrm{d}E\, \hat{p}(E) = 1\ .$$

Finally, we must consider the state of prior knowledge. Prior to measuring $\langle E\rangle$ it is natural to imagine all energies $\{E_i\}$ to be present in equal proportions, $q(E_i) \equiv q_i = 1/M$. This is an "equal weights" hypothesis, or state (10.18). Therefore, the unknown $p(E)$ obeys maximum entropy. Solution (10.28) becomes

$$\hat{p}(E) = \exp(-1 + \mu + \lambda_1 E)\ . \tag{10.30}$$

The parameters μ and λ_1 are found by making this solution satisfy the two preceding constraint equations. The result is

$$\mu = 1 - \ln(kT)\ , \quad \lambda_1 = -(kT)^{-1}\ ,$$

so that by substitution

$$\hat{p}(E) = (kT)^{-1}\,\mathrm{e}^{-E/kT}\,. \tag{10.31}$$

This is the Boltzmann distribution!

In fact, the Boltzmann distribution *is known to be the true probability law* for the given physical situation. This lends credence to the estimation procedure we have used, including the "equal weights" hypothesis.

It also shows that the Boltzmann distribution has the special properties we derived for maximum entropy solutions. That is, it is the single *most probable* law, also the smoothest, most equi-probable and consequently most random law consistent with a single average energy constraint on the particles.

As a corollary, if Nature required some additional constraint on the particles, such as a maximum permissible E or a second moment for E, a *different* probability law $p(E)$ would result by use of solution (10.28). So that, *a forteriori*, since the Boltzmann law is observed to be true *there must be only the one constraint* $\langle E\rangle$ on the particles.

10.4.12 When Does the Maximum Entropy Estimate Equal the True Law?

Because the approach is based upon sound, mathematical argument, it seems reasonable that sometimes the maximum entropy estimate will coincide with the true probability law for the process. We saw an example of this at (10.31) describing the Boltzmann distribution. Can a general principle be stated in this regard?

Other examples of coincidence exist. For example, let us briefly reconsider the speckle phenomenon of Sect. 5.9. There we found on the basis of a detailed, physical model for the diffuser and for light propagation that the probability density $p_I(x)$ for intensity obeys

$$p_I(x) = \langle I\rangle^{-1} \exp(-x/\langle I\rangle)\,. \tag{10.32}$$

Note that this law is defined by but one free parameter $\langle I\rangle$, the average intensity. What if, instead, we chose to use maximum entropy to estimate $p_I(x)$, with the constraint $\langle I\rangle$ *the only information* we had on hand about the speckle process?

We now note the mathematical similarity of this problem to the derivation of result (10.31). The answer is then once again an exponential law,

$$p_I(x) = \langle I\rangle^{-1} \exp(-x/\langle I\rangle)\,. \tag{10.33}$$

Noting that the results (10.32) and (10.33) are identical, we once again have an example of convergence between maximum entropy and the truth.

A third example of such convergence occurs for a Poisson event sequence. At Ex. 6.1.16, the probability density $p(t)$ was found for a gas molecule lasting a time $t - t_0$ before making a collision, after last colliding at t_0. With $\langle t\rangle$ the average time $\langle t - t_0\rangle$ between collisions, the solution was

$$p(t) = \langle t\rangle^{-1} \exp[-(t - t_0)/\langle t\rangle]\,. \tag{10.34}$$

If we had instead ignored the Poisson nature of the problem and simply sought the $\hat{p}(t)$ which has maximum entropy subject to a mean time $\langle t \rangle \equiv \langle t - t_0 \rangle$ between collisions, result (10.34) would once again have resulted!

Ignoring the Poisson nature of the preceding problem and simply tracking collisions is a statistical mechanics view of the problem. In fact, maximum entropy generally gives the true answer in problems of classical statistical physics. This is because *maximum entropy can only give exponential answers* (10.28), and the distribution laws of statistical mechanics, including (10.32)–(10.34), are precisely of this form. However, in non-classical, quantum problems the general answer is not an exponential form, and the maximum entropy principle (10.25) generally gives the wrong answer (see also Ex. 17.3.5).

How does incomplete or erroneous knowledge of the constraints affect maximum entropy solutions $\hat{p}(x)$? These will cause a maximum entropy estimate that errs, but in a smooth way. With incomplete knowledge, there are fewer constraints upon $H_{\max}$. With fewer constraints upon $H_{\max}$, it increases as with any maximum; hence by Sect. 10.4.10, $\hat{p}(x)$ becomes smoother. This is a desirable property for the estimate $\hat{p}(x)$, since the opposite tendency of artifact peaks leads to faulty assertions on bias, or preferred values, in x.

10.4.13 Maximum Probable Estimates of Optical Objects

Optical objects consist of photons radiating from resolution cells or pixels. Suppose that during a fixed exposure time, m_1 photons have left resolution cell 1, m_2 have left cell 2, etc. across the entire object. Suppose further that N photons in all have so radiated, and N is known by means of conservation of energy into the image. Then form

$$m_i/N \equiv p_i \ , \quad i = 1, \ldots, M \ . \tag{10.35}$$

Let the optical object be defined by its relative occurrences $\{p_i\}$. The aim is to estimate $\{p_i\}$.

Since the object is defined by a law $\{p_i\}$, it might at first be thought that the answer is estimation principle (10.17). However, in derivation of (10.17) it was assumed that the sole degeneracy present is that of event order, as in (10.11). That kind of degeneracy is a property of discrete, particle-like events. Photons, on the other hand, obey a different kind of degeneracy since they are not particles. This is described below.

Let us consider the kinds of information the user will have at hand to aid in his estimate of object $\{p_i\}$. These might be called "direct" and "indirect" forms of information. The given direct information is image data values $\{F_k\}$. These obey (10.12), where now the kernel $f_k(i)$ is the point spread function $s(y_k; y_i)$ [compare (10.12) with the imaging equation (2.21)].

The given indirect information is the user's expectations about the object. This brings in *prior probabilities* $\{q_i\}$, as in the preceding dice problem. A value q_i is defined as the probability that a given photon will radiate somewhere from within cell i. The user's expectations about the object take the form of a probability law

$p(q_1, \ldots, q_M)$ defining an "object class." For example, the particular form (10.18) for this law represents the case where the user believes that the object class consists of but one object, which is uniformly gray. This might occur, e.g., in astronomy where the major source of light is known to be interstellar gas.

With these identifications made, the development of the MP principle for estimating optical objects is now analogous to that of Sect. 10.4.1 for estimating occurrence rates. The only difference lies in the particular degeneracy factor to be used. The degeneracy in (10.17) was due to different orderings of the die rolls. By contrast, photons have degeneracy due to many physical causes *that includes* such orderings.

The question we must address, then, is: What is the degeneracy of an optical object $\{m_i\}$, or equivalently, in how many ways W_{opt} can an object defined by cell occurrences $\{m_i\}$ physically occur? From the field of quantum optics, it is known [10.4] that each resolution cell has z "bins" or degrees of freedom that any given photon may occupy. Also, photons are indistinguishable, so any number may occupy a bin. The number z of bins depends upon factors such as the exposure time, temporal coherence time, detector aperture area, and resolution cell area. The number of ways W_i that m_i such photons may occupy z bins in a cell then obeys the Bose-Einstein factor [10.4]

$$W_i = \frac{(m_i + z - 1)!}{m_i!(z-1)!} \,. \tag{10.36}$$

Finally, as the cells are independent of one another, the object $\{m_i\}$ may be formed in

$$W_{\text{opt}} = \prod_{i=1}^{M} \frac{(m_i + z - 1)!}{m_i!(z-1)!} \tag{10.37}$$

different ways.

Accordingly, in analogy with the development (10.15) through (10.17), the general MP principle for estimating an optical object $\{p_i\}$ is[4]

$$\begin{aligned} &W_{\text{opt}}(m_1, \ldots, m_M) \int \cdots \int \mathrm{d}q_1 \ldots \mathrm{d}q_M q_1^{m_1} (1-q_1)^z \ldots \\ &q_M^{m_M}(1-q_M)^z p(q_1, \ldots, q_M) = \text{maximum} \,, \quad m_i \equiv N p_i \,. \end{aligned} \tag{10.38}$$

The particular estimation principle to use will depend upon the object class $p(q_1, \ldots, q_M)$ at hand. We consider one such next.

---

[4] The principle (10.38) permits a general state of prior knowledge to be present, through $p(q_1, \ldots, q_m)$. If the object is known *a priori* to be a transparency consisting, e.g., of a black letter A against a clear background, certain cells *a priori* will radiate more photons than others. The integral factor in (10.38) takes this into account through input $p(q_1, \ldots, q_M)$, and builds this into the estimates $\{p_i\}$. Hence, if pixel i lies in the clear portion of the transparency, the tendency $p_i \cong 1$ *will be present in the estimate*. This is beneficial. The extra factors $(1-q_i)^z$ are required by normalization over the $\{m_i\}$.

10.4.14 Case of Nearly Featureless Objects

It is interesting to consider the special case (10.18) of *bias toward a uniformly gray object*. This arises in astronomy, for example, when most of the object is constant background radiation due to interstellar dust. Using (10.18) in the general principle (10.38), taking its logarithm and ignoring irrelevant constants, the principle becomes

$$-\sum_i m_i \ln m_i + \sum_i (m_i + z - 1)\ln(m_i + z - 1) = \text{maximum}\,,$$

$$m_i \equiv Np_i\,. \tag{10.39}$$

Note that this is not a principle of maximum entropy, as it *was* for the corresponding case (10.20) involving die events (and particle models resembling die events). In particular, for the case of $z = 1$ degree of freedom per cell, the left-hand side is identically zero. There is now no estimation principle at all; all objects are equally MP!

It is also interesting to consider the limiting cases $m_i \gg z$ and $m_i \ll z$. These describe very bright objects and very dim objects, respectively. In the bright object case,

$$\begin{aligned}(m_i + z - 1)\ln(m_i + z - 1) &= (m_i + z - 1)\ln\left[m_i\left(1 + \frac{z-1}{m_i}\right)\right]\\ &\simeq (m_i + z - 1)\left(\ln m_i + \frac{z-1}{m_i}\right)\\ &\simeq (m_i + z - 1)\ln m_i + z - 1\,.\end{aligned}$$

Hence principle (10.39) becomes

$$-\sum_i m_i \ln m_i + \sum_i (m_i + z - 1)\ln m_i = \text{maximum} - M(z-1)$$

or simply

$$\sum_i \ln m_i = \text{maximum} \tag{10.40}$$

after ignoring irrelevant constants. This is *Burg's* estimation principle [10.5]. It was originally used in processing seismic data, but more recently was used upon astronomical images [10.6].

In the opposite limit of a dim object,

$$\begin{aligned}(m_i + z - 1)\ln(m_i + z - 1) &= (m_i + z - 1)\ln\left[(z-1)\left(\frac{m_i}{z-1} + 1\right)\right]\\ &\simeq (m_i + z - 1)\left(\ln(z-1) + \frac{m_i}{z-1}\right)\\ &\simeq (m_i + z - 1)\ln(z-1) + m_i\,.\end{aligned}$$

Now the MP principle (10.39) becomes

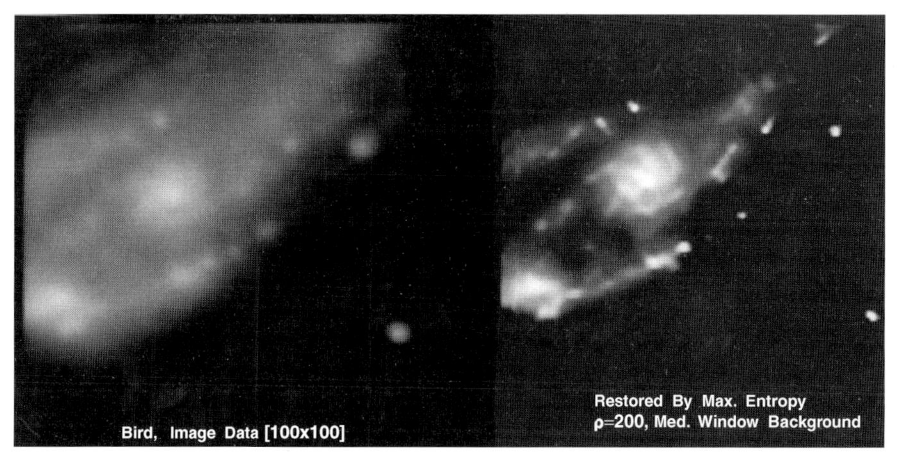

Fig. 10.3. (*left*). A galactic image blurred by atmospheric turbulence; (*right*) its restoration by maximum entropy. (Image courtesy of Kitt Peak National Observatory. Restoration by the author.)

$$-\sum_i m_i \ln m_i + \ln(z-1)\sum_i (m_i + z - 1) + \sum_i m_i = \text{maximum} .$$

Since $\sum m_i = N$, a constant, we get once again

$$-\sum_i m_i \ln m_i = \text{maximum} \tag{10.41}$$

or maximum entropy.

These are important results, since many optical objects fall into the bright or dim object category. For example, planetary astronomical objects have a very small m_i/z ratio in the visible and infrared regions. Also, bright radio astronomical objects often have a very large m_i/z ratio.

These observations perhaps explain *why* maximum entropy- and Burg-estimation work well on certain astronomical images. It has long been known by workers in the field that these estimators do a fine job on objects consisting of isolated impulses against a uniform (or nearly uniform) background. We see that these are precisely the featureless objects required above for optimum MP estimation! Of course, the impulses must be relatively sparse, for the object to be nearly featureless.

A typical maximum entropy estimate of an astronomical object of this type is shown in Fig. 10.3. Note the emergence of spiral arms out of the original blur in the central region.

Having read the poem of the student of Dr. Jaynes at the beginning of this chapter, the reader might now want to hear the professor's reply to it:

MAXIMUM ENTROPY REVISITED

There's a fine old branch of science coming back now to the fore,
And if you learn its principles you'll need to grope no more.
For though others may be blinded (which can make them twist and turn),
You'll be certain that you see whatever is there to discern.

This is no cut-and-try device, empirical ad hockery,
But a principle of reasoning, of proven optimality.
For prior information, which empiricism spurns,
Has proved to be the pivot point on which decision turns.

The procedure's very simple (it is due to Willard Gibbs).
First define your prior information, without telling any fibs.
On this space of possibilities, a constraint is then applied
For every piece of data, until all are satisfied.

If the states you thought were possible, in setting up this game,
And the real possibilities in Nature are the same,
Then LO! you see before you (there is nothing more to grind)
Reproducible connections that experimenters find.

But the principles of logic are the same in every field,
And regardless of your ignorance, you'll always know they yield
What your information indicates; and (whether good or bad)
The best predictions one could make, from data that you had.

Notwithstanding the excellent poetry, our experience with the concept has been equivocal. It is often an excellent numerical procedure of estimation, on one hand, but certainly is not as optimal as the poetry suggests. See Sect. 10.4.12, and the references [10.3, 10.4] and [10.7]. Also see the specifically non-MaxEnt solutions in Exs. 10.1.10–10.1.16 below.

Exercise 10.1

10.1.1 Based upon simply the occurrence of its individual letters, compute the probability of acquiring the following thought:

I enjoy working these problems

The individual letter probabilities are as follows [10.8]:

Letter	Probability	Letter	Probability	Letter	Probability
A	0.086	J	0.001	S	0.061
B	0.014	K	0.004	T	0.105
C	0.028	L	0.034	U	0.025
D	0.038	M	0.025	V	0.009
E	0.131	N	0.071	W	0.015
F	0.029	O	0.080	X	0.002
G	0.020	P	0.020	Y	0.020
H	0.053	Q	0.001	Z	0.001
I	0.063	R	0.068		

The preceding sentence had 26 letters. What is the most probable sentence of 26 letters, and how probable is it? Again, assume the *only* constraint to be the individual letter probabilities.

10.1.2 a) Verify the degeneracy formula (10.9) for the case $M = 2$, $m_1 = 3$, $m_2 = 1$, by actually listing the event sequences.

b) Verify the degeneracy formula (10.37) for the case $M = 2, m_1 = 3, m_2 = 1$, with $z = 2$ degrees of freedom, again by listing the event sequences.

10.1.3 ***Honors problem.*** We showed in Sect. 9.6 that if a particular event occurs m times out of a total of N trials, and if m is assumed to be the most likely number of occurrences of the event, then the probability of the event is m/N. Show by an analogous derivation that the principle of maximum likelihood for a datum m_i in (10.10) leads to the same result, that $p_i = m_i/N$.

10.1.4 Probabilities $p_i, i = 1, \ldots, M$ have an entropy H. Suppose the particular probability p_k is less than another of these probabilities, p_j. The curve of $p_i, i = 1, \ldots. M$ would be made smoother if we perturbed p_k by adding a small amount Δp to it and simultaneously perturbed p_j by subtracting Δp from it. Compute the resulting change in H, treating Δp as a differential. Is the change in H positive or negative?

10.1.5 An unknown occurrence law $p(x)$ is known to have a first moment m and a variance σ^2. An estimate $p(x)$ is formed using the principle of maximum entropy. Show that the estimate is a normal law with mean m and variance σ^2. Find the resulting $H_{\max}$.

10.1.6 Nothing is known about the statistics of a RV x except that x is confined to a finite interval $|x| \leq x_0$. An estimate $\hat{p}(x)$ may nevertheless be formed using the principle of maximum entropy. Show that by this procedure the estimate is *uniformly random* over interval $|x| \leq x_0$. How does the degree of ignorance about x, i.e., $H_{\max}$, vary with the size of x_0?

10.1.7 In Fig. 10.4, a filled cone of photons leaves point 0 and strikes the screen S distance D away. Assume a photon travels in a straight line from point 0 to the screen, striking it at radius r, as shown.

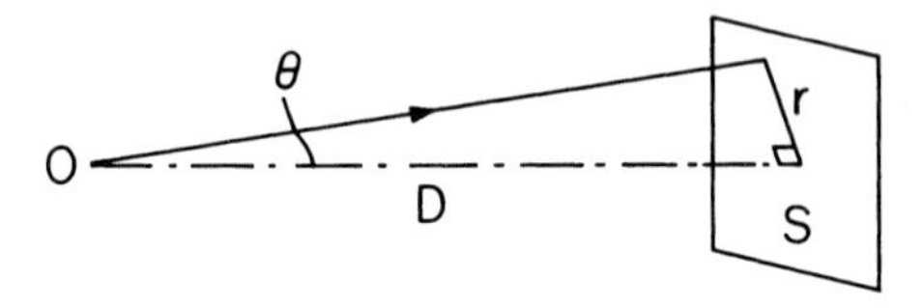

Fig. 10.4. If $p_\Theta(\theta)$ is formed according to maximum entropy, what is $p_R(r)$?

The angular distribution $p_\Theta(\theta)$ of source photons is unknown, except for the knowledge that $0 \leq \theta \leq \theta_0$, where θ is a polar angle and $\theta_0 < \pi/2$.

(a) Find the maximum entropy estimate of $p_\Theta(\theta)$. (Note equivalence to Ex. 10.1.6.)

(b) Find the corresponding probability law $p_R(r)$ for photon radial position on S. (Do not assume r/D to be small.) Why is it not a uniform law?[5]

10.1.8 Figure 10.5 shows a slit located a distance of 1 (arbitrary unit) from point 0. Photons radiate from the slit according to a probability law $p(x)$, with coordinate x along the slit. The mean $\langle \ln r^2 \rangle$ is known where r is a photon's distance from point 0. This mean is used as a constraint when forming an estimate $\hat{p}(x)$ according to maximum entropy. Show that the estimate obeys a Cauchy shape $(a/\pi)(a^2 + x^2)^{-1}$. What is a in terms of the known $\langle \ln r^2 \rangle$?

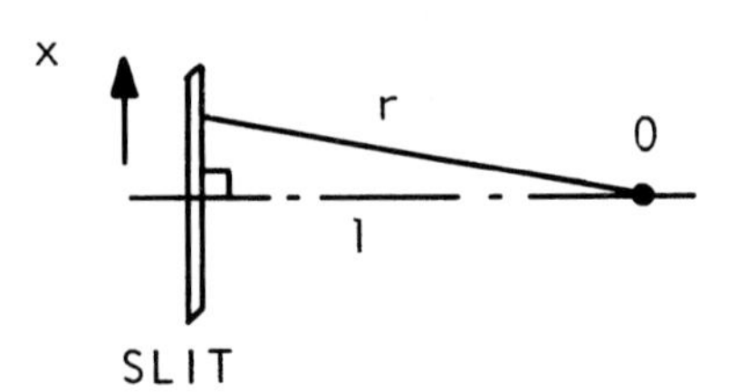

Fig. 10.5. Geometry giving rise to a Cauchy probability law as the maximum entropy estimate

10.1.9 ***Honors problem.*** The binomial probability law (6.5) can be derived from the principle of maximum entropy. Let a Bernoulli trial sequence $c_1, c_2, \ldots, c_N \equiv \mathbf{c}$ with

$$c_n = \begin{cases} 1 & \text{if success} \\ 0 & \text{if failure} \end{cases} \tag{10.42}$$

have m successes (or values) 1. The probability of success is p. Then the maximum entropy estimate of the joint probability of $c_1, \ldots, c_N$, denoted as $P(\mathbf{c})$, obeys

$$P(\mathbf{c}) = p^m (1-p)^{N-m} , \tag{10.43}$$

[5] This is an open-ended question that relates to *Bertrand's paradox* [10.9]: What probability law describes "complete" randomness?

if it is given that

$$\langle m \rangle = Np \tag{10.44}$$

with $\langle m \rangle$ the average value of m. Multiplication by degeneracy factor (6.4) then gives the binomial law. Prove (10.43), using the principle (10.24) with joint summation over $c_1, \ldots, c_N$ replacing the single summation over i. *Hint:* By definition (10.42),

$$m = \sum_{n=1}^{N} c_n \, .$$

Also note the mathematical identities

$$\sum_{\mathbf{c}} \exp\left(\lambda \sum_{n=1}^{N} c_n\right) = \sum_{m=0}^{N} \binom{N}{m} e^{\lambda m} = (1 + e^{\lambda})^N$$

and

$$\sum_{\mathbf{c}} c_m \exp\left(\lambda \sum_{n=1}^{N} c_n\right) = e^{\lambda} \sum_{\mathbf{c}} \exp\left(\lambda \sum_{n=1}^{N-1} c_n\right) = e^{\lambda}(1 + e^{\lambda})^{N-1}$$

for any m.

10.1.10 Show that if all prior probabilities $\{q_i\}$ are equally probable over (0, 1), defining the MacQueen–Marschak state of "maximum prior ignorance," the MP estimation principle (10.17) collapses into a constant, independent of the unknowns $\{p_i\}$. *Hint:* The Dirichlet integral obeys

$$\int_{\sum q_1 = 1} \cdots \int \mathrm{d}q_1 \ldots \mathrm{d}q_M q_1^{m_1} \ldots q_M^{m_M} \mathrm{Rect}(q_1 - 1/2) \ldots \mathrm{Rect}(q_M - 1/2)$$

$$= \frac{1}{(N + M - 1)!} m_1! \ldots m_M! \tag{10.45}$$

10.1.11 ***Negative prior knowledge.*** An optical object is to be estimated. As prior knowledge, the observer knows only that the object *is not the Mono Lisa* (for example), or a close approximation to it. Let $\{R_i\}$ describe $\{p_i\}$ for the Mona Lisa. Then this state of knowledge is defined by a class of objects

$$p(q_1, \ldots, q_M) = (1 - \varepsilon)^{-1} \left[\prod_{i=1}^{M} \mathrm{Rect}(q_i - 1/2) - \varepsilon \prod_{i=1}^{M} \delta(q_i - R_i) \right] \tag{10.46}$$

Here, ε is a small constant that measures the amount of spread in pictures about the Mona Lisa that are also known *to not be* present. This defines a "missing class" of pictures. (Note that the first product-term in (10.46) states that *all other* pictures are equally probable *to be present*.) Show that when (10.46) is substituted into estimation principle (10.38), for the dim object case $m_i \ll z$ the principle becomes one of *minimum* cross-entropy

$$-\sum_{i=1}^{M} p_i \ln(p_i / R_i) = \text{minimum} \, . \tag{10.47}$$

For the case R_i = constant, all i, this is the exact opposite of maximum entropy! See also the next problem.

10.1.12 The "bias" of a coin is the probability p_1 that it will yield a head on a given flip. A bagful of coins originally has every bias present in equal proportions. But then, all coins whose biases lie between 0.2 and 0.3 *are removed.* Next, the observer randomly draws a coin from the bag, and flips it $N = 10$ times. What is the most probable number m_1 of heads to result? [*Hint:* This defines a binary $M = 2$ estimation problem where the prior knowledge is "negative," as in problem (11). Accordingly, the MP solution p_1 obeys (10.47). The answer is rather startling.]

10.1.13 ***Empirical prior knowledge.*** Perhaps the fairest way of determining the prior $p(q_1, \ldots, q_M)$ is *to actually observe* a typical member of the class for its physical occurrences $\{M_i\}$. For example, consider the unknown die problem. Let the die be a member of a class of dice, one of which is rolled L times, with the number of outcomes $\{M_i\}$ for events $i = 1, \ldots, 6$ duly noted. *Can knowledge of empirical data* $\{M_i\}$ *define the prior* $p(q_1, \ldots, q_M)$?

Note that here the prior is actually a conditional probability $p(q_1, \ldots, q_M|\{M_i\})$, and by Bayes' rule (2.22)

$$p(q_1, \ldots, q_M|\{M_i\}) = \frac{P(M_1, \ldots, M_M|\{q_i\})p_0(q_1, \ldots, q_M)}{\int \cdots \int dq_i \ldots dq_M P(M_1, \ldots, M_M|\{q_1\})p_0(q_1, \ldots, q_M)} . \tag{10.48}$$

The right-hand quantities are known. Probability $P(M_1, \ldots, M_M|\{q_i\})$ obeys the multinomial law (10.15). Also, $p_0(q, \ldots, q_M)$ defines the prior probability *before* observing the empirical data (a "pre-prior"). It is reasonable to assume uniformity (10.21) for this law. Then the denominator of (10.48) may be evaluated using identities (10.15) and (10.45). Show that the resulting prior obeys

$$p(q_1, \ldots, q_M|\{M_i\}) = \frac{(L+M-1)!}{(M-1)!} \frac{1}{M_1! \ldots M_M!} q_1^{M_1} \ldots q_M^{M_M} ,$$
$$L \equiv \sum_i M_i , \quad N \equiv \sum_i m_i , \quad \text{all } 0 \leqq q_i \leqq 1 . \tag{10.49}$$

This is basically the multinomial probability law (10.10) once again, despite a reversal in the roles played by unknowns and data. Note that with no empirical data ($M_i = 0$ all i, $L = 0$) the prior reverts to the "pre-prior" assumption of uniformity. This is consistent.

10.1.14 Continuation from problem (13). Let us find the probability law for unknown occurrences $\{m_i\}$, based upon knowledge of the $\{M_i\}$. Show by substitution of (10.49) into general principle (10.17) that

$$P(m_1, \ldots, m_M) = \frac{M!(L+M-1)!}{(L+M+N-1)!} \frac{1}{M_1! \ldots M_M!} \frac{(M_1+m_1)! \ldots (M_M+m_M)!}{m_1! \ldots m_M!} , \tag{10.50}$$
$$m_i \equiv Np_1 .$$

10.1.15 Continuation from problem (14). By taking the natural logarithm of (10.50) and using Stirling's approximation (10.19), show that the MP estimation principle for an *empirical prior* is

$$-\sum_i m_i \ln m_i + \sum_i (M_i + m_i) \ln(M_i + m_i) = \text{maximum} . \tag{10.51}$$

The first sum by itself represents maximum entropy. However, the net principle is generally different from maximum entropy.

10.1.16 Continuation from problem (15). Examine the two limiting cases of empirical data:

a) ***Meagre data.*** Here all $M_i \ll m_i$. Show that MP principle (10.51) becomes

$$+\sum_i M_i \ln m_i = \text{maximum} , \qquad m_i \equiv N p_i . \tag{10.52}$$

If all M_i are equal, this is a Burg-type of estimator (10.40). In general, the unequal M_i act as empirical weights in (10.52), selectively emphasizing different p_i in the sum. Hence, the more prior observations M_i there are of event i, the higher the estimated p_i will be. This is consistent.

b) ***Compelling data.*** Here all $M_i \gg m_i$. Show that MP principle (10.51) now becomes

$$\sum_i m_i \ln(M_i/m_i) = \text{maximum} . \tag{10.53}$$

If in particular all M_i are equal, how is this result consistent with the derivation of (10.20)? *Hint:* What does the law of large numbers (2.8) say about the spread in probabilities $\{q_i\}$ if all $\{M_i\}$ are very large?

10.1.17 ***Derivation of Bose–Einstein probability law (Honors problem)*** [10.10]. The geometric law (3.57) of quantum optics is actually an MP estimate (10.17).[6] Consider N photons trapped within an enclosure whose walls are kept at temperature T The photons are being absorbed and re-radiated by harmonic oscillators within the walls, and at equal rates – a state of thermal equilibrium exists (hence the name "thermal" photons). These behave like classical particles, in their statistical properties, and hence obey the particle-like estimator (10.17), where m_i is now the number of photons in a given energy state E_i. Since these originate in harmonic oscillators,

$$E_i = (i + 1/2)h\nu , \quad i = 0, 1, \ldots , \tag{10.54}$$

where h is Planck's constant and ν is the light frequency (see any book on quantum mechanics). The problem is to find the MP probability $p(E_i) \equiv p_i = m_i/N$ that a photon will be in energy state E_i.

The data is knowledge of the average energy $\langle E \rangle$ over all states, due to the known temperature T:

$$\langle E \rangle = h\nu \left(\frac{1}{e^{h\nu/kT} - 1} + \frac{1}{2} \right) . \tag{10.55}$$

Parameter k is the Boltzmann constant.

[6] Physically, an MP estimate is the most probable, and therefore most frequently *observed*, answer.

The prior knowledge is that q_i, the value of p_i prior to knowing $\langle E \rangle$, equals g_i, where Ng_i is a degeneracy factor representing the a priori number of different photons having the same energy E_i. This is supposed to be known with certainty, as part of a physical model, so that

$$p(q_1, \ldots, q_M) = \delta(q_1 - g_1) \ldots \delta(q_M - g_M) \tag{10.56}$$

[compare with (10.18)]. Parameter M is the total number of different energy states, taken to be very large.

a) Show that use of the prior knowledge (10.56) in estimator (10.17) results in an estimator

$$\frac{N!}{\prod_i m_i!} \prod_i g_i^{m_i} = \text{maximum} , \quad m_i = Np_i . \tag{10.57}$$

b) Show that use of Stirling's approximation (10.19) leads to an estimator

$$-\sum_i m_i \ln(m_i/g_i) = \text{maximum} , \quad m_i \equiv Np_i . \tag{10.58}$$

This is called a principle of maximum "cross-entropy" (compare with minimum cross-entropy expression (10.47)).

c) Show that when constraints on total number N of photons and average energy $\langle E \rangle$ are attached as Lagrange constraints, the solution for p_i is

$$p_i = (g_i/N)\,\mathrm{e}^{-1+\mu+\lambda E_i} . \tag{10.59}$$

We have to determine the constants μ and λ.

Consider the simplest case, where all

$$g_i = 1/M . \tag{10.60}$$

This is the case where one "mode" of radiation is present a priori.

(d) Show that in order for the MP estimate (10.59) of p_i to have a mean energy given by (10.55), it must be that

$$\lambda = -1/kT . \tag{10.61}$$

(*Hint:* Use $\langle E \rangle = \sum_i E_i p_i / \sum_i p_i$.)

(e) Show that in order for (10.59) to obey normalization, it must be that

$$(1/MN)\,\mathrm{e}^{-1+\mu} = \frac{1 - \mathrm{e}^{-h\nu/kT}}{\mathrm{e}^{-h\nu/2kT}} . \tag{10.62}$$

Substitution of (10.60) to (10.62) into (10.59) yields the celebrated Bose-Einstein law

$$p_n = (1 - \mathrm{e}^{-h\nu/kT})\,\mathrm{e}^{-nh\nu/kT} . \tag{10.63}$$

See the related Exercise 10.1.19.

10.1.18 ***Derivation of multinomial degeneracy factor.*** The multinomial degeneracy factor (10.9) may be expressed formally as the product of binomial degeneracies

$$\frac{N!}{m_1!(N-m_1)!} \cdot \frac{(N-m_1)!}{m_2!(N-m_1-m_2)!} \cdot \frac{(N-m_1-m_2)!}{m_3!(N-m_1-m_2-m_3)!} \ldots \tag{10.64}$$

Explain what each factor represents, and therefore show how the multinomial factor arises out of purely binomial considerations.

10.1.19 ***Alternative derivation of Negative Binomial Probability law.*** The law (5.88) arises out of the *geometric* law (3.57) as follows. For $M = 1$, (5.88) is a geometric law on laser speckle

$$P_1(n) = qp^n\ , \quad q = 1 - p\ .$$

Now reconsider the case of general M. Let cell m contain n_m speckles. Then the total number n obeys

$$n = \sum_{m=1}^{M} n_m\ ,$$

where each n_m is a geometric RV.

Since the speckles are distributed independently among the cells, each n_m in the sum also is independent. *Show*, then, that $P(n)$ is negative binomial. *Hint:* The characteristic function for a negative binomial law is

$$\varphi(\omega) = \left[\frac{q}{1 - p\exp(\mathrm{j}\omega)}\right]^M \tag{10.65}$$

Comment: Although this derivation has been carried through for laser speckle, it holds as well for the situation where *n photons* are distributed randomly among z degrees of freedom in a radiation field. The situation is analogous for, by (10.63), the probability law $P_1(n)$ for photons within one degree of freedom is geometric; and, any number of photons may crowd into one degree of freedom (a defining property of photons). These were precisely the properties we assumed above for laser speckles within cells. *Photons therefore follow a $P(n)$ law which is also negative binomial, z replacing M in the law.* See [10.4] and [10.11] for further details, including the full, quantum mechanical derivation.

10.1.20 ***MP as a degenerate MAP case.*** The aim of the general problem of Sect. 10.4 is to estimate occurrence rates p_i or, equivalently, the occurrences m_i themselves. The latter can be absorbed within the framework of MAP estimation (Sect. 17.2.3), if we regard them as unknown MAP parameters a_i that are to be estimated. However, the data values F_k of the occurrence problem cannot so readily be regarded as data values y_k of the MAP problem. This is because MAP requires the latter to be generally random according to some well-defined probability law. By comparison, our data F_k are formed from the parameters m_i via the deterministic Eqs. (10.12). Therefore our data *have no random components.* Hence the likelihood law governing our data is an infinitely narrow Dirac delta function. As in the overall MAP procedure, the parameters m_i can be solved for by differentiation, but not quite as in the "MAP equation" (17.54) per se. *What would be the revised MAP equation?*

11. The Chi-Square Test of Significance

We previously noted the widespread occurrence in nature of the *chi-square probability law*. It was also found to describe the random behavior of the sample variance. Here we shall show how the chi-square law may be used to describe the state of "significance" of a set of data.

Significant data are defined to be improbably far from consistency with pure chance. For example, if a coin is flipped 10 000 times and heads occurs 8000 times, the data are significant, that is, significantly far from consistency with $P(\text{head}) = P(\text{tail}) = 1/2$. This implies that a *physical effect* is present that is forcing heads to predominate. The coin has been "fixed" by some physical procedure, such as rounding off the rim profile on the head side. In general, when it is not known whether a hypothetical physical effect exists, data from the effect may be tested for their significance.

This idea may be applied to the investigation of any brand new physical phenomenon. A reasonable first step is to show that the effect exists. Let us restrict attention to deterministic effects, like the coin-fixing effect above. For example, consider the effect "force equals mass times acceleration." If this effect exists, then repetition of an experiment where mass and acceleration are systematically varied such that their product remains fixed should yield output data consisting of force values that are constant. *They should not be random*, except to the extent that measurement error creeps into the data.

If the force data are close enough to being constant, we say that the data are "significant." They probably show the existence of the physical effect, since they lack randomness. It therefore behooves us to devise a test of randomness, or significance, of given data.

To show that data are significant, we shall resort to a common trick of statistics. This may aptly be called the *statistical indirect method of proof*, in analogy to the ordinary indirect method. This is to propose a probability law (the chance hypothesis) to have given rise to the data, and then to show that on this basis the data represent an event way out *on the tail* of the law. Since this is an improbable event, the data and the law are inconsistent: one must be rejected, on the simple grounds that very unlikely events don't tend to happen (the ML principle stated in a negative way). Since by hypothesis the data are correct, the probability law (chance hypothesis) must be rejected. Consequently, the data are significant.

One way to carry through such an approach is to form a χ^2 statistic, as shown next.

11.1 Forming the χ^2 Statistic

Consider an experiment, such as rolling a die, which has M possible event outcomes. Let the events be independent, with known probabilities $P_i, i = 1, 2, \ldots, M$. A large number N of trials of the basic experiment are performed, with m_i the observed number of events i, $i = 1, 2, \ldots, M$. For example, with $N = 10$ die rolls, the occurrences $m_i = (2, 3, 1, 2, 2, 0)$ of the six possible die values might result. This constitutes *one* set of $\{m_i\}$, and hence one overall experiment as far as they are concerned.

With $\{P_i\}$ *fixed*, and the experiment repeated, a new set of $\{m_i\}$ will be obtained. These will in general differ from those of the first experiment. Hence, the $\{m_i\}$ are actually random variables. Consider one such set of data $\{m_i\}$.

We want to know whether the observed $\{m_i\}$ *are consistent with the presumed* $\{P_i\}$. If they are *not*, we call the experiment "interesting" or "significant." By this we mean that the observed $\{m_i\}$ could not reasonably have occurred as a random outcome of the $\{P_i\}$. They are so far from agreeing with the $\{P_i\}$ as to be inconsistent with the $\{P_i\}$.

And so, we want to devise a test of the $\{m_i\}$ data that *will measure* their departure from consistency with the $\{P_i\}$. *In fact, consistent data would obey* $m_i \cong NP_i$ *for N observed outcomes*, by the law of large numbers (2.8).

This suggests forming a statistic

$$\Lambda \equiv \sum_{i=1}^{M} (m_i - NP_i)^2 \,. \tag{11.1}$$

Notice that Λ is large if and only if the data $\{m_i\}$ are inconsistent with the $\{P_i\}$. Hence, Λ is a measure of the significance of data $\{m_i\}$. Other possible measures exist (Sect. 11.4).

It turns out that a better measure of significance is

$$\chi^2 \equiv \sum_{i=1}^{M} \frac{(m_i - NP_i)^2}{NP_i} \,, \tag{11.2}$$

called the *chi-square* statistic. Since the $\{P_i\}$ are merely constants, this is closely related to form (11.1).

Imagine, then, that a die is rolled $N = 10$ times with resulting occurrences $\{m_i\}$. These are substituted into (11.2) along with a known set of probabilities $\{P_i\}$ to yield a value χ_0^2. If χ_0^2 is large, we want to state that the data $\{m_i\}$ are significant. But, how large is *large*?

Observe that χ^2 is a random variable, since it is formed in (11.2) from random occurrences $\{m_i\}$. (All this means is that the $\{m_i\}$, and hence χ^2, will change randomly from one set of 10 die rolls to another.) Hence, the cumulative probability $P(\chi^2 \geqq \chi_0^2)$ may be computed. This number gives us a handle on deciding whether or not χ_0^2 is significantly large. If it is, $P(\chi^2 \geqq \chi_0^2)$ *will be small*. That is, χ_0^2 is so large that if the 10 die rolls were repeated over and over, in only a small fraction of cases would the ensuing χ^2 values exceed value χ_0^2.

In particular, if

$$P(\chi^2 \geqq \chi_0^2) = 0.05 \tag{11.3}$$

or less the data are said to be *significant at the 5% level.* Sometimes the 1% level is used instead, in a more stringent test.

The evaluation of $P(\chi^2 \geqq \chi_0^2)$ is called a *one-tail chi-square test.* It graphically represents the area of one tail of the probability density $p_{\chi^2}(x)$, as illustrated in Fig. 11.1a. If the number χ_0^2 is far enough out on the tail of $p_{\chi^2}(x)$, then $p(\chi^2 \geqq \chi_0^2)$ will be smaller than 0.05, and the verdict is "significance."

A more strict test of significance would be the "two-tail" test, of

$$P(\chi^2 \geqq \chi_0^2 \text{ or } \chi^2 \leqq 2\langle\chi^2\rangle - \chi_0^2) = 0.05\ . \tag{11.4}$$

This tests whether χ_0^2 is *improbably far* from (on either side of) its mean. This is illustrated by the two shaded areas in Fig. 11.1b.

To compute either probability (11.3) or (11.4), by identity (3.1) we have to know the probability density for χ^2. This is in fact chi-square, as the label suggests. The proof follows.

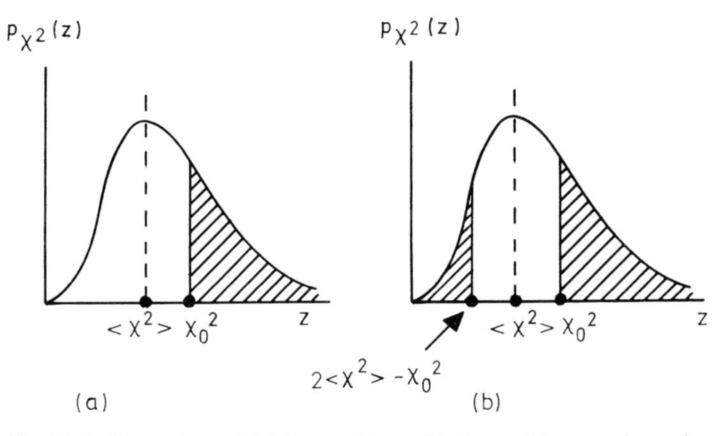

Fig. 11.1. Comparison of (a) "one-tail" and (b) "two-tail" test regions, shown shaded

11.2 Probability Law for χ^2 Statistic

We shall show that RV χ^2 is the sum of squares of N Gaussian RV's, as is the standard form (5.54) for a chi-square RV. Hence χ^2 is chi-square.

Consider a single random variable m_i, such as the number m_2 of die outcomes 2 in N rolls. What are its statistics? Note first that $P(m_i) \neq P_i$. That is, for a fair die value i, $P_i = 1/6$ whereas $P(m_i)$ represents the relative occurrence of value m_i when the experiment consisting of N die rolls *is repeated over and over.*

To get a handle on m_i, we can represent it as a sum

$$m_i = \sum_{l=1}^{N} n_{li}\ , \quad N \text{ large}\ , \tag{11.5}$$

formed over trials $l = 1, 2, \ldots, N$ of a single experiment. Quantity n_{li} is a "counter," i.e.,

$$n_{li} \equiv \begin{cases} +1 \text{ if trial } l \text{ has outcome } i \\ \;\;0 \text{ if trial } l \text{ does not have outcome } i \,. \end{cases}$$

Now the n_{li} are RV's, since roll outcomes are random. They also are binary (+1 or 0), and independent of l, since roll outcomes are independent. Also, each n_{li} is governed by the *same* binary probability law

$$\begin{aligned} P(n_{li} = 1) &= P_i \\ P(n_{li} = 0) &= 1 - P_i \end{aligned} \tag{11.6}$$

independent of trial l. (Note the absence of any l-dependence in the righthand quantities). Finally, assume that N is large.

Then (11.5) has the form of a sum of many, statistically independent and identically distributed RV's, so that m_i *obeys the central limit theorem, and is normal* (Sect. 4.26).

Now, from (11.5), the mean of this RV obeys

$$\langle m_i \rangle = N \langle n_i \rangle = N[1 \cdot P_i + 0 \cdot (1 - P_i)] = NP_i \,,$$

a constant. Therefore, a new random variable

$$x_i' \equiv m_i - NP_i$$

would be Gaussian, as would another choice

$$x_i \equiv \frac{m_i - NP_i}{(NP_i)^{1/2}} \,. \tag{11.7}$$

Hence the statistic (11.2) is actually

$$\chi^2 = \sum_{i=1}^{M} x_i^2 \tag{11.8}$$

with each of the x_i Gaussian. This is almost in the standard form (5.54) for a chi-square RV, but not quite because: (a) the variances obey $\sigma_{X_i}^2 = 1 - P_i$ and hence are not equal (Ex. 11.1.5); and (b) the $\{x_i\}$ are not all independent. The former may be shown from properties (11.5) through (11.7). The latter follows from the fact that

$$\sum_{i=1}^{M} m_i = N \,.$$

Combining this with (11.7) leads to the constraint

$$\sum_{i=1}^{M} x_i P_i^{1/2} = 0 \,. \tag{11.9}$$

This shows that, instead, only $M - 1$ of the $\{x_i\}$ are independent.

If (11.9) is solved for x_M, and this is substituted into (11.8), there results an expression for χ^2 that is quadratic, with cross-terms $x_i x_j$. By a process of completing the square, it is then possible to define a new set of $M-1$ independent random variables y_i that are linear combinations of the x_i, whose sum of squares equals χ^2, and which have the required constant variance. In fact, $\sigma_Y^2 = 1$. In [11.1], this procedure was carried through for the specific cases $N = 2$ and 3.

Thus,

$$\chi^2 = \sum_{i=1}^{M-1} y_i^2 , \tag{11.10}$$

with each y_i independent. This *is* of the form (5.54), so χ^2 is indeed a chi-square random variable, obeying

$$p_{\chi^2}(z) = 2^{-(M-1)/2}\{\Gamma[(M-1)/2]\}^{-1} z^{(M-3)/2} \exp(-z/2) . \tag{11.11}$$

This has $M-1$ degrees of freedom.

We may now use either criterion (11.3) or (11.4) of significance. For example, the one-tail statistic is

$$P(\chi^2 \geqq \chi_0^2) = \int_{\chi_0^2}^{\infty} \mathrm{d}z \, p_{\chi^2}(z) . \tag{11.12}$$

This gives a numerical value to the probability that an experimental value χ_0^2 for χ^2 will be exceeded. If this P is small, *then the data were significant.*

Tables of the integral (11.12) exist for discrete values of its free parameters χ_0^2 and M. See, for example, Appendix B, reprinted from [11.2]. In this table, $\nu \equiv M-1$, the number of degrees of freedom for the problem. Alternatively, (11.12) may be evaluated on the computer by numerical integration; see, for example, [11.3].

11.3 When is a Coin Fixed?

This simple example beautifully illustrates the χ^2 test: A coin was flipped 5000 times and heads appeared 2512 times. Is this datum significant, i.e., is it significantly far from consistency with a law $P_{\text{head}} = P_{\text{tails}} = 1/2$? Notice that here "significance" means biased away from "fairness." We want to know if the coin was "fixed," for example, by having its rim edge rounded off on one side.

We have $N = 5000$, $M = 2$, $P_1 = P_2 = 1/2$, $m_1 = 2512$, and $m_2 = 2488$. Then, $NP_1 = NP_2 = 2500$, and by (11.2),

$$\chi_0^2 = \frac{(2512-2500)^2}{2500} + \frac{(2488-2500)^2}{2500} = 0.115 .$$

Appendix B shows that for value $\chi_0^2 = 0.115$ and $M = 2$ (in the table $\nu = 1$), $P(\chi^2 > 0.115) = 0.73$. This is not very small. Hence value 0.115 is *consistent with chance* ($P_1 = P_2 = 1/2$). The datum is *not* significant; the coin is probably fair, and we have no justification for suspecting foul play.

Appendix B also shows a *required value* of $\chi_0^2 = 3.84$ to get $P(\chi^2 > \chi_0^2) = 0.05$ when $\nu = 1$ degree of freedom. This *would be* significant on the 5% level. Data that would cause this large a χ_0^2 are $m_1 = 2570$ and $m_2 = 2430$ for the coin-flipping problem.

11.4 Equivalence of Chi-Square to Other Statistics; Sufficient Statistics

Let us restrict attention to binary problems, such as coin problems. In these cases there is only $M - 1 = 1$ degree of freedom, the value m_1. This is because of constraints

$$m_1 + m_2 = N , \quad P_1 + P_2 = 1 , \tag{11.13}$$

on m_2 and P_2. Since there is only one RV to consider, the situation is vastly simplified. Let us define a general statistic based on RV m_1, N and the probability P_1 of chance to test against. Call it

$$f(m_1, N, P_1) \equiv y . \tag{11.14}$$

For example, for $y = \chi^2$,

$$f = (m_1 - NP_1)^2/NP_1P_2 , \quad P_2 = 1 - P_1 . \tag{11.15}$$

We show next that an observed value m_1^0 of m_1 *will have the same significance level* $P(y \geqq y_0)$, *regardless of the choice of statistic* f, provided that when (11.14) is solved for m_1, it has only one root.

The proof is quite elementary. Each event m_1 corresponds uniquely to an event y via (11.14). In particular, let m_1^0 correspond to y_0.

Then

$$P(y) = P(m_1) \tag{11.16}$$

We may merely sum both sides of (11.16) over corresponding events for which $y \geqq y_0$, obtaining

$$P(y \geqq y_0) = P(m_1 \geqq m_1^0) . \tag{11.17}$$

Now the right-hand side is simply a sum over the binomial probability law (6.5), since the event sequence is binary:

$$P(y \geqq y_0) = \sum_{M_1=m_1^0}^{N} P_N(m_1) , \tag{11.18}$$

where

$$P_N(m_1) = \binom{N}{m_1} P_1^{m_1}(1 - P_1)^{N-m_1} . \tag{11.19}$$

The right-hand side is thus independent of the choice of statistic f, so $P(y \geqq y_0)$ is as well. Q.E.D.

This result is intriguing. It states that *any statistic f that has a single root will yield significance on the same level as by simply computing the probability $P(m \geqq m_1^0)$ that the observed occurrences $m_1 = m_1^0$ will be exceeded.* The only test of significance that really matters is $P(m_1 \geqq m_1^0)$, in this binary situation. This result has an intuitive appeal as well.

How, then, does the χ^2 statistic relate to this result? If choice (11.15) for f is made, inversion for m_1 yields the double root

$$m_1 = NP_1 \pm \sqrt{NP_1 P_2 y}\ . \tag{11.20}$$

This violates the single-root premise that led to result (11.18). However, it can easily be shown that in this case $P(y \geqq y_0)$ is simply the sum over *two portions* of the binomial probability law $P_N(m_1)$,

$$P(y \geqq y_0) \equiv P(\chi^2 \geqq \chi_0^2) = \sum_{m_1=0}^{NP_1-\sqrt{NP_1P_2\chi_0^2}} P_N(m_1) + \sum_{m_1=NP_1+\sqrt{NP_1P_2\chi_0^2}}^{N} P_N(m_1)\ . \tag{11.21}$$

Hence, remarkably enough, in the binary case the chi-square test can be accomplished by merely summing over the binomial probability law.

This is of fundamental importance. Note that the right-hand side of (11.21) is the probability

$$P(m_1 \geqq m_1^0 \text{ or } m_1 \leqq 2NP_1 - m_1^0)\ ,$$

by identities (11.20) with $y_0 = \chi_0^2$. Hence, *the chi-square test is equivalent to the simple probability that the observed number of occurrences m_1^0 is improbably far from its mean NP_1*. This is a "two-tail" test, as in Fig. 11.1b.

In summary, for the binary situation *all* "single-root" statistics of general form (11.14) lead to a "single-tail" test (11.18) on the binomial law, while the "double-root" chi-square statistic leads to a "two-tail" test (11.21) on the binomial law. All these tests are really equivalent or *sufficient*, as it is up to the user in any case as to whether a single- or a double-tail test will be employed. Hence, we say that the chi-square is a "sufficient statistic" in the binary case, as are all statistics of the form (11.14).

11.5 When Is a Vote Decisive?

We return to binary use of the chi-square test, keeping in mind its sufficiency in such cases.

The faculty of a certain graduate department met to vote on whether to grant tenure to a certain professor. The vote was 19 to 15 in favor of tenure. The presider at the meeting declared that this was a "decisive" vote, but one professor objected to this judgment on the grounds that if but three voters had changed their minds the decision would have been reversed. Who was right?

Let us analyze the situation from the standpoint of a presider who is not a voter, and who wishes to judge the "mood" (yes or no) of the faculty. To him, each vote decision (yes or no) is ideally not predictable before the vote is made. Each vote decision is then a random variable with two possible outcomes. The outcomes are the data that will help him to decide on the true mood and on its decisiveness.

Suppose the observer had wanted to judge the "bias" of an unknown coin, instead of the "mood" of the faculty. That is, he wanted to know the actual probability p of a head. He has the following evidence for making the judgment on "bias": A group of N people flip, in turn, the same coin once each. From the N outcomes as data he can test for significant departure from a "fair" coin $P = 1/2$, for example. The χ^2 test may be used for this purpose, as we did previously.

Now he returns to the "mood" determination. The observer's evidence, or data, here are the N independent votes made. These data correspond to the N independent coin flips made above. The coin flips were samples from the one probability law governing the coin's "bias." Likewise, here, *the N votes may be regarded as samples from the one probability law governing the faculty's "mood" overall.* A decisive mood would be indicated by significant departure from a $P = 1/2$ state. On this basis, he may use the χ^2 test to judge the "mood," or "bias," from a $P = 1/2$ state.

Here $N = 19 + 15 = 34$, $M = 2$, $P_1 = P_2 = 1/2$, $m_1 = 19$, and $m_2 = 15$. Then $NP_1 = NP_2 = 17$, and

$$\chi_0^2 = \frac{(19-17)^2}{17} + \frac{(15-17)^2}{17} = 0.47 \, .$$

We decide to use a one-tail test, with significance at 5%. Appendix B shows $P(\chi_0^2 > 0.47) = 0.48$. Value 0.47 is not overly large, and is consistent with equanimity $(P_1 = P_2 = 1/2)$. Working backward, to yield a $P(\chi^2 > \chi_0^2) = 0.05$ would have required a $\chi_0^2 \cong 4.0$, or a vote of *23 to 11* on the issue. Hence a 2:1 majority would have been needed to characterize the vote as decisive!

11.6 Generalization to *N* Voters

It is interesting to generalize the last calculation. Suppose N people vote, with two alternatives possible. What margin of vote beyond $N/2$ is needed to define a decisive vote? Since $M = 2$, this still requires a $\chi_0^2 = 4.0$ for confidence at the 0.05 level. Let $v =$ necessary vote for one alternative. Then by (11.2) we have to solve

$$\frac{2(v - N/2)^2}{N/2} = 4.0$$

for v. The solution is simply

$$v = N/2 \pm \sqrt{N} \, . \tag{11.22}$$

Hence, *the votes must be at least as extreme as* $N/2 \pm \sqrt{N}$. Examples follow.

If $N = 100$ (number of U.S. Senators), this requires a 60:40 vote. A two-thirds majority *is, in fact, required* in the Senate. (Could the founding fathers have known some statistics?)

If instead $N = 10^8$ (voters in the U.S. presidential election), a margin of 10 000 votes ensures a decisive vote! This is somewhat surprising, as it means that even the closest U.S. presidential elections have been decisive! Also, perhaps this result justifies the simple majority rule that applies for electoral college votes, although it really offers even better justification for a majority rule based on *direct* votes.

11.7 Use as an Image Detector

We found in Sect. 7.4 how to create a photon statistical image on the computer. Images as in Fig. 7.5 were used in the following study.

Suppose an observer is told that a noisy image he has received is either random noise (as defined below) or a statistical image of *any object*. The observer has difficulty in making the judgment because there are a limited number of photons in the picture, and this causes the image of a real object to *visually* resemble pure noise. *Is there a way for him to make an educated guess as to which of the two alternatives is true?*

The chi-square test provides a way. Suppose, as in Fig. 7.5, the data consist of $M = 20$ photon-number counts ($m_i, i = 1, \ldots, 20$) across the image. These are given to be formed independently, by the placement of N photons in all ($L = N$ in the figure). The hypothesis we want to test is that these $\{m_i\}$ are consistent with pure noise, the latter specified as *uniformly random* probabilities $P_i = 1/20$, $i = 1, \ldots, 20$, for photon position across the image. Since all the $\{P_i\}$ are equal, the noise image would tend to be flat, or gray in appearance.

Accordingly we form a χ^2 statistic (11.2)

$$\chi_0^2 \equiv \sum_{i=1}^{20} (m_i - N/20)^2/(N/20) \tag{11.23}$$

for data-departure from an ideal gray state $P_i = 1/20$. If χ_0^2 *is sufficiently large*, we surmise that the data are *significantly far* from the gray state to indicate that *noise is not present: a real object is, instead.*

These ideas were tested, by once again creating Monte Carlo images of the rectangle object in Fig. 7.5 using a cosine-bell spread function (7.5). *Thus, in these tests a real object was always present.* Since we know "ground truth," we can judge the effectiveness of the proposed method.

From the noisy images previously generated in Fig. 7.5, the higher the photon count L (or N) is, the more the image approaches its limiting form of a raised cosine; hence, the less the image resembles random noise. This would make the proposed test (11.23) more apt to be correct for N large. We therefore want to see *how small* N can be to still allow the test to reach the correct conclusion.

We arbitrarily decide to adopt a 5% confidence level of acceptance, and a one-tail test. This means that we agree to accept the hypothesis of a real object, if the χ_0^2 value obtained is so large that $P(\chi^2 \geqq \chi_0^2) = 0.05$ or less. We use the χ^2 test with $M - 1 = 19$ degrees of freedom.

Figure 11.2 shows some typical results of the experiments. Each plot is a Monte Carlo created image due to the number N of photons indicated beneath it. Also

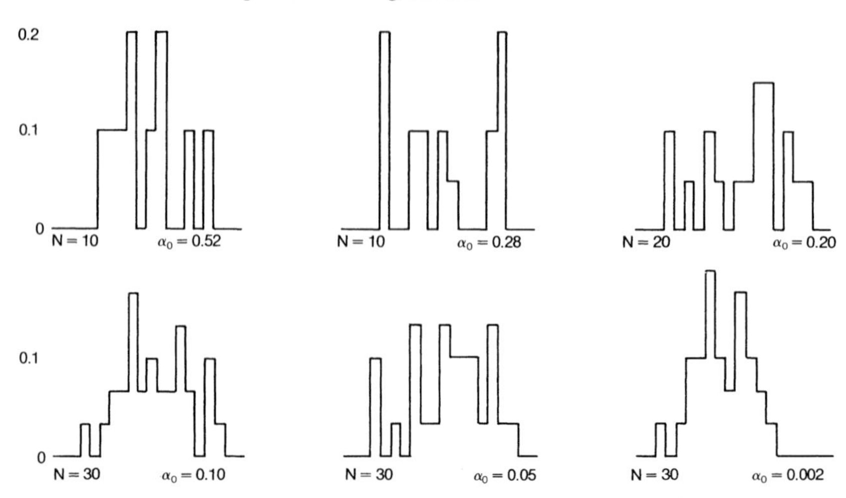

Fig. 11.2. Significance tests on six noisy images

indicated there is the value of $P(\chi^2 > \chi_0^2) \equiv \alpha_0$ corresponding to the χ_0^2 (not shown) for that image. Remember that the value $\alpha_0 = 0.05$ is critical.

The images are shown in order of decreasing α_0 from left to right, and downward. The top left image is so noisy that 52% of the time random noise would give rise to its value χ_0^2 or larger. It is not significant. The bottom right image shows a strong enough resemblance to the raised cosine image of Fig. 7.5 that its probability of occurrence consistent with pure noise is nearly infinitesimal, $\alpha_0 = 0.002$. It is very significant. The in-between curves are at intermediate stages of development away from randomness and toward the cosine image. Note that the bottom middle curve is the first to be significant on the 5% level.

Regarding total photon number N required for accurate discrimination, we found that for $N = 10$ a significance level $\alpha_0 = 0.05$ was seldom attained, whereas for $N = 30$ it nearly always was attained. This is the kind of tendency we expected, since for a great number of photons the image is so regular that there is no problem in distinguishing it from random noise.

This was, of course, merely one example of use of the χ^2 test upon image data. More generally, the test may be used as a vehicle for deciding upon the presence of any *hypothetical signal* $\{NP_i\}$ within a given noisy image. Such signal hypotheses as weak stars, tumors, submarine wakes, and diseased crops are potential candidates for the test. The test might prove particularly useful in medical diagnostic imagery and in aerial surveillance, where decisions on "significant" departure from standard images are, again, crucial.

Exercise 11.1

11.1.1 A given die is suspected of being "loaded," i.e., rigged so that certain numbers come up more often than for a fair die having $P_i = 1/6, i = 1, \ldots, 6$. The die is tested by rolling it 6000 times. The roll counts $m_i, i = 1, \ldots, 6$ are (987, 1013,

1100, 1120, 800, 980). Are these data significant at the 5% level in comparison with a fair die hypothesis?

11.1.2 Radio waves from outer space are being examined for possible signs of "intelligence." The number of occurrences of each of 10 intensity intervals is recorded as follows:

Intensity interval	Occurrence	Intensity interval	Occurrence
0–1	128	5–6	130
1–2	94	6–7	65
2–3	110	7–8	88
3–4	21	8–9	130
4–5	150	9–10	84

Let us define "intelligence" as inconsistency with pure chance, $P_i = 1/10$, $i = 1, \ldots, 10$. Do these data show intelligence at the 5% level of significance?

11.1.3 Suppose a picture is taken of a cell that is suspected of being malignant. It has been proposed [11.4] that such a cell be diagnosed by comparing its occurrence rate of pixel densities against: (i) the average histogram of densities for normal cells, and then (ii) that for malignant cells. Let each comparison be made using a chi-square test. Then each results in a significance value $P(\chi^2 \geqq \chi_0^2)$. The diagnosis would be in favor of the higher significance value, in this case, since we want to test for *agreement with* a given histogram (chance hypothesis). Suppose that the following data are given (these are hypothetical values):

Density	Occurrence rate for cell	Normal histogram value	Malignant histogram value
0–0.25	15	0.01	0.00
0.25–0.50	22	0.03	0.01
0.50–0.75	38	0.10	0.08
0.75–1.00	81	0.15	0.20
1.00–1.25	95	0.15	0.10
1.25–1.50	101	0.20	0.06
1.50–1.75	52	0.15	0.05
1.75–2.00	25	0.10	0.11
2.00–2.25	18	0.07	0.20
2.25–2.50	16	0.02	0.10
2.50–2.75	8	0.01	0.05
2.75–3.00	2	0.01	0.04

(a) Is the given cell malignant or not, and with what significance?

(b) Actually, this test is not sensitive enough. It ignores a priori information about correlations in density from one pixel to the next. In particular, what basic premise of the derivation in Sect. 11.2 is violated by a cell *picture?*

11.1.4 An examination is given, and the occurrence of grades is as follows:

Grade	Occurrence	"Fair probability"
A	5	0.20
B	10	0.40
C	70	0.25
D	40	0.10
F	10	0.05

A "fair" examination might be defined as one for which the majority of students do fairly well, as indicated by the probabilities in the third column above. Did the grade occurrences show significant departure from fairness?

11.1.5 Show that the variances of the RV's $\{x_i\}$ in (11.8) obey $\sigma^2_{X_i} = 1 - P_i$.

11.1.6 It was found at (11.21) that the two-tailed *binomial* statistic is equivalent to use of the chi-square test on the same data. Show, then, that (11.21) gives the same answer as by chi-square to the coin problem of Sect. 11.3. *Hint:* Note that in a "fair coin" case the two sums in (11.21) become identical, and that the resulting sum may be evaluated by use of the error function formula (6.48).

11.1.7 Occurrence data $\{m_i\}$ are to be best fit by a constant level $\overline{m}$. Show that if the criterion-of-fit used is a minimum χ^2 value, then $\overline{m}$ is the rms data level, and not the mean level.

11.1.8 ***Chi-square test on the populations of the tribes of Israel.*** On the basis of modern demographics one would expect the populations of the twelve tribes of Israel to be normally distributed. The actual populations as cited in the Bible are as follows:

$$\begin{bmatrix} \text{Asher} & 41,500 \\ \text{Benjamin} & 35,400 \\ \text{Dan} & 62,700 \\ \text{Ephraim} & 40,500 \\ \text{Gad} & 45,650 \\ \text{Issacher} & 54,400 \\ \text{Judah} & 74,600 \\ \text{Manasseh} & 32,200 \\ \text{Naphtali} & 53,400 \\ \text{Reuben} & 46,500 \\ \text{Simeon} & 59,300 \\ \text{Zebulun} & 57,400 \end{bmatrix}$$

Do a chi-square test on the hypothesis that the populations follow a normal law with a mean equal to the sample mean, and a variance equal to the same variance, of these population figures. Use a bin size of 10 000. There are then $M = 5$ different event intervals bracketing the populations, hence $\nu = 4$ degrees of freedom to the problem. Integrate over each population span of 10 000 using the error function formula (4.62) to form the hypothesis probabilities P_i, $i = 1, \ldots, 5$. Show that the result is a value of $\chi_0^2 = 0.93$, for which Appendix B gives an $\alpha = 0.92$. Hence, we cannot reject the normal hypothesis. On the contrary, the data agree with the hypothesis with 92% confidence.

11.1.9 ***Chi-square test on the universal physical constants.*** We previously concluded on theoretical grounds (Sect. 5.14) that the universal physical constants should obey a *reciprocal* probability law $p(y) = a/y$ over a finite range of y-values. If the numbers y vary over a finite range of values 10^{-b} to 10^{+b} then a new variable $\log_{10} y \equiv x$ obeys a uniform law

$$p_X(x) = \frac{1}{2b} \text{ for } -b \le x \le b \tag{11.24}$$

Derive this uniform law from the reciprocal law by the Jacobian approach of Chap. 5.

The law (11.24) may be used as a test hypothesis for representing *the known* physical constants.

One representative table [11.5] of the constants, in *cgs* units, is as follows:

Quantity	**Magnitude**	$\mathbf{Log_{10}}$	$(-30, -18)$	$(-18, -6)$	$(-6, 6)$	$(6, 18)$	$(18, 30)$
Velocity of light c	2.99×10^{10}	$+10$				X	
Gravitational G	6.67×10^{-8}	-7		X			
Planck constant h	6.63×10^{-27}	-26	X				
Electronic charge e	4.80×10^{-10}	-10		X			
Mass of electron m	9.11×10^{-28}	-27	X				
Mass of 1 amu	1.66×10^{-24}	-24	X				
Boltzmann constant k	1.38×10^{-16}	-16		X			
Gas constant R	8.31×10^{7}	$+8$				X	
Joule equivalent J	4.19×10^{0}	0			X		
Avogadro number N_A	6.02×10^{23}	$+24$					X
Loschmidt number n_0	2.69×10^{19}	$+19$					X
Volume gram-molecule	2.24×10^{4}	$+4$			X		
Standard atmosphere	1.01×10^{6}	$+6$				X	
Ice point	2.73×10^{2}	$+2$			X		
Faraday $N_A e/c$	$9.65 \times 10^{+3}$	$+4$			X		

From the range of exponent values in the table, a reasonable value for the maximum absolute exponent value to ever be encountered (in *cgs* units) is $b = 30$. The $M = 5$ indicated equal-length event intervals $(-30, -18), (-18, -6), \ldots, (+18, +30)$ for the exponent events cover all exponents over the range $(-b, +b)$. Each individual number event is assigned to an appropriate interval as indicated by an X. Adding up the number of X's vertically gives the number m_i of events in each interval i. The table shows that these have the values $(m_1, m_2, m_3, m_4, m_5) = (3, 3, 4, 3, 2)$. This is the data to test against the hypothesis (11.24), which corresponds to probabilities $P_i = 1/5, i = 1 - 5$ for these discrete intervals.

Using the chi-square test, show that the result is an $\alpha = 0.95$. The hypothesis that the logarithms of the fundamental constants are uniformly distributed is well-confirmed by the data.

Although the test was performed upon numbers in one particular set of units (*cgs*), the distribution (11.24) obeyed by the numbers is actually *independent of choice of units*. See Sect. 5.14. If the numbers in the table were expressed in any other consistent set of units they would likewise give an α close to 0.95 for the hypothesis of a flat law on the exponents.

The theoretical median of the probability law on the constants is *unity*; see Ex. 9.2.7. Again, this is independent of choice of units. The sample median of the numbers in the table is 2.73×10^2, not that far from the theoretical value considering that the numbers range over 51 orders of magnitude.

12. The Student t-Test on the Mean

This is a famous example of how statistical methods can be contrived to overcome a problem of inadequate (in fact, missing) information.

Suppose, as in Sect. 9.1, N values of the speed of light c are determined by N independent experiments, and the sample mean

$$\bar{c} \equiv N^{-1} \sum_{n=1}^{N} c_n \tag{12.1}$$

is taken. The $\{c_n\}$ are the experimental values of c. The purpose of taking the sample mean is to obtain an improved estimate of c, the true value of the speed of light. The latter is also presumed to be the theoretical mean of the probability law whose outputs are the $\{c_n\}$. Thus we assume

$$c_n = c + e_n \ , \tag{12.2}$$

where

$$\langle c_n \rangle = c \ . \tag{12.3}$$

Experimental apparatus having this property are called physically *unbiased*. By Eqs. (12.2) and (12.3),

$$\langle e_n \rangle = 0 \ . \tag{12.4}$$

Since the experiments are independent, we also assume the $\{e_n\}$ to be independent. *We want to know how accurately the sample mean* (12.1) *represents the true mean* c.

This type of problem is widely encountered. For example, a noisy image is scanned with a finite aperture in order to beat down speckle (Sect. 5.10). How accurate a version of the signal image is this scanned image? We shall continue to use the particular notation $\{c_n\}$, while simultaneously addressing the more general problem.

This problem is of course directly answerable *if the variance* σ_c^2 in the $\{c_n\}$ *outputs is known*. By the result (9.5), the mean-square error $\sigma_{\bar{c}}^2$ in the sample mean (12.1) obeys simply

$$\sigma_{\bar{c}}^2 = \sigma_c^2 / N \ . \tag{12.5}$$

Suppose, on the other hand, that σ_c^2 *is not known*. This is a situation of ignorance regarding the accuracy in the experimental data. When does such a situation arise?

12.1 Cases Where Data Accuracy is Unknown

If an experiment is of the "quick-and-dirty" variety, where large errors are expected, the user can pretty well estimate the expected error σ_c^2 in the data. However, in the opposite case where a highly refined experiment is performed, the errors are liable to be so small as to be difficult to estimate with any degree of assurance. For example, they may easily be off by a factor of 2 or more. With this amount of uncertainty, it might be inadvisable to use the estimate of the error in any quantitative way, such as in (12.5).

This problem would also arise when experimental data $\{c_n\}$ are published but the reader has limited knowledge of the accuracy in the data, because either (i) they were simply not published or (ii) the reader suspects that some sources of error may have been overlooked by the experimenter.

Then there are some data that by their nature preclude knowledge of their accuracy. The phenomenon giving rise to the data is so poorly understood, or so rare, as to preclude a reliable figure for σ_c. An example of the former is the yearly rate c_n of UFO reports, $n = 1, 2, \ldots$. The latter is exemplified by the yearly discovery rate of new fundamental particles.

Finally, there are physical problems where the unknown quantity per se defines an unknown σ. For example, consider the problem where the underlying signal c represents an *unknown* transmittance level on a grainy photographic emulsion. Photographic noise σ_c is transmittance dependent (6.12), so that σ_c is again unknown. If a finite aperture scans the emulsion, each output of the aperture is a sample mean. The problem of estimating the error $\sigma_{\bar{c}}$ in this sample mean is then precisely our problem once again.

12.2 Philosophy of the Approach: Statistical Inference

This is a good example of a problem in missing information. If the sample $\bar{c}$ is formed from given data $\{c_n\}$, can the expected error in $\bar{c}$ somehow be estimated when the intrinsic fluctuation σ_c in the data is not known? Since σ_c is missing information, the estimate must be independent of this quantity. This at first seems an impossible task, even in principle.

However, could a replacement for knowledge of the variance σ_c^2 be observation of the *sample variance* in the data? Let us define this as a quantity S_0^2 where

$$S_0^2 \equiv N^{-1} \sum_{n=1}^{N} (c_n - \bar{c})^2 . \tag{12.6}$$

Note that this S_0^2 is not quite the sample variance (9.41) S^2.

One way to use S_0 is to attempt to replace the exact (12.5) by

$$\sigma_{\bar{c}}^2 = \lim_{N\to\infty} S_0^2/N . \tag{12.7}$$

But this formula, although correct, is not very useful, since by hypothesis we want to know $\sigma_{\overline{c}}^2$ for *finite* N. A new tactic must be sought.

Let us try now the basic method of statistics called *statistical inference*. (This closely resembles the indirect method of proof described in Chap. 11.) By this method, a hypothesis is made about the correct value of the mean c. The hypothesis is that $c = \hat{c}$, a test number.

Next, regard the individual data values $\{c_n\}$ *to be formed as random departures from the "true" value* $\hat{c}$*, as in* (12.2) *with* $\hat{c}$ *replacing* c. Then the sample mean $\overline{c}$ of these data ought to tend to equal $\hat{c}$. Thus, if the test hypothesis $\hat{c}$ is correct, the difference $(\overline{c} - \hat{c})$ should tend statistically to be small.

Conversely, too large a difference $(\overline{c} - \hat{c})$ implies that the test hypothesis $\hat{c}$ is wrong. In fact, the difference $(\overline{c} - \hat{c})$ is known numerically, since $\overline{c}$ is the sample mean and $\hat{c}$ is a test number. Hence *if* $(\overline{c} - \hat{c})$ *is observed to be overly large, the hypothesis* $\hat{c}$ *is rejected*. The grounds for this rejection are that the sample mean $\overline{c}$ contradicts the assumption that $\hat{c} = c$. Now, what do we mean by "overly large"?

There is, in fact, a certain, small probability that $\hat{c}$ was the correct value c, and the observed large number $(\overline{c} - \hat{c})$ occurred as a purely random fluctuation from $\hat{c}$. Even larger fluctuations could have occurred, again purely as a random fluctuation from $\hat{c} = c$. If such fluctuation actually occurred, resulting in $\overline{c}$, we would be wrong to *reject* the hypothesis that $\hat{c} = c$. How are we to know, then, when to reject the hypothesis? The answer is that we can never know with perfect assurance. We must be satisfied to make a guess that is *probably correct*.

The probability that a value $(\overline{c} - \hat{c})$ will be exceeded, where $\hat{c} = c$, may be computed (see below). If this is very small, say 0.05 or less, then we have grounds for rejecting the hypothesis $\hat{c}$. The grounds are that the observed sample mean $\overline{c}$ is improbably far away numerically from the "true" value $\hat{c}$. We say, then, that we reject the hypothesis with 0.05 (5%) confidence. Often 1% confidence level is used instead.

What hypothesis value $\hat{c}$ will be *accepted* instead? The particular number $\hat{c} \equiv \overline{c}$ can never be rejected, since then $(\overline{c} - \hat{c}) = 0$ identically, and the probability of a fluctuation occurring that is greater than or equal to 0 is exactly 50% (see below). Of course, 50% is not a reasonable confidence level for rejection. In a similar way, values of $\hat{c}$ near to $\overline{c}$ will be accepted. Finally, when either value

$$\hat{c} = \overline{c} \pm \Delta c \tag{12.8}$$

is tested, it will define a just large enough fluctuation $(\overline{c} - \hat{c}) = \pm\Delta c$ to be rejected.

The answer to the error estimation problem is then the following: Sample mean $\overline{c}$ is correct to $\pm\Delta c$ with confidence 5% (or whatever).

12.3 Forming the Statistic

The statistic we proposed to use for testing the hypothesis $\hat{c}$ was $(\overline{c} - \hat{c})$. Any quantity proportional to this statistic would also suffice, by the same reasoning. We use this fact below to form a statistic that is proportional to $(\overline{c} - \hat{c})$ and that does not require knowledge of the unknown σ_c (Sect. 12.1).

The procedure outlined above requires knowing the probability that a value $(\overline{c} - \hat{c})$ will be exceeded. In turn, this requires knowing the probability density for $(\overline{c} - \hat{c})$. This is found next.

By (12.1–3),

$$\overline{c} - \hat{c} = N^{-1} \sum_{n=1}^{N} c_n - \hat{c} = c - \hat{c} + N^{-1} \sum_{n=1}^{N} e_n = N^{-1} \sum_{n=1}^{N} e_n \,, \tag{12.9}$$

since the statistic is tested under the assumption that the test hypothesis $\hat{c}$ is correct, $\hat{c} = c$.

Let us assume that the errors $\{e_n\}$ are Gaussian random variables, in addition to the previous assumptions of zero mean and independence. By the central limit theorem of Sect. 4.26 this is very often the case.

Then from (12.9) the statistic $(\overline{c} - \hat{c})$ is a Gaussian random variable whose variance is $N^{-2}(N\sigma_c^2)$ or $N^{-1}\sigma_c^2$. Also, the related statistic

$$x \equiv (\sqrt{N}/\sigma_c)(\overline{c} - \hat{c}) \tag{12.10}$$

must be a Gaussian random variable with unit variance (independent of σ_c). The trouble with this statistic is that to form it requires knowledge of the quantity σ_c^2, which we have been assuming to be unknown (Sect. 12.1).

How about, then, replacing division by σ_c in (12.10) with division by quantity S_0 defined in (12.6)? By (9.51), S_0 is a reasonable approximation to σ_c with N large. Hence, form a new statistic

$$y \equiv (\overline{c} - \hat{c})/S_0 \,. \tag{12.11}$$

This quantity is formed independent of an assumption of knowing σ_c. But is the probability density for y also independent of σ_c? This is shown next to be the case, basically because both numerator and denominator of (12.11) are proportional to σ_c, which cancels.

From the form of defining relation (12.6) the random variable S_0^2 is proportional to a sum of squares of N random variables $(c_n - \overline{c})$. By (12.1) these random variables are not all independent. In fact, by (9.49), the RV $(N-1)S^2/\sigma_c^2$ is standard chi-square with $N-1$ degrees of freedom. Then, since $S_0^2 = (N-1)S^2N^{-1}$, we may represent

$$S_0^2 = (\sigma_c^2/N)z \,, \tag{12.12}$$

where z is a chi-square RV with $N-1$ degrees of freedom. We can now combine results.

Consider the new statistic very closely related to y,

$$t \equiv \sqrt{N-1}(\overline{c} - \hat{c})/S_0 \,. \tag{12.13}$$

Note that t is formed from quantities that only depend upon the data $\{c_n\}$. By (12.10) and (12.12), t is also an RV obeying

$$t = \sqrt{N-1}\,x/\sqrt{z}\,. \tag{12.14}$$

We found before that x is Gaussian with unit variance and z is chi-square with $N-1$ degrees of freedom. Then, neither the formation (12.13) of the t-statistic nor the random variable t defined in (12.14) depends in any way upon the unknown quantity σ_c. This solves the problem. We have found a statistic that permits a decision to be made about the likelihood of a test hypothesis $\hat{c}$, without requiring knowledge of the data accuracy σ_c!

The size of a number t, and more precisely, its probability of being that large or larger, now determines our decision on the acceptance of an hypothesis $\hat{c}$. Accordingly, we must establish the probability density for t.

12.4 Student's t-Distribution: Derivation

The derivation of probability law $p_T(t)$ is a simple exercise in the use of the Jacobian approach of Sect. 5.13.1. Accordingly, we form two output RV's t, u according to transformation laws

$$t = \sqrt{N-1}\frac{x}{\sqrt{z}}$$
$$u \equiv z$$

with the last defining a helper variable u. This is a *one-root* case, as can easily be shown (recall that $\sqrt{z}$ corresponds to $\sqrt{S_0}$, a standard deviation and therefore positive). Then the first term of (5.18) gives an output density

$$p_{TU}(t,u) = |J(x,z/t,u)|\,p_{XZ}(x,z).$$

Computing $J = (N-1)^{-1/2}u^{1/2}$ and using the fact that x and z are independent, gives

$$p_{TU}(t,u) = (N-1)^{-1/2}u^{1/2}p_X(tu^{1/2}(N-1)^{-1/2})\,p_Z(u). \tag{12.15}$$

With the known forms for $p_X(x)$ and $p_Z(z)$, this becomes

$$\begin{aligned} p_{TU}(t,u) = {} & (2\pi)^{-1/2}(N-1)^{-1/2}u^{1/2}\exp[-t^2u/2(N-1)] \\ & \cdot 2^{-(N-1)/2}[\Gamma(N-1)/2]^{-1}u^{(N-3)/2}\exp(-u/2). \end{aligned}$$

Combining like powers of u and integrating out over all positive values of u, we get the marginal answer

$$p_T(t) = \frac{\Gamma(N/2)}{\Gamma(N/2-1/2)}[(N-1)\pi]^{-1/2}\left(1+\frac{t^2}{N-1}\right)^{-N/2}, \tag{12.16}$$

which we set out to find.

This is known as *Student's distribution*, after the mathematician Student (pseudonym of W.S. Gosset), who contrived it in 1908 to solve the very problem of this chapter [12.1]. The number N of data is also called the number of "degrees of freedom" in the distribution.

12.5 Some Properties of Student's t-Distribution

This distribution depends upon t only through t^2. Hence, $p_T(t)$ is even in t and

$$\langle t \rangle = 0 \,. \tag{12.17}$$

The variance may be found by directly integrating out $t^2 \cdot p_T(t)$. The result is

$$\sigma_T^2 = \frac{N-1}{N-3} \,, \tag{12.18}$$

or unit variance for N moderately large.

Unless N is very small (4 or less), $p_T(t)$ is very close to a Gaussian curve with unit variance. To show this, from (12.16) the t-dependence is

$$\left(1 + \frac{t^2}{N-1}\right)^{-N/2} .$$

Calling this y,

$$\lim_{N\to\infty} \ln y = \lim_{N\to\infty} -\frac{N}{2} \ln\left(1 + \frac{t^2}{N-1}\right) = -\frac{t^2}{2} \lim_{N\to\infty} \frac{N}{N-1} = -t^2/2 \,.$$

Consequently,

$$\lim_{N\to\infty} p_T(t) \propto \mathrm{e}^{-t^2/2} \,, \tag{12.19}$$

a Gaussian with unit variance.

To show how close Student's distribution is to a Gaussian, it is plotted in Fig. 12.1 for the case $N = 3$, along with the unit variance Gaussian.

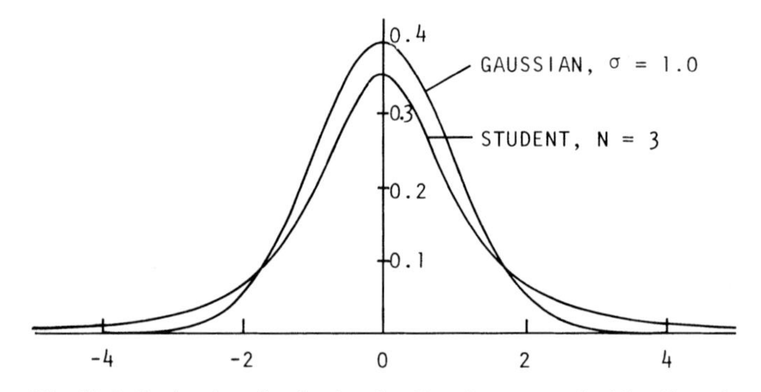

Fig. 12.1. Student's t-distribution for $N = 3$ compared with a Gaussian curve of unit variance

Tables of the integral

$$P(|t| \leq t_0) \equiv 2 \int_0^{t_0} \mathrm{d}t \; p_T(t) \tag{12.20}$$

exist, as functions of t_0 and N (the only free parameters). For example, see Appendix C, from [12.2]. Here, $\nu \equiv N - 1$ and $A \equiv P$ in (12.20). Or, the integral may easily be evaluated numerically, since the integrand is of a simple, rational form.

12.6 Application to the Problem; Student's t-Test

In summary, the procedure we have derived to test an hypothesis $\hat{c}$ is as follows: Form a value of $t = t_0$ from the N data $\{c_n\}$ and the hypothesis $\hat{c}$, via

$$t_0 = (N-1)^{1/2}(\overline{c} - \hat{c})/S_0\,, \tag{12.21}$$

where $\overline{c}$ is the sample mean (12.1) and S_0^2 is defined at (12.6). Next, look up the corresponding value of $P(|t| \leqq t_0)$ in tables of the Student distribution. Then $1 - P(|t| \leqq t_0)$ is the probability that the value t_0 will be exceeded, based on the assumption that $\hat{c}$ is correct. If this probability is *less than* a prescribed confidence level, say 0.05, then the value $\hat{c}$ *is rejected* as a possible value of c. In words, the normalized fluctuation (12.21) in the sample mean from the test mean is too large to be reasonably expected. Conversely, if $1 - P$ exceeds the confidence level, then the test $\hat{c}$ is accepted as a possible value of c. In this manner, a range $\pm\Delta c$ of acceptable $\hat{c}$ is constructed, and $\pm\Delta c$ is the estimated uncertainty in the sample mean $\overline{c}$.

12.7 Illustrative Example

In an attempt to reduce image noise, a slowly varying image is scanned with a small aperture. The output of the aperture at each of its positions is the sample mean over the data within it. The standard deviation σ_i in the image at any one point is unknown.

Suppose that, at any position in the image, the aperture contains four image values, as in Table 12.1. The output image value is the sample mean of these four numbers, $\overline{i} = 2.0$. By how much is this liable to be in error from the true value of the image at this aperture position? That is, if the true $i = 2.0 \pm \Delta i$, what is Δi?

Table 12.1. Image values within aperture at one scan position.

n	i_n
1	2.0
2	3.0
3	1.0
4	2.0

It we assume that each

$$i_n = i + e_n \tag{12.22}$$

where i is the unknown signal value and the $\{e_n\}$ are independent and Gaussian, then we may use Student's t-test to decide on any hypothesis $\hat{i}$ for i.

Let us adopt a 5% confidence level. For the case $N = 4$ at hand, Appendix C shows that a value $t_0 = 3.182$ just barely flunks the t-test, i.e., its probability of being exceeded is less than 0.05. To see what test hypothesis $\hat{i}$ this corresponds to, from Table 12.1, $S_0^2 = 0.5$ and $\bar{i} = 2.0$. Hence, by (12.21), for any test $\hat{i}$

$$t_0 = \sqrt{3}(2.0 - \hat{i})/\sqrt{0.5} \ .$$

Setting this equal to 3.182, we find this to result from a just-rejected hypothesis

$$\hat{i}_{\text{rej}} = 2.0 - 1.3 \ .$$

Since $p_T(t)$ is even in t, the same t_0 would result from an $\hat{i}$ on the other side of the mean, or

$$\hat{i}_{\text{rej}} = 2.0 + 1.3 \ .$$

We conclude that for the data at hand, the estimated signal $\hat{i}$ can be known no more accurately than

$$\hat{i} = 2.0 \pm 1.3 \ . \tag{12.23}$$

Or, the sample mean is probably in error by ± 1.3 or less. This is for a confidence level 5%. A lower confidence level would permit even wider latitude in acceptable $\hat{i}$ values, i.e., more uncertainty than ± 1.3 in the sample mean.

We emphasize that this calculation has been based solely upon four data numbers, without requiring knowledge of the theoretical accuracy σ_i in each data value. Herein lies the power of the approach.

It is instructive to form the range of estimates $\hat{c}$ of the mean c as predicted by the t-statistic (12.21),

$$\hat{c} = \bar{c} \pm \frac{S_0}{\sqrt{N-1}} t_0 \ , \tag{12.24a}$$

and *compare this* with the range as predicted by the standard error expression (12.5),

$$\hat{c}_0 \approx \bar{c} \pm \frac{\sigma_c}{\sqrt{N}} \ . \tag{12.24b}$$

First, both show spreads about the same number, the sample mean $\bar{c}$. Second, the spreads themselves may be conveniently compared since, once $N \gtrsim 10$, S_0 becomes a fairly good estimate of the true standard deviation, $S_0 \approx \sigma_c$ [see (9.52)], and t_0 hardly varies with N: see Appendix C. The Appendix also shows that, for the most useful range of confidence level values, i.e. $0.02 \le A \le 0.20$, the value of t_0 varies only between the specific numbers 1.3 and 2.3. This means that, in *the absence* of knowledge of σ_c, which provided the motivation for this chapter, (12.24a) gives an estimated spread that is somewhere between 1.3 and 2.3 times that which would be made in *the presence* of knowledge of σ_c. We conclude that the t-statistic fulfills its major aim admirably.

12.8 Other Applications

The t-test has other uses as well. These are developed and applied in problems 4) through 9) below. It may be used to decide whether two specimens of data probably (at some level of confidence) have the same mean. This allows a decision to be made on whether the two specimens arose from identical physical causes. For example, two samples of uniformly exposed film are given, and we want to test whether they are probably pieces of the same film (Ex. 12.1.8).

Or, the failure rates over the years are given for brand X and for Y, and from these we wish to determine whether brand Y is an improvement over brand X (Ex. 12.1.9).

Or, two weak stars are observed to have nearly equal brightness. How long a time exposure is needed to have confidence at 5% that they are not equal? There are potentially many uses of the t-test in optical research.

Exercise 12.1

12.1.1 The radio wave occurrence values of Ex. 11.1.2 have a certain sample mean and sample variance. What are these? Assume that the occurrences have a common mean. Find the spread in estimates of the mean at a confidence level of 5%. If the data in Ex. 11.1.2 were significant on the basis of a chi-square test, is the preceding calculation justified? Explain.

12.1.2 A sample of six transistors is chosen from a lot, all of whose members were manufactured under identical conditions. Their hours to failure were noted as (40, 38, 42, 44, 36, 41). Another transistor failed after only 32 hours of use. It is not known whether this transistor belongs to the same lot as the other six. What is the strongest confidence with which we can infer that it does not belong? (Recall that a strong confidence means a low confidence level.)

12.1.3 In a certain course, a student's consecutive test scores were (100, 82, 92, 74, and 88). On the next test he received a grade of 50. He explained his poor showing by stating that his mind was unusually dull owing to a significant lack of sleep the night before the exam. With what confidence can we accept his statement, at least to the extent that his grade score was uncharacteristic of its predecessors?

12.1.4 ***The t-statistic may be used to test the hypothesis that two data sets have the same mean.*** The hypothesis is verifiable from the closeness of their respective sample means. If it is verified, one concludes that *the two sets of data arose from the same cause*, for example that both are photographic density values from the same piece of film. This and the next three problems deal with the derivation of the statistic that tests out this hypothesis.

Let subscript 1 describe the first set of data, subscript 2 the second. Let these both have the same standard deviation σ_c, the assumption being that if the two data sets are from different causes this will be reflected only in different means, c_1 and c_2. The derivation is analogous to that of the ordinary t-statistic in Sect. 12.3. Accordingly, form

$$x \equiv \frac{(\overline{c}_1 - \overline{c}_2) - (c_1 - c_2)}{\sigma_{\overline{c}_1 - \overline{c}_2}}, \tag{12.25}$$

where $\sigma_{\overline{c}_1 - \overline{c}_2}$ is the standard deviation in $(\overline{c}_1 - \overline{c}_2)$. Show that this x is analogous to x in (12.10), that is, it is Gaussian with unit variance.

12.1.5 Continuing the development, quantity $N_1 S_1^2/\sigma_c^2$ is chi-square with $N_1 - 1$ degrees of freedom, and similarly for the subscript 2 quantity. Show that

$$z^2 \equiv z_1^2 + z_2^2 \equiv N_1 S_1^2/\sigma_c^2 + N_2 S_2^2/\sigma_c^2 \tag{12.26}$$

is also chi-quare, with

$$\nu = N_1 + N_2 - 2 \tag{12.27}$$

degrees of freedom.

12.1.6 In analogy to (12.14), we form

$$t \equiv \sqrt{\nu}\, x/\sqrt{z}\,. \tag{12.28}$$

Show that this t must obey a Student probably law with $N_1 + N_2 - 2$ degrees of freedom.

12.1.7 By combining (12.25) through (12.28) and finding $\sigma_{\overline{c}_1 - \overline{c}_2}$, show that

$$t = \frac{(\overline{c}_1 - \overline{c}_2) - (c_1 - c_2)}{(N_1 S_1^2 + N_2 S_2^2)^{1/2}} \left(\frac{N_1 + N_2 - 2}{1/N_1 + 1/N_2} \right)^{1/2}. \tag{12.29}$$

This statistic may be used to test hypotheses about the size of $c_1 - c_2$, the difference in the intrinsic means of the two data sets. Notice that the remaining right-hand quantities are experimental quantities derived from the data. This allows t to once again be computed purely from the data. Then Appendix C may be used to see whether t is large enough to reject the hypothesis at the required confidence level.

The user may want to try out a *finite* test value $c_1 - c_2$ if it has previously been shown that the test hypothesis $c_1 = c_2$ is *not* reasonable. It may then be of interest to know how large a difference $c_1 - c_2$ *is* reasonable.

The following two problems may be attacked by use of the t-statistic (12.29).

12.1.8 Two small pieces of uniformly exposed film are found. These are scanned on a densitometer, with the following outputs:

Film 1	Film 2
2.1	2.6
2.8	2.5
2.4	2.8
2.6	2.4
	2.9
	3.0

Based on such limited information, should we conclude that they are actually parts of the same piece of film? With what confidence?

12.1.9 Product X suffered from too high a failure rate. It was improved, and renamed Y. A consumer's testing agency decided to evaluate the degree of improvement, by observing the time to failure of samples of product X and of product Y. Since the products were expensive, only a limited number of samples could be taken. These were as follows:

X (hours)	Y (hours)
11	18
14	11
8	9
12	7
9	14
	10
	15

(a) With what confidence may they conclude that there was an improvement?

(b) The expected gain in product hours $c_Y - c_X$ can be bracketed. Using a 5% confidence level, find the upper and lower bounds to the gain. The lower bound in particular represents a conservative estimate of the amount of gain such a limited amount of data can imply. (*Hint:* Leave $c_Y - c_X$ as an unknown in (12.29); 5% confidence level and the given number of degrees of freedom imply that the absolute value of the resulting t must be less than the value given in the tables.)

12.1.10 Confidence levels on a sample mean may also be established in a naive way. Since all the data in this chapter are assumed to be normal, a test value $\hat{c}$ represents a point at distance $\hat{c} - c$ from the true mean c on the normal curve. The mean and the variance of the curve may be estimated as sample values from the data. Then the curve *is known*, so that the area beyond value c is also known. This represents the probability that a value greater than or equal to $\hat{c}$ can occur, consistent with the given mean and variance. If this probability is less than or equal to a required confidence level, the value $\hat{c}$ is rejected as an estimate of c.

In this way, establish the range of permissible values for i in Table 12.1, using a confidence level of 5%. Compare with the result (12.23). Why are they different? Why is the approach naive, particularly for N small?

13. The F-Test on Variance

Suppose we are given two sets of data, each of which is a random sample of some RV. We saw in Chap. 12 how the t-test may be applied to these data to decide whether the two RVs have the same mean; see (12.29). In turn, equality implied the same physical cause for the RV's.

Here we consider the analogous problem of deciding upon the equality of the *two variances*. This is sometimes a more sensitive test of the hypothesis that two sets of data arose from the same cause.

As the t-statistic was the difference of two sample means, define the F-statistic as the ratio of the two sample variances,

$$f \equiv S_1^2/S_2^2 \tag{13.1}$$

where subscript denotes data set. S^2 is the sample variance (9.41). We next show that, as with the t-statistic, the F-statistic obeys a probability law which does not depend upon knowledge of σ_c^2, the intrinsic variance of either set.

We assume, as in Sect. 12.3, that both sets of data are normally distributed. *The test hypothesis shall be that the two sets have equal variances* σ_c^2, and therefore arose from the same cause. (Remarkably, no assumption need be made about the means. The probability law for f will turn out to be independent of them!)

13.1 Snedecor's F-Distribution; Derivation

From (9.49) the RV $(N-1)S^2/\sigma_c^2 \equiv z$ is standard chi-square with $N-1$ degrees of freedom. Alternatively, $S^2 = \sigma_c^2(N-1)^{-1}z$, where z is a chi-square RVwith $N-1$ degrees of freedom. Then (13.1) becomes

$$f = \frac{N_2-1}{N_1-1}\frac{z_1}{z_2}, \tag{13.2}$$

where z_1, and z_2 are independently chi-square, with N_1-1 degrees of freedom and N_2-1 degrees of freedom, respectively.

The density $p_F(f)$ may now be found, using the same approach as in Sect. 12.4 for the t-statistic. Accordingly, we first form the joint law

$$p_{Z_1Z_2}(z_1, z_2) = p_{Z_1}(z_1)p_{Z_2}(z_2) \tag{13.3}$$

by independence. Using standard form (11.11) for the chi-square law,

$$p_{Z_1 Z_2}(z_1, z_2) = A_1 A_2 z_1^{(N_1-3)/2} z_2^{(N_2-3)/2} \mathrm{e}^{-(z_1+z_2)/2} \tag{13.4}$$

with coefficients

$$A_i \equiv 2^{-(N_i-1)/2} / \Gamma\left[\frac{1}{2}(N_i - 1)\right], \quad i = 1, 2$$

Now regard (13.2) as a transformation of RV z_1 to RV f. Using the helper variable approach of Sect. 5.13, let the new RV $u = z_2$. Then the Jacobian $\mathrm{J}(z_1, z_2 / f, u)$ is easily found to be $r\,u$, with

$$r \equiv (N_1 - 1)/(N_2 - 1) \; . \tag{13.5}$$

There is a single root (z_1, z_2) for each couplet (f, u). Then by transformation law (5.18)

$$p_{FU}(f, u) = p_{Z_1 Z_2}(r f u, u) \cdot r u \; .$$

Combining this with (13.4), we find

$$p_{FU}(f, u) = A_1 A_2 r u (r f u)^{(N_1-3)/2} u^{(N_2-3)/2} \mathrm{e}^{-(r f u + u)/2} \tag{13.6}$$

as the joint law.

The marginal density $p_F(f)$ of interest is found by simply integrating over u, with result

$$p_F(f) = K_{12} f^{(N_1-3)/2} (1 + r f)^{-(N_1+N_2)/2+1} \, ,$$
$$K_{12} \equiv \frac{\Gamma[\frac{1}{2}(N_1 + N_2 - 2)]}{\Gamma[\frac{1}{2}(N_1 - 1)]\Gamma[\frac{1}{2}(N_2 - 1)]} r^{(N_1-1)/2} \; . \tag{13.7}$$

This is called *Snedecor's F-distribution* [13.1] with $N_1 - 1$ and $N_2 - 1$ degrees of freedom.

13.2 Some Properties of Snedecor's F-Distribution

Because $p_F(f)$ is a rational function (13.7) of f, its moments may easily be found by direct integration. These obey

$$m_k = \left(\frac{N_2 - 1}{N_1 - 1}\right)^k \frac{\Gamma\left(\frac{N_1-1}{2} + k\right) \Gamma\left(\frac{N_2-1}{2} - k\right)}{\Gamma\left(\frac{N_1-1}{2}\right) \Gamma\left(\frac{N_2-1}{2}\right)}, \quad k = 1, 2, \ldots \; . \tag{13.8}$$

In particular,

$$m_1 = \left(\frac{N_2 - 1}{N_2 - 3}\right), \quad N_2 > 3 \; . \tag{13.9}$$

Strangely enough, this does not depend upon N_1.

The shape of $p_F(f)$ is shown in Fig. 13.1 for one particular case N_1, N_2. It qualitatively resembles the chi-square law (Fig. 5.6).

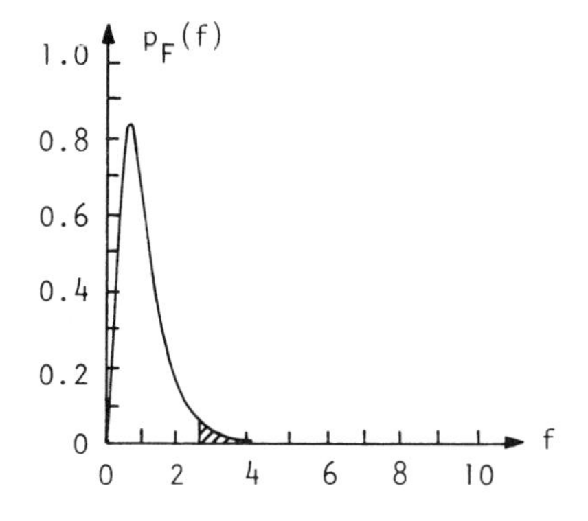

Fig. 13.1. Snedecor's F-distribution for $N_1 = 11$, $N_2 = 21$. The shaded tail region has area 0.025

13.3 The F-Test

One aim of this gambit has been to show that the F-statistic does not depend upon knowledge of σ_c, neither in forming the statistic nor in the probability law obeyed by the statistic. This is shown by (13.1), (13.7), respectively. The statistic (13.1) depends solely upon the data values, and the probability law (13.7) depends solely upon the number of data values in each of the two sets. This permits the F-test, described next, to be widely applicable.

We may now describe how the test of the hypothesis of equal variances is carried out. The test is another example of statistical inference (Sect. 12.2).

If the hypothesis is correct, the observed value of f, call it f_0, should be reasonably close to 1. Hence, if f_0 is instead large it is unlikely that the two variances are equal, and we reject the hypothesis. But how large is too large? As in the chi-square and t-tests, we decide upon a confidence level, say 5%, and reject the hypothesis if f_0 is so large that its probability of being equalled or exceeded falls below 5%. A more strict test would use 1%. Letting α represent a general confidence level, we reject the hypothesis if

$$P(f \geqq f_0) \equiv \int_{f_0}^{\infty} \mathrm{d}f\, p_F(f) \leqq \alpha\,. \tag{13.10}$$

For example, with an α of 0.025, any f_0 lying within the shaded tail of Fig. 13.1 would cause rejection of the hypothesis.

Note that if f_0 is instead small, this also implies that the hypothesis of equal variances is incorrect. For these values $f_0 < 1$ the test would be

$$P(0 \leqq f \leqq f_0) \equiv \int_0^{f_0} \mathrm{d}f\, p_F(f) \leqq \alpha\,.$$

However, this test need never be made. The area under $p_F(f)$ for $0 \leqq f \leqq f_0$ with $f_0 < 1$ must equal the area under $p_F(f')$ for $1/f_0 \leqq f' \leqq \infty$ with N_1 and N_2 interchanged. This is ultimately because any event $1/f$ is identical with an event f when the names of the two data sets are interchanged. Hence, if $f_0 < 1$ the user need only interchange the names of the two sets, obtain then $f_0 > 1$, and use test (13.10).

The numerical value of the integral in (13.10) only depends upon f_0, N_1 and N_2. Tables of the integral exist, for combinations of these parameters. For example, see Appendix D, reprinted from [13.2]. Here, a value $\nu \equiv N - 1$ and $Q \equiv \alpha$.

13.4 Illustrative Example

Let us regard the two images in Tables 12.1 and 13.1 as candidates for the test. Are they pieces of the same image? Do they obey the hypothesis of equal variances? We have $N_1 = N_2 = 4$, $S_1^2 = 2/3$, $S_2^2 = 1/6$, and so $f_0 = 4$. Looking in Appendix D we see that for cases $\nu_1 = \nu_2 = 3$, at $\alpha = 0.25$ ($Q = 0.25$ in the table) an f_0 of 2.36 would just reject the hypothesis. Since our f_0 of 4 *exceeds* 2.36, the hypothesis is being rejected with higher confidence than 25% (i.e., a lower α). Proceeding to lower α, we see that a value $\alpha = 0.10$ would require an f_0 of 5.39. Since our f_0 is not this large, we cannot reject at this stricter level of confidence. Hence, our confidence level for just rejecting the hypothesis lies somewhere between values 10% and 25%. Note that this level of confidence is not significant (by convention, 5% or lower). Therefore, we cannot reject the hypothesis that the two data sets have a common cause. They may.

Table 13.1. As in Table 12.1, but now less fluctuation.

n	i_n
1	2.0
2	2.5
3	1.5
4	2.0

In the preceding work, we have used the F-statistic in a binary way, either to accept or to reject the hypothesis of common cause. In the following application, a continuous mapping of f_0 values into output (gray level) values is performed. This imaginative use of the F-test may be unorthodox, but it will be seen to fit well the groundrules and aims of the particular problem.

13.5 Application to Image Detection

In photographic imagery, the variance of noise increases with the signal density level. Suppose an image scene is either empty, consisting of only photographic noise caused by a background signal level, or has somewhere superimposed upon the background a feature of interest. The feature, if it is present, is uniform with an unknown signal level of its own. Can the scene be tested for the hypothesis that the feature is present? This problem is, for example, encountered in medical diagnostic imagery, where tumors are to be detected.

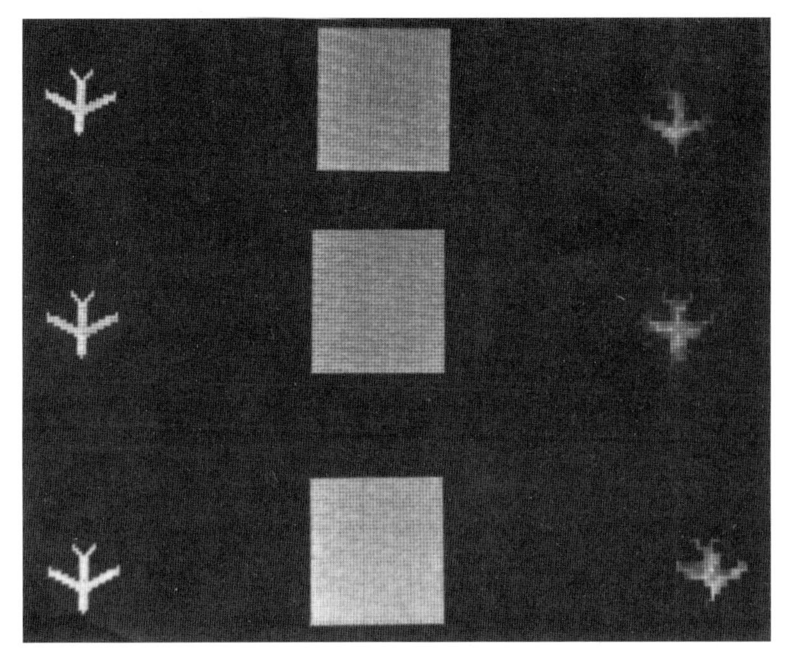

Fig. 13.2. Three demonstrations (rows) of F-test feature extraction. In each row, the feature on the right was extracted from the image in the center! The true feature is shown on the far left. Signal-to-noise is unity in all cases. (Illustrations courtesy of Professor Peter Bartels, University of Arizona)

Imagine the image to be subdivided into subareas or pixels, each containing a finite number of data values. We can test each subarea for the presence of the feature based on the fact that where the feature is present, its signal adds to that of the background, increasing the local variance of noise. Hence, the pertinent statistic is the ratio of the sample variance for the subarea to that within an image subarea where the feature *is known not to be present*. If this value f_0 is near 1, we reject the hypothesis that the foreground feature exists there. This test is performed at each pixel subarea in sequence over the picture.

Bartels and *Subach* [13.3] have carried through this test by computer simulation. In order to create gray-tone pictures, at each pixel subarea they generate a gray value linear with the local value of $P(f > f_0)$. This technique is given the name *significance probability mapping*. Their test feature was a white object (Fig. 13.2) and the background was black. Hence, a high $P(f > f_0)$ signifies acceptance of the "background only" hypothesis, or a high probability of local blackness. Therefore, the higher P was, the blacker the pixel was made.

Each of the three rows of photos in Fig. 13.2 contains three images. On the left is the test feature, superimposed somewhere within the square background field shown in the middle. Both the feature plus background, and the background, were generated with signal level equal to standard deviation. Hence, the signal-to-noise ratio S/N in the simulated data is unity in all three cases. The cases differ in the amount of noise used, and in the position of the feature in the background.

A pixel subarea size of $N_1 = N_2 = 200$ data points was chosen. This defines the pixel size in the right-hand images. The F-test was performed upon each image in the second column of the figure, subarea by subarea, as described previously. The right-hand images are the corresponding outputs of the test. These indicate a very good job of feature extraction, considering what the unaided eye can discern in the middle images!

Note that the finite subarea size defines the blur spot size in the outputs. The result is a tradeoff between sensitivity of detection and spatial resolution, since the F-test is more discriminating as N_1 and N_2 increase. Nevertheless, that the method even works with a signal-to-noise ratio of unity is quite remarkable, and holds promise for future applications.

Exercise 13.1

13.1.1 In Ex. 12.1.8 we used the t-test to decide whether two film pieces were actually part of the same emulsion. That test was based upon the sample means. Now apply the F-test to that data. Compare the t-test and F-test confidence levels at which the hypothesis of equality is just rejected. Which was the more decisive test?

13.1.2 A good optical flat will have very little variation in optical thickness across its surface. Flats of type A were manufactured with loose tolerances and gave rise to thickness measurements (across one typical sample) as in column A below. These are departures from the ideal flat surface.

After complaints from numerous customers, the manufacturer claimed to have improved the product, and as proof gave the typical thickness measurements as given in column B below. How decisively can we reject the hypothesis of equality, i.e., reject the premise that no improvement was made?

A	B
2.1λ	1.11λ
2.5	0.9
1.8	0.8
2.8	1.2
1.5	1.0
	0.9

These two sets of data could also be compared on the basis of their means. Assume that absolute light transmission is not a problem for a given customer. Would it make sense for him to test for equality of mean thickness of optical flats as a criterion of quality ?

13.1.3 The larger in diameter a star is, the smaller is its rms fluctuation of light (Ex. 4.3.12). Stars A and B were observed to have the following light outputs (arbitrary flux units):

A	B
8.5	6.8
8.1	7.2
7.8	8.0
8.0	8.1
9.1	
7.9	

What confidence do we have that the two stars have the same diameter?

14. Least-Squares Curve Fitting – Regression Analysis

The aim of a least-squares approach is to estimate the parameters that define a known, or hypothetical, physical law. The estimated parameters are those that make the law fit a given set of data points in the least-squares sense. Probably the reader has carried through a least-squares curve fit more than once during his career. But, there is actually a great deal more to this problem than accomplishing the fit.

The data points must inevitably suffer from noise. Hence, the parameters of the fit must suffer from error. *If so, what is this error?* Aside from accuracy, the *significance of the individual parameters* must also be questioned – are all needed, or can some be dropped without significantly reducing the accuracy of the fit? Even more basically, the functional form presumed for the curve may be in question.

The approach called *regression analysis* permits such questions to be answered in a direct way. Better yet, we shall see that it permits a "best" experimental program of tests to be defined a priori, such that the estimated parameters of the least-squares fit *will* have minimum error.

14.1 Summation Model for the Physical Effect

Suppose that the data points plotted in Fig. 14.1 are given. These are samples from an unknown, continuous physical law which we want to find. The law is an unknown curve which fits the data in some sense. But because of inevitable noise in the data, and because of their finite spacing, there are an infinity of possible curves which may fit them. At first, we do not even know whether to force the curve to pass through all the points, or to "miss" them by some criterion that promotes smoothness. This is a basic problem faced by all curve-fitting procedures.

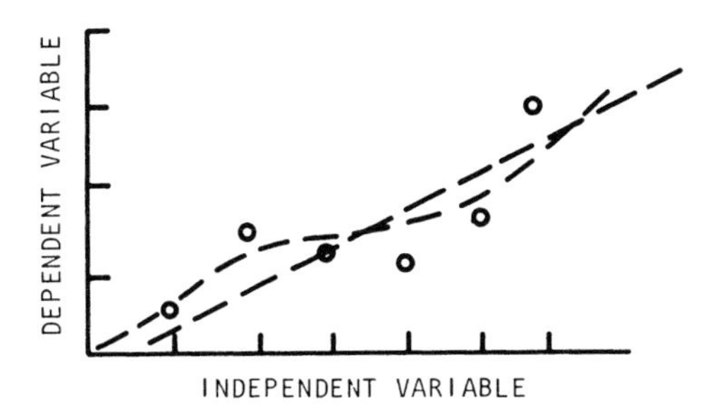

Fig. 14.1. Given the experimental points o, which continuous curve "best" fits them?

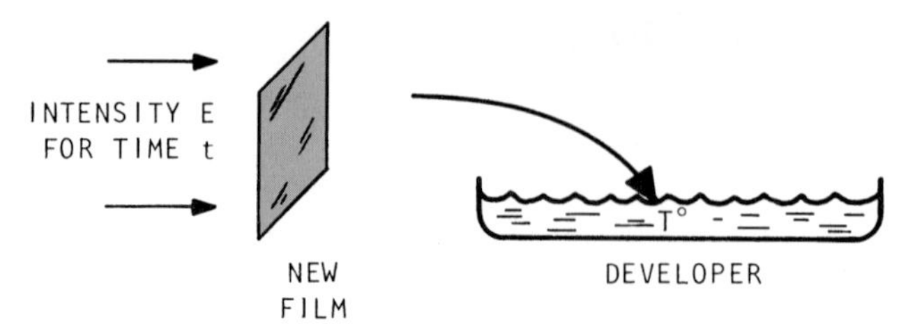

Fig. 14.2. Establishing the sensitometry of a new film. The film is exposed to light intensity E for time t, and then is developed in a solution at temperature T, resulting in a density D. Many values of factors (t, E, T) are so used. How may the dependence of density D upon factors (t, E, T) be found? With what accuracy? Are all the factors significant?

More generally, the problem is multidimensional. Suppose that a new type of film is discovered. We want to find its development characteristics. Hence, an experimental procedure is devised, by which film density D is measured for different values of exposure time t, light level E, and developer temperature T (Fig. 14.2). For example, by testing five different samples of the film, the data in Table 14.1 are obtained. How can a functional relation $D(t, E, T)$ be formed from these data? Experimental variables such as t, E and T are called "factors" in statistics.

Table 14.1. Data for the unknown film

D	t [min]	E [m Cd s]	T [°C]
2.6	18.0	1.0	20.0
2.0	12.0	1.0	21.1
1.6	10.0	0.8	18.8
0.7	8.0	0.1	19.6
0.2	3.0	0.07	20.8

In general, these problems are of the following type: An output or dependent variable y depends upon M independent factors $x_1, \ldots, x_M$ in an unknown way $y(x_1, \ldots, x_M)$. We want to establish $y(x_1, \ldots, x_M)$ by observing N data "points" $(y_n, x_{1n}, \ldots, x_{Mn})$, $n = 1, \ldots, N$, resulting from N repetitions of the experiment (Note: the left-hand member m of each subscript in x_{mn} denotes the factor name x_m, while the right-hand member denotes the experiment number n). See Table 14.2.

Returning to the film problem, the observer might guess that the unknown dependence $D(t, E, T)$ is of an additive or *summation form*

$$D = \alpha_0 + \alpha_1 t + \alpha_2 E^{0.8} + \alpha_3 T^2 + \text{noise}\,, \tag{14.1}$$

where the weights $\{\alpha_n\}$ need to be found. This is the special case where the functional dependence upon factors t, E and T is expressible as a weighted sum of functions of t, E, and T, respectively. Note that the unknown weights or coefficients $\{\alpha_n\}$

Table 14.2. Data in the general case

Experiment No.	y	x_1	x_2		x_M
1	y_1	x_{11}	x_{21}	$\cdots$	x_{M1}
2	y_2	x_{12}	x_{22}	$\cdots$	x_{M2}
$\vdots$	$\vdots$	$\vdots$	$\vdots$	$\cdots$	$\vdots$
N	y_N	x_{1N}	x_{2N}	$\cdots$	x_{MN}

contribute *linearly* to D. This will be the type of model assumed for the physical effect. This model may be greatly generalized, as follows.

The model effect need not be *explicitly* linear in unknown coefficients $\{\alpha_n\}$. The left-hand quantity in (14.1) may instead be *any function* of the data so long as it is computable. Therefore, if the model dependence were initially nonlinear in the $\{a_n\}$, as in

$$D = \alpha_0 \mathrm{e}^{-\alpha_1 t} \cdot E^{\alpha_2} \cdot T^{-\alpha_3} \,,$$

its logarithm could be taken to bring the right-hand side into the summation form (14.1) once again.

As a further generalization some of the terms in (14.1) may be different functions of *the same* factor. For example, note the two terms in factor t, in a model

$$D = \alpha_0 + \alpha_1 t + \alpha_2 t^2 + \alpha_3 E^{0.8} + \alpha_4 T^2 + \text{noise} \,.$$

By estimating α_1 and α_2, an unknown functional dependence of D upon t is estimated, through Taylor series. Obviously, more terms may be added to the series, although the noise limits the extent to which this procedure may be pushed.

In general, then, it will be assumed that the unknown dependence $y(x_1, \ldots, x_M)$ is of a summation form

$$y = \alpha_0 + \alpha_1 f_1(x_1) + \cdots + \alpha_M f_M(x_M) + u \,. \tag{14.2}$$

Quantity y is either the data directly, or some known function of the data. The functions $\{f_m(\cdot)\}$ and factors $\{x_m\}$ are specified. The problem is to find the coefficients $\{\alpha_m\}$ of the model. Quantity u is the inevitable noise that will accompany any measurement y. Its presence will, of course, degrade calculation of the $\{\alpha_m\}$. To proceed further, we need a model for the noise.

14.2 Linear Regression Model for the Noise

Assume the experiment to be repeated N times, with outputs $\{y_n\}$. Then by (14.2), each y_n obeys

$$y_n = \alpha_0 + \alpha_1 f_1(x_{1n}) + \cdots + \alpha_M f_M(x_{Mn}) + u_n \,, \quad n = 1, \ldots, N \,. \tag{14.3}$$

Each noise value u_n may now be regarded as a distinct RV. Accordingly, y_n is also random, and is in fact a *stochastic process*. We shall next define a model for this process. As usual, key parts of the model will be denoted by dots •.

Purely on the grounds of simplicity, at first, let the noise contribute *additively* to the process, and be Gaussian. Hence, regarding

$$\alpha_0 + \alpha_1 f_1(x_{1n}) + \cdots + \alpha_M f_M(x_{Mn}) \equiv y_n^s \tag{14.4a}$$

as the signal or true value of the output, we postulate that

$$y_n = y_n^s + u_n , \quad \text{where} \tag{14.4b}$$

- The noise u_n is independent of the signal y_n^s.
- The noise is zero mean and Gaussian, with variance σ_n^2.

But by the additivity assumption, it is reasonable to expect all variances $\{\sigma_n^2\}$ to be equal.

- Hence, let $\sigma_n^2 = \sigma^2$, all n.

Then, as a consequence of (14.4b):

- The observables $\{y_n\}$ are statistically independent, and are distributed

$$y_n = N(y_n^s, \sigma^2) . \tag{14.4c}$$

These assumptions constitute what is called a *linear regression model.* Observe that it is linear in the coefficients $\{\alpha_m\}$, but not necessarily in the independent factors $\{x_m\}$; e.g., see (14.1). This makes it widely applicable to physical problems.

To further clarify the model, consider again the unknown film problem. Imagine one set of factors t, E and T (e.g., $t = 12.0$ min, $E = 1.0$ m Cd s, $T = 21.1$ °C) to be repeated over and over experimentally. Then the outputs D_n will not be constant, but will randomly vary about a signal level D_n^s. Moreover, the variance in outputs D_n would be the same if any other fixed set of experimental parameters t, E and T were so repeated. Of course, this describes additive noise, as the model prescribes.

The unknowns of the model are the $\{\alpha_m\}$ and σ^2. The latter is of secondary importance, of course; primarily we want the $\{\alpha_m\}$ However, knowledge of σ^2 will allow us to tackle the allied problem of determining which among the $\{\alpha_m\}$ are truly *significant contributors* to the effect y. This question is considered in detail in Sect. 14.6.

How useful is the model? Will it often apply to physical phenomena? We have already discussed the generalized nature of the data $\{y_n\}$ and functions $f_n(\cdot)$. The assumption of additive noise should likewise be discussed. In many physical situations, the measured quantity y_n will indeed suffer from additive noise: examples include cases where (i) y_n is detected light intensity which is bright enough that the major detection error is due to thermal or Johnson noise in the detection circuit; and (ii) y_n is detected so accurately that it only suffers from roundoff error as a form of noise. Important counterexamples also exist, however, notably (i) y_n is detected light intensity where the light is of *low* enough intensity that Poisson fluctuations dominate the detection error; then the noise variance *depends upon* signal level, in fact equalling it (Chap. 6); and (ii) y_n is detected photographic density; then the variance depends upon signal level through the Selwyn root law (6.12). Hence, *the additivity assumption of the regression model must be used with caution.* In particular, the very example of an "unknown film," which we have used to exemplify concepts in this chapter, may violate this very premise! (Although, since it is "unknown" and possibly unconventional, it may not.)

14.3 Equivalence of ML and Least-Squares Solutions

The system of equations (14.3) is usually "tall," $N > M + 1$, where N is the number of equations and $M + 1$ is the number of unknowns $\{\alpha_m\}$. This is because almost invariably there are a lot more data outputs $\{y_n\}$ at hand than there are known (or, suspected) physical causes $\{x_m\}$. It would appear, then, that these equations are overconstrained, and therefore *do not* have a solution $\{\alpha_m\}$ satisfying all of them. But this overlooks the presence of noise values $\{u_n\}$ in the equations. Because these *also* are unknown, there are in total $M + N + 1$ unknowns, compared with N equations. Hence, the system is actually *under* constrained. There are an infinity of possible solutions $\{\hat{\alpha}_m\}$, $\{\hat{u}_n\}$ (one of which is, of course, the *correct* one $\{\alpha_m\}$, $\{u_n\}$). Which candidate $\{\hat{\alpha}_m\}$ should be chosen?

Let us seek the maximum likelihood ML solution for the $\{\alpha_m\}$. By (14.4c), each y_n is normally distributed with mean value y_n^s. Combining this with the independence of the $\{y_n\}$ implies that the joint probability of the $\{y_n\}$ must be a simple product of normal laws

$$p(y_1, \ldots, y_N | \{\alpha_m\}) = (\sqrt{2\pi}\sigma)^{-N} \exp\left[-\frac{1}{2\sigma^2}\sum_{n=1}^{N}(y_n - y_n^s)^2\right] . \tag{14.5}$$

This is conditional upon the $\{\alpha_m\}$, since each y_n^s is a known function (14.4a) of the fixed $\{\alpha_m\}$. Eq. (14.5) is the likelihood law for the problem.

The maximum likelihood principle states that the data $\{y_n\}$ that occurred are the ones that are a priori most likely to occur through choice of the $\{\alpha_m\}$. This choice of solution $\{\hat{\alpha}_m\}$ will then maximize (14.5), or more simply, its logarithm

$$\ln p = -N \ln(\sqrt{2\pi}\sigma) - \frac{1}{2\sigma^2}\sum_{n=1}^{N}(y_n - y_n^s)^2 . \tag{14.6}$$

Note that the first term is a constant, and the second can only decrease $\ln p$. Then the solution $\{\hat{\alpha}_m\}$ obeys

$$\sum_{n=1}^{N}(y_n - \hat{y}_n^s)^2 \equiv e = \text{minimum} , \tag{14.7}$$

where $\hat{y}_n^s$ is the estimate of y_n^s based on the solution $\{\hat{\alpha}_m\}$. This is often called a "least-squares estimate." Explicitly putting in the dependence of $\hat{y}_n^s$ upon the $\{\hat{\alpha}_m\}$ via (14.4a), we obtain

$$e \equiv \sum_{n=1}^{N}\left[y_n - \hat{\alpha}_0 - \sum_{m=1}^{M}\hat{\alpha}_m f_m(x_{mn})\right]^2 = \text{minimum} . \tag{14.8}$$

Hence, *the maximum-likelihood estimate* $\{\hat{\alpha}_m\}$ *is also the least-squares estimate.* Such a correspondence is, of course, not universal. Tracing its roots backward through the derivation, we see that it originated in the model assumption of additive, Gaussian noise. Hence, with this type of noise the least-squares curve is the *most likely* curve to satisfy the data. This gives it some real meaning.

Notice that if the summation assumption (14.2) *really is* correct, the minimum achieved in (14.8) is precisely

$$e = \sum_{n=1}^{N} \hat{u}_n^2 , \tag{14.9}$$

where $\hat{u}_n$ is the estimate of u_n. But this is the total estimated squared noise over all the experiments. By seeking the minimum in (14.8), we are therefore stating that the noise tends to be small and to contribute minimally over many experiments. At least, we found it maximum likely to do so.

14.4 Solution

This is obtained in the usual way, by setting $\partial/\partial\hat{\alpha}_m$, of (14.8) equal to zero for each of $m = 0, 1, \ldots, M$. The mth equation so obtained is

$$\sum_{n=1}^{N} f_m(x_{mn}) \left[y_n - \hat{\alpha}_0 - \sum_{k=1}^{M} \hat{\alpha}_k f_k(x_{kn}) \right] = 0 .$$

This equation is linear in unknowns $\{\hat{\alpha}_k\}$. Moreover, the system of equations is square, so there is now a *unique solution*

$$\hat{\alpha} = ([X^T][X])^{-1}[X]^T y . \tag{14.10}$$

Notation T denotes the transpose. Matrix $[X]$ is formed as

$$[X] \equiv \begin{pmatrix} 1 & f_1(x_{11}) & f_2(x_{21}) & \cdots & f_M(x_{M1}) \\ 1 & f_1(x_{12}) & f_2(x_{22}) & \cdots & f_M(x_{M2}) \\ \vdots & \vdots & \vdots & & \vdots \\ 1 & f_1(x_{1N}) & f_2(x_{2N}) & \cdots & f_M(x_{MN}) \end{pmatrix} \tag{14.11a}$$

and $\boldsymbol{y}$, $\boldsymbol{\alpha}$ are vectors

$$\boldsymbol{y} \equiv (y_1 y_2 \ldots y_N)^T$$
$$\boldsymbol{\alpha} \equiv (\hat{\alpha}_0\, \hat{\alpha}_1 \ldots \hat{\alpha}_M)^T . \tag{14.11b}$$

Computer implementation of the solution (14.10) is quite easy, since many standard algorithms exist for taking the inverse and transpose operations.

14.5 Return to Film Problem

The data in Table 14.1 were input into solution (14.10). The resulting $\{\hat{\alpha}_m\}$ when substituted into model (14.1) predict the dependence

$$\hat{D} = -0.429 + 0.100t + 1.033E^{0.8} + 0.0005T^2 \tag{14.12}$$

for the estimated film density $\hat{D}$. The relation is reasonable regarding the signs of the coefficients, since this correctly implies that D increases with t, E, and T.

14.6 "Significant" Factors; The *R*-Statistic

From the very small size of the term $0.0005T^2$ compared to the other terms in (14.12), $\hat{D}$ (and hence D) may not really depend upon T^2. Or equivalently, the dependence may be *insignificant*. To check out this hypothesis, we need a measure of the strength of the contribution of T^2 to $\hat{D}$. Essentially, we want to test the hypothesis that factor T^2 does not belong in the model dependence (14.12) for $\hat{D}$; or, that $\alpha_3 = 0$. What measurable quantity will be sensitive to its presence in the analysis?

The most direct approach is to simply obtain new data of an appropriate kind: experimentally fix factors t and E while varying only factor T. The response of the outputs D to T may then be directly observed. The *correlation coefficient* between T and D could be computed and used as a measure of significance. In particular, it could be tested against the hypothesis that there is no correlation (*no significance*) (Ex. 14.1.10). This approach would, however, require new data to be taken. The following approach has the advantage of using the *given* data in Table 14.1.

If $\hat{D}$ *really does* depend upon factor T^2 (as well as t and E), then a solution (14.10) which includes α_3 will fit the data $\{D_n\}$ better, in the sense of a smaller residual error (14.8), than if it does not include α_3. The fit with one or more parameters α_n fixed at zero cannot be better than the best unrestricted fit.

This suggests that we first form the solution (14.10) including T^2 and α_3 in the analysis; yielding the minimized residual error e of (14.8). Then again solve the problem, now ignoring factor T^2 (equivalently, setting $\alpha_3 = 0$). This yields a residual error e_1. The difference

$$e_1 - e$$

is then a measure of the worsening of the fit due to ignoring factor T^2. If the difference is big enough, T^2 is indeed a significant factor.

However, $(e_1 - e)$ by itself cannot be the complete test. For, each of e_1 and e tend to be large if the intrinsic noise (14.9) in the data is large. Hence, the difference can also be large on this basis alone, irrespective of the significance of T^2. Now, each noise value u_n of course depends upon the size of σ. Hence, the relation between e_1 and σ, and between e and σ, must be found. If, e.g., it turns out that both e_1 and e are proportional to σ^2, a statistic $(e_1 - e)/\sigma^2$ would be independent of the size of σ, and hence a better measure of significance alone.

Any statistic must use known numbers as inputs. However, often σ will not be known, or will be known too crudely for test purposes. By contrast, the closely related quantity $\hat{\sigma}$, the estimate of σ based upon the data, is always known. Also, it is directly proportional to e, as shown next. We therefore replace our search for the relation between e and σ, with one for the relation between e and $\hat{\sigma}$.

Residual error e is by (14.7) the sum-of-squares of terms $(y_n - \hat{y}_n^s)$. Since each y_n is, by (14.4c), a normal RV with variance σ^2, it would appear from (5.54) that e is chi-square. Indeed, e would be chi-square and with exactly N degrees of freedom (d.o.f.) if the estimated signal values $\hat{y}_n^s$ were fixed numbers. However, instead they depend linearly upon the $\{y_n\}$ through the $(M+1)$ parameters $\{\alpha_m\}$; see (14.4a,b). It results that only $(N-M-1)$ terms $(y_n - \hat{y}_n^s)$ are *linearly* independent. This implies, of course, that not all the N RV's $(y_n - \hat{y}_n^s)$ are *statistically* independent d.o.f.s. But, how many are? Would they also number $(N-M-1)$?

We encountered this kind of problem before, when the probability law for the sample variance S^2 was sought (Sect. 9.10.2). The N factors there were $(c_n - \overline{c})$, with $\overline{c}$ a *single* parameter depending linearly upon the $\{c_n\}$. This resulted in $(N-1)$ d.o.f. In fact, this was an example of a general rule: *The number of d.o.f. in a sum* $\sum z_n^2$, *where each* $z_n = N(0, \sigma^2)$, *equals the number of linearly independent terms among the* $\{z_n\}$. This is a correspondance between the distinct concepts of *linear* independence and *statistical* independence.

By this rule, e is chi-square with $(N-M-1)$ d.o.f. Then, by (5.55c)

$$\langle e \rangle = (N-M-1)\sigma^2 .$$

This shows that if $e/(N-M-1)$ is regarded as an estimate of σ^2, it will be an unbiased estimate. We conclude that $e/(N-M-1)$ is a valid estimate, and so form

$$\hat{\sigma}^2 = e/(N-M-1) . \tag{14.13}$$

We may now return to the main problem, in fact a generalization of it. Suppose that we wish to test *the significance of the last K factors in* $\{x_m\}$; for example, both factors $E^{0.8}$ and T^2 in (14.12). Accordingly, we first form a solution (14.10) allowing all $\{\alpha_m\}$ to be present; this results in a residual error e. Next, form a solution (14.10) ignoring the last K factors in $\{x_m\}$ (operationally, lop the last K columns off the matrix $[X]$, solve for the $\{\hat{\alpha}_m\}$ only out to value $m = M - K$); call the resulting residual error e_1. From (14.13), $e/(N-M-1) \equiv \hat{\sigma}^2$ will be a good estimate of σ^2. It is also reasonable that, *if the last K factors are insignificant*, ignoring them in the solution will produce no better or worse estimate of σ^2. Hence, as well $e_1/(N-M-1+K) \equiv \hat{\sigma}_1^2$ will be a good estimate of σ^2. Cross-multiplying each of the two preceding equations, then taking their difference, gives

$$(e_1 - e)/K \simeq \hat{\sigma}_1^2 ,$$

so that ratio

$$\frac{\hat{\sigma}_1^2}{\hat{\sigma}^2} \simeq \frac{(e_1 - e)/K}{e/(N-M-1)} = \frac{N-M-1}{K}\left(\frac{e_1 - e}{e}\right) \equiv R \tag{14.14}$$

will tend toward unity. Accordingly, *the closer statistic R is to unity*, the more acceptable is the test hypothesis that *the last K factors are insignificant.*

Statistic R is seen to be proportional to the statistic $(e_1 - e)$ proposed at the outset. Division by e has the effect of cancelling out all dependence upon the unknown σ^2. Recall that this was one major requirement of the sought statistic.

Statistic R also much resembles the F-statistic of Chap. 13. In fact, it will have an F-distribution. This is discussed next.

We observe from (14.14) that the statistic R is linear in e_1/e, where e_1 has been shown to be chi-square with $N - M - 1 + K$ d.o.f. and e chi-square with $N - M - 1$ d.o.f. The quotient of two chi-square RV's was shown in Chap. 13 to result in an F-distributed RV. In fact, the same result follows here, although not quite as simply [14.2]: The probability law $p_R(r)$ for R follows an F law (13.7) with K and $(N - M - 1)$ degrees of freedom.

With this known, the significance of the K factors may be tested to any degree of confidence α. If r_0 is the computed value of R by (14.14), then if r_0 is near unity $P(r \geqq r_0)$ should be large. Hence *we accept a hypothesis of insignificance* with confidence α if

$$p(r \geqq r_0) \equiv \int_{r_0}^{\infty} \mathrm{d}r\, p_R(r) \geqq \alpha \,. \tag{14.15}$$

The table of Appendix D may be used for evaluation of this integral. Use $\nu_1 = K$ and $\nu_2 = N - M - 1$, the degrees of freedom for this problem, and $Q = \alpha$.

14.7 Example: Was T^2 an Insignificant Factor?

The preceding analysis arose from a suspicion that the factor T^2 is an insignificant contributor to $\hat{D}$ in (14.12). We found that the answer is to perform a test (14.14, 15) of significance. Here $K = 1$ factor is in question, out of $M + 1 = 4$ total, with $N = 5$ data (Table 14.1). The solution (14.10) including the questionable factor T^2 in $[X]$ yields a value $e = 0.38 \times 10^{-3}$ by (14.8). The solution obtained after dropping the factor yields a value $e_1 = 0.14 \times 10^{-2}$. Hence, by (14.14) $r_0 = 2.79$, far from unity, but is it improbably far? Using Appendix D with $\nu_1 = K = 1$ and $\nu_2 = N - M - 1 = 1$, we find a confidence level $Q = \alpha$ somewhere between 0.25 and 0.5 for the hypothesis that factor T^2 is insignificant. This is not improbable enough to reject the hypothesis. Hence, T^2 probably *is* insignificant.

Does this mean that density D does not depend upon temperature T in any manner? If true, this is actually a much more definitive statement than what was proven. Although factor T^2 was shown to be insignificant, perhaps some other function of T *is* significant. The user must test out all feasible model dependences upon T for significance before he can unequivocally state that D intrinsically does not depend upon T.

Let us now check for the significance of some other factor in the model dependence (14.12), for example $E^{0.8}$. From its appreciable size compared to the other terms in (14.12), it appears that this factor is highly significant. Accordingly, we perform the R-test now upon it. Including factor $E^{0.8}$ in the analysis yields a value $e = 0.38 \times 10^{-3}$; excluding it yields an $e_1 = 0.25$, much higher; it appears that the factor will be significant. The resulting r_0 is a colossal 653.9. By Appendix D a confidence level α lying near 0.025 results. This is strong confidence. Hence, we confidently reject the hypothesis that $E^{0.8}$ is insignificant. It is indeed significant.

14.8 Accuracy of the Estimated Coefficients

An estimate $\hat{\alpha}_m$ must be partly in error, because the data $\{y_n\}$ suffer from additive noise $\{u_n\}$. The variance of this error is found in this section.

14.8.1 Absorptance of an Optical Fiber

To be specific, let us consider the following problem in experimental measurement (given as an M.S. degree project at the Optical Sciences Center of the Univ. of Arizona). The optical fiber of Fig. 14.3 suffers light absorptance at a position x along the fiber which is proportional to the intensity $i(x)$ at x. If the proportionality constant is α (called the absorptance coefficient) it follows that

$$i(x) = i_0 \mathrm{e}^{-\alpha x} . \tag{14.16}$$

Suppose that initial intensity i_0 is unknown. Then the model (14.16) has two unknown parameters, i_0 and α, to be estimated. This model has so far ignored the presence of noise. Let this be of a "multiplicative" form, so that the net model is

$$i(x) = i_0 \mathrm{e}^{-\alpha x} n(x) , \tag{14.17}$$

$n(x)$ the noise at x. *The student was asked to best estimate coefficient α, given a length L of the fiber and a light detector to measure intensity.*

The ground rules were as follows: The fiber could be cut at will, so that $i(x)$ could be measured at any x. In all, N measurements of i could be made at positions $\{x_n\}$. What would be the best values $\{x_n\}$ of length to use, on the basis of smallest expected error in the estimated α? Should they be *uniformly spaced,* for example? Let us use regression analysis to solve this problem (Note: The student did not know regression analysis, and decided on the basis of intuition that uniform spacing is optimal. This turns out to be incorrect[1]).

The model (14.17) is not linear in the unknown α, and therefore does not, at first, obey the regression model (14.2). However, taking the logarithm of (14.17) produces

$$\ln i(x) = \ln i_0 - \alpha x + \ln n(x) . \tag{14.18}$$

This is linear in the unknown α (and in the unknown $\ln i_0$), so we are encouraged to try to put it in the form of a regression model.

If measurements $\{i_n\}$ of intensity are made at the points $\{x_n\}$, (14.18) becomes

$$\ln i_n = \ln i_0 - \alpha x_n + \ln n_n \tag{14.19}$$

[1] But, this story has a happy ending. The student's advisor supposed that the optimum solution *had* uniform spacing. The student got his degree.

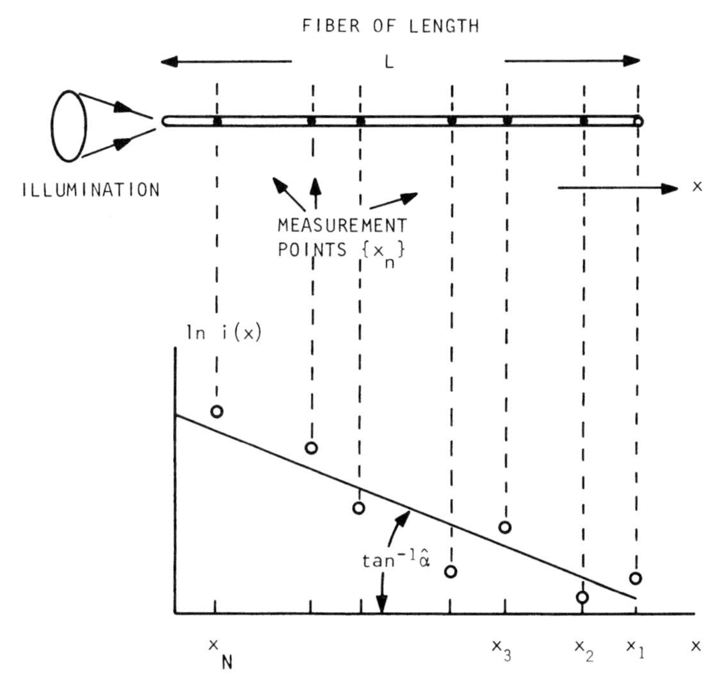

Fig. 14.3. It is desired to ML estimate the absorptance α of an optical fiber. An estimate $\hat{\alpha}$ is based upon measurements of intensity $i(x)$ at positions $x_1, \ldots, x_N$ (in that order). A measurement i_n requires cutting the fiber at point x_n and placing a light detector there. The estimate $\hat{\alpha}$ is the slope of the least-squares straight line through the points $(\ln i_n, x_n)$. What is $\hat{\alpha}$, and what is its error $\sigma^2_{\hat{\alpha}}$? What measurement positions $\{x_n\}$ minimize $\sigma^2_{\hat{\alpha}}$?

where $n_n \equiv n(x_n)$. This equation is now in the form of a linear regression model

$$y_n = \alpha_0 + \alpha_1 f_1(x_n) + u_n \tag{14.20}$$

if the following identifications are made: $\{y_n\} = \{\ln i_n\}$ comprise the data, $\alpha_0 \equiv \ln i_0$, $\alpha_1 \equiv -\alpha$, factor $f_1(x_n) \equiv x_n$, the nth position, and noise $u_n \equiv \ln n_n$. Finally, the noise u_n must be additive and Gaussian. The ramification is that the *physical* noise n_m must be log-normal (its logarithm is normal).

Since the problem is one of linear regression, the solution $\hat{\alpha}_0, \hat{\alpha}_1$ obeys (14.10). But since the log-intensity data $\{y_n\}$ suffer from noise $\{u_n\}$, the solution will be in error. We return to the general problem to find these errors.

14.8.2 Variance of Error in the General Case

This has a very simple answer, which is obtained by straightforward matrix multiplications. Start from solution (14.10), which in terms of a matrix $[S]$ defined as

$$[S] \equiv [X^T][X] \tag{14.21}$$

obeys

$$\hat{\boldsymbol{\alpha}} = [S]^{-1}[X]^T \boldsymbol{y} . \tag{14.22}$$

Take the expectation. Then since $\langle \boldsymbol{y} \rangle = \boldsymbol{y}^s$, by (14.4c), $\langle \hat{\boldsymbol{\alpha}} \rangle = \boldsymbol{\alpha}$ exactly. Hence $\hat{\boldsymbol{\alpha}}$ is an unbiased estimate. Then the variance of *error* in an $\hat{\alpha}_m$, is directly $\sigma^2_{\hat{\alpha}_m}$. We seek this next.

Multiplication of (14.22) on the left by $[S]$ produces

$$[S]\hat{\boldsymbol{\alpha}} = [X]^T \boldsymbol{y} . \tag{14.23}$$

Taking the expectation of both sides produces

$$[S]\langle \hat{\boldsymbol{\alpha}} \rangle = [S]\boldsymbol{\alpha} = [X]^T \boldsymbol{y}^s \tag{14.24}$$

by (14.4c) and the fact that $\boldsymbol{\alpha}$ is unbiased. Subtracting (14.24) from (14.23) yields

$$[S](\hat{\boldsymbol{\alpha}} - \boldsymbol{\alpha}) = [X]^T \boldsymbol{u} \tag{14.25}$$

by (14.4b). Let the elements of $[S] \equiv s_{ij}$, those of $[X]^T \equiv x^T_{ij}$. Then identity (14.25) is explicitly

$$\sum_j s_{ij}(\hat{\alpha}_j - \alpha_j) = \sum_j x^T_{ij} u_j \tag{14.26a}$$

and

$$\sum_{j'} s_{i'j'}(\hat{\alpha}_{j'} - \alpha_{j'}) = \sum_{j'} x^T_{i'j'} u_{j'} \tag{14.26b}$$

for two output equations numbered i and i'. Multiplying the two yields

$$\sum_{j,j'} s_{ij} s_{i'j'} (\hat{\alpha}_j - \alpha_j)(\hat{\alpha}_{j'} - \alpha_{j'}) = \sum_{j,j'} x^T_{ij} x^T_{i'j'} u_j u_{j'} .$$

This is quadratic in the $\{\hat{\alpha}_j\}$, almost in the form of variances $\sigma^2_{\alpha_j}$, as required.

Taking the expectation yields

$$\sum_{j,j'} s_{ij} s_{i',j'} \rho_{\hat{\alpha}}(j, j') = \sum_{j,j'} x^T_{ij} x^T_{i'j'} \rho_u(j, j') \tag{14.27}$$

in terms of autocorrelation coefficients

$$\begin{aligned} \rho_{\hat{\alpha}}(j, j') &\equiv \langle (\hat{\alpha}_j - \alpha_j)(\hat{\alpha}_{j'} - \alpha_{j'}) \rangle , \\ \rho_u(j, j') &\equiv \langle u_j u_{j'} \rangle . \end{aligned} \tag{14.28}$$

We need to solve (14.27) for unknowns $\rho_{\hat{\alpha}}(j, j)$ since these equal $\sigma^2_{\hat{\alpha}_j}$, the quantities we want. Matrix methods come to the rescue.

In view of the additive noise model, we have

$$\rho_u(j, j') \equiv \langle u_j u_{j'} \rangle = \sigma^2 \delta_{jj'} . \tag{14.29}$$

Hence, (14.27) becomes

$$\sum_{j,j'} s_{ij} s_{i'j'} \rho_{\hat{\alpha}}(j, j') = \sigma^2 \sum_j x_{ij}^T x_{i'j}^T . \tag{14.30}$$

In matrix form this is

$$[S][\Gamma_{\hat{\alpha}}][S] = \sigma^2 [S] \tag{14.31}$$

where $[\Gamma_{\hat{\alpha}}]$ is defined as the matrix of coefficients $\rho_{\hat{\alpha}}(j, j')$ and is called the *covariance matrix*. Identity (14.21) was also used.

We want $\sigma^2_{\hat{\alpha}_m}$. But this is simply the mth item down the diagonal of $[\Gamma_{\hat{\alpha}}]$. Hence we want to solve (14.31) for $[\Gamma_{\hat{\alpha}}]$. Multiplying on the left and right by $[S]^{-1}$ yields the solution

$$[\Gamma_{\hat{\alpha}}] = \sigma^2 [S]^{-1} . \tag{14.32}$$

In this way, not only do we know $\sigma^2_{\hat{\alpha}_j}$, but also all cross-dependencies $\rho_{\hat{\alpha}}(j, j')$ among the estimated coefficients. Examples follow.

14.8.3 Error in the Estimated Absorptance of an Optical Fiber

Matrix $[X]$ for this problem is, by (14.11, 19, 20)

$$[X] = \begin{pmatrix} 1 & x_{11} \\ 1 & x_{12} \\ \vdots & \vdots \\ 1 & x_{1N} \end{pmatrix} \tag{14.33}$$

where x_{1n} denotes the nth position x along the fiber. Since there is only one variable x we drop the first subscript in x_{1n} for simplicity. Then by definition (14.21)

$$[S] = \begin{pmatrix} 1 & 1 & 1 & \cdots & 1 \\ x_1 & x_2 & x_3 & \cdots & x_N \end{pmatrix} \begin{pmatrix} 1 & x_1 \\ 1 & x_2 \\ 1 & x_3 \\ \vdots & \vdots \\ 1 & x_N \end{pmatrix} = \begin{pmatrix} N & \sum x_n \\ \sum x_n & \sum x_n^2 \end{pmatrix} , \tag{14.34}$$

all sums over n. We need $[S]^{-1}$ in order to use (14.32):

$$[S]^{-1} = \left[N^{-1} \sum x_n^2 - (N^{-1} \sum x_n)^2 \right]^{-1} \begin{pmatrix} \sum x_n^2/N^2 & -\sum x_n/N^2 \\ -\sum x_n/N^2 & 1/N \end{pmatrix} .$$

Then from (14.32) and the identifications beneath (14.20),

$$\sigma^2_{\widehat{\ln i_0}} = \sigma^2 \frac{\sum x_n^2/N^2}{\sum x_n^2/N - (\sum x_n/N)^2} \quad \text{and} \tag{14.35a}$$

$$\sigma^2_{\hat{\alpha}} = \sigma^2 \frac{1/N}{\sum x_n^2/N - (\sum x_n/N)^2} . \tag{14.35b}$$

The last quantity in particular was sought by this analysis. It may be used for finding what $\sigma_{\hat{\alpha}}$ to expect given a set of measurement positions $\{x_n\}$ along the fiber.

It also can be used in reverse, to define those $\{x_n\}$ which *will* minimize the error $\sigma_{\hat{\alpha}}$. That is, *it can be used for designing an optimized experimental program which will have minimal error in the output.*

For example, let us seek the positions $\{x_n\}$ which minimize the error $\sigma_{\hat{\alpha}}$ in (14.35b). It is equivalent to solve

$$\sum x_n^2/N - \left(\sum x_n/N\right)^2 = \text{maximum} .$$

This is equivalent to requiring

$$\sum_{n=1}^{N}(x_n - \overline{x})^2 P(x_n) = \text{maximum} \tag{14.36}$$

where

$$\overline{x} \equiv N^{-1} \sum_{n=1}^{N} x_n P(x_n)$$

and $P(x_n)$ denotes the relative number of times position x_n is used. Thus, by (14.36) we want to maximize the spatial variance in positions $\{x_n\}$, where all x_n obey $0 \leqq x_n \leqq L$, through choice of an occurrence function $P(x_n)$. This is equivalent to Ex. 3.1.25. Hence the answer is:

$$\begin{aligned} &N/2 \text{ readings } \{x_n\} \text{ at } x_n = 0 \\ &N/2 \text{ readings } \{x_n\} \text{at } x_n = L . \end{aligned} \tag{14.37}$$

That is, *intensity should be measured only at the extreme ends of the fiber* (and not as in Fig. 14.3!). This makes sense intuitively.[2] The estimated α is the estimated slope of the least-squares straight-line through the data points $\{\ln i_n, x_n\}$; and a slope is best defined by points that are maximally far apart.

Note that the answer (14.37) does not necessarily minimize the error $\sigma^2_{\widehat{\ln i_0}}$ in the *other* estimated quantity $\widehat{\ln i_0}$; see (14.35a). Of course, this does not bother us because we only wanted to estimate α. However, in other problems where many parameters are to be estimated, it would be a further complication. Some compromise state of accuracy among the estimates would have to be decided upon, before solving for the optimum $\{x_{mn}\}$.

Exercise 14.1

14.1.1 Suppose that a phenomenon obeys the bizarre model form

$$y = \sin[\alpha_0 \exp(\alpha_1 x_1^2 - \alpha_2/x_2)]^{1/2} .$$

Coefficients $\alpha_0, \alpha_1, \alpha_2$ are to be estimated by varying the experimental factors x_1 and x_2. Despite its very nonlinear form, can this model be transformed into a linear regression model ?

[2] But only after knowing the solution.

14.1.2 The Boltzmann law

$$N_n = N_0 \exp(-E_n/kT) , \quad n = 1, \ldots, N$$

may be used to estimate the Boltzmann constant k, if the occupation numbers $\{N_n\}$ of energy states $\{E_n\}$ are observed. Temperature T is fixed and known, but parameter N_0 is an unknown. What optimum set of $\{E_n\}$ will minimize the error $\sigma_{\hat{k}}$ if all $\{E_n\}$ lie in interval $0 \leqq E_n \leqq E_{\max}$?

14.1.3 Consider the optical fiber problem of Sect. 14.8. What set of positions $\{x_n\}$ will minimize the error $\sigma_{\widehat{\ln i_0}}$ in the estimate of $\ln i_0$?

14.1.4 If the error $\sigma_{\widehat{\ln i_0}}$ is known, what is $\sigma_{\widehat{i_0}}$? (*Hint:* Because of (14.5, 10) the $\{\hat{\alpha}_n\}$ are jointly normal with means $\{\hat{\alpha}_n\}$ and correlation matrix $[\Gamma_{\hat{\alpha}}]$ given by (14.32). Then any single RV $\hat{\alpha}_n$ is normal with mean α_n and variance $\sigma^2_{\hat{\alpha}}$.

14.1.5 ***Weighted least-squares approach.*** Suppose that a phenomenon is of the form (14.4a, b), where each

$$u_n = N(0, \sigma_n^2)$$

and the σ_n^2 are generally *unequal*. That is, the variance in noise u_n depends upon the particular values $\{x_{mn}\}$ for that experiment. Show that the ML solution for the $\{\alpha_n\}$ (Sect. 14.3) now becomes a *weighted* least-squares solution. Intuitively, what do these weights accomplish?

14.1.6 ***Calibration of a radiometer.*** A radiometer has an output voltage y_n due to input exposure x_n obeying a linear relation

$$y_n = \alpha x_n + u_n ,$$

where u_n is noise. It is desired to estimate the sensitivity α of the instrument from a sequence of measurement points $\{x_n, y_n\}$. The ML solution $\hat{\alpha}$ is sought. Assume that the noise u_n follows the linear regression model, with variance σ^2.

(a) Show that $\hat{\alpha} = \sum x_n y_n / \sum x_n^2$. Is this solution as sensitive to errors in small values y_n as it is to large?

(b) Find the dependence of the error $\sigma_{\hat{\alpha}}$ upon exposures $\{x_n\}$. What set of $\{x_n\}$ will minimize $\sigma_{\hat{\alpha}}$ and hence produce the best estimate of α?

(c) Assume now that the noise obeys the Poisson-like property $\sigma_n^2 = b\langle y_n \rangle$, with b constant. Also, the noise has zero mean. The ML solution $\hat{\alpha}$ is again sought (Sect. 14.3). Show that now $\hat{\alpha} = \sum y_n / \sum x_n$, or $\hat{\alpha} = \overline{y}/\overline{x}$ in terms of these sample means. Is this solution more equitable in its sensitivity to large and small values y_n than was (a) ?

14.1.7 ***Atmospheric turbulence.*** In the presence of long-term turbulence (exposure times exceeding 0.01 s), the optical transfer function is (Fig. 8.4)

$$T(\omega) = T_0(\omega) \exp[-3.44(\lambda R\omega/r_0)^{5/3}] .$$

Here $T_0(\omega)$ is the intrinsic transfer function of the lens system in use, λ is the light wavelength, R is the distance through the turbulent atmosphere from object to image, and r_0 is an "equivalent optical pupil size" due to atmospheric seeing.

Assume that all quantities are known, except for r_0 which it is desired to estimate. Show how to cast this problem as one of linear regression; solve for the estimate $\hat{r}_0$ in terms of the given quantities; and derive an expression for the standard deviation in quantity $\hat{r}_0^{-5/3}$. Assuming that this standard deviation is small, find that in $\hat{r}_0$ itself.

(*Hint:* If $y = \hat{r}_0^{-5/3}$, then $dy = (-5/3)\hat{r}_0^{-8/3}\, d\hat{r}_0$.) The use of what set of frequencies $\{\omega_n\}$ in the test program will minimize the uncertainty in $\hat{r}_0$?

14.1.8 ***Image restoration.*** The imaging equation (8.51) for a one-point object (say, a star) located at a known x_0 is

$$i(x_n) = \alpha s(x_n - x_0) + u_n\,, \qquad n = 1, \ldots, N\,,$$

where $i(x_n)$ is the image intensity at a known position point x_n, α is the unknown object value, s is the point spread function and u_n is additive, Gaussian noise. Restoration of the object is then seen to be a one-parameter problem in linear regression.

We want to know how the answer $\hat{\alpha}$ and its estimated error $\sigma_{\hat{\alpha}}$, depend upon the chosen image points $\{x_n\}$. These are often at the user's discretion.

(a) Show that the estimated object obeys

$$\hat{\alpha} = \left[\sum_n i(x_n)s(x_n - x_0)\right] \Big/ \left[\sum_n s(x_n - x_0)^2\right].$$

(b) Show that the variance of error in the estimate $\hat{\alpha}$ obeys

$$\sigma_{\hat{\alpha}}^2 = \frac{\sigma^2}{\sum_n s(x_n - x_0)^2}$$

where σ^2 is the variance in the image noise.

(c) Then what choice of image points $\{x_n\}$ will yield a most-accurate estimate $\hat{\alpha}$? Is this physically realizable? What do the answers to parts (a) and (b) become under these circumstances?

14.1.9 ***Lens design.*** Linear regression analysis allows for the identification of insignificant contributors to an observable. Let that observable be Strehl intensity, defined as the value of the point spread function at its central maximum. Let the contributors (or factors) be the radii and refractive indices of a lens system which forms the point spread function. Not all of these factors will be equally significant. Knowledge of small significance for a lens parameter, e.g., will permit a loose manufacturing tolerance on that parameter, and hence a saving of money.

Suppose a ray-trace program is used to establish the following values of Strehl for the indicated radii $\{r_i\}$ and refractive indices $\{n_i\}$:

Strehl intensity	r_1	n_1	r_2	n_2	r_3
0.95	3.8	1.3	5.1	1.7	2.3
0.90	3.6	1.5	5.4	1.5	2.5
0.88	3.5	1.7	4.6	1.4	2.7
0.71	2.5	1.2	3.6	1.4	2.9
0.63	2.6	1.1	4.7	1.3	3.3
0.50	2.3	1.3	5.1	1.3	3.7
0.42	2.1	1.5	5.0	1.2	3.9

From the behavior in the table, it appears that systematically decreasing r_1 tends to decrease the Strehl. Therefore, r_1 should be significant. Likewise, r_3 appears to systematically affect the Strehl, as does n_2. However, parameters n_1 and r_2 do not show consistent effects upon Strehl.

Test these suspicions by letting

$$\text{Strehl} = \alpha_1 r_1 + \alpha_2 n_1 + \alpha_3 r_2 + \alpha_4 n_2 + \alpha_5 r_3$$

and finding the $\{\alpha_n\}$. Also, establish the significance of each factor.

14.1.10 ***Significance measured by correlation.*** An alternative way of judging the significance of a factor x to an output variable y is through the *measured correlation* r between x and y. While holding the other factors fixed, we observe the outputs $\{y_n\}$ due to the particular values $\{x_n\}$ of factor x, and compute the statistic

$$r \equiv N^{-1}\sum_{n=1}^{N}(x_n - \overline{x})(y_n - \overline{y})/s_1 s_2\,, \quad \overline{x} \equiv N^{-1}\sum_{n=1}^{N} x_n\,, \tag{14.38}$$

$$\overline{y} \equiv N^{-1}\sum_{n=1}^{N} y_n\,, \quad s_1^2 \equiv N^{-1}\sum_{n=1}^{N}(x_n - \overline{x})^2\,, \quad s_2^2 \equiv N^{-1}\sum_{n=1}^{N}(y_n - \overline{y})^2\,.$$

Statistic r is noted to be a sample version of the correlation coefficient ρ defined at (3.47).

Our working principle is that the significance of x to y is defined by the correlation ρ between x and y. If $\rho \neq 0$ there is significance. Therefore *we form a hypothesis, that x and y do not correlate or $\rho = 0$*, and seek to use the computed r as a test of the hypothesis. As usual, we shall need the probability law for the statistic.

Let the $\{x_n\}$ be *chosen* as samples from a normal law. This may be easily done: see Sect. 7.1 on Monte Carlo methods. Then if outputs $\{y_n\}$ do not correlate with the $\{x_n\}$, by the linear regression model values $\{x_n, y_n\}$ will jointly obey a probability law which is a *product* of normal laws.

Under these circumstances, the probability law $p_R(r)$ is uniquely defined by (14.38) [14.2]. Using this known law, the associated statistic

$$t \equiv (N-2)^{1/2}[r^2/(1-r^2)]^{1/2} \tag{14.39}$$

may be found to obey a *Student t-distribution* with $\nu = N - 2$ degrees of freedom. We see from the form of (14.39) that if r is near 1 (suggesting that $\rho > 0$ or there *is* significance) statistic t is large. Thus, the test: Reject the hypothesis of no significance, with confidence α, if t_0 is so large that

$$\int_{r_0}^{\infty} \mathrm{d}r\, p_R(r) = \int_{t_0}^{\infty} \mathrm{d}t\, p_T(t) < \alpha \,. \tag{14.40}$$

Statistic t_0 is the function (14.39) of a value r_0 computed from the data by (14.38).

The test quantitatively works as follows. If r is close to 1, by (14.39) t_0 is very large. Then the integrals in (14.40) are very small. This indicates strong confidence in significance (strong confidence in rejecting the hypothesis of no significance). This makes sense, since if r is close to 1, ϱ is also probably close to 1.

Note that the factor $(N-2)^{1/2}$ in (14.39) brings the effect of data sparsity into play. If r is close to 1, but N is small, the net t_0 can still be small, causing (now) the hypothesis of no significance to instead be accepted. The test is stating that although r was large, we cannot have high confidence that $\varrho > 0$ since computation of r was based only upon a small number N of data.

Apply this test of significance to the following density D versus developer temperature T data (factors t and E were held fixed):

D	T[°C]	D	T[°C]
1.8	21.1	2.3	21.8
2.1	21.5	1.4	19.3
1.6	19.2	0.8	19.0
1.8	20.9	3.0	22.0

With what confidence α can we reject the hypothesis that T was not significant to D?

14.1.11 ***Orthogonalized factors (Honors problem).*** There is an alternative way of solving the least-squares problem of this chapter. It is most easily explained for a particular problem, the unknown film of Sect. 14.1, but may readily be generalized.

Observe from Table 14.1 that the correlation coefficient *between factors t* and E may be computed. Since the factors are *purposeful, known numbers*, the correlation coefficient measures the inadvertent correlation in t and E merely due to the user's choice of numbers. Most likely, this will be nonzero. The same can be said about the model functions $f_1(t)$ and $f_2(E)$ of these factors.

By contrast, it is possible that certain linear combinations of $f_1(t)$, $f_2(E)$ and $f_3(T)$

$$\begin{aligned} \psi_1(t,E,T) &\equiv a_{11} f_1(t) + a_{12} f_2(E) + a_{13} f_3(T) \\ \psi_2 &\equiv a_{21} f_1(t) + a_{22} f_2(E) + a_{23} f_3(T) \\ \psi_3 &\equiv a_{31} f_1(t) + a_{32} f_2(E) + a_{33} f_3(T) \end{aligned} \tag{14.41}$$

will have *zero correlation*, i.e., obey orthogonality

$$\sum_{n=1}^{N} \psi_i(t, E, T)_n \psi_j(t, E, T)_n = \sigma_i^2 \delta_{ij} . \tag{14.42}$$

Notation $(t, E, T)_n$ means that t, E, and T are evaluated at the nth experiment (nth row of numbers in Table 14.1). Functions $\{\psi_i(t, E, T)\}$ are orthogonalized versions of the experimental factors. These may be regarded as more fundamental to the formation of density $D(t, E, T)$ than the dependent factors $f_1(t)$, $f_2(E)$ and $f_3(T)$. In fact, D may be found directly in terms of the $\{\psi_i\}$.

But first, *what coefficients* $\{\alpha_{ij}\}$ *will accomplish orthogonality* (14.42)? Denote

$$(t, E, T) \equiv (x_1, x_2, x_3)$$

and

$$(t, E, T)_n \equiv (x_{1n}, x_{2n}, x_{3n})$$

for the nth experiment. These are all known numbers. Then by this notation the equations of (14.41) are

$$\psi_i(x_{1n}, x_{2n}, x_{3n}) = \sum_{k=1}^{3} a_{ik} f_k(x_{kn}) , \quad i = 1, 2, 3$$

and the requirement (14.42) becomes

$$\sum_{n=1}^{N} \sum_{k=1}^{3} a_{ik} f_k(x_{kn}) \sum_{l=1}^{3} a_{jl} f_l(x_{ln}) = \sigma_i^2 \delta_{ij}$$

$$\sum_{k=1}^{3} a_{ik} \sum_{l=1}^{3} a_{jl} \rho_f(k, l) = \sigma_i^2 \delta ij \tag{14.43}$$

after rearranging the sums, where

$$\varrho_f(k, l) \equiv \sum_{n=1}^{N} f_k(x_{kn}) f_l(x_{ln})$$

is the correlation coefficient between columns in Table 14.1. Let the matrix of coefficients $\rho_f(k, l)$ be denoted as $[\Gamma_f]$.

Comparing the form of (14.43) with that of (14.30), we recognize that it may be cast in the matrix form

$$\boldsymbol{a}_i^T [\Gamma_f] \boldsymbol{a}_j = 0 \quad \text{for} \quad i \neq j . \tag{14.44}$$

Vector $\boldsymbol{a}_i$ has elements $a_{ij} = 1, 2, 3$.

The well-known solution to problem (14.44) is that the $\boldsymbol{a}_i$ satisfy an eigenvalue equation

$$[\Gamma_f] \boldsymbol{a}_i = \lambda_i \boldsymbol{a}_i \tag{14.45a}$$

where

$$\boldsymbol{a}_i^T \boldsymbol{a}_j = \delta_{ij} . \tag{14.45b}$$

Eigenvalues λ_i may be found by noting that

$$\boldsymbol{a}_i^T[\Gamma_f]\boldsymbol{a}_j = \boldsymbol{a}_i^T\lambda_i\boldsymbol{a}_i$$

by (14.45a),

$$= \lambda_i$$

by (14.45b),

$$\equiv \sigma_i^2$$

by (14.43). Hence

$$\lambda_i = \sigma_i^2 \,. \tag{14.46}$$

In summary, *the coefficients needed in* (14.41) *to make* ψ_i *orthogonal are eigenfunctions of the correlation matrix of the experimental factors.* Also, *the* σ_i^2 *are the eigenvalues of the same matrix.*

Suppose, now, that we want to represent the density D *not* directly in terms of the t, E and T but in terms of the *orthogonalized factors* $\psi_1(t, E, T)$, $\psi_2(t, E, T)$, and $\psi_3(t, E, T)$.

(a) Show that if

$$\hat{D}(t, E, T)_n \equiv \sum_{i=1}^{3} b_i\psi_i(t, E, T)_n \,, \tag{14.47}$$

then the coefficients $\{b_i\}$ must be formed as

$$b_i = \sigma_i^{-2}\sum_{n=1}^{N} D(t, E, T)_n\psi_i(t, E, T)_n \,.$$

The simplicity of this solution is the main advantage of orthogonalizing the factors. However, as a practical matter it requires the prior solution to an eigenvalue problem (14.45a). This is no easier to solve numerically than it is to accomplish the matrix inversion required of the ordinary least-squares solution (14.10). "There is no free lunch."

(b) Show that the solution (14.47) is also a least-squares solution, satisfying

$$\sum_{n=1}^{N}\left[D(t, E, T)_n - \sum_{i=1}^{3} b_i\psi_i(t, E, T)_n\right]^2 = \text{minimum} \tag{14.48}$$

through choice of the $\{b_i\}$.

(c) Is the solution (14.47) for $\hat{D}$ in terms of orthogonalized factors the same as (14.10) in terms of the original factors? Prove that it is. *Hint:* Consider first a case $M = 2$. Note that

$$\begin{aligned} &b_1[a_{11}f_1(x_{1n}) + a_{12}f_2(x_{2n})] + b_2[a_{21}f_1(x_{1n}) + a_{22}f_2(x_{2n})] \\ &= (b_1a_{11} + b_2a_{21})f_1(x_{1n}) + (b_1a_{12} + b_2a_{22})f_2(x_{2n}) \end{aligned}$$

so that (14.48) goes into the form (14.8).

In the main, this development is of value as a bridge between the least-squares problem and the principal components problem of the next chapter. Whereas the former orthogonalizes the factors describing one set of data points, the latter orthogonalizes many sets of data points. It will become evident in the next chapter that the same eigenvalue problem results in either case.

15. Principal Components Analysis

A concept that is closely related to linear regression (preceding chapter) is *principal components* [15.1]. Linear regression addressed the question of how to fit a curve to one set of data, using a minimum number of factors. By contrast, *the principal components problem asks how to fit many sets of data with a minimum number of curves*. The problem is now of higher dimensionality. Specifically, can each of the data sets be represented as a weighted sum of a "best" set of curves? Each curve is called a "principal component" of the data sets.

As usual, the manner by which "best" is defined is crucial in defining the answer. By "best" we shall mean orthogonal, as in linear regression by orthogonalized factors (Ex. 14.1.11). Since the latter problem and principal components also share analogous aims, it should not be surprising that they will have the same kind of mathematical solution. This is the diagonalization of a covariance matrix.

These points will be clarified below, as usual by recourse to a specific problem in optics.

15.1 A Photographic Problem

Let us consider how to best represent a set of (say) 101 photographic Hurter-Driffield (H–D) curves. An H–D curve of an emulsion is a plot of its photographic density D response to log exposure $\log E$ values. For convenience, let $\log E = x$. A typical H–D curve is shown in Fig. 15.1 (see also Ex. 6.1.25). These curves do not vary that much, at least qualitatively, from one emulsion to the next. It seems reasonable, then, to seek a *smallest set* of H–D curves that would incorporate among them all the essential features of the given 101 curves. Orthogonal curves have such properties. Thus the problem:

From a test sample $D_j(x)$, $j = 1, \ldots, J$ of H–D curves for $J = 101$ different types of film, can a much smaller set of orthogonal H–D curves $\hat{D}_k(x), k = 1, \ldots, K$, $K \ll J$, somehow be formed?

Toward this end, let us assume a subdivision x_n, $n = 1, \ldots, N$, of points to adequately define each curve. Then, form a sample mean (*across* the sample)

$$\overline{D}_j \equiv N^{-1} \sum_{n=1}^{N} D_j(x_n) \quad \text{for} \quad j = 1, \ldots, J \,, \tag{15.1}$$

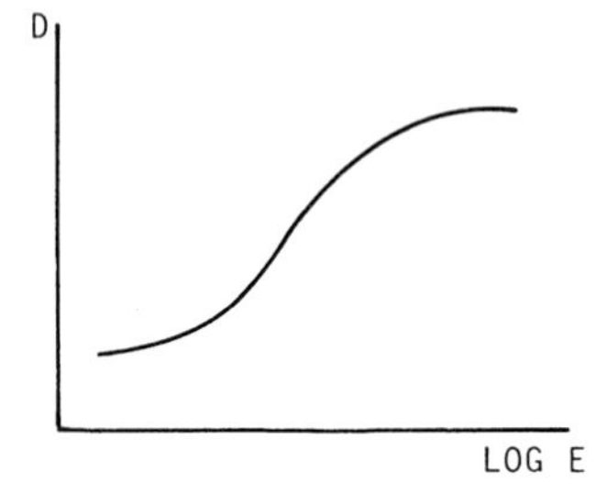

Fig. 15.1. An H–D curve for an emulsion typically has a low response region on the left, followed by a rise region of nearly constant slope, and then a saturation region to the right

and a sample covariance matrix

$$\overline{C}(i, j) \equiv N^{-1} \sum_{n=1}^{N} D_i(x_n) D_j(x_n) - \overline{D}_i \overline{D}_j \tag{15.2}$$

for all i and j. (Sample averages will always be denoted as barred quantities.) These comprise the data for the problem.

15.2 Equivalent Eigenvalue Problem

Next, define the principal components $\{\hat{D}_k(x)\}$ to be a certain linear combination of the data curves $\{D_j(x)\}$, as in

$$\hat{D}_k(x_n) \equiv \sum_{j=1}^{J} u_k(j) D_j(x_n) \,. \tag{15.3a}$$

For later use, summing n produces

$$\overline{\hat{D}_k} = \sum_{j=1}^{J} u_k(j) \overline{D}_j \,. \tag{15.3b}$$

Coefficients $\{u_k(j)\}$ are fixed by the requirement that *the* $\{\hat{D}_k(x)\}$ *curves are to be orthogonal.* This is again in the sense of a sample average

$$\overline{\hat{D}_k \hat{D}_l} \equiv \overline{\hat{D}_k}\,\overline{\hat{D}_l} \text{ for } k \neq l \tag{15.4}$$

where

$$\overline{\hat{D}_k \hat{D}_l} \equiv N^{-1} \sum_{n=1}^{N} \hat{D}_k(x_n) \hat{D}_l(x_n) \tag{15.5}$$

and

$$\overline{\hat{D}_k} \equiv N^{-1} \sum_{n=1}^{N} \hat{D}_k(x_n) \,.$$

Note that orthogonality (15.4) also means zero sample correlation.

Next, we bring the unknowns $\{u_k(j)\}$ into the requirement (15.4), (15.5) by the use of (15.3a). After a rearrangement of summations,

$$\overline{\hat{D}_k \hat{D}_l} = \sum_i \sum_j u_k(i) u_l(j) N^{-1} \sum_n D_i(x_n) D_j(x_n) . \tag{15.6}$$

Next, by definition (15.2),

$$\overline{\hat{D}_k \hat{D}_l} = \sum_i \sum_j u_k(i) u_l(j) [\overline{C}(i, j) + \overline{D}_i \overline{D}_j] .$$

And by identity (15.3b),

$$\overline{\hat{D}_k \hat{D}_l} = \sum_i \sum_j u_k(i) u_l(j) \overline{C}(i, j) + \overline{\hat{D}}_k \overline{\hat{D}}_l . \tag{15.7}$$

Compare this with the orthogonally requirement (15.4). The $\{u_j(i)\}$ must obey

$$\sum_i \sum_j u_k(i) u_l(j) \overline{C}(i, j) = 0 \quad \text{for} \quad k \neq l . \tag{15.8}$$

In vector notation, this is a requirement that

$$\boldsymbol{u}_k^T [\overline{C}] \boldsymbol{u}_l = 0 \text{ for } k \neq l , \tag{15.9}$$

T denoting transpose.

The well known solution to this problem is that u_k must satisfy the eigenvalue equation

$$[\overline{C}] \boldsymbol{u}_k = \overline{\sigma}_k^2 \boldsymbol{u}_k , \tag{15.10a}$$

where

$$\boldsymbol{u}_k^T \boldsymbol{u}_l = \delta_{kl} . \tag{15.10b}$$

Hence, *the required $u_k(i)$ are the orthogonal eigenvectors $\boldsymbol{u}_k$ of the sample covariance matrix $[\overline{C}]$*. There are J such eigenvectors in all.

For later use, (15.10a, b) in summation form are

$$\sum_i \overline{C}(i, j) u_k(i) = \overline{\sigma}_k^2 u_k(j) \tag{15.11a}$$

and

$$\sum_i u_k(i) u_l(i) = \delta_{kl} \tag{15.11b}$$

respectively. Also, using (15.11a, b) and definition (15.5) in (15.7) yields

$$N^{-1}\sum_n \hat{D}_k(x_n)\hat{D}_l(x_n) = \overline{\sigma}_k^2\delta_{kl} + \overline{\hat{D}_k}\,\overline{\hat{D}_l}\,, \quad \text{all } k, l\,. \tag{15.12}$$

15.3 The Eigenvalues as Sample Variances

The eigenvalue equation (15.10a) will always have real, positive eigenvalues $\overline{\sigma}_k^2$, $(k = 1, \ldots, J)$ because by (15.2) matrix $[\overline{C}]$ is symmetric and otherwise well behaved. These eigenvalues are usually numbered in order of descending size, with

$$\overline{\sigma}_1^2 \geqq \overline{\sigma}_2^2 \geqq \cdots \geqq \overline{\sigma}_J^2 \geqq 0\,. \tag{15.13}$$

But, what significance do they have *vis-a-vis* principal components $\{\hat{D}_k(x)\}$? This may be found by use of Eq. (15.12) with (now) $k = l$,

$$\overline{\sigma}_k^2 = \overline{\hat{D}_k^2} - (\overline{\hat{D}_k})^2\,. \tag{15.14}$$

Or an *eigenvalue* $\overline{\sigma}_k^2$ *represents the sample variance of fluctuation across the* kth *principal component.* Hence, because of the ordering property (15.13), the first principal component will be highly fluctuating, the second less so, etc., until a principal component appears that has negligible fluctuation. This defines (effectively) the last principal component K. Usually $K \ll J$, the original number of data curves, representing a great compression of data. This is clarified in Sect. 15.5.

15.4 The Data in Terms of Principal Components

It is possible to express any data curve $D_j(x)$ in terms of the principal components $\{\hat{D}_k(x)\}$. Let

$$D_j(x_n) = \overline{D}_j + \sum_{k=1}^{J} a_k(j)[\hat{D}_k(x_n) - \overline{\hat{D}_k}]\,, \tag{15.15}$$

coefficients $\{a_k\}$ to be determined. These must be dependent upon the particular curve j to be fit, and so are denoted $\{a_k(j)\}$. Transposing $\overline{D}_j$ to the left side, multiplying by $N^{-1}[\hat{D}_i(x_n) - \overline{\hat{D}_i}]$, and summing over n yields

$$\begin{aligned} N^{-1} &\sum_n [\hat{D}_i(x_n) - \overline{\hat{D}_i}][D_j(x_n) - \overline{D}_j] \\ &= \sum_k a_k(j) N^{-1} \sum_n [\hat{D}_i(x_n) - \overline{\hat{D}_i}][\hat{D}_k(x_n) - \overline{\hat{D}_k}] \\ &= \sum_k a_k(j)[\overline{\sigma}_k^2\delta_{ik} + \overline{\hat{D}_i}\,\overline{\hat{D}_k} - \overline{\hat{D}_i}\,\overline{\hat{D}_k} - \overline{\hat{D}_i}\,\overline{\hat{D}_k} + \overline{\hat{D}_i}\,\overline{\hat{D}_k}] \end{aligned}$$

by (15.12) and definition of the sample (barred) averages. Then

$$a_i(j) = \overline{\sigma}_k^{-2}(\overline{\hat{D}_i D_j} - \overline{\hat{D}_i}\,\overline{D_j}) . \tag{15.16}$$

This answer will also *least-squares fit* the right-hand expansion (15.15) to the left-hand side.

The coefficient $a_i(j)$ is seen from (15.16) to measure the correlation between the data curve D_j to be fit and the ith principal component $\hat{D}_i$. If they are independent, $a_i(j) = 0$. We now consider how many terms in series (15.15) need be used in practice.

An alternative to (15.15) exists for expressing the data in terms of the principal components. See Ex. 15.1.2.

15.5 Reduction in Data Dimensionality

In practice, terms in (15.15) beyond a small order $k = K$ may be ignored, $K \ll J$. The reasoning is as follows. Because of the descending values (15.13) of standard deviation, once an order $k = K$ is reached, the corresponding $\hat{D}_k$ is nearly constant, obeying

$$\hat{D}_k(x_n) = \overline{\hat{D}_k} + \varepsilon(x_n) , \quad \varepsilon(x_n) \text{ small} . \tag{15.17}$$

Then the error $e(x_n)$ due to ignoring this term in (15.15) is simply

$$e(x_n) = a_K(j)\varepsilon(x_n) . \tag{15.18}$$

Combining this result with (15.16) yields

$$e(x_n) = \frac{N^{-1}\sum_n[\overline{\hat{D}_K} + \varepsilon(x_n)]D_j(x_n) - \overline{\hat{D}_K}\,\overline{D_j}}{N^{-1}\sum_n[\overline{\hat{D}_K} + \varepsilon(x_n)]^2 - (\overline{\hat{D}_K})^2}\varepsilon(x_n) \tag{15.19}$$

by virtue of (15.14) and (15.17), or

$$e(x_n) = \frac{N^{-1}\sum_n \varepsilon(x_n)D_j(x_n)}{2N^{-1}\overline{\hat{D}_K}\sum_n \varepsilon(x_n)}\varepsilon(x_n) \tag{15.20}$$

after some cancellation, and ignoring second-order terms $\varepsilon(x_n)^2$ in the denominator. Since the numerator is second-order in $\varepsilon(x_n)$ while the denominator is first-order, the quotient is vanishingly small.

Hence, principal components analysis allows a very compact series representation for a given set $\{D_j(x_n)\}$ of data. *The required number K of terms equals the order number of the first negligible variance $\overline{\sigma}_k^2$ or eigenvalue.* In practice, this is usually very small, the order of 10 or less, drastically reducing the dimensionality of the data.

15.6 Return to the H–D Problem

For example, consider the photographic H–D case. *Simonds* [15.2] found that value $K = 4$ sufficed to describe all 101 H–D curves in the test sample! This is a highly useful result. Originally, a given film j was specified by 10 density numbers $D_j(x_n)$, $n = 1, \ldots, 10$. By contrast, use of series (15.15) would permit it to be specified by $K = 4$ coefficients $a_k(j)$, $k = 1, \ldots, 4$. This means that *in practice a roll of film need have only four numbers printed upon its box in order to completely specify its* H–D *development characteristics.*

The principal component curves $\hat{D}_k(x)$, $(k = 1, \ldots, 4)$ should permit high variability among their superpositions. Only in this way could they stand a chance of fitting the entire sample of 101 data H–D curves. For example, Fig. 15.2 shows the range of H–D curves that were generated by the use of coefficients $\{a_k\}$ listed beneath it.

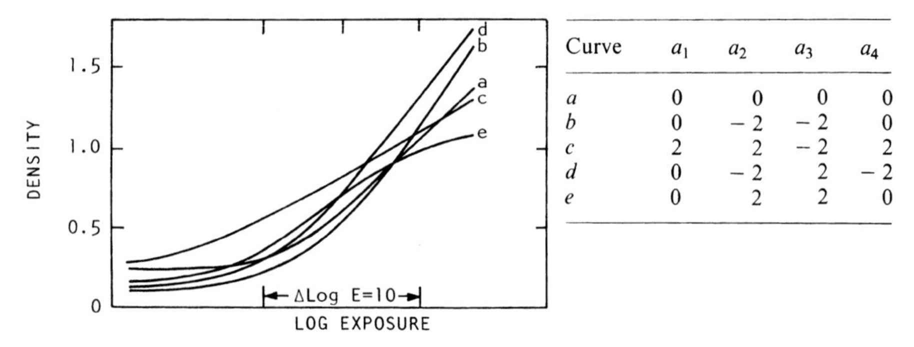

Curve	a_1	a_2	a_3	a_4
a	0	0	0	0
b	0	− 2	− 2	0
c	2	2	− 2	2
d	0	− 2	2	− 2
e	0	2	2	0

Fig. 15.2. Five H–D curves generated by adding the principal component H–D curves together in various amounts, as listed to the right

15.7 Application to Multispectral Imagery

The proceeding analysis has taken its sample averages across one-dimensional curves. Two-dimensional images may be handled in a completely analogous way, the averages now going across *and down* each image [15.3]. For example, consider the case where images of the Earth's surface are taken in different spectral windows. Each image is of the same ground area, but has different intensity values at corresponding points because of the wavelength dependence of the reflectance and emittance of plants and soils. For example, Fig. 15.3 shows six spectral images of the same farmland in Indiana. Although these images show important differences (vegetation appears dark in visible wavelengths but bright in the infrared), they share similarities as well: for example, all share a common geometry of boundaries.

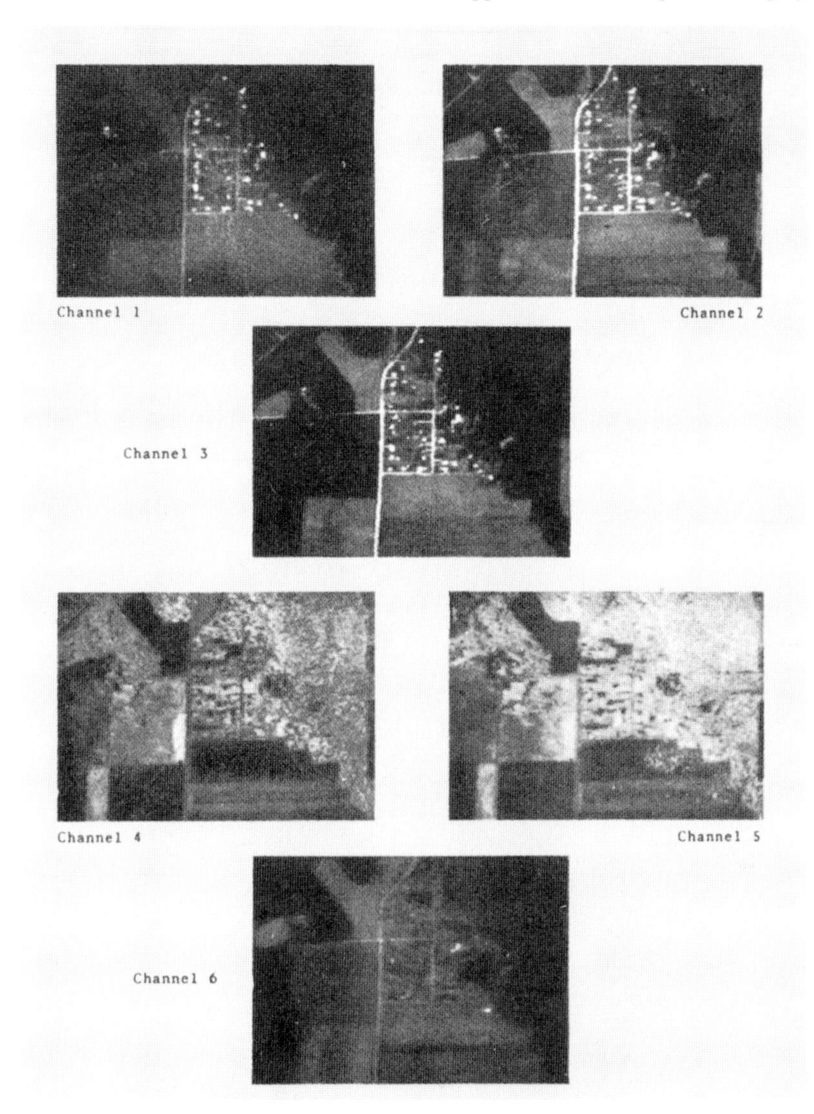

Fig. 15.3. Six spectral images from an airborne camera [15.3]

Hence, we might expect that a principal components analysis of the six will uncover a smaller number than 6 of truly independent images.

A principal components analysis was carried through. The eigenvalues or sample variances were $\overline{\sigma}_k^2 = 3210, 931.4, 118.5, 83.88, 64.00$, and 13.40. This shows that the first principal component will have much more variation in intensity (that is, more features) than the second, the second much more than the third, etc. The last three are so small as to be ignorable for most applications.

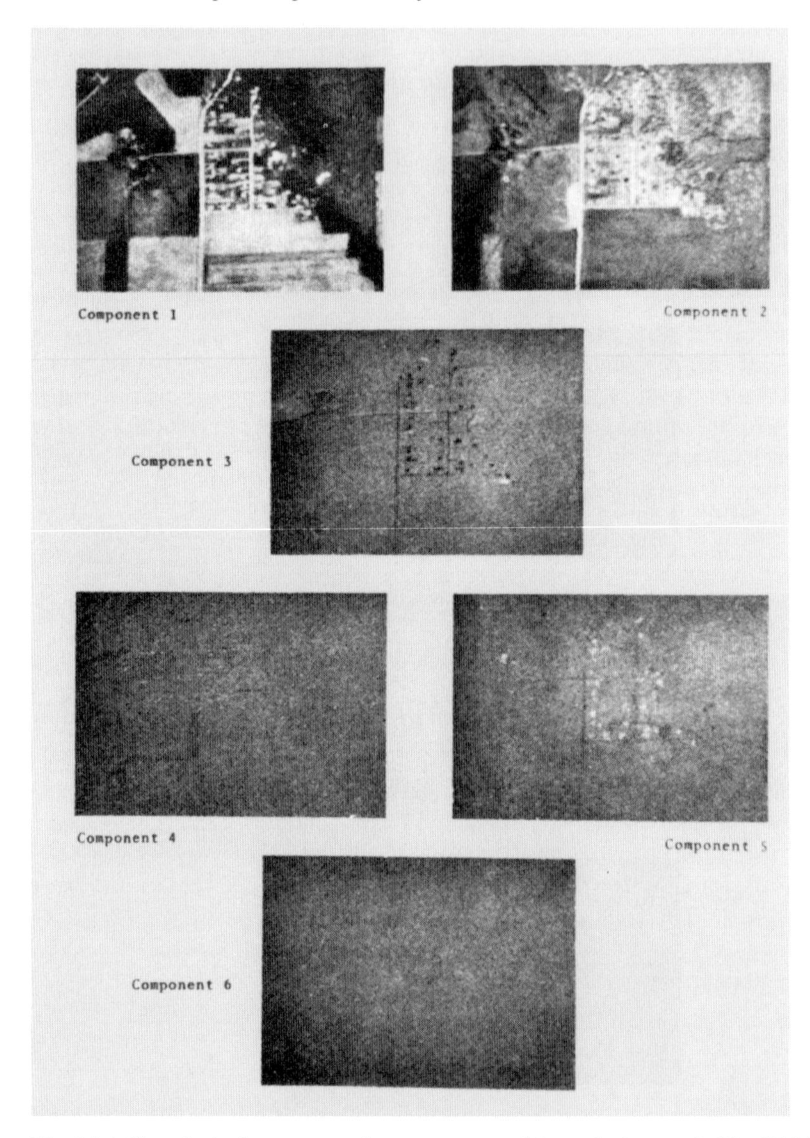

Fig. 15.4. Six principal component images computed from the images in Fig. 15.3 [15.3]

The principal component images are shown in Fig. 15.4. As was expected from the variances, the first two components are much richer in features than the remaining four. It would appear that the effective number of components is here $K = 2$ or 3.

What use can be made of these two or three principal components? First of all, for purposes of feature extraction it is easier and quicker to work with three images than with six. Also, any of the six candidate wavelength-images could be reconstructed as a linear superposition (15.15) over the components. But, even further, perhaps *in-between wavelengths* could likewise have their images of the same ground area so constructed, each with its own coefficients. This would suppose the original six discrete wavelengths to form a *basis set for the entire continuum* of frequencies emitted by the scene. This property would lend even more importance to the two or three principal component images. Its presence is obviously dependent upon slow and smooth changes, between spectral windows, of the intensity at corresponding pixels in the scene.

15.8 Error Analysis

There will inevitably be error in the inputs $\overline{C}(i, j)$ to the eigenvalue problem (15.10a) since the $\overline{C}(i, j)$ are computed via (15.2) from *data* curves $\{D_i(x)\}$. The outputs of the problem, eigenvalues

$$\overline{\sigma}_k^2 \equiv \lambda_k \tag{15.21}$$

and eigenvectors $\boldsymbol{u}_k$, will therefore suffer from error. We shall pay particular attention to the errors in the eigenvalues. Errors in the eigenvectors are discussed in [15.4].

For simplicity, it shall be assumed that errors $\Delta C(i, j)$ are differentially small. Then, if all errors $\Delta C(i, j)$ are simultaneously present, how will they contribute to the resulting error $\Delta\lambda_k$ in any one of the eigenvalues? At first, this seems a formidable problem, since λ_k is formed from *all* the $\overline{C}(i, j)$ as the kth root of a polynomial equation

$$\det \begin{bmatrix} \overline{C}(1,1)-\lambda_k & \overline{C}(1,2) & \cdots & \overline{C}(1,J) \\ \overline{C}(2,1) & \overline{C}(2,2)-\lambda_k & \cdots & \overline{C}(2,J) \\ \vdots & \vdots & \vdots & \vdots \\ \overline{C}(J,1) & \overline{C}(J,2) & \cdots & C(J,J)-\lambda_k \end{bmatrix} = 0\,. \tag{15.22}$$

(This, of course, is the requirement that the homogeneous set of equations has nontrivial roots.) The form of (15.22) shows that error $\Delta\lambda_k$ must be a function *of all* the input errors $\Delta\overline{C}(i, j)$. The problem seems very complicated. But consider the following alternative argument.

Because of identity (15.21) we are really searching for the error in a sample variance $\overline{\sigma}_k^2$. By (15.12), the latter is equivalently formed from N input values $\hat{D}_k(x_n)$, $n = 1, \ldots, N$. And we found in Sect. 9.10.2 that the error in a sample variance based on N inputs (that are independently normal) is proportional to the true variance; see (9.51). The latter would translate into an answer

$$\sigma_{\lambda_k} = \left(\frac{2}{N-1}\right)^{1/2} \sigma_k^2 \tag{15.23}$$

to our problem, where σ_k^2 is the *true* variance of the kth principal component $\hat{D}_k$. If we now assume that $\overline{\sigma}_k^2 = \sigma_k^2 + 0(1/N^{1/2})$, which is consistent with (15.23), then to lowest order in $(1/N)$ the latter becomes

$$\sigma_{\lambda_k} = \left(\frac{2}{N-1}\right)^{1/2} \overline{\sigma}_k^2 ,$$

or

$$\sigma_{\lambda_k} = \left(\frac{2}{N-1}\right)^{1/2} \lambda_k \tag{15.24}$$

by (15.21). This simple and useful answer has been found by heuristic means. A real proof of (15.24) due to *Girshick* [15.5] follows.

Imagine errors $\mathrm{d}\overline{C}(i, j)$ of differential size to be present. The differential of (15.11a) shows that these cause errors $\mathrm{d}u_k(i)$ and $\mathrm{d}\lambda_k$ in the eigenvectors and eigenvalues, respectively, obeying

$$\sum_i \overline{C}(i, j)\,\mathrm{d}u_k(i) + \sum_i u_k(i)\,\mathrm{d}\overline{C}(i, j) = \lambda_k\,\mathrm{d}u_k(j) + u_k(j)\,\mathrm{d}\lambda_k . \tag{15.25}$$

Multiplying this by $u_m(j)$ and summing over j produces simply

$$\mathrm{d}\lambda_k = \sum_{i,j} u_k(i)u_k(j)\,\mathrm{d}\overline{C}(i, j) \tag{15.26}$$

after using (15.11a, b), setting $m = k$, and noticing that the lefthand-most terms cancel from each side of the equation. Remarkably, the unknown errors $\mathrm{d}u_k(j)$ have dropped out.

However, what we want is $\langle \mathrm{d}\lambda_k\,\mathrm{d}\lambda_p \rangle$. Accordingly, we take the version of (15.26) with index k replaced by p, and multiply it by (15.26). After taking expectations,

$$\left\langle \mathrm{d}\lambda_k\,\mathrm{d}\lambda_p \right\rangle = \sum_{i,j}\sum_{i',j'} u_k(i)u_k(j)u_p(i')u_p(j') \left\langle \mathrm{d}\overline{C}(i, j)\,\mathrm{d}\overline{C}(i', j') \right\rangle . \tag{15.27}$$

To proceed further, we have to know the statistics of changes $\mathrm{d}\overline{C}(i, j)$. Assume that these are multivariate normal, as at (3.48). Under these conditions, it can be shown [15.4] that

$$\left\langle \mathrm{d}\overline{C}(i, j)\,\mathrm{d}\overline{C}(i', j') \right\rangle = (N-1)^{-1}[\overline{C}(i, i')\overline{C}(j, j') + \overline{C}(i, j')\overline{C}(j, i')] . \tag{15.28}$$

[Note the particular case $i = i' = j = j'$, which agrees with the expression (9.51) for the error in a sample variance.]

The answer is obtained by substituting (15.28) into (15.27), and using the identity

$$\sum_{i,j} u_k(i)u_p(j)\overline{C}(i, j) = \lambda_k\delta_{pk} . \tag{15.29}$$

[This identity is obtained by combining Eqs. (15.11a, b).] The use of (15.29) collapses four sums into terms $\lambda_k\delta_{pk}$, giving an answer

$$\left\langle \mathrm{d}\lambda_k\,\mathrm{d}\lambda_p \right\rangle = 0 \quad \text{for} \quad k \neq p , \tag{15.30a}$$

$$\left\langle \mathrm{d}\lambda_k^2 \right\rangle = 2(N-1)^{-1}\lambda_k^2 . \tag{15.30b}$$

The latter is identical to the heuristic result (15.24).

Hence, in summary, if the noise in data $\{D_j(x_n)\}$ is such that the resulting covariance values $\overline{C}(i, j)$ obey multivariate normal statistics, and if the noise is small enough, then:

(i) the resulting errors in eigenvalues $\{\lambda_k\}$ do not correlate, and
(ii) the standard deviation of error in an eigenvalue obeys relation (15.24). Note that the latter defines a relative error e, which obeys

$$e \equiv \sigma_{\lambda_k}/\lambda_k = \left(\frac{2}{N-1}\right)^{1/2}. \tag{15.31}$$

Hence, *the relative error in an eigenvalue is purely a function of the number of samples across each given curve* $D_j(x_n)$. This is a very useful result, since it does not require explicit knowledge of either the $\overline{C}(i, j)$ values or their uncertainties.

Exercise 15.1

15.1.1 Two sensitometry curves $D_j(x)$, $j = 1, 2$, $x = \ln E$, are obtained:

x	D_1	D_2
0.0	2.2	1.9
1.0	2.6	1.5
2.0	2.1	1.8
3.0	3.0	2.5
4.0	4.0	4.0
5.0	5.1	6.0
6.0	5.5	8.0
7.0	5.8	8.5
8.0	6.0	8.6

(a) Find the associated 2×2 covariance matrix for the D's.
(b) Setting its determinant equal to zero, find roots λ_j, $j = 1, 2$.
(c) Find its eigenvectors $u_k(j)$, $k = 1, 2$; $j = 1, 2$.
(d) Find its principal components $\hat{D}_j(x_n)$, $j = 1, 2$; $n = 1, \ldots, 9$. Plot these, and compare with the plotted data $D_j(x)$, $j = 1, 2$.
(e) Using (15.14), find the sample variance across each principal component. Do these equal the roots λ_1, λ_2 previously obtained?
(f) Estimate the relative error in each sample variance.
(g) Perturb the covariance matrix (preserving its symmetry about the diagonal) by small numbers. Solve again for the roots. Do the changes in the roots agree with the estimated relative errors in part (f)?

15.1.2 There is a simple alternative to (15.15) for expressing the data in terms of the principal components. In (15.3a) *regard* x_n *as a fixed number*. Show that (15.3a) can then be represented as a matrix equation $\hat{\boldsymbol{D}} = [\boldsymbol{u}]\boldsymbol{D}$ where matrix $[\boldsymbol{u}]$ has a general element $u_k(j)$ at (row, column) position (k, j). Then multiplying on the left by the inverse matrix $[\boldsymbol{u}]^{-1}$ gives $[\boldsymbol{u}]^{-1}\hat{\boldsymbol{D}} = \boldsymbol{D}$.

Using the fact that $[\boldsymbol{u}]$ is unitary, obeying $[\boldsymbol{u}]^{-1} = [\boldsymbol{u}]^T$, the transpose, show that each datum $D_k(x_n)$ may be reconstructed from the principal components $\hat{D}_j(x_n)$ as

$$D_k(x_n) = \sum_j u_k(j)\hat{D}_j(x_n) \,. \tag{15.32}$$

16. The Controversy Between Bayesians and Classicists

A mathematical science ordinarily admits of two (or more) possible routes to a solution. This is also true in statistics, with one notable exception. There are *two fundamentally different approaches* to solving certain statistical problems. These are the Bayesian and Classical[1] approaches. Both camps have strong and vocal partisans, and both approaches give correct answers within their frameworks. As is usual in great controversies, the argument arises out of a difference in point of view.

The point of view regards what can be legitimately assumed about an unknown probability law that gave rise to given data. The Classical view is that *nothing should be assumed about that law.* Any analysis should be based entirely upon the given data. The answer should be "distribution free," that is, free of any assumption about the unknown distribution. This is a completely objective point of view. Indeed, if the answer can be found in this way the Bayesians agree with this point of view. There is no argument.

However, what should be done when formulating an answer requires use of the unknown probability law? The Classical view is to give up on that statement of the problem, perhaps replacing it with a soluble one having a less-ambitious goal. The Bayesian view, by contrast, is *to permit a subjective estimate of the unknown probability law.* Any "reasonable" estimate of the unknown law, called now the "prior probability," is okay. For example, a maximum entropy estimate (Chap. 10) is within the realm of possibilities. This viewpoint is justified, say the Bayesians, on the grounds that (a) the overall method of science is empirical, and not deductive, so that (b) an investigator will always want to justify *his own* hypothesis on the basis of the data. Therefore, (c) he should use, along with the data, any relevant opinions he had *before* acquiring the data. For example, if $p_0(x)$ is the unknown prior probability, and if the investigator knows that x lies in interval (a, b) but knows nothing else about its fluctuation, he might well take the viewpoint of maximum entropy that $p_0(x)$ is uniform over (a, b) (Ex. 10.1.6).

As further justification for this approach, it often happens that the Bayesian answer to the problem depends only weakly, if at all, upon the choice of the prior. Such will be the case in an illustrative example below. This usually happens, though, only when the number N of data are very large. As *Weber* [16.1] stated, "when the empirical evidence is compelling, a priori personal probability is irrelevant...."

[1] The reader may already surmise a bias on the part of this author, and indeed all others who describe the opposition viewpoint as "Classical." To paraphrase the late J.R. Oppenheimer, "The word 'Classical' means only one thing in science: it's wrong!"

However, when the data are sparse and resultingly the answer does depend upon the choice of $p_0(x)$, the Bayesian approach comes under strong attack. It may be, after all, that the observer is lacking some vital prior knowledge which would strongly change his choice of the prior. Any solution under these circumstances must be viewed from this perspective, and hence with skepticism, say the Classicists. Specific cases are examined below.

The Bayesian approach takes its name from its necessary use of Bayes' rule (2.22).[2] Consider, for simplicity, a one-datum test of a hypothesis. The hypothesis will be accepted if it is known that P(hypothesis | datum) is large. Bayes' theorem is used to construct the required P(hypothesis | datum) from the usually computable P(datum | hypothesis), along with a usually non-essential normalization constant P(datum), and P(hypothesis). The latter is called a *prior probability*, since it must be known prior to seeing the datum. By (2.21),

$$P(\text{hypothesis}|\text{datum}) = \frac{P(\text{datum}|\text{hypothesis})\,P(\text{hypothesis})}{P(\text{datum})} . \tag{16.1}$$

Hence, different estimates of P(hypothesis) lead to different estimates of P(hypothesis | datum) and hence, to different decisions on the truth of the hypothesis. Of course, this is an overly simple problem, using but one piece of data. We show next how the Bayesian approach may be applied to a problem we already solved classically, in Sect. 9.9.

16.1 Bayesian Approach to Confidence Limits for an Estimated Probability

We found in Chap. 9 that the maximum-likelihood estimate $\hat{p}$ of a probability is $p = m/N$, when m out of N observed events obey the event in question. As in Sect. 9.9, we now want to establish confidence limits for the estimate. To review, this means that if p is fixed and the N total events are repeated over and over, each giving rise to a new m and hence a new estimate $\hat{p}$, in $\alpha\%$ of these estimates an error ε_0 is tolerable.

The Bayesian approach differs immediately from the Classical. Whereas the Classical (9.26) framed the requirement as

$$P(p - \varepsilon_0 \leqq \hat{p} \leqq p + \varepsilon_0) = \alpha , \tag{16.2}$$

the Bayesian criterion is

$$P(\hat{p} - \varepsilon_0 \leqq p \leqq \hat{p} + \varepsilon_0) = \alpha . \tag{16.3}$$

Notice that these requirements are really saying the same thing: the probability that the absolute error $|p - \hat{p}|$ in the estimate is less than ε_0 must equal α. However, they give rise to different approaches. Criterion (16.2) requires that we know the probability law on $\hat{p}$, Eq. (9.25); whereas criterion (16.3) requires the probability

[2] Perhaps so as not to be led into temptation, one well-known Classicist never uses the rule!

law on p. It is the latter quest which ultimately requires use of Bayes' rule and, hence, the prior probability on the unknown p. To avoid such cumbersome notation as $P(p)$, let

$$p = x \; . \tag{16.4}$$

What (16.3) requires then, is $p(x|m)$, the probability density on x given m observations of the event. The unknown density $p(x|m)$ has the following meaning. Suppose a coin of unknown bias (probability x of a head) is randomly selected from a bagful of coins containing all biases. If the coin is flipped N times, and m heads occur, we are tempted to say that $x = m/N$. However, this is only the maximum likelihood value of x (see Sect. 9.6). Actually, *any bias between* 0 *and* 1 *could be present.* The probability of each such bias x is defined by $p(x|m)$. This is found next, as we continue the Bayesian approach.

16.1.1 Probability Law for the Unknown Probability [16.2]

Using Bayes' rule (2.23),

$$p(x|m) = \frac{P(m|x)\, p_0(x)}{P(m)} \; . \tag{16.5}$$

(Note the discrete probabilities of m, capitalized by convention.) This is used because we know $P(m|x)$ and can make "educated guesses" at the form of the probability law $p_0(x)$. The latter is a "prior probability," since it is the probability of the hypothesis x prior to seeing the datum.

Since the sequence of events under observation is binary, each either contributing to m or not, they describe Bernoulli trials. Hence, by (6.5)

$$P(m|x) = \binom{N}{m} x^m (1-x)^{N-m} \; . \tag{16.6}$$

Also, the denominator in (16.5) is simply a constant (with respect to x) which guarantees normalization for $p(x|m)$. It results that

$$p(x|m) = \frac{p_0(x) x^m (1-x)^{N-m}}{\int_0^1 \mathrm{d}y\, p_0(y) y^m (1-y)^{N-m}} \; . \tag{16.7}$$

To further evaluate this requires an assumption on the form of the prior probability $p_0(x)$. Here is the Bayesian dilemma directly, for it seems obvious that the form of the answer $p(x|m)$ will vary with our assumption on $p_0(x)$. However, for this problem at least, such will not be the case if N is large enough. This will be shown below.

16.1.2 Assumption of a Uniform Prior

Since x is p, an unknown probability, necessarily $0 \leqq x \leqq 1$. Assuming that there is no reason to expect a preferred value of x over that range, it is reasonable to construct

$$p_0(x) = \begin{cases} 1 \text{ for } 0 \leqq x \leqq 1 \\ 0 \text{ otherwise} \end{cases} . \tag{16.8}$$

This is, of course, the maximum entropy solution (10.28) as well (Ex. 10.1.6).

Substituting (16.8) into (16.7), the denominator integrates out analytically, and we have

$$p(x|m) = (N+1)\binom{N}{m} x^m (1-x)^{N-m} . \tag{16.9}$$

This is called *Bayes' distribution,* or the Beta distribution. It is easily shown by direct differentiation of (16.9) that the maximum probable value of x obeys

$$\hat{x}_{MAP} = m/N . \tag{16.10}$$

The subscript denotes the terminology "maximum a posteriori" probability. See Sect. 17.2.3 for more on this kind of estimate.

It is interesting that the maximum-probable value m/N differs from the mean-probable value, which from (16.9) obeys

$$< x >= (m+1)/(N+2). \tag{16.11}$$

We note parenthetically that this is as valid an estimate of p as is (16.10). The choice depends upon whether the mean or the mode value (MAP value) of the curve $p(x|m)$ is deemed more important by the user.

All moments of the law (16.9) may easily be found, by using identity (16.31) below.

16.1.3 Irrelevance of Choice of Prior Statistic $p_0(x)$ if N is Large

Some typical forms for Bayes' distribution are shown in Fig. 16.1. These have the interesting tendency to become ever more sharply concentrated about the MAP value

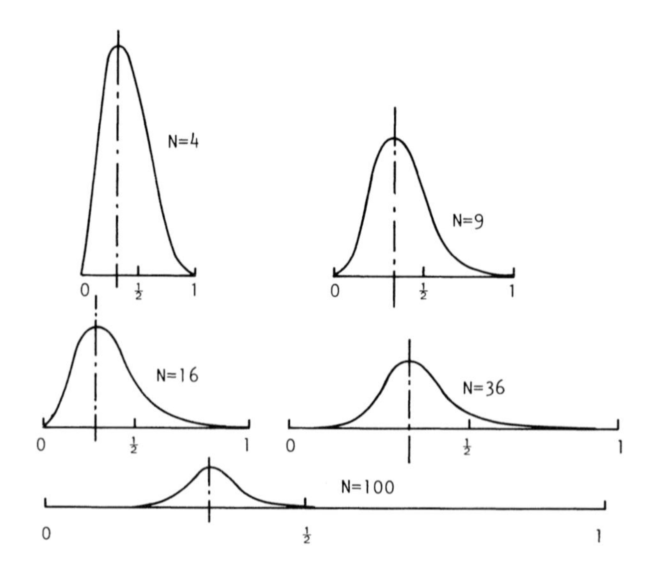

Fig. 16.1. Bayes' distribution with $m/N = 1/3$

m/N as N increases (note the positions of abscissa $1/2$ on the curves). This is, of course, a verification of the law of large numbers (2.8). But it also suggests that the function

$$x^m(1-x)^{N-m}$$

when multiplied by any other function $p_0(x)$ will also become ever-more concentrated, provided $p_0(x)$ has some reasonable properties: (i) non-zero at m/N, and (ii) non-singular elsewhere.

The upshot is that the general expression (16.7) for $p(x|m)$ asymptotically approaches an invariant form, regardless of choice of the prior $p_0(x)$, as $N \to \infty$. This fact is sometimes called *Bayes' theorem,* or the second law of large numbers. The limiting form is discussed next.

16.1.4 Limiting Form for *N* Large

As suggested by the curve shapes in Fig. 16.1, as $N \to \infty$[3]

$$x \to N(\hat{p}, \hat{p}(1-\hat{p})/N)\,, \quad \hat{p} \equiv m/N\,. \tag{16.12}$$

The variance $\hat{p}(1-\hat{p})/N$ is, of course, known from (9.21). Result (16.12) is, once again, independent of the choice of the prior $p_0(x)$. It is also very useful in practical problems, as follows.

16.1.5 Illustrative Problem

We now reconsider the problem at the end of Sect. 9.9, previously solved using the Classical approach. Will the Bayesian approach give the same numerical answer?

How many trials N must be made to have at least 99% probability of correctness for the inference that $0.38 \leqq p \leqq 0.42$, if the estimate $\hat{p} = m/N = 0.4$?

We anticipate that the answer N will be large. This allows us to use the asymptotic result (16.12). Then

$$P(0.38 \leqq x \leqq 0.42) = \frac{1}{\sqrt{2\pi}\sigma}\int_{0.38}^{0.42} \mathrm{d}x\, \mathrm{e}^{-(x-0.4)^2/2\sigma^2} \equiv 0.99\,,$$

where

$$\sigma^2 \equiv 0.4(1-0.4)/N\,.$$

This reduces to

$$\mathrm{erf}\left(\frac{0.02}{\sqrt{2}\sigma}\right) = 0.99\,,$$

with solution (Appendix A)

$$N = 4050 \text{ trials}\,.$$

[3] Expand $\ln[x^m(1-x)^{N-m}]$ about point $x = m/N$ in a *second-order* Taylor series.

This agrees with the answer found by using the Classical approach in Sect. 9.9. Noting that N is large, we see that this is a case when the sheer size of N makes the choice of the prior $p_0(x)$ irrelevent, so that Classical and Bayesian approaches give the same answer.

16.2 Laplace's Rule of Succession

This problem will show that when the number of data are small, the Bayesian answer quite critically depends upon the choice made for the prior $p_0(x)$. Furthermore, the particular choice of a uniform prior will no longer seem so unerringly reasonable in all cases.

Basically, the problem is as follows. A coin of *unknown fairness* is flipped N times. The flips are independent, the unknown probability p of a head remaining constant during the flips. All N flips have the outcome head. What is the probability that the next m flips will also be all heads?

This problem may be used as a model for many others [16.3]. For example, a basketball team about which you know nothing wins its first N games in a row. What are appropriate odds against it winning the next m? The case $m = 1$ is usually of immediate interest. Obviously, the experiment of flipping a coin may be replaced by any experiment having independent binary outcomes (see Ex. 16.2.1 for the extension to trinary, etc., outcomes).

16.2.1 Derivation

As at (16.4), let the unknown bias (probability of a head) p of the coin be represented by the generic RV x. Since $x = p$ and p is a probability,

$$0 \leq x \leq 1. \tag{16.13}$$

Each flip of the coin constitutes a trial. Let B denote the event that there are N successes on the first N trials, and A denote the event that there are m successes on the next m trials. Our problem is to compute the probability $P(A|B)$.

Obviously, *if the coin has a known bias* $x = x_0$ then the direct answer is, for the m independent trials,

$$P(A|B) = x_0^m. \tag{16.14}$$

Notice that this follows as well from the definition (2.11) of a conditional probability,

$$P(A|B) \equiv P(A, B)/P(B) , \tag{16.15}$$

since here

$$P(A, B) \equiv P(A, B|x) = x^{N+m} \tag{16.16}$$

and

$$P(B) \equiv P(B|x) = x^N . \tag{16.17}$$

The division once again gives $P(A|B) = x^m \equiv x_0^m$ as the particular value.The provisional knowledge of the bias x_0 takes the form of the indicated *conditional dependence* upon x in the notation. This will prove useful.

In our problem x is of course not known. In the absence of such specific information, let x instead be known to be *randomly chosen* from a given probability law $p_0(x)$. To define such a law, we can again resort to the "bag of coins" model described before in Sect. 16.1. This model fits many physical scenarios. Hence, there is a bag of coins present, where each coin has randomly a bias x, and the frequency with which x occurs in the bag is $p_0(x)$. Subsequently, one coin is randomly selected from the bag, and is flipped repeatedly to produce all the trials we are considering.

The entire experiment must of course obey causality. Thus, the random coin selection occurs *prior to* observation of the datum N. For this reason $p_0(x)$ is called a *prior probability law*.

Despite the more vague state of knowledge $p_0(x)$ in place of x per se, we can still use definition (16.5) to form $P(A|B)$. The proviso is that equations (16.16) and (16.17) be altered to mirror the new state of knowledge. Using the partition law (3.7), (16.16) becomes

$$P(A, B) = \int_0^1 \mathrm{d}x P(A, B|x) p_0(x) = \int_0^1 \mathrm{d}x\, x^{N+m} p_0(x) , \tag{16.18}$$

and (16.17) becomes

$$P(B) = \int_0^1 \mathrm{d}x\, P(B|x) p_0(x) = \int_0^1 \mathrm{d}x\, x^N p_0(x) . \tag{16.19}$$

The finite limits follow from the limited range (16.13) of possible bias values. The answer (16.15) is then

$$P(A|B) = \frac{\int_0^1 \mathrm{d}x\, x^{N+m} p_0(x)}{\int_0^1 \mathrm{d}x\, x^N p_0(x)} . \tag{16.20}$$

To proceed further we have to specify the form of the prior probability law $p_0(x)$. However, oftentimes the observer cannot do this with any degree of confidence. The law existed in the past of the data, and there may be no physical model for what transpired then. What can be done?

The Bayesian answer is to proceed on the basis of a state of ignorance defined by a *hypothetical shape* of the prior law. Since probabilities are subjective, and our aim is to compute a probability, this idea seems reasonable. One computes probabilities on the basis of, or relative to, what one knows. For example, the hypothesis (16.8) of a *uniform law*,

$$p_0(x) = 1, \quad 0 \le x \le 1 \tag{16.21}$$

defines a state of *maximum ignorance* about x. At the other extreme, the hypothesis of a Dirac delta function,

$$p_0(x) = \delta(x - x_0) \tag{16.22}$$

defines a state of *minimum ignorance* about x. The bias has a known value x_0. An in-between state of ignorance is provided by a triangular law (3.20),

$$p_0(x) = Tri(2x - 1)\ . \tag{16.23}$$

Here there is *a tendency* to choose a fair coin $x = 1/2$, but other levels of fairness are still possible, with smaller probabilities.

Of course each of the choices (16.21)–(16.23) must yield a different prediction (16.20) for $P(A|B)$. In this sense there are many Laplace's Rules of succession. This again returns us to the subjective nature of probability values.

With the hypothesis (16.22) of definite knowledge of the bias used in (16.20), the sifting property of the delta function gives back the answer $P(A|B) = x_0^m$ we computed at (16.14). This is expected since definite knowledge of x is defined by a Dirac-delta law (16.22).

With the hypothesis (16.21) of maximum ignorance, (16.20) gives a simple quotient of integrals

$$P(A|B) = \frac{\int_0^1 \mathrm{d}x\, x^{N+m}}{\int_0^1 \mathrm{d}x\, x^N}\ . \tag{16.24}$$

The integrals are easily evaluated, giving

$$P(A|B) = \frac{N+1}{N+m+1}\ , \tag{16.25}$$

the *Laplace rule of succession.*

Note that for the special case A of 1 success on the next trial, since the trials are independent $P(A|B)$ represents an estimate $\hat{x} \equiv \hat{p}$ of the simple occurrence of a head. We shall also denote this $\hat{p}$ as $P(1|N)$.

Example. An unknown coin is flipped 10 times, yielding 10 heads. What's the probability of obtaining a head on the next flip? In (16.25), $N = 10$, $m = 1$. Then $P(1|10) = 11/12$. This seems reasonable. Certainly, with 10 heads in a row we have high suspicion that the coin is far from a fair one ($p = 1/2$).

However, consider the following counter-example. The coin is flipped but once, and gives a head. Then (16.25) predicts that $P(1|1) \equiv \hat{p} = 2/3$. This is not at all reasonable. If a coin has probability 2/3 of giving a head, it is very biased. It does not seem appropriate to make such an extreme estimate $\hat{p}$ based upon only one piece of data, the occurrence of a head. In fact, we know that in real life the estimate would be wrong. That is, if all the coins in the world are flipped once, about half will give a head. Yet, if the latter are flipped once more, certainly nothing like two out of every three will again be heads. Where has the derivation of (16.25) gone astray?

The problem lies in the choice of a uniform prior (16.21). Although uniformity may be a correct choice in other cases, it certainly is incorrect for the coin case. For it states that a randomly selected coin is as likely to have a bias of 1 (heads only) as a bias of 1/2 (fairness); whereas we know from experience that *the vast majority of coins are fair.* Hence, our prior knowledge is very different from the choice (16.21) of equanimity. Instead, it should be closer to

$$p_0(x) = \delta(x - 1/2) \tag{16.26}$$

which is (16.22) for the case of *only* fair coins. A compromise would be to construct $p_0(x)$ as triangular (16.23) about a state of fairness. The resulting Laplace rule of succession is as follows.

Exercise 16.1

Show that the use of a triangular prior probability law (16.23) in (16.20) gives an answer

$$P(A|B) = \frac{(N+1)(N+2)}{(N+m+1)(N+m+2)} \cdot \left(\frac{1-2^{-N-m-1}}{1-2^{-N-1}}\right). \tag{16.27}$$

(*Hint:* Each integral is conveniently evaluated as a sum of two integrals, with limits going respectively from 0 to 1/2 and from 1/2 to 1. Each integrand is still a simple power of x and, hence, easy to integrate.)

16.2.2 Role of the Prior

Again considering the special case A of 1 success on the next trial, (16.27) estimates $\hat{p}$ based on N successes as

$$P(1|N) \equiv \hat{p} = \left(\frac{N+1}{N+3}\right)\left(\frac{1-2^{-N-2}}{1-2^{-N-1}}\right). \tag{16.28}$$

For N small or moderate this is quite different from $(N+1)/(N+2)$ predicted by (16.25). For example, reconsidering the case $N = 1$, (16.28) gives $\hat{p} = 7/12$. This is much closer to fairness ($p = 1/2$) and more reasonable than the result 2/3 previously obtained for the uniform prior.

Is there *an optimum* prior $p_0(x)$ to use for the coin problem? The optimum prior should be the true histogram of coin biases as they occur naturally. This could be obtained in principle by testing every coin in existence for its bias $p \equiv x$, and building up the histogram $p_0(x)$ from these events. Although this is unrealistic in practice, at least it is definable. The coin problem is a metaphor for a wide class of statistical problems. In some, the *proper prior law* is too nebulous even to be defined. In these cases the estimated probability $P(A|B)$ becomes very strongly subjective and, so, has reduced value for widespread application.

In summary, we have found that the Bayesian answer will often depend upon the choice made for the prior. (See also Ex. 16.2 below.) Conversely, for certain problems, when N is very large the Bayesian answer goes over into the Classical (as in Sect. 16.1.5) independent of the form chosen for the prior. Then the empirical evidence is so compelling as to make the choice of subjective probability irrelevant.

16.2.3 Bull Market, Bear Market

An interesting application of the Laplace rule of succession is to the daily valuation of the stock market. A bull market is characterized by a long string of days for which the valuation rises virtually every day; a bear market is one for which the market valuation falls virtually every day. (To get a steady trend, one might use a 30-day moving average in place of the daily valuation.) Does this have something to do with the Laplace rule (16.25)?

The Laplace rule tends to predict long trends. If the rule is obeyed, then the longer the market is in (say) a bull market phase the more probable it is that it will remain so on the next day of trading. Or, conversely for a bear market. In fact this is empirically confirmed by the long durations of many bull- or bear markets. On the coarsest scale, use of a 20-*year* moving average shows a definite bull market in operation where the rise is about 11% per year. But, of course much shorter-range bull and bear markets occur as well.

Perhaps such behavior is effectively predicted by the model we used for arriving at the Laplace rule – successive independent tosses of a single coin with an unknown fixed bias x. In application to the market, let x represent the probability of a rise in market valuation on a given day with use of a given averaging window. The value of x may be taken to characterize *the prevailing mood of the participants,* the buyers and sellers. If they are optimistic about the economy then $x > 1/2$, there is a positive bias, and there is a tendency toward a bull market; or, if they are pessimistic then $x < 1/2$ and there is a tendency toward a bear market. If they are neutral then $x \approx 1/2$, the coin is essentially "fair", and the market tends to trade within a fixed range. (This is also the model used in Sect. 11.5 for deciding if a "vote" is decisive.) Does the model have any support in economic theory?

According to the "random walk" (Sect. 4.16) model of the market, the market is "efficient" [17.32, 17.33]. That is, market price alone reflects all of the prevailing economic forces of the day. Only the empirical data – current and past values of the market – really matter for purposes of predicting its future behavior. No other information (book values, price/earnings ratios, etc.) about specific stock values matters. Then any day's market price is independent of its predecessors, and any value of the bias x is *a priori* equally probable. If so, one is in a state of maximum ignorance as to bias toward either bull or bear behavior. The law $p_0(x)$ is flat.

In fact these agree with the model assumptions in Sect. 16.2.1 for derivation of the Laplace law. The model assumption of independent trials is implied by the random walk model. Also, a flat prior law (16.21) is mirrored in the above assumption of maximum ignorance.

Thus, by the Laplace model, each new rise of the market adds to the probability of a subsequent rise, and hence to the extension of a bull market. However, every bull market eventually ends. This seems to refute the model. But in fact the model may be extended to include random termination events, whereby the old coin of bias $x > 1/2$ is discarded, and replaced with a new coin of bias $x < 1/2$ or $x \approx 1/2$. Then at each such termination occurrence an existing bull market would suddenly change to a bear market or a fixed-range market. A key ingredient of the extended model would be the probability law on the length N of a bull market, subject to the length of the given moving average window. Likewise for a bear market, or for a fixed-range market. These could be obtained from past behavior of the market. However, the usual caveat must be stated: there is no guarantee that these laws will be obeyed in the future.

Exercise 16.2

16.2.1 ***Use of Bayes or beta "prior" in the Laplace problem.*** The observer may have the following *prior knowledge* (prior to the observed N heads) about the unknown probability x for the given coin. The coin was first randomly picked from a bag containing biases x in equal portions, and was flipped M times, with m_1 heads observed.

Thus, the observer has a rough estimate of x at the outset, prior to the Laplace part of the overall experiment when N heads in a row are observed. Now what should be his estimate of the probability of the next m flips being heads?

Show that the answer is

$$P(A|B) = \frac{(M+N+1)!(m_1+N+m)!}{(M+N+m+1)!(m_1+N)!} \tag{16.29}$$

with special case

$$P(1|N) = \hat{p} = \frac{m_1+(N+1)}{M+(N+2)} . \tag{16.30}$$

Note the reasonableness of (16.30). When the number N of Laplace data is small, the answer $\hat{p}$ is biased toward simply m_1/M, the a priori MAP estimate of x. Or, when the number of Laplace data is large, the answer is biased toward the Laplace estimate (16.25) with $m = 1$.

Hint: Use the Bayes' density $p(x|m_1)$ given by (16.9) as the prior law $p_0(x)$ in (16.20). Note the identity

$$\int_0^1 \mathrm{d}x x^b (1-x)^a = \frac{a!b!}{(a+b+1)!} . \tag{16.31}$$

16.2.2 ***Laplace rule for a trials sequence of general order r.*** The derivation in Sect. 16.2.1 is for a sequence of binary events *head* or *tail*, *win* or *loss*, etc. However, some sequences are intrinsically of higher order than binary. For example, a die gives rise to events ranging from numbers 1 to 6. This defines an events sequence of order $r = 6$.

Suppose that a die of unknown biases is rolled N times, with N outcomes (say) 2. Now what is $P(1|N)$, the probability that it will roll one more 2 on *the next* trial?

This can be solved from two different viewpoints. Although both are correct from their vantage points, the results do not agree!

Viewpoint (1): Each event is still basically binary, i.e., a 2 or "not 2". This viewpoint ignores the $r = 6$ nature of the events sequence. On this basis, the problem has already been solved in Sect. 16.2.1. The answer is $P(1|N) = (N+1)/(N+2)$. What is the significance of assuming a uniform prior law $p_0(x)$ now?

Viewpoint (2): Each event *is known to be* of order r. This information should not be ignored. Hence the derivation of Sect. 16.2 should be redone on this basis.

Thus as at (16.17) $P(N|x_1, \ldots, x_r) = x_2^N$, so that

$$P(N) = \int_{\sum x_i = 1} \cdots \int \mathrm{d}x_1 \cdots \mathrm{d}x_r \, x_2^N = \frac{N!}{(N+r-1)!}. \tag{16.32}$$

The integral is called "Dirichlet's integral". The integration is subject to the given normalization constraint, and the resulting region of integration is called a "simplex region".

Show in the same way that

$$P(N, 1) = \frac{(N+1)!}{(N+r)!}$$

Show that the preceding results imply that

$$P(1|N) = \frac{N+1}{N+r}. \tag{16.33}$$

For a die, then, the answer is $P(1|N) = (N+1)/(N+6)$. This differs considerably from the answer from *Viewpoint (1)*, which was $(N+1)/(N+2)$. Why do the two answers differ? Suppose that you had to bet on the occurrence of the event 2 on the next roll of the die. The two answers now differ widely. Discuss why you would choose one over the other. Is this an example of Bertrand's paradox?

17. Introduction to Estimation Methods

The general aim of statistics is to infer something about the source that gave rise to given data. An ultimate aim is to find *exact* numerical values for the source. There is a large body of literature on this subject, called *estimation theory*. Estimation is both an art and a science – typical of most subjects in statistics. Hence, there are passionate devotees of one approach or another, to a given problem, especially economic problems. Personally, we like the following motivation for the subject: If Adam and Eve were expelled from Paradise for attempting to Know, then the least we can do, having lost our chance at earthly Paradise, is to find out just how well we *can* Know.

Hence, our aims in this chapter are to establish optimal methods of estimating, and the minimal errors that such methods entail. Optical examples will be given wherever possible, as usual.

Two general types of estimation problems will be pursued: *parameter estimation*, and *probability law estimation* (see also Chap. 10). It will be seen, in Sect. 17.3, that an excellent method of estimating probability laws grows out of a detailed model of measurement, which postulates that the Fisher information level in any measurement results from a flow of the information from the observed phenomenon into the data. The carriers of the information are photons or any other probe particles that are used to illuminate the object. This approach is shown to derive the Schrodinger wave equation and other probabilistic laws of physics.

The subject of parameter estimation neatly divides, at the outset, into two distinct areas. Consider an unknown *parameter* a that is to be estimated from given data $y_1, \ldots, y_N = \boldsymbol{y}$. Parameter a may usefully be regarded as either deterministic (but unknown) or random. Either viewpoint leads to good estimates. It will be seen that the deterministic viewpoint is the random viewpoint in the presence of zero prior information about a.

In general we will be seeking a function $\hat{a}(\boldsymbol{y})$ of the data $\boldsymbol{y}$, called an *estimator*, that best approximates a in some sense. As might be expected, depending upon the "sense" used, different estimators will generally result. However, in diametric opposition, it will also be seen that an optimum estimator by one specific criterion (the quadratic "cost" function) will also be optimum for a very wide class of cost functions.

17.1 Deterministic Parameters: Likelihood Theory

There are circumstances for which parameter a cannot profitably be regarded as random. This is when a probability law $p_0(a)$ cannot be known, e.g., because the scenario for a is ill-defined, or because a is a one-time event.

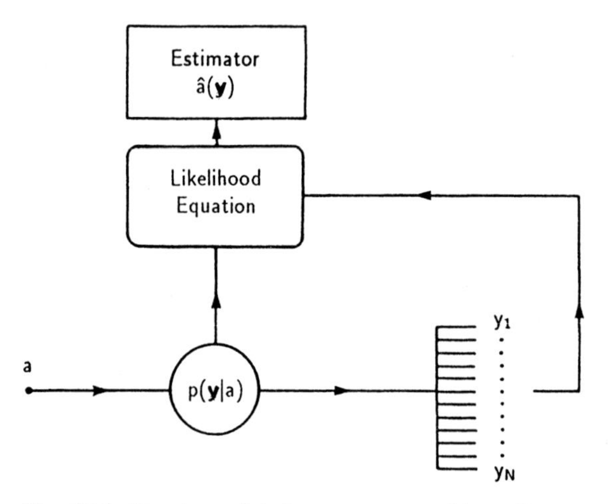

Fig. 17.1. The deterministic parameter problem. Parameter a is fixed but unknown. Then a causes data $\boldsymbol{y}$ through random sampling of likelihood law $p(\boldsymbol{y}|a)$. Finally, $p(\boldsymbol{y}|a)$ and $\boldsymbol{y}$ are used to form the estimator $\hat{a}(\boldsymbol{y})$ via the likelihood equation

Let the data $\boldsymbol{y}$ arise from a via a known probability law $p(\boldsymbol{y}|a)$. A model for this process is shown in Fig. 17.1. It is traditional to call $p(\boldsymbol{y}|a)$ the *likelihood law.*

Consider the spatial statistics in intensities $\boldsymbol{y}$ due to an extended thermal light source of uniform intensity a. These can obey likelihood laws

$$p(\boldsymbol{y}|a) = \prod_{n=1}^{N} \mathrm{e}^{-a} a^{y_n} / y_n! \quad \text{(Poisson)} \tag{17.1}$$

or

$$p(\boldsymbol{y}|a) = \prod_{n=1}^{N} (\sqrt{2\pi}\sigma)^{-1} \exp[-(y_n - a)^2/2\sigma^2] \quad \text{(Normal)}\,. \tag{17.2}$$

Note that these are simple product laws; it is assumed that there is no correlation between adjacent data values. From Chap. 6, (17.2) is the special case of (17.1) when the light source is very bright.

17.1.1 Unbiased Estimators

The general problem we consider in Sect. 17.1 is how to estimate the parameter a based upon knowledge of the data $\boldsymbol{y}$ and the likelihood law $p(\boldsymbol{y}|a)$. The estimate is denoted as $\hat{a}$. It is generally some function $\hat{a}(\boldsymbol{y})$ of the data, and is called an "estimator function" of the data.

All averages are with respect to the likelihood law $p(\boldsymbol{y}|a)$. Hence the mean value of an estimator is

$$\langle \hat{a}(\boldsymbol{y}) \rangle \equiv \int \mathrm{d}\boldsymbol{y}\, \hat{a}(\boldsymbol{y}) p(\boldsymbol{y}|a) \,. \tag{17.3}$$

(In this and subsequent integrals, undefined limits mean infinite limits $\int_{-\infty}^{\infty}$.)

Depending upon the functional form of $\hat{a}(\boldsymbol{y})$, this can have values

1. a, in which case we call the estimator "unbiased"
2. $a + b$ or a/b, where b is a known constant, independent of a
3. $a + b$ or a/b, where b is an unknown constant that depends on a.

Of the three cases, the unbiased one is by far the most important to consider. Although there are exceptions [see (9.65) or (16.11)], most good estimators are unbiased, i.e., correct on average. This is the analog of unbiased experimental apparatus, such as a device for measuring the speed of light c that has error at each determination of c, but whose average over many determinations approaches c (Sect. 9.2).

Case 2 allows for conversion to an unbiased estimate, since b is known. Where the mean has value $a + b$, simply subtract b from each estimate $\hat{a}$; where the mean has the value a/b, multiply each estimate by b [see (9.65) *et. seq.*].

Cases 3 do not generally allow for recourse to an unbiased estimate, since the bias is unknown.

17.1.2 Maximum Likelihood Estimators

The principle of maximum likelihood states that data that occur presumably have had maximum probability of occurring (Sect. 9.6). Hence, given the likelihood law $p(\boldsymbol{y}|a)$ and fixed data $\boldsymbol{y}$, a must have the property that it maximized the likelihood of the data,

$$p(\boldsymbol{y}|a) = \text{maximum} \,. \tag{17.4}$$

The a that satisfies this condition is called the "maximum likelihood estimator" $\hat{a}_{\mathrm{ML}}$. In general, its value will depend upon data $\boldsymbol{y}$, and so it is denoted as the function $\hat{a}_{\mathrm{ML}}(\boldsymbol{y})$.

Many likelihood laws, such as (17.1) or (17.2), simplify when their logarithms are taken. Since the logarithm of a quantity monotonically increases with the quantity, principle (17.4) is equivalent to

$$\ln p(\boldsymbol{y}|a) = \text{maximum} \,. \tag{17.5}$$

Finally, if $\hat{a}_{\mathrm{ML}}$ occurs at an interior point of the permissible range of a, then it occurs at a point of zero derivative, or

$$\left. \frac{\partial \ln p(\boldsymbol{y}|a)}{\partial a} \right|_{a=\hat{a}_{\mathrm{ML}}} = 0 \,. \tag{17.6}$$

This is traditionally called the *likelihood equation.* Its inputs are $p(\boldsymbol{y}|a)$ and $\boldsymbol{y}$. Its output is $\hat{a}_{ML}(\boldsymbol{y})$ (Fig. 17.1). Examples follow.

Exercise 17.1

17.1.1 Show that either likelihood law (17.1) or (17.2), when substituted into the likelihood equation (17.6), yields the same ML estimate

$$\hat{a}_{ML} = \frac{1}{N}\sum_{n=1}^{N} y_n \equiv \overline{y}\,, \tag{17.7}$$

the sample mean.

17.1.2 Show that use of a log-normal likelihood law

$$p(\boldsymbol{y}|\ln a) = \prod_{n=1}^{N} \frac{1}{\sqrt{2\pi}\sigma} \exp[-(y_n - \langle \ln a \rangle)^2/2\sigma^2] \tag{17.8}$$

in likelihood equation (17.6) yields as an ML estimate

$$\hat{a}_{ML} = \exp(\overline{y})\,. \tag{17.9}$$

Comment: Note that in all the above cases the estimator depended upon the data through solely its sample mean. This is not always the case. However, when an estimator depends upon some well-defined function $f(\boldsymbol{y})$ of the data, that function is called a *sufficient statistic*. This is in the sense that knowing $f(\boldsymbol{y})$ alone is "sufficient" to make the estimate. The individual data *per se* do not have to be known. Sufficient statistics are usually some linear combination of the data.

17.1.3 Maximum likelihood theory is extendable to multiple parameter estimation, in the obvious way. Hence, if a vector $a_1, a_2, \ldots, a_N \equiv \boldsymbol{a}$ of unknowns exists, instead of a single likelihood equation (17.6) we have N likelihood equations

$$\left.\frac{\partial \ln p(\boldsymbol{y}|\boldsymbol{a})}{\partial a_n}\right|_{a_n=\hat{a}_{ML}^{(n)}} = 0\,. \tag{17.10}$$

Apply this approach to estimating a_1 and a_2, when the likelihood equation is

$$p(\boldsymbol{y}|a_1, a_2) = \prod_{n=1}^{N} \frac{1}{\sqrt{2\pi a_2}} \exp[-(y_n - a_1)^2/2a_2]\,. \tag{17.11}$$

This is the problem of jointly estimating a mean and a variance from a random sample of the normal law. the sample mean.

17.1.4 The solution to ML problems is not always in closed form. A transcendental equation may result. Consider the problem of determining the position a of a slit from a discrete sample $\boldsymbol{y}$ of photon positions in its diffraction pattern. The likelihood law, from (3.16), has the form

$$p(\boldsymbol{y}|a) = \prod_{n=1}^{N} \operatorname{sinc}^2(y_n - a) \ . \tag{17.12}$$

Show that the likelihood equation becomes

$$\sum_{n=1}^{N} (y_n - a)^{-1} = \sum_{n=1}^{N} \cot(y_n - a) \ , \tag{17.13}$$

transcendental in a. Work out the solutions for the cases $N = 1$, $N = 2$. These are surprisingly simple.

17.1.5 The image restoration problem (Sect. 8.12) may be addressed by ML theory. Assume that additive, Gaussian noise is present. Then the likelihood law is

$$p(\boldsymbol{y}|\boldsymbol{a}) = c \prod_{n=1}^{N} \exp\left[-\left(y_n - \sum_{m=1}^{N} a_m s_{mn} \right)^2 \Big/ 2\sigma^2 \right] \ , \tag{17.14}$$

where s_{mn} are sampled values of the point spread function, $\boldsymbol{y}$ are the image data values and $\boldsymbol{a}$ are the unknown object values. Using the multidimensional ML approach (17.10), show that

$$\hat{\boldsymbol{a}}_{\mathrm{ML}} = [S]^{-1} \boldsymbol{y} \ ,$$

where matrix $[S]$ has as its elements the s_{mn} values.

17.1.3 Cramer–Rao Lower Bound on Error

In this section we derive a truly important, fundamental result of estimation theory. The question addressed is, if *any* unbiased estimator $\hat{a}(\boldsymbol{y})$ is allowed (not necessarily ML), can the mean-square error in the estimate be made arbitrarily small? Or, is there some finite lower limit on error? We shall see that the latter is true. This has far-reaching consequences, both to estimation theory and to the broader, physical world around us [17.1]. We next calculate the lower limit on estimation error, following *Van Trees* [17.2].

Assume that the estimator is unbiased. Then by (17.3)

$$\int \mathrm{d}\boldsymbol{y} \, p(\boldsymbol{y}|a)[\hat{a}(\boldsymbol{y}) - a] = 0 \ . \tag{17.15}$$

Differentiate with respect to a:

$$\begin{aligned} &\int \mathrm{d}\boldsymbol{y} \frac{\partial}{\partial a} \{p(\boldsymbol{y}|a)[\hat{a}(\boldsymbol{y}) - a]\} = 0 \\ &= \int \mathrm{d}\boldsymbol{y} \frac{\partial p(\boldsymbol{y}|a)}{\partial a} (\hat{a} - a) - \int \mathrm{d}\boldsymbol{y} \, p(\boldsymbol{y}|a) = 0 \ . \end{aligned} \tag{17.16}$$

By normalization, the second integral is 1. Also,

$$\frac{\partial p(\boldsymbol{y}|a)}{\partial a} = \frac{\partial \ln p(\boldsymbol{y}|a)}{\partial a} p(\boldsymbol{y}|a)$$

identically. Then (17.16) becomes

$$\int d\boldsymbol{y} \frac{\partial \ln p(\boldsymbol{y}|a)}{\partial a} p(\boldsymbol{y}|a)[\hat{a}(\boldsymbol{y}) - a] = 1 .$$

Preparing for use of the Schwarz inequality, we factor the integrand and square the whole equation:

$$\left[\int d\boldsymbol{y} \left(\frac{\partial \ln p(\boldsymbol{y}|a)}{\partial a}\sqrt{p(\boldsymbol{y}|a)}\right)\left(\sqrt{p(\boldsymbol{y}|a)}[\hat{a}(\boldsymbol{y}) - a]\right)\right]^2 = 1 . \tag{17.17}$$

By the Schwarz inequality, the left-hand side (LHS) obeys

$$\text{LHS} \le \int d\boldsymbol{y} \left(\frac{\partial \ln p(\boldsymbol{y}|a)}{\partial a}\sqrt{p(\boldsymbol{y}|a)}\right)^2 \int d\boldsymbol{y} \left(\sqrt{p(\boldsymbol{y}|a)}[\hat{a}(\boldsymbol{y}) - a]\right)^2$$

or

$$\text{LHS} \le \int d\boldsymbol{y} p(\boldsymbol{y}|a) \left(\frac{\partial \ln p(\boldsymbol{y}|a)}{\partial a}\right)^2 \int d\boldsymbol{y} p(\boldsymbol{y}|a)[\hat{a}(\boldsymbol{y}) - a]^2 .$$

But the second integral defines mean-square error ε^2 in the estimate. Also, the first integral is called the *Fisher information,* and is designated as I. Hence,

$$\text{LHS} \le I\varepsilon^2 .$$

But by (17.17) LHS $= 1$. Hence,

$$\varepsilon^2 \ge 1/I , \tag{17.18}$$

where

$$I \equiv \int d\boldsymbol{y} p(\boldsymbol{y}|a) \left(\frac{\partial \ln p(\boldsymbol{y}|a)}{\partial a}\right)^2 . \tag{17.19}$$

Hence, no matter how ingenious the estimation approach, its mean-square error cannot be less than the value given by (17.18). Notice that this result depends solely upon the form of the likelihood law $p(\boldsymbol{y}|a)$, which is presumably known. Hence, it is usually quite straightforward to compute the error bound. This bound is commonly called the "Cramer–Rao", after its co-discoverers.

To get a feel for *how* the likelihood law $p(\boldsymbol{y}|a)$ influences the lower bound, consider the case of additive, Gaussian noise corrupting $N = 1$ data value,

$$y = a + x , \quad x = N(0, \sigma^2) .$$

The likelihood law is the case $N = 1$ of (17.2),

$$p(y|a) = (\sqrt{2\pi}\sigma)^{-1} \exp[-(y-a)^2/2\sigma^2] .$$

Following the prescription (17.19) we form

$$\frac{\partial \ln p(y|a)}{\partial a} = \sigma^{-2}(y - a) ,$$

so that

$$
\begin{aligned}
I &= \int \mathrm{d}y \, p(y|a)\sigma^{-4}(y-a)^2 \\
&= \sigma^{-4} \cdot \sigma^2 = \sigma^{-2} \,.
\end{aligned}
$$

Hence by (17.18)

$$\varepsilon^2 \geq \sigma^2 \,.$$

In other words, the broader the law $p(y|a)$ is, the greater (worse) is the lower bound. This makes sense, in that a broad law $p(y|a)$ implies large fluctuations in y from a, so that a ought to be harder to estimate.

The reader can easily show, by analogous steps, that use of the N-dimensional law (17.2), i.e., the case of many data values, leads to a result

$$\varepsilon^2 \geq \sigma^2/N \,.$$

The right-hand side looks familiar. Comparing this result with (9.5), we see that the problem of Sect. 9.2 and the present problem are identical, and that we had actually found, in Sect. 9.2, the processor $\hat{a}(\boldsymbol{y})$ that *achieves* the lower error bound. This was the sample mean, (9.1),

$$\hat{a}(\boldsymbol{y}) = \overline{y} \,. \tag{17.20}$$

Going one step further, we found at (17.7) that estimator (17.20) is the ML estimator. This leads us to ask, is this result general? Will the ML estimator always work to achieve the Cramer–Rao lower bound?

17.1.4 Achieving the Lower Bound

Recall that the Schwarz inequality is nothing more than the relation

$$(\boldsymbol{A} \cdot \boldsymbol{B})^2 \leq |\boldsymbol{A}|^2 \, |\boldsymbol{B}|^2 \quad \text{or} \quad \left(\sum_n A_n B_n \right)^2 \leq \sum_n A_n^2 \sum_n B_n^2 \tag{17.21}$$

between vectors $\boldsymbol{A}$ and $\boldsymbol{B}$. The equality sign holds when $\boldsymbol{A}$ is parallel to $\boldsymbol{B}$, i.e., when

$$A_n = kB_n \,, \qquad k = \text{constant} \tag{17.22}$$

for all components n. Now we had from (17.17)

$$A_n \equiv A(\boldsymbol{y}) = \frac{\partial \ln p(\boldsymbol{y}|a)}{\partial a} \sqrt{p(\boldsymbol{y}|a)}$$

and

$$B_n \equiv B(\boldsymbol{y}) = [\hat{a}(\boldsymbol{y}) - a]\sqrt{p(\boldsymbol{y}|a)} \,.$$

Consequently, by condition (17.22), equality will hold when

$$\frac{\partial \ln p(\boldsymbol{y}|a)}{\partial a} = k(a)[\hat{a}(\boldsymbol{y}) - a] \,. \tag{17.23}$$

The constant k can, in general, be a function of a.

Hence, if the derivative $\partial/\partial a$ of the logarithm of the likelihood law can be put in the form of the right-hand side of (17.23), the lower bound of error can be achieved by an estimator. And, the estimator becomes immediately apparent. The estimator is then called an "efficient estimator". Examples will be shown below. Conversely, if the law $p(\boldsymbol{y}|a)$ is such that it cannot be put in the form of (17.23), then the lower error bound cannot be achieved by an unbiased estimator.

We next find the estimator that, in general, satisfies condition (17.23) *if* $p(\boldsymbol{y}|a)$ can be placed in the form (17.23). Starting with (17.23), we evaluate it at $a = \hat{a}_{\mathrm{ML}}$ and see if an equality results. The LHS gives

$$\left.\frac{\partial \ln p(\boldsymbol{y}|a)}{\partial a}\right|_{a=\hat{a}_{\mathrm{ML}}} \equiv 0$$

according to definition (17.6). The RHS gives

$$k(\hat{a}_{\mathrm{ML}})[\hat{a}(\boldsymbol{y}) - \hat{a}_{\mathrm{ML}}] \,.$$

Hence, if

$$\hat{a}(\boldsymbol{y}) = \hat{a}_{\mathrm{ML}}$$

the RHS is zero as well, and an equality does result.

We conclude that if $p(\boldsymbol{y}|a)$ can be placed in the form (17.23), then the estimator that satisfies (17.23) will always be the maximum likelihood estimator. More succinctly, if an efficient estimator exists, it will be $\hat{a}_{\mathrm{ML}}$.

17.1.5 Testing for Efficient Estimators

Assume, as usual, that likelihood law $p(\boldsymbol{y}|a)$ is known. It is easy, of course, to form the ML estimator from this law [via (17.6)]. But, as we found, the estimate will not be efficient unless relationship (17.23) can be shown to hold. As an example, suppose that $p(\boldsymbol{y}|a)$ is the normal law (17.2). By direct operation upon (17.2),

$$\begin{aligned}\frac{\partial \ln p(\boldsymbol{y}|a)}{\partial a} &= \frac{\partial}{\partial a}\left[-\frac{1}{2\sigma^2}\sum_n (y_n - a)^2\right] \\ &= \frac{1}{\sigma^2}\sum_n (y_n - a) = \frac{N}{\sigma^2}\left(\frac{1}{N}\sum_n y_n - a\right),\end{aligned}$$

which *is* in the form (17.23) if we identify

$$\begin{aligned}k(a) &= N/\sigma^2 \,, \\ \hat{a}(\boldsymbol{y}) &= \frac{1}{N}\sum_n y_n \,.\end{aligned}$$

However, this is still not enough to conclude that the estimator achieves the lower bound. The estimator must also be shown to be unbiased. Taking the expectation of the last equation gives

$$\langle \hat{a}(\boldsymbol{y}) \rangle = \frac{1}{N} \sum_n \langle y_n \rangle = \frac{1}{N} Na = a \, ,$$

by (17.2). Hence the estimator is indeed unbiased. We conclude that the estimator is efficient in this case.

The Poisson scenario (17.1) may be checked as well. We get

$$\begin{aligned} \frac{\partial \ln p(\boldsymbol{y}|a)}{\partial a} &= \frac{\partial}{\partial a} \left[-Na + (\ln a) \sum_n y_n - \sum_n \ln y_n! \right] \\ &= -N + \frac{1}{a} \sum_n y_n \, . \end{aligned}$$

The $1/a$ factor bothers us, since by (17.23) we need linearity in a. However, this can be achieved by factoring out the quantity $N/a = k(a)$ to yield

$$\frac{\partial \ln p(\boldsymbol{y}|a)}{\partial a} = \frac{N}{a} \left(\frac{1}{N} \sum_n y_n - a \right) \, .$$

Once again the estimator is the sample mean. The sample mean is, moreover, unbiased, since the mean of each Poisson RV y_n is a. We conclude that the Poisson case also permits efficient estimation.

Do all cases permit efficient estimation? A counter example is as follows. Reconsider the problem of estimating the position a of a slit from observation of photon positions in its diffraction pattern. For simplicity take the case of $N = 1$ photon position. By (17.12),

$$\frac{\partial}{\partial a} \ln p(y|a) = \frac{\partial}{\partial a} \left[2 \ln \frac{\sin(y-a)}{y-a} \right] = 2 \left[\frac{1}{y-a} - \cot(y-a) \right]$$

after some manipulation. This is clearly not of the form (17.23). Therefore, an unbiased efficient estimate does not exist in this case.

The case

$$p(y|a) = c \sin^2[n\pi(y-a)/L]$$

represents the likelihood law for a freefield quantum mechanical particle observed to be at position y in a box of length L. Given y, the aim is to estimate the position a of the box in the field. The reader can easily show, as in the preceding case, that an unbiased efficient estimator for a does not exist.

17.1.6 Can a Bound to the Error be Known if an Efficient Estimator *Does Not* Exist?

We found that in order to know if the lower error bound in (17.18) *is met*, we check if the likelihood law $p(\boldsymbol{y}|a)$ at hand obeys condition (17.23). If it does, and if the resulting estimator $\hat{a}(\boldsymbol{y})$ is unbiased, then the error in the estimator *achieves its lower bound* of $1/I$. The estimator is then called *efficient*. But, what can we do if the test (17.23) is not passed? Obviously, now the lowest possible error $1/I$ cannot be achieved. The error must be larger. How much larger, is ascertained next.

As in Sect. 3.18.7 we start with a helper vector $\boldsymbol{x}$, form its outer-product matrix $\langle xx^{\mathrm{T}}\rangle$, where the mean value of each element in the matrix is taken, and then use the latter's positive-definiteness to get the sought inequality. The steps are as follows [17.9]:

Define an $(N+1)$ dimensional vector

$$\boldsymbol{x} \equiv \begin{bmatrix} \hat{a}(\boldsymbol{y}) - a \\ p^{-1}\partial p/\partial a \\ p^{-1}\partial^2 p/\partial a^2 \\ \cdot \\ \cdot \\ p^{-1}\partial^N p/\partial a^N \end{bmatrix} \tag{17.24}$$

The size of N will be discussed below. From this vector form the mean outer product matrix

$$\langle \boldsymbol{x}\boldsymbol{x}^{\mathrm{T}}\rangle = \begin{bmatrix} \langle(\hat{a}-a)^2\rangle & \langle(\hat{a}-a)\frac{1}{p}\frac{\partial p}{\partial a}\rangle & \langle(\hat{a}-a)\frac{1}{p}\frac{\partial^2 p}{\partial a^2}\rangle & \cdots \\ \langle(\hat{a}-a)\frac{1}{p}\frac{\partial p}{\partial a}\rangle & \langle(\frac{1}{p}\frac{\partial p}{\partial a})^2\rangle & \langle\frac{1}{p^2}\frac{\partial p}{\partial a}\frac{\partial^2 p}{\partial a^2}\rangle & \cdots \\ \langle(\hat{a}-a)\frac{1}{p}\frac{\partial^2 p}{\partial a^2}\rangle & \langle\frac{1}{p^2}\frac{\partial p}{\partial a}\frac{\partial^2 p}{\partial a^2}\rangle & \langle(\frac{1}{p}\frac{\partial^2 p}{\partial a^2})^2\rangle & \cdots \\ \vdots & \vdots & \vdots & \end{bmatrix} .$$

The matrix is symmetric by construction. Beyond element (2, 2) it consists of a sub matrix $[B]$ whose general element is

$$B_{ij} \equiv \left\langle \frac{1}{p^2}\frac{\partial^i p}{\partial a^i}\frac{\partial^j p}{\partial a^j} \right\rangle, \quad i, j = 1, ..., N . \tag{17.25}$$

$[B]$ is called the *Bhattacharyya matrix* in honor of its inventor [17.9]. It remains to evaluate the elements of the top row (or left column, by symmetry).

Element $(1,1)$ is just ϵ^2. As in (3.106), element $(1,2) = (2,1) = 1$. Element $(1,3)$ is evaluated by twice differentiating $\partial/\partial a$ the condition of unbiasedness (17.15). Differentiating (17.15) gives (17.16). Differentiating (17.16) gives an equation

$$\int \mathrm{d}\boldsymbol{y}(\hat{a}-a)\frac{\partial^2 p}{\partial a^2} - \int \mathrm{d}\boldsymbol{y}\frac{\partial p}{\partial a} - \int \mathrm{d}\boldsymbol{y}\frac{\partial p}{\partial a} = 0 . \tag{17.26}$$

But since $\int \mathrm{d}\boldsymbol{y}\,\partial p/\partial a = \partial/\partial a \int \mathrm{d}\boldsymbol{y} p = (\partial/\partial a)\,1 = 0$ by normalization of p, the two right-hand integrals in (17.26) are zero, so that the left-hand integral is identically zero. But this integral is identically the element $(1,3)$. Hence, $(1,3) = (3,1) = 0$. In the same way, the general element $(1,k) = (k,1) = 0$, $k = 4, 5, \ldots, N+1$.

The resulting matrix $\langle \boldsymbol{x}\boldsymbol{x}^{\mathrm{T}}\rangle$ is

$$\langle \boldsymbol{x}\boldsymbol{x}^{\mathrm{T}} \rangle = \begin{bmatrix} \epsilon^2 & 1 & 0 & \cdot\cdot & 0 \\ 1 & B_{11} & B_{12} & \cdot\cdot & B_{1N} \\ 0 & B_{21} & B_{22} & \cdot\cdot & B_{2N} \\ \cdot & \cdot & \cdot & & \cdot \\ \cdot & \cdot & \cdot & & \cdot \\ 0 & B_{N1} & \cdot & \cdot\cdot & B_{NN} \end{bmatrix}$$

Since $\langle \boldsymbol{x}\boldsymbol{x}^{\mathrm{T}} \rangle$ is formed as an outer product, it is positive definite. Then its determinant is positive or zero. Taking the determinant by expanding in cofactors along the top row gives $\epsilon^2 \det[B] - 1 \cdot Cof B_{11} \geq 0$, or

$$\epsilon^2 \geq \frac{Cof\, B_{11}}{\det[B]} \equiv [B]_{11}^{-1}\,, \tag{17.27}$$

the $(1, 1)$ element of the inverse matrix to $[B]$. The lower bound to the error is $[B]_{11}^{-1}$.

How large is this lower bound? Taking $N = 2$ shows the trend for the bound. Matrix $[B]$ is now only 2×2, so that $Cof\ B_{11} = B_{22}$. Also, $\det[B] = B_{11}B_{22} - B_{12}^2$. The result is that

$$\epsilon^2 \geq \frac{B_{22}}{B_{11}B_{22} - B_{12}^2} \equiv \frac{B_{11}B_{22}}{B_{11}(B_{11}B_{22} - B_{12}^2)} = \frac{1}{B_{11}} + \frac{B_{12}^2}{B_{11}(B_{11}B_{22} - B_{12}^2)}\,.$$

Now, $B_{11} \equiv \langle (\frac{1}{p}\frac{\partial p}{\partial a})^2 \rangle = \langle (\frac{\partial \ln p}{\partial a})^2 \rangle \equiv I$. Thus the first right-hand term $1/B_{11} = 1/I$ is the Cramer–Rao lower error bound (17.18). Further considering the second term, the error bound is then the Cramer–Rao bound plus a term which represents an improvement (increase) in the bound *if it is positive*. Since the matrix $[B]$ is positive definite the determinant of any of its principal submatrices is positive or zero. Thus, $B_{11}B_{22} - B_{12}^2 \geq 0$. Hence the second term *is* positive, and represents an improvement in the bound. Taking higher values of N increases the bound further.

17.1.7 When can the Bhattacharyya Bound be Achieved?

As with the Cramer–Rao bound, there is a class of likelihood laws such that the Bhattacharyya bound is *achievable*. This is found as follows. From the way (17.27) was formed, if $\det\langle \boldsymbol{x}\boldsymbol{x}^{\mathrm{T}} \rangle = 0$ the lower bound $[B]_{11}^{-1}$ is achieved. It is shown in Ex. 3.2.13 that, if the top element of $\boldsymbol{x}$ is a linear combination of the elements beneath, then the top row of the matrix $\langle \boldsymbol{x}\boldsymbol{x}^{\mathrm{T}} \rangle$ is likewise a linear combination of the rows beneath. This is a condition for a zero determinant, as was required. Hence from (17.24), if

$$\hat{a}(\boldsymbol{y}) - a = \sum_{i=1}^{N} C_i \frac{1}{p}\frac{\partial^i p}{\partial a^i}, \quad p \equiv p(\boldsymbol{y}|a), \ \ C_i = const\,. \tag{17.28}$$

then the lower bound is achieved. Notice that for $N = 1$ this reverts to the Cramer–Rao requirement (17.23) for efficiency. For $N = 1$ the entire Bhattacharyya approach goes over into the Cramer–Rao; see Ex. 17.2.4. The overall Bhattacharyya approach thereby represents a generalization of the Cramer–Rao.

Knowledge of the *least upper bound* to an unknown represents the strongest statement of information about a bound, since it confines the unknown to the smallest possible interval. The following strategy for obtaining a least upper bound to the error ϵ is suggested by the preceding. If the Cramer–Rao requirement (17.23) is obeyed, then the desired bound is defined by the case $N = 1$. But, if (17.23) is not obeyed then one should try a case $N = 2$ in (17.28). If (17.28) is obeyed, then that defines the bound. If it isn't obeyed, then $N = 3$ should be tried; etc. Since the bound increases with N, this procedure defines the smallest one, our aim. If no value of N satisfies (17.28), then the largest practicable value should be used in (17.27) to obtain *a finite, but unattainable,* upper bound.

17.2 Random Parameters: Bayesian Estimation Theory

Up till now, the unknown parameter a has been assumed to be fixed and deterministic. However, there are situations where a is randomly chosen from a known probability law $p_0(a)$. This is called the *prior probability law* since its action occurs prior to formation of the data. (Examples of prior probability laws were given in Chap. 16.) The data are now formed according to the model shown in Fig. 17.2. How can the additional information $p_0(a)$ be used to improve the estimate of a? This is the question addressed by Bayesian estimation theory.

The *posterior probability law* $p(a|\boldsymbol{y})$ describes the probable values of a when the data are known. If this law were known, any number of key points a along the curve $p(a|\boldsymbol{y})$ could be used as appropriate estimates of a. (These points are defined by a *cost function,* taken up below.)

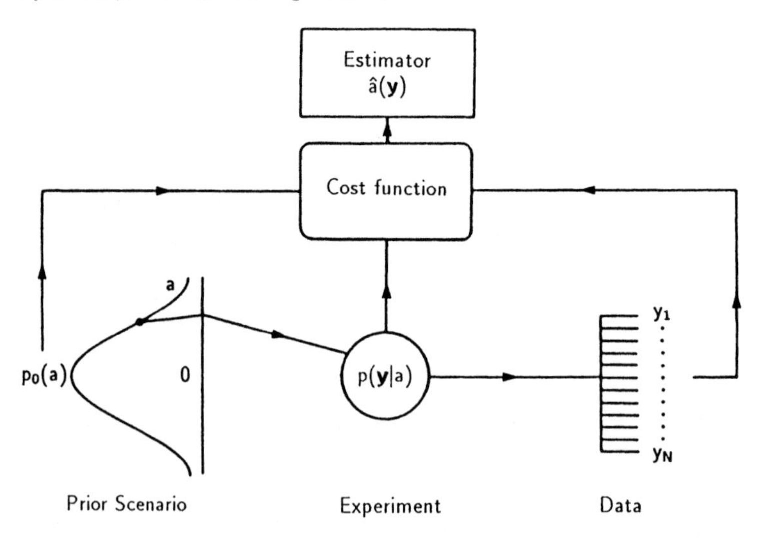

Fig. 17.2. The random parameter problem. Parameter a is chosen randomly from a prior statistic $p_0(a)$. Then a causes data $\boldsymbol{y}$ through random sampling of likelihood law $p(\boldsymbol{y}|a)$. Finally $p_0(a)$, $p(\boldsymbol{y}|a)$, $\boldsymbol{y}$, and the cost function form the estimator $\hat{a}(\boldsymbol{y})$

Since both $p(\boldsymbol{y}|a)$ and $p_0(a)$ are now known, we can actually form the posterior probability law. By Bayes' rule,

$$p(a|\boldsymbol{y}) = \frac{p(\boldsymbol{y}|a)p_0(a)}{p(\boldsymbol{y})} \tag{17.29a}$$

where

$$p(\boldsymbol{y}) = \int \mathrm{d}a\, p(\boldsymbol{y}|a)p_0(a) \,. \tag{17.29b}$$

A typical posterior probability law is shown in Fig. 17.3. Notice that it is simply a one-dimensional curve, in a; the $\boldsymbol{y}$ are fixed. The appropriate estimator to use is a subjective matter, and is defined by a *cost function* [17.2], described next.

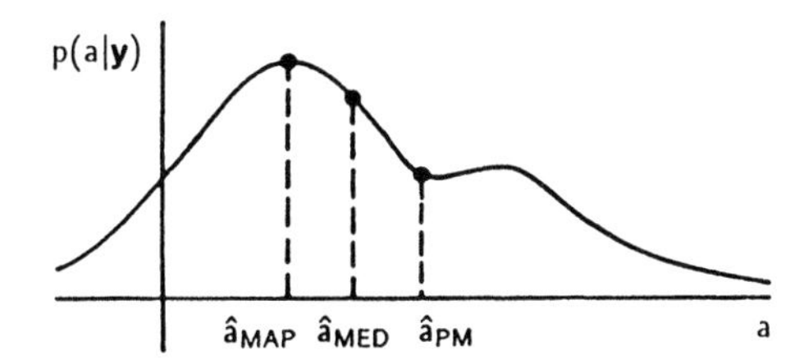

Fig. 17.3. A posterior probability law and some candidate estimators

17.2.1 Cost Functions

The appropriate a value to choose from a $p(a|\boldsymbol{y})$ curve depends upon the cost to the user of making errors of estimation. First-degree cost yields one value of a, second-degree a different one, etc. This is quantified as follows.

Let $C(a - \hat{a}(\boldsymbol{y}))$ designate the cost function appropriate to the user. Notice that it is a function of the error

$$e = a - \hat{a}(\boldsymbol{y}) \,. \tag{17.30}$$

Typical cost functions are shown in Fig. 17.4. A quadratic cost function penalizes larger errors more severely than does, e.g., the absolute cost function. As an opposite effect, the step cost function penalizes larger errors exactly as much as smaller errors, provided these exceed a critical value.

Estimator $\hat{a}$ is defined by minimizing *average cost*. The result is that the quadratic cost function, since it penalizes large errors extra strongly, tends to produce estimates $\hat{a}$ whose largest errors occur very infrequently. However, this means that small-to-midsize errors are, to a degree, ignored by the minimization procedure. Hence, these tend to grow in frequency. If relatively frequent midsize errors are intolerable to the user, he should consider use of a different cost function.

By constrast with the quadratic, the absolute cost function penalizes large errors less, with a stronger emphasis on minimizing the small-to-midsize errors. The result is estimation errors that tend to be very small, except for some very large ones occasionally. The user has to decide if he can tolerate occasional, very large errors. If not, a different cost function should be used.

Finally, the step cost function penalizes large errors equally with smaller ones, provided these exceed the threshold error in use. The result is a situation that is more extreme than the absolute error case. The emphasis is upon producing most often the errors that are less than the threshold in size, i.e., the tiniest errors, but at the expense of allowing the larger-than-threshold errors to occur relatively more frequently, and with about equal frequency.

The step cost function has many applications. It describes situations where "a miss is as good as a mile", i.e., where all depends upon getting the error less than the threshold value e_0. The penalty is zero, i.e., the goal is accomplished, if the error is less than e_0, but the goal is thwarted if e exceeds e_0 by a factor of 2, or by a factor of 100. Two idealized examples show up the situation.

Two airplanes are engaging in a dogfight. Each has missiles with proximity fuses. If airplane A can get his missile within distance e_0 from airplane B, he will win. However, if his missile misses airplane B by more than e_0 not only is B not eliminated but probably A is shot down by B's missile. Hence, to A, the result is success (no penalty) if distance e_0, or less, can be achieved, or disaster if e_0 is exceeded by *any amount.*

A similar situation occurs when a cancer cell is being irradiated by a focused energy beam. If the beam misses the cancer by less than a distance e_0 it will be destroyed. But if the miss exceeds e_0 by any amount, the cancer survives, to the detriment of the patient. The patient loses his life whether the miss is by $2e_0$ or $100e_0$.

17.2.2 Risk

Once the user decides on a cost function, his goal is to achieve the estimator $\hat{a}$ that *minimizes average cost.* The average cost is called *risk* R,

$$R \equiv \langle C(a - \hat{a}(\boldsymbol{y})) \rangle = \iint \mathrm{d}a\, \mathrm{d}\boldsymbol{y}\, C(a - \hat{a}(\boldsymbol{y}))\, p(a, y) = \text{minimum} \,. \quad (17.31)$$

This is sometimes called "risk aversion" in economic applications. Note that the averaging is over all random aspects of the problem, i.e., input a and data $\boldsymbol{y}$. An

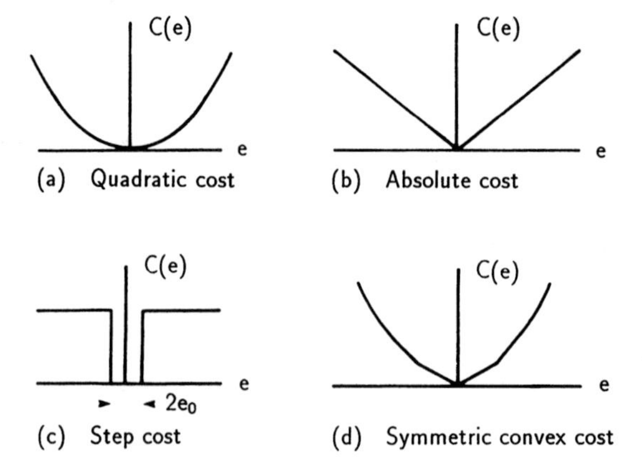

Fig. 17.4. Cost functions $C(e)$ that are commonly used

estimator $\hat{a}(\boldsymbol{y})$ that achieves the minimization of risk is called a *Bayes estimator.* We next find the Bayes estimator for the four cost functions shown in Fig. 17.4.

Quadratic Cost Function. Use of a quadratic cost function (Fig. 17.4a) in (17.31) gives

$$R = \int \mathrm{d}\boldsymbol{y}\, p(\boldsymbol{y}) \int \mathrm{d}a [a - \hat{a}(\boldsymbol{y})]^2 p(a|\boldsymbol{y}) \,. \tag{17.32}$$

The identity

$$p(a, \boldsymbol{y}) = p(\boldsymbol{y}) p(a|\boldsymbol{y})$$

was used, in order to bring in the posterior law whose particular point $\hat{a} = a$ is sought.

We seek to minimize R through choice of $\hat{a}$. Since $p(\boldsymbol{y})$ is positive and independent of $\hat{a}$, while the inner integral K of (17.32) is positive and depends on $\hat{a}$, R is minimized by minimizing the inner integral alone

$$\int \mathrm{d}a [a - \hat{a}(\boldsymbol{y})]^2 p(a|\boldsymbol{y}) \equiv K = \text{minimum} \,.$$

We assume the minimum to be at an interior point, so that

$$\frac{\partial K}{\partial \hat{a}} \equiv 0 = -2 \int \mathrm{d}a [a - \hat{a}(\boldsymbol{y})] p(a|\boldsymbol{y}) = 0 \,. \tag{17.33}$$

This is linear in $\hat{a}$, and so allows for direct solution,

$$\hat{a}(\boldsymbol{y}) = \int \mathrm{d}a\, a p(a|y) \equiv \hat{a}_{\mathrm{PM}}(\boldsymbol{y}) \,. \tag{17.34}$$

The normalization property

$$\int \mathrm{d}a\, p(a|\boldsymbol{y}) = 1 \tag{17.35}$$

was also used.

Result (17.34) shows that the quadratic cost function leads to the mean of the posterior probability law as the estimator. This is called the *posterior mean,* and is designated as $\hat{a}_{\mathrm{PM}}(\boldsymbol{y})$ (Fig. 17.3).

To verify that solution point (17.34) defines a minimum in the risk, and not a maximum, we form $\mathrm{d}^2 K / \mathrm{d}\hat{a}^2$. From (17.33) this has value $+2$, once identity (17.35) is used. Hence, the solution point defines minimum risk, as we wanted.

Absolute Cost Function. Use of the absolute cost function (Fig. 17.4b) in (17.31) gives

$$R = \int \mathrm{d}\boldsymbol{y}\, p(\boldsymbol{y}) \int \mathrm{d}a \left| a - \hat{a}(\boldsymbol{y}) \right| p(a|\boldsymbol{y}) \,. \tag{17.36}$$

By the same reasoning as above, we need only minimize the inner integral

$$K = \int \mathrm{d}a \left| a - \hat{a}(\boldsymbol{y}) \right| p(a|\boldsymbol{y}) \equiv \text{minimum} \,. \tag{17.37}$$

To allow differentiation $\mathrm{d}/\mathrm{d}\hat{a}$ we need to get rid of the absolute value sign. This is accomplished by judicious use of the integration limits,

$$K = -\int_{-\infty}^{\hat{a}(\boldsymbol{y})} \mathrm{d}a[a - \hat{a}(\boldsymbol{y})]p(a|\boldsymbol{y}) + \int_{\hat{a}(\boldsymbol{y})}^{\infty} \mathrm{d}a[a - \hat{a}(\boldsymbol{y})]p(a|\boldsymbol{y}) \,. \tag{17.38}$$

Note the Leibnitz's rule

$$\frac{\mathrm{d}}{\mathrm{d}x}\int_{g(x)}^{h(x)} \mathrm{d}y f(x, y) = f[x, h]h' - f[x, g]g' + \int_{g(x)}^{h(x)} \mathrm{d}y \frac{\mathrm{d}f(x, y)}{\mathrm{d}x} \tag{17.39}$$

where primes denote derivatives $\mathrm{d}/\mathrm{d}x$ Then from (17.38)

$$\frac{\mathrm{d}K}{\mathrm{d}\hat{a}} = -\int_{-\infty}^{\hat{a}(\boldsymbol{y})} \mathrm{d}a(-1)p(a|\boldsymbol{y}) + \int_{\hat{a}(\boldsymbol{y})}^{\infty} \mathrm{d}a(-1)p(a|\boldsymbol{y}) = 0 \,,$$

or

$$\int_{-\infty}^{\hat{a}(\boldsymbol{y})} \mathrm{d}a p(a|\boldsymbol{y}) = \int_{\hat{a}(\boldsymbol{y})}^{\infty} \mathrm{d}a p(a|\boldsymbol{y}) \,. \tag{17.40}$$

This defines the median $\hat{a}(\boldsymbol{y})$ of the curve $p(a|\boldsymbol{y})$. Hence, the absolute cost function leads to the median of the posterior probability law as the estimator. Denote this as $\hat{a}_{\mathrm{MED}}(\boldsymbol{y})$ (Fig. 17.3).

Step Cost Function. Use of the step cost function (Fig. 17.4c) in (17.31) gives

$$\begin{aligned} R &= \int \mathrm{d}\boldsymbol{y} p(\boldsymbol{y}) \int \mathrm{d}a \left[1 - \mathrm{Rect}\left(\frac{a - \hat{a}}{e_0}\right)\right] p(a|\boldsymbol{y}) \\ &= \int \mathrm{d}\boldsymbol{y} p(\boldsymbol{y}) \left[1 - \int_{\hat{a}-e_0}^{\hat{a}+e_0} \mathrm{d}a p(a|\boldsymbol{y})\right] \end{aligned} \tag{17.41}$$

after using normalization (17.35) once again. This is minimized when the inner integral is maximized. The latter is simply the area under a rectangular slice of $p(a|\boldsymbol{y})$, where the rectangle is centered on the point $a = \hat{a}$. Recalling that we are free to choose $\hat{a}$, i.e., to slide the rectangle along the curve $p(a|\boldsymbol{y})$, we slide the rectangle to the position a for which the largest rectangular area occurs. If the width $2e_0$ of the rectangle is small, this will simply be where $p(a|\boldsymbol{y})$ has its maximum in a.

Hence, the step cost function leads to an estimator $\hat{a}(y)$ which is the maximum probable value of a, given $\boldsymbol{y}$ (Fig. 17.3). This estimator is called the *maximum a posteriori* estimate, or MAP, since it is formed posterior to knowledge of $\boldsymbol{y}$. It is designated as $\hat{a}_{\mathrm{MAP}}$.

In practice, the MAP estimator is usually the easiest to find, among the three estimators considered so far. This is because its solution requires (usually) the mere differentiation of a known function $p(a|\boldsymbol{y})$, whereas the other estimators require its integration.

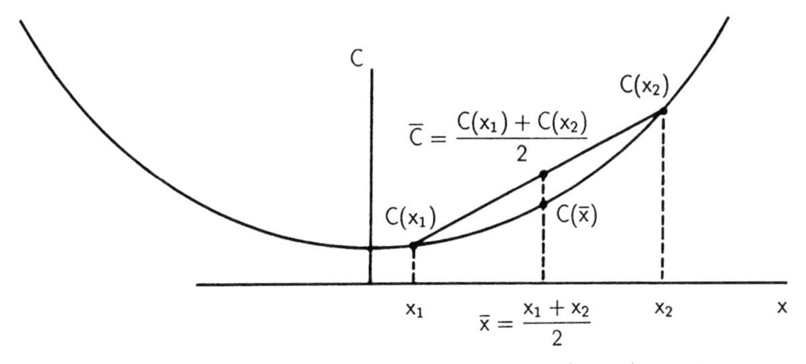

Fig. 17.5. For a convex cost function, by inspection $C\left(\frac{x_1+x_2}{2}\right) \leq \frac{C(x_1)+C(x_2)}{2}$.

Symmetric Convex Cost Function. Consider a cost function as in Fig. 17.4d. The convexity property means that the chord connecting any two points on the curve will lie on or above the curve. Obviously such a cost function includes (a) quadratic cost and (b) absolute cost as special cases (but not (c) step cost). Hence it is a very general cost function. Let the cost function also be symmetric,

$$C(a - \hat{a}) = C(\hat{a} - a)\ . \tag{17.42}$$

Assume, finally, that the posterior law $p(a|\boldsymbol{y})$ is *symmetric about its posterior mean.*

$$p(\hat{a}_{PM} - \epsilon|\boldsymbol{y}) = p(\hat{a}_{PM} + \epsilon|\boldsymbol{y}),\ \text{any}\ \epsilon\ . \tag{17.43}$$

This is a quite specialized property, not true of a general posterior law. As will be seen, however, it gives special meaning to the posterior mean point a_{PM}. *The estimator that minimizes the convex, symmetric cost function is then the posterior mean estimator.* This is shown next.

First, we need a preliminary property established. In Fig. 17.5, the x-value for point $\overline{C}$ must lie halfway between x_1 and x_2, since it is on the *straight line* chord connecting $C(x_1)$ and $C(x_2)$. Then it lies somewhere along a vertical line through $\overline{x} = (x_1 + x_2)/2$. But since the curve is convex, it must lie *above* point $C(\overline{x})$. Hence, the important inequality

$$C\left(\frac{x_1 + x_2}{2}\right) \leq \frac{C(x_1) + C(x_2)}{2} \tag{17.44}$$

is established.

The rest of the proof is simple algebraic manipulation. Let $\hat{a}(\boldsymbol{y})$ be the estimator that minimizes risk for the symmetric cost function. As before, we merely have to minimize the inner integral (17.32),

$$\int \text{d}a\, C(a - \hat{a}(\boldsymbol{y}))p(a/\boldsymbol{y}) \equiv R(\hat{a}|\boldsymbol{y}) \tag{17.45}$$

called the *conditional risk.* First, replace RV a by a new RV

$$z \equiv a - \hat{a}_{PM}(\boldsymbol{y}) \; . \tag{17.46}$$

Suppress the argument $\boldsymbol{y}$ of $\hat{a}_{\mathrm{PM}}(\boldsymbol{y})$. Then from (17.46)

$$a - \hat{a} = z + \hat{a}_{\mathrm{PM}} - \hat{a} \; .$$

Also, of course

$$p_Z(z|\boldsymbol{y})\,\mathrm{d}z = p(a|\boldsymbol{y})\,\mathrm{d}a \tag{17.47}$$

by (5.4). Then (17.45) becomes

$$R(\hat{a}|\boldsymbol{y}) = \int \mathrm{d}z\, C(\hat{a}_{\mathrm{PM}} - \hat{a} + z)\, p_Z(z|\boldsymbol{y})$$

$$= \int \mathrm{d}z\, C(\hat{a} - \hat{a}_{\mathrm{PM}} - z)\, p_Z(z|\boldsymbol{y})$$

by symmetry property (17.42) for the cost function,

$$= \int \mathrm{d}z\, C(\hat{a} - \hat{a}_{\mathrm{PM}} + z)\, p_Z(-z|\boldsymbol{y})$$

by merely replacing the integration variable z by $-z$,

$$= \int \mathrm{d}z\, C(\hat{a} - \hat{a}_{\mathrm{PM}} + z)\, p_Z(z|\boldsymbol{y}) \tag{17.48}$$

by symmetry property (17.43) for the posterior law,

$$= \int \mathrm{d}z\, C(\hat{a}_{\mathrm{PM}} - \hat{a} - z)\, p_Z(z|\boldsymbol{y})$$

by symmetry property (17.42) for the cost function,

$$= \int \mathrm{d}z\, C(\hat{a}_{\mathrm{PM}} - \hat{a} + z)\, p_Z(z|\boldsymbol{y}) \; , \tag{17.49}$$

again by replacing z by $-z$ and using symmetry property (17.43) for the posterior law. Next, we add (17.48) and (17.49). This gives

$$R(\hat{a}|\boldsymbol{y}) = \int \mathrm{d}z\, p_Z(z|\boldsymbol{y})[\tfrac{1}{2}C(\hat{a} - \hat{a}_{\mathrm{PM}} + z) + \tfrac{1}{2}C(\hat{a}_{\mathrm{PM}} - \hat{a} + z)] \; . \tag{17.50}$$

Observe that the $[\cdot]$ part is in the form of the right-hand side of inequality (17.44), where we identify

$$x_1 = \hat{a} - \hat{a}_{\mathrm{PM}} + z \; ,$$
$$x_2 = \hat{a}_{\mathrm{PM}} - \hat{a} + z \; .$$

Then by (17.44) the $[\cdot]$ part is greater than $C(z)$, or $C(a - \hat{a}_{\mathrm{PM}})$ by (17.46). Then from (17.50),

$$R(\hat{a}|\boldsymbol{y}) \geq \int \mathrm{d}z\, p_Z(z|\boldsymbol{y}) C(a - \hat{a}_{\mathrm{PM}}) \; .$$

Once again using identity (17.47), we have

$$R(\hat{a}|\boldsymbol{y}) \geq \int \mathrm{d}a \; p(a|\boldsymbol{y})C(a - \hat{a}_{\mathrm{PM}}) \; .$$

But by definition (17.45), the right-hand side is precisely $R(\hat{a}_{\mathrm{PM}}|\boldsymbol{y})$, the conditional risk associated with a posterior mean estimate. Hence

$$R(\hat{a}|\boldsymbol{y}) \geq R(\hat{a}_{\mathrm{PM}}|\boldsymbol{y}) \; . \tag{17.51}$$

The estimator that minimizes the convex, symmetric cost function is the posterior mean. The proviso is that the likelihood law must be symmetric about the posterior mean. This limitation is discussed next.

Limitation in Scope. When is the posterior law symmetric about $\hat{a}_{\mathrm{PM}}$? By Bayes' theorem (17.29a), this is when the product

$$p(\boldsymbol{y}|a)p_0(a) \tag{17.52}$$

of the likelihood law and prior probability law has this symmetry. This will be the case when, e.g., the likelihood law is normal (17.2), and $p_0(a)$ is either flat, exponential, or normal. (In these cases, completing the square in the exponent will lead to a normal, symmetric law.)

However, counter-examples abound. For example, a likelihood law of the Poisson form (17.1) is not symmetric about its mean, particularly for small N. It would take a quite contrived shape $p_0(a)$ to render the product (17.52) symmetric. The log-normal law, χ^2-law, Rayleigh law, and many others are also nonsymmetric about their means. Hence, these too would not obey the required symmetry, unless multiplied by very contrived $p_0(a)$ shapes.

As a practical matter, though, if the products are close to being symmetric, then theorem (17.51) will be close to being true. Estimator $\hat{a}_{\mathrm{PM}}$ will then be close to optimum, i.e., *suboptimum*. This gives important additional meaning to the posterior mean estimate $\hat{a}_{\mathrm{PM}}$.

17.2.3 MAP Estimates

We found that the MAP estimator $\hat{a}_{\mathrm{MAP}}$ follows from a step cost function. The MAP estimate is the value of a that maximizes the posterior law $p(a|\boldsymbol{y})$. If a is known to lie *within* an interval $a_0 \leq a \leq a_1$, then by elementary calculus $\hat{a}_{\mathrm{MAP}}$ is the solution to

$$\left.\frac{\partial \ln p(a|\boldsymbol{y})}{\partial a}\right|_{a=\hat{a}_{\mathrm{MAP}}(\boldsymbol{y})} = 0 \; , \tag{17.53}$$

called the *MAP equation.* (As before, we use the fact that a maximum in $\ln p$ lies at the same value a as a maximum in p directly.)

In most cases the posterior law must be inferred from knowledge of the likelihood and prior laws, in which case Bayes' rule (17.29a) must be used. Then the MAP equation (17.53) becomes

$$\left[\frac{\partial \ln p(\boldsymbol{y}|a)}{\partial a} + \frac{\partial \ln p_0(a)}{\partial a}\right]_{a=\hat{a}_{\mathrm{MAP}}(\boldsymbol{y})} = 0\,. \tag{17.54}$$

Note that $p(\boldsymbol{y})$ has dropped out, after taking $\partial/\partial a$, since it is not a function of a.

In the special case of a flat prior law $p_0(a) = \text{constant}$, the MAP equation (17.54) becomes the likelihood equation (17.6). Hence $\hat{a}_{\mathrm{MAP}}$ and $\hat{a}_{\mathrm{ML}}$ are the same when the prior law is flat. This gives an alternative meaning to the maximum likelihood principle.

Examples of the use of MAP equation (17.54) follow next, and after Ex. 17.2.

All-normal Case. Consider the case (17.2) of a normal likelihood law, and a Gaussian prior law $p_0(a)$,

$$a = N(0, \mu^2)\,. \tag{17.55}$$

The MAP equation is

$$\frac{\partial}{\partial a}\left[-\frac{1}{2\sigma^2}\sum_n (y_n - a)^2 - a^2/2\mu^2\right]_{a=\hat{a}_{\mathrm{MAP}}} = 0\,.$$

The solution is

$$\hat{a}_{\mathrm{MAP}} = \frac{\overline{y}}{1 + \sigma^2/(N\mu^2)}\,. \tag{17.56}$$

Recall that $\hat{a}_{\mathrm{ML}} = \overline{y}$ for this case (17.2); see Ex. 17.1.1. Hence $\hat{a}_{\mathrm{MAP}}$ is not the same as $\hat{a}_{\mathrm{ML}}$ in this case.

Exercise 17.2

17.2.1 Consider a normal likelihood law (17.2) and an exponential prior law

$$p_0(a) = \frac{1}{\langle a \rangle}\exp(-a/\langle a \rangle)\,. \tag{17.57}$$

Show that

$$\hat{a}_{\mathrm{MAP}} = \overline{y} - \frac{\sigma^2}{N\langle a \rangle}\,.$$

Is this estimator unbiased? Discuss the meaning to $p_0(a)$ and $\hat{a}_{\mathrm{MAP}}$ of cases $\langle a \rangle \to 0$, $\langle a \rangle \to \infty$.

17.2.2 Consider the problem of estimating star density a from an independent observation of N nearest-neighbor distances $\boldsymbol{y}$. According to (6.58), the likelihood law is Rayleigh

$$p(\boldsymbol{y}|a) = \prod_{n=1}^{N} 2\pi a y_n \exp(-\pi a y_n^2)\,.$$

Show that if the prior law $p_0(a)$ is flat, then

$$\hat{a}_{\text{MAP}} = \hat{a}_{\text{ML}} = \frac{N}{\pi \sum_n y_n^2} \,. \tag{17.58}$$

This is a non-intuitive result. Is the estimator unbiased?

Show that if the prior law is exponential (17.57), the solution for $\hat{a}_{\text{MAP}}$ closely resembles (17.58).

17.2.3 ***Estimating a universal physical constant by MAP.*** Suppose that a universal physical constant, of unknown value a, is measured N times. The measurements $\boldsymbol{y}$ suffer from additive Gaussian noise $N(0, \sigma^2)$. How should a be estimated? Ordinarily the sample mean of the data are taken as the estimate. But, is this a maximum probable estimate?

Recall that the physical constants obey a reciprocal probability law (9.98). This, then, defines a prior probability law $p_0(a)$ for the constant. This is a real effect, providing extra information about the value of the constant, and so should be built into its estimate. Of course a MAP estimate does just that. Note in particular that since $p_0(a)$ is large at small a, *the MAP estimate should favor small estimates* in some sense.

From the preceding, the likelihood law $p(\boldsymbol{y}|a)$ is a product (17.2) of normal laws. Coupled with our knowledge (9.98) that $p_0(a) = (2a \ln b)^{-1}$, the MAP equation (17.54) becomes

$$\frac{\partial}{\partial a}\left\{-\sum_n \left[\ln(\sqrt{2\pi}\,\sigma) + \frac{(y_n - a)^2}{2\sigma^2}\right] - \ln(2a\ln b)\right\}_{a=a_{MAP}} = 0\,.$$

This assumes that the answer will lie within the known range $(10^{-b}, 10^{b})$ of values of a. If it doesn't, then the approach is inconsistent, and some other way of achieving the maximum must be found, e.g., a simplex method. Assume that it lies within the interval.

Carrying through the indicated differentiation, show that the MAP solution obeys

$$\hat{a}_{MAP} = \frac{\bar{y}}{2}\left[1 + \sqrt{1 - \left(\frac{4}{N}\right)\left(\frac{\sigma}{\bar{y}}\right)^2}\,\right]. \tag{17.59}$$

Show that a minus sign before the radical is another solution for extremizing the posterior probability, but the extremum is a minimum and not the maximum as is desired. *Hint*: Merely take a second derivative and examine its sign, in the usual way.

Notice that, since the radical is generally ≤ 1, the MAP answer (17.59) is generally *less than* the sample mean $\bar{y}$. This *downward departure* from $\bar{y}$ can be substantial. An example is the case of a newly discovered constant, where N might be quite small and/or the level of noise σ appreciable relative to $\bar{y}$. Naively taking the average of the measurements would then give estimates that are significally too large. Of course this is because just using the sample mean ignores *the bias toward small a* shown in the prior probability law. This is important information.

However, in the opposite case of a large number N of measurements and/or a small noise level σ, the estimator $\hat{a}_{\text{MAP}} = \bar{y}$. The naive sample mean is now MAP as well. This is also the ML answer. *Why?*

It may be noted from (9.5) that $\sigma^2/N = \sigma_{\bar{Y}}^2$, the mean-square error *of the sample mean* from a. Using this in (17.59), we find that in cases where $\sigma_{\bar{Y}}$ is small relative to $\bar{y}$

$$\hat{a}_{MAP} \approx \bar{y}\left[1 - \left(\frac{\sigma_{\bar{Y}}}{\bar{y}}\right)^2\right] \approx \bar{y}\left[1 - \left(\frac{\sigma_{\bar{Y}}}{a}\right)^2\right].$$

Show the first approximation by expanding out the radical by Taylor series in (17.59). The second follows from the fact that $\bar{y} \approx a$ under these circumstances. Then with a fixed, small $\sigma_{\bar{Y}}$ the downward shift from $\bar{y}$ of the MAP estimator is large for *small* physical constants a, and small for *large* physical constants. Why is this a reasonable result?*Hint*: Look at the shape of the curve $p_0(a)$.

17.2.4 ***Cramer–Rao inequality derived from Bhattacharyya bound.*** The Cramer–Rao inequality (17.18) can be derived by the approach of Sect. 17.1.6. Show that in the case $N = 1$ the inequality (17.27) becomes (17.18). This avoids use of the Schwarz inequality in derivation of (17.18). Furthermore, show that in this case the efficiency condition (17.28) goes over into the Cramer–Rao efficiency condition (17.23).

17.2.5 ***Fisher information in a joint measurement: Possibility of perfect processing.*** Suppose there are two Fisher variables (y_1, y_2) formed by data equations

$$y_1 = a + x_1, \quad y_2 = a + x_2 .$$

Variables x_1, x_2 are noise fluctuations that obey a joint Gaussian probability density with a common variance σ^2 and a general correlation coefficient ρ. Quantity a is regarded as the unknown parameter of the problem.

(a) Show by direct calculation using (17.19) that the Fisher information in joint data readings (y_1, y_2) about a obeys

$$I = \frac{2}{\sigma^2}\left(\frac{1}{1+\rho}\right).$$

(b) Discuss the behavior of the information as σ^2 and ρ range over their possible values. Notice that when $\rho = 0$ the information is exactly double that of a single Gaussian variable [in (3.99)], and as ρ increases from value zero the information level monotonically drops. Why?

(c) Cases where $\rho < 0$ are intriguing because they give information values that are larger than in the uncorrelated case! Why should negatively correlated noise convey more information than uncorrelated noise? Remember that the ultimate significance of any information value I is its affect upon predicted error via the Cramer–Rao relation (17.18). The specific answer arises out of taking the ultimate limit $\rho = -1$. Then the information $I \to \infty$. By (17.18) this means that there is an estimator function $\hat{a} = \hat{a}(y_1, y_2)$ that achieves an error $e^2 = 0$! In fact this estimator function is easy to find. Find it. *Hint:* Given that $\rho = -1$, in the presence of a noise fluctuation x_1 what must x_2 be?

Particle Tracking; The Scintillation Camera. A fascinating problem in Bayesian estimation is provided by a camera whose image requires MAP estimation for its formation (Fig. 17.6a). In the scintillation camera, the image is built up of individual γ-ray arrivals. However, as shown in the figure, the γ-ray positions x that form the image are not observed directly. Instead, two photomultipliers A, B register numbers n_1, n_2, respectively, of secondary photons that are emitted from the phosphorescent screen that the γ-ray strikes.

The problem is to estimate the image coordinate x on the screen based on photon counts n_1, n_2 as data. Hence, an estimator $\hat{x}(n_1, n_2)$ is sought. Coordinate y (out of the page) can be estimated from two other photomultipliers C, D placed alongside A, B respectively, so that A, B, C, D form four quadrants about the origin of image space (Fig. 17.6b).

A mathematically identical problem occurs in airplane tracking. With the airplane very far away its image is a point spread function consisting of photon counts $(n_1, n_2, n_3, n_4) \equiv \boldsymbol{n}$ in the four quadrants surrounding an origin of coordinates (Fig. 17.6b). From the data $\boldsymbol{n}$ it is desired to estimate the coordinates x, y of the airplane in the image field.

We show a one-dimensional analysis. The two-dimensional problem is attacked analogously. Or, each of the x, y can be separately estimated as one-dimensional problems, with slightly less accuracy.

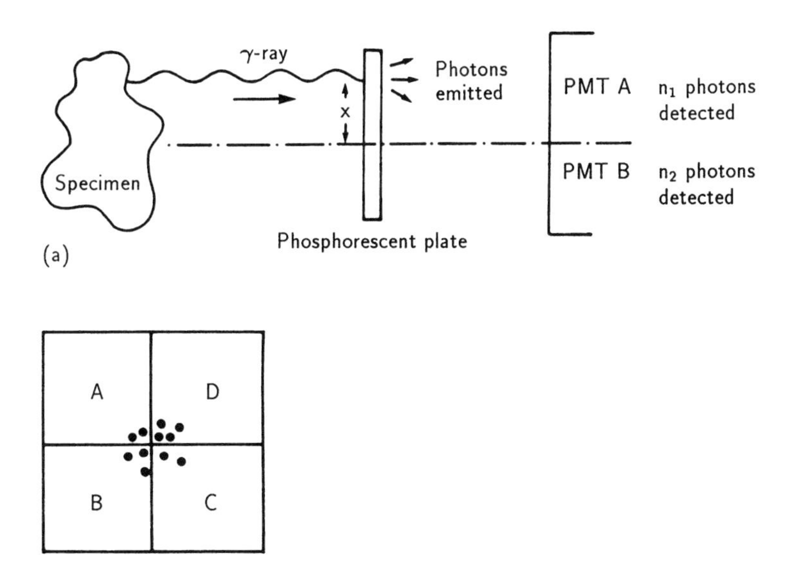

Fig. 17.6. (a) Scintillation camera problem (one dimensional version). From data n_1, n_2, estimate coordinate x. (b) Two-dimensional tracking problem. Four quadrant detectors A, B, C, D collect $\boldsymbol{n} = (2, 3, 2, 4)$ photons, respectively. The problem is to estimate the origin position of the point spread function that formed this data

Taking a Bayesian approach, we must first form the likelihood law $P(\boldsymbol{n}|x)$. Physically, counts n_1, n_2 are independent Poisson, so that

$$P(\boldsymbol{n}|x) = P(n_1|x)P(n_2|x)\ ,$$

where

$$P(n_i|x) = \mathrm{e}^{-\lambda_i}\lambda_i^{n_i}/n_i!\ , \quad \lambda_i = \langle n_i|x\rangle\ , \quad i = 1, 2\ . \tag{17.60}$$

Note that each λ_i is the mean count, for x fixed. Each $\lambda_i = \lambda_i(x)$ must be known, since we anticipate that any estimator $\hat{x}$ will be a function of the λ_i values. In practice, the λ_i values may be found by a process of *calibration.* A pinhole source is placed at successive x values. At each x, the longterm average counts $\langle n_i|x\rangle = \lambda_i$, $i = 1, 2$ are observed. Assume, then, that each function $\lambda_i(x)$ is known.

We next have to form the prior probability law $p_0(x)$. This will depend upon the particular application. In the airplane tracking problem, once the airplane has been located, subsequent tracking should see close to a Gaussian law for $p_0(x)$. In the scintillation camera problem, if a known internal organ is being observed then $p_0(x)$ should be the average intensity image for organs of that type.

To keep the analysis simple, let us assume that the prior law is flat over a unit interval, $p_0(x) = \mathrm{Rect}(x - 1/2)$. We seek $\hat{x}_{\mathrm{MAP}}$, which is the same as $\hat{x}_{\mathrm{ML}}$ in this case. Given that the likelihood law is independent Poisson (17.60), the MAP equation is now

$$\frac{\partial}{\partial x}\left\{\ln\left[\mathrm{e}^{-\lambda_1(x)}\frac{\lambda_1(x)^{n_1}}{n_1!}\mathrm{e}^{-\lambda_2(x)}\frac{\lambda_2(x)^{n_2}}{n_2!}\right]\right\} = 0\ . \tag{17.61}$$

This becomes

$$\frac{\mathrm{d}\lambda_1(x)}{\mathrm{d}x}\left[\frac{n_1}{\lambda_1(x)} - 1\right] + \frac{\mathrm{d}\lambda_2(x)}{\mathrm{d}x}\left[\frac{n_2}{\lambda_2(x)} - 1\right] = 0\ . \tag{17.62}$$

To solve this for the root $\hat{x}_{\mathrm{MAP}}$, some functional form must be assumed for $\lambda_i(x)$, $i = 1, 2$. For simplicity, assume that these are linear,

$$\lambda_1(x) = mx, \quad \lambda_2(x) = m(1 - x), \quad 0 \le x \le 1\ .$$

(Then $\lambda_1(x) + \lambda_2(x) =$ constant, which is reasonable by conservation of energy.) Substituting these forms into (17.62) immediately yields a solution

$$\hat{x}_{\mathrm{MAP}} = \frac{n_1}{n_1 + n_2}\ . \tag{17.63}$$

Note that this is nonlinear in data n_1, n_2. This solution is commonly used in tracking problems. Its use in a simulation with a medical thyroid "phantom" is shown in Fig. 17.7.

MAP Image Restoration. Image formation is a stochastic process whereby an object intensity function $o(x)$ is blurred via convolution with a point spread function $s(x)$, and is further corrupted by the addition of noise; see (8.51). The aim of image restoration is to recover the object $o(x)$ from observed data related to the image function $i(x)$. We consider a problem where the data consists, in fact, of *finite photon*

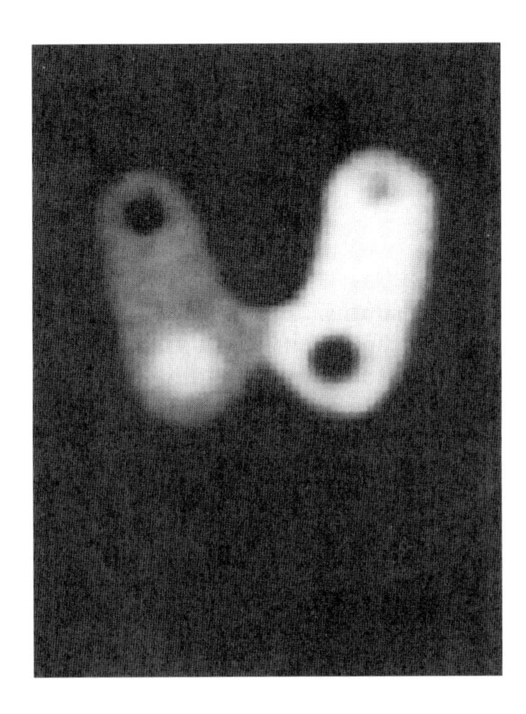

Fig. 17.7. Estimated γ-ray positions $\hat{x}_{\text{MAP}}$ form the image of a medical phantom thyroid [17.7]

counts $n_1, \ldots, n_N \equiv \boldsymbol{n}$ in N detectors covering image space (Fig. 17.8). This is the case when an array of solid state detectors is used to pick up the data. Such detectors are capable of counting photons. This image acquisition approach is currently state-of-the-art in astronomy, space research, and earth surveillance systems.

Regard the object as a discrete sequence $o_1, \ldots, o_N \equiv \boldsymbol{o}$ (Fig. 17.8) of intensities, where each o_n value is the average value of $o(x)$ over a resolution length Δx. The latter is presumed to be the user's goal in acquired resolution. Then the imaging equation (ignoring the noise) becomes a sum

$$i_m = \sum_{n=1}^{N} o_n s_{mn} \, , \quad m = 1, \ldots, N \tag{17.64}$$

where $s_{mn} \equiv s(x_m - x_n)$. Hence, the problem has now been entirely discretized. This aids in its analysis.

The noise in the image data $\boldsymbol{n}$ is presumed to be entirely due to finiteness of duration of exposure. Hence, each n_m is a random sample from a probability value i_m given by (17.64). It follows that

$$\langle n_m \rangle = K i_m \, , \tag{17.65}$$

with K a constant of the image-forming apparatus.

We seek a MAP estimator $\hat{\boldsymbol{o}}_{\text{MAP}}(\boldsymbol{n})$. Hence, we again need a likelihood law $P(\boldsymbol{n}|\boldsymbol{o})$ and a prior probability law $p_0(\boldsymbol{o})$. Let the light be ordinary ambient light, i.e.,

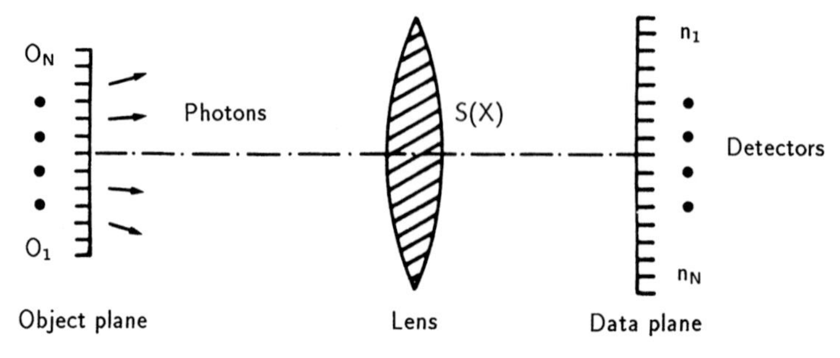

Fig. 17.8. The image restoration problem, when a photon-counting array of detectors is used to acquire the data

thermal, and for simplicity assume it to be linearly polarized. Then, by Ex. 10.1.19, the likelihood law is negative binomial,

$$P(\boldsymbol{n}|\boldsymbol{o}) = \prod_{m=1}^{N} \frac{(n_m + z - 1)!}{n_m!(z-1)!} p_m^{n_m}(1 - p_m)^z \tag{17.66}$$

where

$$p_m = \frac{\langle n_m \rangle}{\langle n_m \rangle + z} \,. \tag{17.67}$$

Note that parameters p_m bring in the unknowns $\boldsymbol{o}$ through combination of (17.64, 65, 67).

Parameter z is the number of degrees of freedom per detector. It is modeled [17.4] as the number of photons that may be crowded into the "detection volume" $\mathcal{V}$ swept out by photons during the finite exposure time T,

$$\mathcal{V} = ATc \,. \tag{17.68}$$

Here A is the area of a detector and c is the speed of light. The effective volume of a photon, by this model, is

$$V = A_{\text{coh}} \tau c \tag{17.69}$$

where A_{coh} is the coherence area and τ is the coherence time (3.76). Hence

$$z \equiv \frac{\mathcal{V}}{V} = \frac{A}{A_{\text{coh}}} \frac{T}{\tau} \tag{17.70}$$

by (17.68, 69).

The estimate $\hat{\boldsymbol{o}}_{\text{ML}}(\boldsymbol{n})$ based on likelihood law (17.66) turns out to be the simple inverse solution $[S]^{-1}\boldsymbol{n}$ given in Exercise 17.1.5. This is not a satisfactory estimator, however; see Fig. 8.11c. The role of the prior probability law $p_0(\boldsymbol{o})$ will be to force Bayesian estimators away from simple inverse solutions. Its use is thus crucial, in this application.

The prior probability law consistent with knowledge of the mean-square value

$$\langle o^2 \rangle = \frac{1}{N} \sum_{n=1}^{N} \langle o_n^2 \rangle \tag{17.71}$$

across the object is a one-sided Gaussian law,

$$p_0(\boldsymbol{o}) = a \exp(-\lambda \sum_n o_n^2) \tag{17.72}$$

according to the principle of maximum entropy (Sect. 10.4.9). The same answer results from alternative use of the principle of *extreme physical information* (see Eq. (17.132)). It is found that parameter λ relates to the known $\langle o^2 \rangle$ as approximately

$$\lambda = N^3 \ln \frac{N(N+1)\langle o^2 \rangle}{2} . \tag{17.73}$$

Now that the likelihood law and prior law are known, these may be used to form a MAP solution. Substituting laws (17.66) and (17.72) into the n-dimensional MAP equation [compare with (17.54)]

$$\left[\frac{\partial \ln P(\boldsymbol{n}|o)}{\partial o_i} + \frac{\partial \ln p_0(\boldsymbol{o})}{\partial o_i} \right]_{o_i = \hat{o}^{(i)}_{\text{MAP}}(\boldsymbol{n})} = 0 , \quad i = 1, \ldots, N , \tag{17.74}$$

yields the MAP estimates. In practice, the N equations that so result must be solved numerically. Figure 17.9 shows one demonstration of the approach [17.3]. The object (top) was blurred laterally with a rectangle point spread function of width 9 pixels, and was made noisy by computer simulation of negative binomial image statistics (17.66). The image that so results is the middle picture. This comprises data $\boldsymbol{n}$ to be fed into the MAP approach (17.74). The output $\hat{\boldsymbol{o}}_{\text{MAP}}(\boldsymbol{n})$ is shown as the bottom picture. Note that both the visual appearance and the quantitative rms error were improved over those of the image data.

17.3 Exact Estimates of Probability Laws: The Principle of Extreme Physical Information

Probability law estimation is the subject of Chap. 10, Ex. 9.2.1 and Sect. 5.14. The overall aim of Chap. 10 is to produce a *good empirical estimate* of the unknown law. This is an estimate based upon given empirical data from the law. It is tacitly assumed that the estimate rarely, if ever, *equals* the unknown law.

By contrast, the approaches of Sect. 5.14 and Ex. 9.2.1 are not data-based but, rather, are logic-based. They aim to construct the law solely by the use of an appropriate *invariance principle*. Also, the goal is a lot more ambitious than to produce an empirical estimate. It is to produce *the exact* probability law! Sometimes it succeeds in this quest, as for example in Ex. 9.2.1 and reference [5.15]. Alternatively, some invariance principles, such as invariance to reciprocation, are inadequate to by themselves produce the unique answer; see (5.107). In cases such as these an

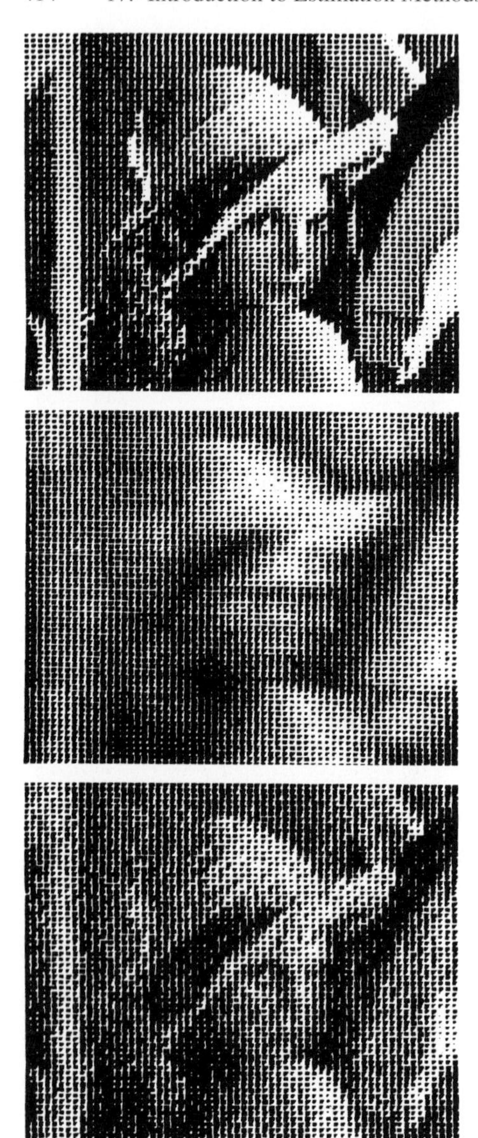

Fig. 17.9. *Top:* object; *middle:* image, blurred laterally by 9 pixels, S/N = 20.4, quantum degeneracy $\langle n \rangle / z = 5$. Root mean square error from object = 0.0061. *Bottom:* MAP restoration using known object variance; rms error is reduced to 0.0043 [17.3]

additional invariance principle is needed to winnow out the right one. In Sect. 5.14, it was invariance to change of units.

As is evident from the preceding examples, such use of an invariance principle is problem-specific: each is attacked by generally a different principle. Then the overall approach would be even more valuable if it were known that *a single invariance principle exists that applies generally, that is, physically.* Can such a principle exist? In fact we will find that it is (17.79), *invariance of an* L^2 *length to unitary transformation*. Mathematically, we will see that it arises out of the L^2 form (17.19) of Fisher information. Why the latter, is discussed in sections to follow.

17.3.1 A Knowledge-Based View of Nature

Nature is full of symmetries or invariances. Is there a most basic one?

The probability laws we know with extreme accuracy are those that describe quantifiable effects: those of physics but, also, of chemistry, genetics, etc. Hence our invariance principle must be capable of deriving the probability laws of a wide range of phenomena. What could all have in common to permit a common approach? An immediate candidate would be the concept of energy, but, energy is undefined in genetics, for example. Our viewpoint is that *the commonality of these phenomena lies in our ability to observe them.*

By the term "physical law" we will mean any quantifiable phenomenon of nature. Data are taken so as to acquire knowledge about nature, such as the position of an electron. In fact, *a physical law is really but an optimally brief way of expressing all the data that ever have been, or will, be taken about a phenomenon.* The use of mathematical symbols and operations allow for the compression. However, measurements are imperfect. So, therefore, must be our knowledge of nature.

Since our knowledge of nature is imperfect, physical laws cannot be absolute truths but, rather, estimates, or models, of nature. Of course, the better an estimate or model is, the more confidence we have that it describes reality; but in any event it describes what *we know* about nature, not nature itself. Thus, the famous wave functions of physics describe what we know about quantum systems, and not necessarily the states of the systems. This ties in with our concept of a probability law (Chap. 2). It describes what the observer knows about the random variable, not the state of the random variable itself.

If the laws we seek are but models, what actually exists? Our premise is that observed data exist. They are not, e.g., the subject of a dream, hallucination or prank. Furthermore, the act of measurement presumes an existent measurer, along with the phenomenon under measurement and the measuring instrument itself. *These four items are the elements of a closed measuring "system"*. In brief, the closed measuring system consists of four existent items:

(i) The phenomenon under measurement.
(ii) The measuring instrument.
(iii) The acquired data.
(iv) The observer.

Anything exterior to this closed system *is taken to be a model of reality*, only existing in our brains. This includes the statements of physical laws defining probability laws, *except at the particular values taken by the observed data.*

We further discuss the last point. A data value is assumed to be a random sample from a given probability law (as in Chap. 7). Such a data point exists, by hypothesis (item (iii) above). However, points that are not observed cannot be presumed to exist, even if they have a high probability of occurrence. Thus, non-observed positional values of a particle are regarded as predictions of the Schroedinger wave equation, and do not necessarily exist. Only when a position is observed can the position be said to exist.

Let us consider the above list in the light of Descarte's adage of existence, "I think, therefore I am". In agreement with Descarte, the observer who measures must think and, therefore, exist. But also, he must use a real measuring instrument and acquire real data from a real phenomenon. Moreover we will find that each *such datum implies a model physical law* defining a probability law. In this way, the items (i)-(iv) somewhat extend the scope of Descarte's adage, to "I measure and therefore both I and it exist", where by "it" is meant a physical law.

The invariance principle we seek follows *from detailed consideration of the manner by which measurements are formed out of physical laws.* Whereas physical laws give rise to data, the opposite is true as well: *analyzing how data arise* will show us how to derive physical laws and predict probabilities. This is the general procedure to be taken in the following sections.

As was noted, the laws we will so establish are not restricted to statements about probability laws. Truths about probability laws are inextricably attached to truths about other aspects of nature. For example, the following peripheral questions will be addressed: *What quantities are "legitimately" measured? What constants must be universal physical constants? What are the ultimate resolution lengths of physics? What is the basic connection between energy and mass; between energy and information? What uncertainty principles should exist?*

The invariance principle we will develop from these considerations is called the principle of *Extreme Physical Information* or EPI. Here we give an introduction to the approach, mainly on the one-dimensional level. The full development, which applies to generally multiple-dimensioned, multiple-particle phenomena, is the subject of the book [17.12].

The starting point to EPI is to analyze *exactly how* measurements are acquired. This requires a *physical interpretation* of measurement. The interpretation will, in return, imply the structure of the probability law that gave rise to the data.

17.3.2 Fisher Information as a Bridge Between Noumenon and Phenomenon

As we saw, physical laws: (a) describe our state of knowledge of given phenomena, and (b) are simply optimally brief encodings of all the data ever taken about a given phenomenon. Now, *data are obtained in answer to a question.* What type of question?

The type of question we assume is, "What is the size of a parameter"? For example, what are the coordinates of a particle? The data are presumed to not be the response to questions of existence, such as, is there a particle there? Such questions do not necessarily perturb the phenomenon, as needed below.

The question initiates the measurement process. Of course the questioner seeks the truth about a measured parameter or, equivalently, seeks the truth about the underlying probability law it is sampled from. Suppose, then, that an object is measured, as in Fig. 17.10.

The observed measurement necessarily suffers from noise, or fluctuation, to some degree. The nature of these fluctuations is of interest. *The physical law we seek is the one that forms the probability law on the fluctuations.* These intrinsic fluctuations are to be distinguished from error fluctuations due to a measuring instrument. The measuring instrument is assumed to add negligable error to its readings. In effect, the error fluctuations are presumed exterior to the closed system (i)-(iv) above. Although a more general theory [17.12] allows the instrument to incur its own fluctuations, here we ignore them in the interests of simplicity.The upshot is that the observed data suffers from fluctuations x *that are purely characteristic of the phenomenon.* These are called "intrinsic fluctuations". No exterior sources of fluctuation enter in.

For example, the intrinsic fluctuations in position x of a particle obey the phenomenon called the Schroedinger wave equation (SWE). Knowledge of the SWE allows a probability amplitude function $\psi(x)$ to be formed, whose modulus-square is the required probability law $p(x)$. In general, we will address probability laws that are the squares of amplitude laws.

In what follows we will use the SWE to exemplify general aspects of the approach.The SWE exists or "arises" whenever an attempt is made to measure the position of a particle. Here the ideal position value a is sought, but, because of the intrinsic randomness in position of the particle, the datum y incurs a random

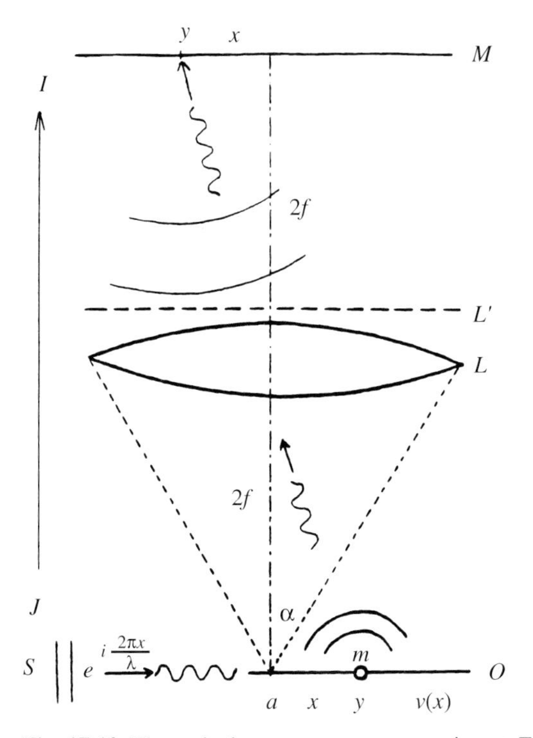

Fig. 17.10. The gedanken measurement experiment. Estimate parameter a (the mean coordinate) from observation of a single coordinate value y

component x, where

$$y = a + x \,. \tag{17.75}$$

It is assumed that *the statistics of its random fluctuations x are independent of the size of the ideal parameter value a*. This is a case of *additive noise* (Sect. 8.9), resulting in shift-invariant statistics (see below). All probability laws obeying shift-invariant statistics will be considered.

It is the probability law on y or on x that is the subject of the SWE. Thus, the SWE arises out of a quest for knowledge about a parameter, a. It also is the answer to a particular question, namely, *where is the particle*? All probability laws that address such questions about the size of observables will be considered. Again, physics is knowledge-based.

Notice that this viewpoint is the reverse of the conventional view that data arise out of the SWE. Of course, the latter is true, but in fact the implication can go either way: the SWE *is equivalent to* its data. It may not be surprising to find, then, that the SWE *may be derived* on the basis of how its data are formed. More generally, a detailed consideration of what goes on during measurement gives a principle from which all physical laws may be derived.

The amount of "truth" in a measurement was found at inequality (17.18) to be defined by its level of Fisher information (17.19). For a single data value y this is

$$I_F(p_a;\alpha) \equiv I = \int \mathrm{d}y\, p_a \left(\frac{\partial \log p_a}{\partial a}\right)^2 \equiv \left\langle \left(\frac{\partial \log p_a}{\partial a}\right)^2 \right\rangle, \quad p_a \equiv p_a(y|a)\,. \tag{17.76}$$

The indicated average is, by definition, with respect to the likelihood law p_a. To review earlier parts of this chapter, quantity a is a parameter that it is desired to estimate by means of a function $\hat{a}(y)$ of the datum y called an "estimator function". The *immediate* significance of I is that it predicts *the smallest possible mean-square error that any estimator can achieve*. That error is $1/I$, by the Cramer–Rao inequality (17.18). Information I has been used for this diagnostic purpose for decades. However, the significance of I is actually much broader. Ultimately, I it provides us with a world-view, i.e., a view of physical reality and how we measure it.

At this point, our net view is that *all physical processes specified by probability laws arise out of a quest for knowledge, in the form of a question about the size of a parameter.* The answer, provided by the measuring system, is a data value (or datum). The amount of "truth" that exists in the datum is measured by its level of Fisher information. Indeed, the Fisher information level in a measurement of position, when used with the Cramer–Rao inequality, implies the Heisenberg uncertainty principle (Ex. 17.3.6). This is a direct expression of the amount of truth in the datum. A physical law is a very compact way of representing all such data. Hence, the amount of truth that exists in the inferred law likewise depends upon the level of Fisher information in the datum. Thus we may conclude:

An invariance principle that aims to derive the physical laws governing the probability laws involved in acquiring knowledge should depend in some fundamental way upon the Fisher information in the data which ultimately form the knowledge.

A corollary is that other information measures (see Sect. 3.18) do not apply here. These do not directly measure the quality of data taken about unknown parameters. For example, the Shannon entropy measures instead the *number of distinguishable messages* that can be transmitted by a channel.

It is interesting that the Cramer–Rao inequality quantifies ideas of the ancient Greek philosopher Plato and the German Immanuel Kant as to what constitutes ultimate knowledge. Their thought is that absolute truth about nature exists, which Kant [17.13] called *the noumenon*. However, the observer can rarely sense it. What he usually senses instead is an imperfect version of it called *the phenomenon*.Thus, man senses phenomena and not the noumena that are his aims. There is then a basic *duality of levels of knowledge* in nature.

This duality of truth will, in fact, provide a foundation for our invariance principle. The two levels of truth represent to man two levels of knowledge. We propose that *the two levels of knowledge – phenomenon and noumenon – carry levels of Fisher information I and J, respectively*. Their difference, $I - J \equiv K$, then represents the difference between what "is" and what "is perceived". On this basis K could be suspected of playing a key role in defining the invariance principle we seek.

17.3.3 No Absolute Origins

The case (17.75) of additive noise has a curious effect upon the Fisher information I. With additive noise, equation (8.47b) gives $p_a(y|a) = p(y-a)$, called *shift-invariant statistics*. Using this in (17.76), after a change of integration variable $y - a \equiv x$ we get

$$I = \int \mathrm{d}x\, p \left(\frac{\partial \log p}{\partial x}\right)^2 = \int \mathrm{d}x \left(\frac{p'^2}{p}\right), \quad p' \equiv \mathrm{d}p/\mathrm{d}x\,. \qquad (17.77)$$

The parameter value a has dropped out! (See Ex. 17.3.1) This is useful because the value of a is, by hypothesis, unknown. If I depended upon an unknown number a it would be difficult to use as a tool for finding $p(x)$. The latter would be parametrized by a.

Notice that the invariance of (17.77) to the value of a *does not* mean that the observable y contains zero information *about a*. The information about a is defined to be the I given by (17.77). What the invariance to a does mean is that *observation of y gives no information about an absolute position in data space.*

As an example, consider a quantum mechanical particle moving in the force field of a simple spring. Imagine some absolute origin of coordinates O to be set up in space, as in Fig. 17.10. Relative to this origin, the spring rest position is value a, and the general position of the particle is y. Then the potential function is $V = Kx^2$ with K the spring constant and coordinate $x = y - a$. The SWE defines a probability amplitude function whose square is the density $p(y-a) = p(x)$, and not $p(y)$ alone. Of course, if the origin of coordinates O is shifted, then both the rest position a and the data value y will shift by the same amount, leaving $x = y - a$ invariant. Thus, the SWE solution $p(x)$ is invariant to such a shift in O as well. It gives no information about absolute position in space.

Thus, in Fig. 17.10, imagine the entire measuring apparatus, including planes O, L, L' and M, to be shifted to the right by some fixed amount. Of course the resulting probability law $p(x)$ must stay the same. If the laboratory is shifted by any amount, an observer in the laboratory sees the same physics irrespective of the shift.

All solutions to the SWE are of this nature. Otherwise, a shift of the origin of coordinates would allow information to be obtained about an absolute position in space. It couldn't, since there is no absolute origin of spatial coordinates in space. The same thing holds for velocities and other quantities. There are no absolute origins, according to "Mach's principle". By the way, although Einstein attributed this statement to Mach, it appears that Mach never said it. It was Einstein's idea.

In corresponding multiple-data cases, where the single parameter a is replaced by a four-vector $\boldsymbol{a}$ of parameters, the independence of I to the size of $\boldsymbol{a}$ means that I remains invariant to Lorentz transformation of the data coordinates. See Sect. 17.3.6.

17.3.4 Invariance of the Fisher Information Length to Unitary Transformation

The form (17.77) is noted to depend upon two functions, $p(x)$ and $p'(x)$. Dependence upon a single function would be more simple and elegant. For example, it would permit the information to be represented as an integral over a single Hilbert space of functions. Also, (17.77) seems to suggest (incorrectly) that I goes to infinity if p approaches zero at any x. These problems are overcome by expressing (17.77) in terms of probability amplitudes q instead of densities p,

$$I = 4 \int \mathrm{d}x q'^2(x), \quad q'(x) \equiv \mathrm{d}q(x)/\mathrm{d}x, \quad p(x) \equiv q^2(x) . \tag{17.78}$$

This trick has an unexpected payoff. Information I *is seen to be a sum of squares over the Hilbert space of functions* $q'(x)$. This is commonly called an L^2 norm or measure of the space. Geometrically, it is simply a squared length.

Now, L^2 *lengths remain fixed under a general class of transformations called the unitary transformation.* Algebraically,

$$J = I \tag{17.79}$$

where I is expressed in one space and J is its expression in another. A unitary transformation is a generalized rotation [17.14]. See, for example, Fig. 17.11.

It is of central importance that the form (17.78) holds regardless of the nature of the data and the shift-invariant phenomenon that caused it. This means that the invariance principle (17.79) is an absolute. It is not a matter of a particular phenomenon or of choice (i.e., is not Bayesian) but, rather, has been forced upon us by the nature of the form (17.78). Hence, invariance to unitary transformation is taken to be *the universal invariance principle* we sought.

As a matter of nomenclature, a scalar quantity like I in (17.78) that depends upon the shape of a function [here $q'(x)$] over its entire range of coordinates x is called *a functional.*

Unitary transformations may be carried through in different spaces. *Each such unitary transformation gives rise to a different law of physics via EPI.* See the examples in Table 17.1. The derivations are in the places cited in the second column. Some of the transformations do not directly define probability laws per se (our primary aim), but rather, the parameters and arguments of the laws. For example, see Sect. 17.3.6.

Table 17.1. Unitary transformations and their resulting physical effects. Efficiency $\kappa = 1$ in all cases

Type of Unitary Transformation	Resulting EPI Effect
Rotation of space-time coordinates	Lorentz transform of special relativity (Sect. 17.3.6)
Fourier transform of Hilbert space of space-time data	Wave equations of quantum mechanics (Sect. 17.3.9)
Functional Fourier transform of space of gravitational metric data	Wheeler-DeWitt equation of quantum gravity [17.12]
Rotation of Hilbert space of gauge fields through "Weinberg angle"	Existence of Higgs bosons; creation of mass [17.11]

17.3.5 Multidimensional Form of *I*

Most transformations involve multiple coordinates, so we segue from (17.78) to the multidimensional form for I,

$$I = \frac{4}{N} \int \mathrm{d}x_1 \cdots \mathrm{d}x_M \sum_{n,m} \left(\frac{\partial q_n}{\partial x_m} \right)^2, \qquad q_n = q_n(x_1, \ldots, x_M) . \tag{17.80}$$

In the double sum, n goes from 1 to N, and m independently from 1 to M. The scalar quantity I is now a functional of many functions q_n. A change in any one of them changes the value of I. Also, the probability density $p(x_1, \ldots, x_M)$ now obeys

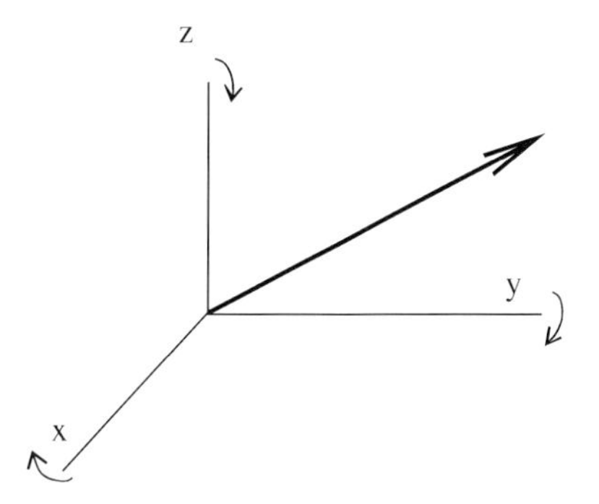

Fig. 17.11. The length of a vector remains the same regardless of rotation of its coordinate axes.

$$p = \frac{1}{N}\sum_n p_n, \quad p_n \equiv p_n(x_1, \ldots, x_M) \equiv q_n^2(x_1, \ldots, x_M) . \tag{17.81}$$

Coordinates $(x_1, \ldots, x_M) \equiv \boldsymbol{x}$ are called "Fisher coordinates".

Form (17.80) is in fact *the average of the trace of the Fisher information matrix* (3.105) in our scenario of shift invariance. I is also called the Fisher information *capacity*, in that I is an upper bound [17.35] to the actual Fisher information for a system whose probability density function obeys (17.81). All such densities $q_n(\boldsymbol{x})$ generally overlap, indicating a general state of mixture or disorder in $\boldsymbol{x}$. However, *if they do not overlap, the information capacity (17.80) becomes directly the Fisher information.* Obviously, the lack of overlap describes a partially ordered system. In effect, the statistical laws of nature arise out of a prior scenario of partial order. This is not to suggest, however, that the derived probability densities are restricted to a class of functions showing partial order in some sense. They are of a general form.

Fisher information capacity (17.80) obeys *a property of additivity* for independent random variables $x_1, \ldots, x_M$. See Ex. 17.3.16. As a corollary, a property of additivity does not uniquely define *entropy* as the appropriate information measure. In fact, many measures of information obey additivity.

17.3.6 Lorentz Transformation of Special Relativity

As indicated in Table 17.1, *each unitary transformation obeying invariance condition (17.79) gives rise to a different law of physics*. Here we apply a unitary transformation to the space of the Fisher coordinates $(x_1, \ldots, x_M) \equiv \boldsymbol{x}$ of the unknown probability law $p(\boldsymbol{x})$. In doing so, we derive *a restriction on the class of Fisher coordinates* that can be treated by the information approach.

Consider the case $M = 4$, with any N. Let a new coordinate system $\boldsymbol{x}'$ be formed linearly from the old one $\boldsymbol{x}$, as

$$\boldsymbol{x}' = [L]\boldsymbol{x}, \quad \boldsymbol{x} = (x_1, \ldots, x_M), \quad \boldsymbol{x}' = (x_1', \ldots, x_M') . \tag{17.82}$$

The matrix $[L]$ is to be real. Let I in the new coordinate system $\boldsymbol{x}'$ be denoted as I'. We require the matrix $[L]$ to preserve the length I. Accordingly, the unitary requirement (17.79) is now

$$I' \equiv J = I$$

under transformation (17.82). The solution for $[L]$ is a member of the Lorentz group of transformations. If, furthermore, the rotation is required to be through an imaginary angle whose tangent is ju/c, with c the speed of light and u the speed with which the coordinate system $\boldsymbol{x}'$ is moving relative to the coordinate system $\boldsymbol{x}$, then the solution for $[L]$ is the Lorentz transformation matrix [17.36]

$$[L] = \begin{bmatrix} \gamma & -\gamma u/c & 0 & 0 \\ -\gamma u/c & \gamma & 0 & 0 \\ 0 & 0 & 1 & 0 \\ 0 & 0 & 0 & 1 \end{bmatrix}, \quad \gamma \equiv \frac{1}{\sqrt{1 - u^2/c^2}} . \tag{17.83}$$

Since the elements of $[L]$ are restricted to real values, factor γ must be real. Then, from the definition (17.83) of γ, speeds u obey $u \le c$. This fixes c as a universal limiting speed. (See also related Ex. 17.3.19.) It is by definition that of light in a vacuum. Hence the speed of light c must be a universal physical constant.

The 4×4 nature of $[L]$ requires that the dimensionality M of coordinate space be fixed at the value $M = 4$ *for all phenomena that are to obey special relativity.*

Another requirement of the unitarity condition (17.79) is that the time, or the space, coordinates must be purely imaginary. Either choice suffices for our purposes. We use the "Bjorken-Drell" choice of imaginary space coordinates,

$$\boldsymbol{x} = (\mathrm{j}\boldsymbol{r},\ ct),\ \boldsymbol{r} \equiv (x, y, z)\ . \tag{17.84}$$

This application of the unitarity requirement (17.79) has not given rise to any particular probability law $p(\boldsymbol{x})$ but, rather, has served to constrain the choice of the Fisher variables $\boldsymbol{x}$ of any law. Any vector $\boldsymbol{x}$ must be a four-vector, and so $M = 4$. Likewise, each of the subscripts of any higher-dimensioned, tensor Fisher coordinates $x_{ij\ldots}$ go from 1 to 4 independently.

This does not, however, rule out the acquisition of a lower-dimensional law such as $p(x)$. Once a four-dimensional law $p(\boldsymbol{x})$ is known from EPI, it may be marginalized to any lower dimension by suitable integrations.

The EPI approach will be based upon the use of I. Given the (now) four-vector nature of the Fisher variables $\boldsymbol{x}$, any probability law that arises out of the use of EPI will likewise obey Lorentz invariance. Such automatic Lorentz invariance characterizes most laws of physics, so it is important to have it built into the approach.

Hence, rigorous use of EPI can only derive laws of physics that define relativistically invariant probability laws. Of course, the *nonrelativistic* limits of these laws can be taken. In fact this is one route to deriving the Schroedinger wave equation (Sect. 17.3.18). But interestingly, EPI also allows the direct derivation of projections of laws into lower-dimensioned spaces. For example, see the EPI derivation of the SWE in Sect. 17.3.19. However, these amount to approximate answers since they are nonrelativistic.

17.3.7 What Constants Should be Regarded as Universal Physical Constants?

Certain constants such as the speed of light c (see preceding), Planck's constant $\hbar$, etc., are regarded as *universal* physical constants. Such constants should not change with the particular boundary conditions of each problem, as ordinary constants do.

Consider a phenomenon where force fields are lacking. Then the invariance principle (17.79) becomes a statement of *an absolute size* for the information in the phenomenon. Also, with no fields present, the information becomes solely a function of one or more constants of the application; e.g., see (17.93). Under these conditions, *the constants that define the information value* must therefore be regarded as *universal* physical constants of the phenomenon.

Although EPI identifies the universal constants, it does not generally assign values to them. Instead, EPI predicts that they obey a reciprocal probability law $1/x$ on their magnitudes x. See Sect. 5.14. This suggests that the constants are random variables and, by implication, randomly formed.

17.3.8 Transition to Complex Probability Amplitudes

Because of the preceding, we consider the case $M = 4$. A *Fourier* transformation is unitary, since Parseval's theorem (8.55) is obeyed. Also, Fourier transforms occur widely in physical phenomena, and these are generally complex. It is therefore useful to re-express the I of (17.80) in terms of *complex amplitude functions* ψ_n that relate to the q_n. The simplest choice is to pack the latter as the real and imaginary parts of the former,

$$\psi_n = \frac{1}{\sqrt{N}}(q_{2n-1} + \mathrm{j}q_{2n}), \quad n = 1, \ldots, N/2 . \tag{17.85}$$

By (17.81) the probability density p is now

$$p = \sum_n \psi_n^* \psi_n . \tag{17.86}$$

By substituting the ψ_n of (17.85) into the r.h.s. of the equation

$$I = 4 \int \mathrm{d}x_1 \cdots \mathrm{d}x_4 \sum_{n,m} \left(\frac{\partial \psi_n^*}{\partial x_m}\right)\left(\frac{\partial \psi_n}{\partial x_m}\right), \quad \psi_n = \psi_n(x_1, \ldots, x_4) , \tag{17.87}$$

we once again get (17.80). Hence the information capacity I alternatively obeys (17.80) or (17.87). Here the sum on n goes from 1 to $N/2$, since by (17.85) there are twice as many q's as ψ's.

Any application of EPI can be made using either form (17.80) or (17.87). The use of complex amplitudes in physics is arbitrary. It often, however, results in simplifications of the theory. For example, if one chooses the Fourier transform as the EPI unitary transformation, from the standpoint of compactness it is clear that a single complex Fourier transform is superior to the use of two purely real cosine and sine transforms.

Students of physics will recognize the integrand of (17.87) as the first terms of the Lagrangian for deriving the Klein–Gordon equation.

17.3.9 Space-Time Measurement: Information Capacity in Fourier Space

The overall approach permits any four-vector $\boldsymbol{x}$ to be the Fisher coordinates; see Sect. 17.3.6.

From this point on, to be definite we assume one particular choice: *the space-time coordinates (17.84).* This means that the subject of the measurement is the space-time value of a particle. Both the space and the time coordinates of the particle are assumed to be random. What physics and what probability law does the particle obey *as a result of the measurement?* (The precise meaning of the last six words will become clear below.)

Any well-defined function has a Fourier transform. Consider the new functions ϕ_n defined by

$$\psi_n(\boldsymbol{r}, t) = \frac{1}{(2\pi\hbar)^2} \int\int \mathrm{d}\boldsymbol{\mu}\, \mathrm{d}E \phi_n(\boldsymbol{\mu}, E)\, e^{\mathrm{j}(\boldsymbol{\mu}\cdot\boldsymbol{r}-Et)/\hbar} \,. \tag{17.88}$$

Each ψ_n is the Fourier transform of a corresponding ϕ_n. The newly introduced constant $\hbar$, Planck's constant, is at this point just regarded as a fixed number. It could conceivably (but won't) vary from problem to problem. However, the four new coordinates $\boldsymbol{\mu}$, E of Fourier space specifically correspond to momentum and energy coordinates. Thus, they have units of mass-velocity and mass-velocity^2, respectively. However, there is at this point in the derivation no known relation (17.92) connecting these coordinates. The relation will be derived below.

The use of form (17.88) in the r.h.s. of (17.87) gives a representation for I in *Fourier space*, of

$$I = \left(\frac{4c}{\hbar^2}\right) \sum_{n=1}^{N/2} \int\int \mathrm{d}\boldsymbol{\mu}\, \mathrm{d}E\, \phi_n^* \phi_n \cdot \left(-\mu^2 + \frac{E^2}{c^2}\right) \equiv J \,, \tag{17.89}$$

the latter by hypothesis (17.79).

17.3.10 Relation Among Energy, Mass and Momentum

By Sect. 17.3.7, since I is an invariant (17.89) *should represent a universal physical constant*. This will have an important physical ramification. This ramification will not be a probability law *per se* , but, the absolute truth we are searching for is not confined to statements about probability laws.

We assumed that *a particle's space time coordinates are measured,* so that the Fisher four-variables are $\boldsymbol{x} = (\mathrm{j}\boldsymbol{r}, ct)$. Let the particle have mass value m.

By the length-preserving property of the unitary transformation (17.88),

$$c \int\int \mathrm{d}\boldsymbol{r}\, \mathrm{d}t \psi_n^* \psi_n = c \int\int \mathrm{d}\boldsymbol{\mu} \mathrm{d}E \phi_n^* \phi_n \,.$$

Summing both sides over n, using (17.86) and, then, the normalization of density p, gives

$$1 = \int\int \mathrm{d}\boldsymbol{\mu}\, \mathrm{d}E P(\boldsymbol{\mu}, E), \quad P(\boldsymbol{\mu}, E) = c \sum_n \phi_n^* \phi_n \,.$$

Then the new quantity P obeys $P \geq 0$ and normalization. This implies that P is a density function in $(\boldsymbol{\mu}, E)$ space.

Then (17.89) becomes an expectation

$$I = \left(\frac{4}{\hbar^2}\right) \left\langle -\mu^2 + \frac{E^2}{c^2} \right\rangle = const \,. \tag{17.90}$$

Since no fields have been presumed present, by Sect. 17.3.7 this I value must be a universal constant. Then, since the two factors in (17.90) are independent of one another, each must be a universal constant. This fixes Planck's constant $\hbar$ as a universal constant.

Likewise, the second factor, the expectation, must be a universal constant. However, the probability law $P(\boldsymbol{\mu}, E)$ necessarily *changes* from one set of boundary conditions to another. This would make the expectation a variable, unless

$$-\mu^2 + \frac{E^2}{c^2} = const. \equiv A^2(m, c) \tag{17.91}$$

where A is some universal constant. In general this can be a function of the known constants of the problem. In the absence of fields, these constants are just the rest mass m of the particle and the speed of light c. (We found the latter to be, in fact, a *universal* constant.) Given the known units for μ^2 or for E^2, by (17.91) those for A^2 must be $mass^2 \cdot velocity^2$. Thus, $A(m, c) \sim mc$. A proportionality constant of unity is chosen to quantitatively define mass (which has not been quantified before this). Using this in (17.91) gives

$$E^2 = c^2\mu^2 + m^2c^4 , \tag{17.92}$$

the famous relativistic equation linking energy, momentum and mass.

17.3.11 Ultimate Resolution

Resolution lengths represent an *ultimate ability to know*. A knowledge-oriented view of physics might be expected to predict these. Substituting (17.92) into (17.90) gives an information

$$I = 4\left(\frac{mc}{\hbar}\right)^2 , \quad \text{where } \frac{\hbar}{mc} = L_c . \tag{17.93}$$

The second equality defines the Compton resolution length L_c. The Compton length is the known quantum mechanical limit of resolution in measuring a particle's position. (Any finer resolution would require, by the Heisenberg uncertainty principle, so large an energy fluctuation as to break up the particle. See Ex. 17.3.5.) Since the Cramer–Rao inequality (17.18) states that $\epsilon^2_{\min} = 1/I$, it follows that (17.93) predicts that $\epsilon_{\min} = L_c/2$.

The latter expresses the Compton resolution length as a minimized error length. This is consistent with its definition as a limiting resolution. A resolution length is an expression of the ultimate ability to know. Hence, EPI succeeds in establishing such a limit in this application.

By Sect. 17.3.7 the information in (17.93) must be a universal constant. Since c and $\hbar$ have already been fixed as such constants, m is now fixed as well. Thus, the rest mass of the measured particle has a universal value.

17.3.12 Bound Information J and Efficiency Constant κ

Eq. (17.79) is a statement of the fact that there are two Fisher information quantitities present, that in the data, with value I, and that in a transform space, with value J. Also, we recall from Sect. 17.3.2 the two levels of knowledge called "phenomenon" and "noumenon". We can now tie these concepts together.

Data information I has to come from somewhere. By the closed aspect of our measuring system (Sect. 17.3.1), this somewhere must be the noumenon. We therefore identify J with the level of information that characterizes the noumenon. We also call J the "bound information", in the sense of being bound to the noumenon [17.19]. In the Kantian framework, then, *J is the level of information present in the noumenon, while I is that in the phenomenon.* This implies that, during measurement, a flow of information

$$J \rightarrow I \tag{17.94}$$

takes place. In Fig. 17.10, it is the indicated flow from the object plane O to the data plane M. The carriers of the information are here photons.

Since the phenomenon is a generally degraded version of the noumenon, there can be no more information in the former than exists in the latter (Ex. 17.3.8),

$$I = \kappa J, \quad 0 \leq \kappa \leq 1 \,. \tag{17.95}$$

This is also in agreement with Kant's philosophy (Sect. 17.3.2). The coefficient $\kappa = I/J$ therefore represents the efficiency with which information is transmitted from the noumenon into the data. Depending upon the nature of the observation, all or some of the existing information reaches the data.

Generally, it is found that when observation can be made on the quantum level then $\kappa = 1$, and (17.95) becomes the full equality (17.79) once more. On the level of information, the phenomenon is then fully the noumenon as well! The state of equivalence is then so complete that the spaces of I and J share *an entangled reality* [17.28, 17.29]. This means that the measurement of a parameter in one space gives knowledge about some parameter in the other. For example, in the particle reaction $e^+ + e^- \rightarrow Z^0 + H^0$ where an electron e^- and positron e^+ combine to give rise to a vector boson Z^0 and a Higgs particle H^0, the latter particles are quantum entangled. It results that a measurement of the position of Z^0 implies an upper bound to the mass of H^0 [17.11]. This provides valuable information about the currently undetected particle H^0. More on κ and J are found in Sect. 17.3.15.

17.3.13 Perturbing Effect of the Probe Particle

Any quantitative measurement of the type we are considering requires the illumination of the subject particle with a probe particle. The latter can be a photon, electron, etc. These carry the information in the transition (17.94). As in Fig. 17.10, the probe particle interacts with the subject (at point y in plane O), thereby picking up information about the ideal position a, passes through a lens or some other focussing instrument, and is then focussed upon a data plane (plane M) where its coordinate y is detected. For simplicity, we are assuming 1:1 magnification in the measurement system; also, the system is assumed ideal in imparting negligable fluctuations of its own to y. In effect, the diameter of the lens is very large. Thus, observation of the coordinate position y gives information I about the true position a of the subject particle; see (17.75). Consequently the phenomenon is the noumenon.

However, owing to the interaction of the subject particle with the probe particle the state of the subject particle is necessarily perturbed. Its amplitude functions $\psi_n(\boldsymbol{r}, t)$ become perturbed, by general amounts $\delta\psi_n(\boldsymbol{r}, t)$. Then by (17.87) the data information level is likewise perturbed, by a general amount δI. Also, by (17.79), the bound information is perturbed. Call this perturbation δJ. How does δI relate to δJ? This will turn out to have an extremely important answer.

17.3.14 Equality of the Perturbed Informations

Recall that a general unitary transformation connects the spaces of informations I and J; see (17.79) and the specific unitary transformation (17.88) of the Fourier variety. We now *derive a variational principle involving I and J that is obeyed in the presence of any unitary transformation.*

For simplicity we work in one dimension x, and with one component amplitude function $q(x)$. Results are later generalized to multiple dimensions and components. Suppose an initial function $q_0(x)$ is perturbed by an amount $\epsilon\eta(x)$, where ϵ is any constant and $\eta(x)$ is any perturbing function. Then the perturbed amplitude function is $q = q_0 + \epsilon\eta$. Differentiating this $\mathrm{d}/\mathrm{d}x$ gives

$$q' = q_0' + \epsilon\eta' \, . \tag{17.96}$$

Using the latter in (17.78) gives a perturbed data information value of

$$I = 4\int \mathrm{d}x\, q'^2 = 4\int \mathrm{d}x\, [q_0' + \epsilon\eta']^2 \, . \tag{17.97}$$

The variation or perturbation in I, called δI, is defined as

$$\delta I \equiv \left(\frac{\partial I}{\partial \epsilon}\right)_{\epsilon=0} \mathrm{d}\epsilon \, .$$

By differentiating (17.97) with respect to ϵ, we therefore get

$$\delta I = 8\,\mathrm{d}\epsilon \int \mathrm{d}x\, q_0'\eta' \, . \tag{17.98}$$

We now turn to J. Represent the unitary transformation of q as $[Lq]$, where L is some linear unitary operator. A transformation

$$Lq \equiv \phi(\mu) \tag{17.98a}$$

is said to define a transform space, of coordinate μ. An example is the Fourier transform

$$Lq = \int \mathrm{d}x\, e^{\mathrm{j}\mu x} q(x) = \phi(\mu). \tag{17.98b}$$

Now, because L is a linear operator it obeys *a distributive law*, so that $[L(q_2 - q_1)] = [Lq_2] - [Lq_1]$. (Again, think of the Fourier transform.)

The defining property of the unitary transformation is that any inner product of functions in x space equals the corresponding inner product in μ space,

$$\int \mathrm{d}x\, f(x)g(x) = \int \mathrm{d}\mu [Lf][Lg]\,.$$

Of course the derivatives of functions f and g are merely another pair of functions, so that the preceding holds for derivatives of the functions as well,

$$\int \mathrm{d}x f'(x)g'(x) = \int \mathrm{d}\mu [Lf'][Lg']\,. \tag{17.99}$$

In the particular case $f = g = q$, (17.97) and (17.99) give

$$I \equiv I[q(x)] = 4\int \mathrm{d}\mu [Lq'][Lq'] = 4\int \mathrm{d}\mu [Lq']^2 \equiv J[\phi(\mu)]. \tag{17.99a}$$

The notation $I[q(x)]$ means that I is ultimately a functional (17.97) of $q(x)$. The last equality is by the use of Eq. (17.98a). As an example, for the Fourier transform case (17.98b) we get

$$J[\phi(\mu)] = 4\int \mathrm{d}\mu\, \mu^2 |\phi(\mu)|^2. \tag{17.99b}$$

Using (17.96) in the second equality (17.99a) gives

$$I = 4\int \mathrm{d}\mu \{L(q_0' + \epsilon\eta')\}^2 \equiv J = 4\int \mathrm{d}\mu \{[Lq_0'] + \epsilon[L\eta']\}^2\,.$$

Information J enters by definition (17.79), and the last equality is again by the distributive property of L. Our perturbation is now in transform- or μ-space, as compared with (17.97) which is in x-space. Differentiating gives directly

$$\frac{\partial J}{\partial \epsilon} = 8\int \mathrm{d}\mu \{[Lq_0'] + \epsilon[L\eta']\}[L\eta']\,.$$

Hence

$$\left(\frac{\partial J}{\partial \epsilon}\right)_{\epsilon=0} = 8\int \mathrm{d}\mu [Lq_0'][L\eta']\,.$$

But by the invariance property (17.99) this becomes

$$\left(\frac{\partial J}{\partial \epsilon}\right)_{\epsilon=0} = 8\int \mathrm{d}x\, q_0'\eta'\,. \tag{17.100}$$

Thus the variation δJ is

$$\delta J \equiv \left(\frac{\partial J}{\partial \epsilon}\right)_{\epsilon=0} \mathrm{d}\epsilon = 8\,\mathrm{d}\epsilon \int \mathrm{d}x\, q_0'\eta'\,.$$

Comparing this with (17.98) shows that

$$\delta I = \delta J\,. \tag{17.101}$$

The two perturbations are equal.

However, in the foregoing I is a functional Eq. (17.97) of the amplitude function $q(x)$, that is $I = I[q(x)]$, while J is a functional Eq. (17.99a) $J = J[\phi(\mu)]$ of the amplitude function $\phi(\mu)$ in the transform space. For example see Eq. (17.99b). Hence Eq. (17.101) really states that

$$\delta I[q(x)] = \delta J[\phi(\mu)]. \tag{17.101a}$$

This is purely a mathematical identity, stating that the unitarity condition $I = J$ is preserved after a perturbation. It holds for *all* functions $q(x)$ that have well-defined unitary (for example, Fourier-) transforms $\phi(\mu)$.

Likewise the equality (17.99a) is another purely mathematical relation. It states the condition $I[q(x)] = J[\phi(\mu)]$ of unitarity between the two spaces. It holds for *any function* $q(x)$ that has a well-defined unitary transformation.

In order to use these purely mathematical relations (17.99a), (17.101a) physically, *we need a physical connection.* In fact in many measurement scenarios the functional $J[\phi(\mu)]$ has a physical meaning. In these, the "coordinate" μ represents the momentum , and $J[\phi(\mu)]$ is the mean-square momentum. Since means can be taken over a choice of spaces (Chap. 5), this permits J to be re-represented as an average over x-space. Thus

$$J[\phi(\mu)] = J[q(x)]. \tag{17.101b}$$

For example, see Eq. (17.112) below.

Hence we now demand Eq. (17.99a) to have the physical significance that it is satisfied by some amplitude function $q(x)$ *in the presence of the physical connection (17.101b).* Notice that these two relations may be used together because (17.99a) holds for *all* well-defined $q(x)$. Combining the two gives the requirement

$$I[q(x)] = J[q(x)], \text{ or, } I[q(x)] - J[q(x)] = 0. \tag{17.101c}$$

Since the two functionals are known, this represents a well-defined problem in the unknown amplitude function $q(x)$. The second condition (17.101c) is called the "EPI zero condition". This by itself leads to solutions for $q(x)$. See Sect. 17.3.19.

Likewise we demand that a function $q(x)$ exist that satisfies Eq. (17.101a) *in the presence of the physical connection (17.101b).* Notice that these two relations may be used together because (17.101a) holds for *all* well-defined $q(x)$. The result is a requirement on $q(x)$ of

$$\delta I[q(x)] = \delta J[q(x)], \text{ or, } \delta\{I[q(x)] - J[q(x)]\} = 0. \tag{17.101d}$$

This does in fact lead to a well-defined solution for $q(x)$, as described in the next section.

The two solutions $q(x)$ to requirements (17.101c) and (17.101d) are generally different. However, the solutions are, of course, the same when *both conditions* (17.101c,d) are required to hold for the given (measured) phenomenon. See the discussion preceding Table 17.2.

The derivation (17.96)-(17.101d) can be readily generalized to the multiple-dimensioned, multiple-component information I given by (17.80) or (17.87). In the latter case, where complex amplitudes $\psi_n(\mathbf{r}, t)$ are used, these merely replace $q(x)$ throughout the preceding. Thus $I[q(x)] \rightarrow I[\boldsymbol{\psi}(\mathbf{r},t)]$, etc.

Eqs. (17.101c) and (17.101d) will turn out to define the dynamics of the system under observation. They are equivalent to the EPI "zero principle" and "extremum principle", respectively.

17.3.15 EPI Variational Principle, and Framework

The second Eq. (17.101d) is $\delta(I - J) = 0$. The first variation of $I - J$ is zero. This means that $I - J$ is an extremum,

$$I - J \equiv K = \text{extrem} . \tag{17.102}$$

Hence, the fact that J is located in a unitary space for I means that *both* $I - J = 0$ [(17.101c)] and $I - J$ is an extremum. *The generalization (17.95) of (17.101c), and (17.102), are the two facets of the EPI principle.* Each element is solved to obtain a solution for the probability amplitudes q_n or ψ_n.

The quantity $I - J$ obviously represents a change (generally, a loss, according to (17.95)) of the information during the information flow (17.94) of the measurement procedure. Thus, (17.102) states this information loss should be an extremum. Since the word "loss" has a negative connotation, we prefer to instead call $I - J$ the "physical information" K (in honor of Kant) for the measurement. Accordingly, (17.102) is a statement of "extreme physical information" (EPI).

The elements of any Lagrangian are functionals I, J and coefficient κ. Functional I always has the same form, (17.80) or (17.87), depending upon whether real or complex amplitudes are required. Operationally, κ and J are found in two different ways, depending upon the nature of the invariance principle.

If the principle is a statement of unitarity, then $\kappa = 1$ and J is known as the representation of I in the transform space. Under these conditions, each of (17.95) and (17.102) are *separately solved* for their probability amplitudes. These solutions do not necessarily agree, since they represent fundamentally different measured objects (e.g., spin 0 vs spin 1/2 particles, in Sects. 17.3.18 and 17.3.19). Some of the resulting EPI results are shown in Table 17.1.

However, if the invariance principle *is not* one of unitarity, e.g., one of continuity of flow, then both κ and J must be solved for, and this is done by demanding both elements of EPI (17.95) and (17.102) to yield *the same solution* for the amplitudes. This is a *self-consistency condition*. Applications of this approach give the EPI outputs listed in Table 17.2. (The student is guided through some of the corresponding EPI derivations in Ex. 17.3.9, 17.3.10 and 17.3.13.)

As indicated in the table, some of the effects (e.g., classical gravity) have an efficiency value of $\kappa = 1/2$, indicating a loss of half of the noumenal level of information. Each of the latter effects also has a more "exact", quantum counterpart (e.g., quantum gravity) for which the value of $\kappa = 1$. A value $\kappa = 1/2$ seems to imply, then, that ignoring quantum effects is equivalent to ignoring half the pre-existing level of Fisher information J. However, why quantum effects should represent precisely 1/2 of the total information is a mystery. Those effects for which κ is at any level of efficiency are also of interest. This implies a phenomenon so pervasive that it comes through into the data no matter what fraction of it is observed. Perhaps this explains why in particular $1/f$ power spectra occur across a wide range of phenomena.

An approximate representation for J is as the mean kinetic energy of the measured object; see Ex. 17.3.12.

Table 17.2. Non-unitary invariance principles and their resulting physical effects

Type of Invariance Principle	Efficiency Value κ	Resulting EPI Effect
Continuity of flow: applied to a vector	1/2	Maxwell's equations (Ex. 17.3.9)
Continuity of flow: applied to a tensor	1/2	Einstein field equations of gravitation [17.12]
Invariance of normalization	1/2	Equation of genetic change (Ex. 17.3.13)
Invariance of normalization	1	Boltzmann energy distribution law [17.12]
Small speed in any frame	1	Maxwell-Boltzmann velocity distribution (Ex. 17.3.10)
Self similarity of power spectrum	Any number on interval (0, 1)	$1/f$ power spectral law [17.12]
Self similarity of magnitudes x	Any number on interval (0, 1)	$1/x$ probability law for mixed phenomena [17.12]

The two facets (17.95) and (17.102) of *EPI provide a definite framework for forming Lagrangians for unknown phenomena.* By comparison, the traditional use of Lagrangians provides minimal framework. An example is in the "minimum Fisher information" approach [17.30]. A later version [17.31] gave a definite approach for finding J, as a mean kinetic energy. However this is now known to only provide an approximate answer; see Ex. 17.3.12. Also, the traditional Lagrangian approach in physics regards the EPI information loss quantity $I - J$, or K, as instead an "action integral". This nomenclature arises out of a viewpoint that emphasizes the dynamics, i.e., particle energy, field strengths, etc., of a given problem, rather than the information aspect as here. By this approach, I is chosen to be of the general Fisher form (17.87) (without calling it that), on the grounds that it can be regarded as a generalized "kinetic energy" term. But, why a kinetic energy in the first place? Basically this is because the kinetic energy is the familiar T term in the action Lagrangian $L = T - V$ of mechanics. The action integrals of all other fields of physics are developed in analogy to this form, including the minus sign. However, there is no *prior rationale* for the $T - V$ form. Of course, it works as a Lagrangian, i.e., gives the correct equations of motion of the particles. But this is a posterior property of the approach, not a prior rationale. In mechanics, when is $T - V$ used other than in Lagrangian/Hamiltonian theory?

Notice that, by comparison, *there is* a prior reason for the choice of Fisher information. It is an integral part of the overall measurement model (Sect. 17.3.16. The difference $I - J$ represents the loss of information suffered by the carriers of information during the measurement process.

Modelling all other physical phenomena after the mechanical Lagrangian also has its pitfalls. This approach becomes tenuous when action integrals *for fields*, rather than particles, are formed. A field does not have a "kinetic energy" in the same sense as a particle. So, the concept has to be generalized. Undoubtedly the generalizations are correct, but, they are only correct in hindsight. For example, in forming the action Lagrangian for the electromagnetic field, a term that is quadratic in the fields, $(E^2 - B^2)$, is chosen. This turns out to be the right choice, in that it leads to Maxwell's equations [17.20a]. But again, this is reasoning from hindsight. There is no convincing *prior* argument for its use. The term $(E^2 - B^2)$ has no stand-alone meaning. True, the resulting Euler–Lagrange solution minimizes the term and, hence, forces E to be as close to B as is possible, but that is a result of *use of* the Lagrangian and not a prior rationale for it.

This conventional approach to constructing the action T term also cannot be applied to phenomena which can be derived by a Lagangian approach, but which *do not utilize energy* in any way. See, for example, Ex. 17.3.13.

Next, we turn to the action terms that are chosen in analogy to the mechanics term V. These are often constructed as a low-order, Taylor expansion of terms from some analogous action integral. Usually, as well, there is a "target" differential equation in mind, and this is used to further guide the construction procedure [17.20a]. The result is, accordingly, approximate in the Taylor expansion, and ad hoc in the choice of analogy. Also, there often are embarrassing "left-over" terms with infinite energy content. These are ignored, through a process of subtraction, and the rest is renormalized appropriately.

What the conventional action approach accomplishes, mainly, is a more elegant starting point for arriving at a desired destination, the differential equation [17.20b]. On the whole, the procedure leaves much to be desired as a real research tool

17.3.16 The Measurement Process in Detail

The EPI framework gives correct answers for estimated probability laws because it describes a physical process. The latter is what happens in detail when a measurement is made. According to preceding sections this is as follows:

(1) The observer decides to measure a four-parameter $\boldsymbol{a}$ or, more generally, a vector of measurements $\boldsymbol{a}$ as in the visual process (which is performed in parallel at a vector of pixels). We continue with the single four-parameter case, for simplicity. The choice of $\boldsymbol{a}$ fixes the Fisher coordinates $\boldsymbol{x}$ of the problem; see (17.75).

(2) The observer selects an appropriate measuring device, consisting of an input port where the subject is located, a source of probe particles to illuminate the subject, and a lens to direct the probe particle into a data space. The overall system is closed, and the measuring device contributes negligable error to $\boldsymbol{y}$. Consequently, the fluctuations in $\boldsymbol{y}$ follow only the intrinsic fluctuations $\boldsymbol{x}$ of the subject; see (17.75).

(3) The observer asks his question: A probe particle is used to illuminate the subject of the measurement. It carries a fraction κ of the pre-existing information J from the noumenon into the data space, where it has the value I. See transition (17.94). This establishes the first half (17.95) of the EPI principle.

(4) The probe particle also perturbs the subject. This in turn perturbs both informations I and J, through functional relation (17.87) and correspondence (17.101d), respectively.

(5) The latter correspondence causes the variational principle (17.102) to physically occur. Both aspects of EPI have therefore occurred.
(6) These define, via Euler–Lagrange solution (17.104) below, the amplitude functions q_n or ψ_n. This solution, a differential equation, is a contrivance of the observer. Although we iconicize the equation with the name "physical law" it is but a mathematical model for the noumenon. All the observer senses are the data. The probability law $p(\boldsymbol{x})$ (17.81) or (17.86) that is defined by the amplitude functions is also a mathematical model.
(7) A fluctuation value $\boldsymbol{x}$ is randomly sampled from the probability law $p(\boldsymbol{x})$ by the measuring instrument. Each component of the datum $\boldsymbol{y}$ is formed via (17.75), and is observed. Observation of $\boldsymbol{y}$, in turn, validates *the reality* of $\boldsymbol{y}$. See the discussion following this list for further on this point.
(8) On the basis of the datum $\boldsymbol{y}$, the observer forms an estimate $\hat{\boldsymbol{a}}(\boldsymbol{y})$ of the required parameter $\boldsymbol{a}$. This can be a mere unit operation $\hat{\boldsymbol{a}}(\boldsymbol{y}) = \boldsymbol{y}$, if so desired.
(9) The estimate $\hat{\boldsymbol{a}}$ is registered as a message. This is in the observer's brain, or possibly in some nonhuman permanent medium such as a bubble chamber, computer memory, etc.
(10) The registered message is observed by a form of intelligence. This can be a human or an intelligent robot. That ends the measurement process.

We discuss further the point made in (7). *We assumed at the outset that the data are real*, i.e. actual outputs of the intended measurement. However, the probability law $p(\boldsymbol{y})$ that EPI generates is *a model* for the noumenon that is under measurement, and does not have an absolute reality. There is no guarantee that the EPI output law p *exactly* holds at $\boldsymbol{y}$ values other than the single data value. One reason is that there is no guarantee that p is a continuous function. For example, the data value $\boldsymbol{y}$ is often a space-time vector. Any measurement of space-time is known to *re-initialize* the law $p(\boldsymbol{y})$ such that it satisfies the measured value [17.15]. Thus, it discontinuously changes the form of the law. When the number $\boldsymbol{y}$ is observed, it is not known if some other $\boldsymbol{y}$ value will be measured at some other time or space value. Each such measurement occurs unpredictably and is of random size. Hence the curve $p(\boldsymbol{y})$ is randomly discontinuous. It results that one can estimate, but cannot know with certainty, $p(\boldsymbol{y})$ at other $\boldsymbol{y}$ values.

We discuss further the need for steps (9) and (10) in the measurement process. Recall that our aim is *to know definitively* that a measurement was made and a corresponding effect $p(\boldsymbol{y})$ occurred. Step (9) distinguishes a pure interaction of particles from an observed event. At an interaction, wave functions are perturbed, as before, and both informations I and J are perturbed, thereby exciting steps (3)–(7) of the process. However, without registration of the interaction there is no evidence that it happened and, so, it is impossible to state with certainty that it occurred.

Likewise, if an intelligent being *does not observe* the registered value of $\boldsymbol{y}$ (step 10), then it is impossible to know with certainty that the sequence (1)–(9) actually occurred. The registered event is evidence for the occurrence of the interaction event. This is in fact the basis of Shannon information theory. But if no one is there to witness the evidence it is impossible to state with certainty that the interaction occurred. ("If a tree falls in the forest, but no one is there to witness it, did the tree really fall?") It can be *estimated* to have occurred, and assuming it did occur, we can *infer* what its physics $p(\boldsymbol{y})$ was (via EPI), but that does not meet our objective of certainty.

It may be noted that a conscious person continually cycles through the measurement process (1)–(10), using his/her senses as the measuring instruments. Each sensed phenomenon gives rise to an acquired amount of Fisher information,which in turn begs further measurement. This cycling continues until the observer loses consciousness, for example by falling asleep. Rather amazingly, a conscious person must maintain a certain *minimal rate of intake of Fisher information*: its absence under artificial conditions called "sensory deprivation experiments" leads to extreme disorientation. It would be interesting to know what the threshold rate is.

17.3.17 Euler–Lagrange Solutions

Once the informations I and J are known, the physical information $K = I - J$ can be formed. Then the EPI extremum condition (17.102) is satisfied by the following procedure. Information K is always expressible as the integral of an information density k called the Lagrangian,

$$K = \int \mathrm{d}\boldsymbol{x}\, k\,, \quad k \equiv k[\psi_n, \psi_{nm}, \psi_n^*, \psi_{nm}^*, \boldsymbol{x}]\,, \quad \psi_{nm} \equiv \frac{\partial \psi_n}{\partial x_m}\,,$$
$$\boldsymbol{x} \equiv \{x_m\} = (x, y, z, t)\,. \tag{17.103}$$

The solution ψ_n to the problem obeys a set of differential equations called the "Euler–Lagrange" equations (Appendix G, Eq. (G.16))

$$\sum_{m=1}^{4} \frac{\mathrm{d}}{\mathrm{d}x_m}\left(\frac{\partial k}{\partial \psi_{nm}^*}\right) = \frac{\partial k}{\partial \psi_n^*}, \qquad n = 1, \ldots, N/2, \,. \tag{17.104}$$

For our problem of four-position measurement, $\boldsymbol{x} = (x, y, z, t)$. The information density k is always constructed as a real function of real coordinates (see examples below).

17.3.18 Free-Field Klein–Gordon Equation

Amplitude functions ψ_n of course depend upon *real* coordinates (x, y, z, t). However, using the coordinates (17.84) in expression (17.87) for I requires derivatives of ψ_n with respect to *imaginary* coordinates such as $\partial/\partial(\mathrm{j}x)$. Then do these derivatives exist? The derivatives are implemented by simply $\partial/\partial(\mathrm{j}x) = (1/\mathrm{j})(\partial/\partial x)$, resulting from a trivial use of the chain rule of differentiation. The derivatives therefore generally exist. The two factors j that result multiply each other, resulting in a functional

I that is negative but depends only upon real coordinates. Also, using (17.93) for J, and multiplying J by the normalization integral $1 = c \int d\boldsymbol{x} \sum_n \psi_n^* \psi_n$, gives a net Lagrangian k of

$$k = 4c \left\{ \sum_n \left[-\sum_{m=1}^{3} \psi_{nm}^* \psi_{nm} + \frac{1}{c^2} \psi_{n4}^* \psi_{n4} - \left(\frac{mc}{\hbar} \right)^2 \psi_n^* \psi_n \right] \right\} \tag{17.105}$$

Its use in (17.104) gives a solution

$$-c^2 \hbar^2 \nabla^2 \psi_n + \hbar^2 \frac{\partial^2 \psi_n}{\partial t^2} + m^2 c^4 \psi_n = 0, \quad \text{where } \nabla^2 \equiv \sum_{m=1}^{3} \frac{\partial^2}{\partial x_m^2} \tag{17.106}$$

defines the Laplacian operator. This is the Klein–Gordon equation (KGE). It describes a particle of mass m and zero spin moving in a free field. Fields may be incorporated into (17.106) in a simple manner, but for brevity we refer the reader to [17.12] for details. Also, see Ex. 17.3.3.

The KGE (17.106) can alternatively be derived by combination of Eqs. (17.88) and (17.92). Eq. (17.88) is multiply differentiated $\partial^2/\partial x^2$ etc., with respect to the space coordinates and, then, $\partial^2/\partial t^2$. Adding the space-coordinate results and subtracting the time-coordinate result then allows (17.92) to be used in the net integrand, showing that it is zero. The result is the KGE. There is often more than one way to derive a dynamical equation. However, the latter approach does not give the Dirac equation of the electron as well. The EPI approach does, as is shown next.

17.3.19 Dirac Equations

A particle of spin 1/2 does not obey the KGE. However, the equation that it obeys can also be found by EPI. We have previously used the extremum facet (17.102) of EPI. If instead the Lagrangian (17.105) is used in *the second EPI* condition (17.95) this takes the form

$$k = 0\,. \tag{17.107}$$

It is found that the Lagrangian k can be factored, so that (17.107) becomes

$$\mathbf{v}_1 \cdot \mathbf{v}_2 = 0\,, \tag{17.108}$$

where

$$\mathbf{v}_1 \equiv \mathrm{j}[\alpha] \cdot \nabla \psi - (mc/\hbar)[\beta]\psi + \mathrm{j}c^{-1}\partial\psi/\partial t \tag{17.109a}$$

and

$$\mathbf{v}_2 \equiv \mathrm{j}[\alpha^*] \cdot \nabla \psi^* + (mc/\hbar)[\beta^*]\psi^* - \mathrm{j}c^{-1}\partial\psi^*/\partial t\,. \tag{17.109b}$$

The indicated matrices are those of Dirac, and the interested reader can consult [17.12] or [17.17] for details on them.

To attain the requirement $k = 0$, either the first, the second, or both, of the factors in (17.108) are zero. Zeroing the first factor means setting $\mathbf{v}_1$ given by (17.109a) equal to zero. This is the Dirac equation for any spin 1/2 particle, e.g. the electron or quark. Setting $\mathbf{v}_2$ given by (17.109b) equal to zero gives the Dirac equation for the corresponding anti particle of spin 1/2.

Finally, setting both $\mathbf{v}_1$ and $\mathbf{v}_2$ equal to zero gives both Dirac equations, implying that both the particle of spin 1/2 and its anti-particle are present. Further, since only one particle was measured these must be bound together, forming what is defined to be a π meson if the spin 1/2 particles are quarks, or a zero-spin state of positronium if the spin 1/2 particles are electrons. In this way, EPI predicts the existence of compound particles out of elementary particles [17.18].

17.3.20 Schroedinger Wave Equation (SWE)

The Schroedinger wave equation may be derived [17.12] as the non-relativistic limit of (17.106). It may also be derived from approximate use of EPI, as follows.

The rigorous use of EPI can only derive relativistic effects, because relativity is built into EPI (see Sect. 17.3.6). This is through the taking of data that are four-vectors.This is why the Klein–Gordon equation (17.106) correctly describes a particle moving at even relativistic speeds. However, EPI has a degree of "robustness" to it, in that it gives close-to correct results even when one-, two- or three-dimensional data are taken. Then, it derives the projection of the correct physical law into the lower-dimensioned measurement space. We next derive the SWE in this way.

Suppose that, contrary to rigorous use of EPI, only the spatial x-coordinate a_x of a particle is measured. Moreover, this is to be at a "given" time value t (again violating a premise of EPI). Accordingly, the amplitude function $\psi(x|t)$ on fluctuations x from a_x is sought. In place of (17.88) we now have a unitary space function $\phi_n(\mu)$ of one dimension, defined as

$$\psi_n(x|t) = \frac{1}{\sqrt{2\pi\hbar}} e^{-\mathrm{j}Et/\hbar} \int \mathrm{d}\mu\, \phi_n(\mu) e^{\mathrm{j}\mu x/\hbar} \,. \tag{17.110}$$

Differentiation of this shows immediately that

$$\frac{\partial \psi_n}{\partial t} = -\mathrm{j}E\psi_n/\hbar \,. \tag{17.111}$$

In place of (17.90) we have

$$I = \frac{4}{\hbar^2}\left\langle \mu^2 \right\rangle \equiv J$$

by conservation of length (17.79). Next we use the specifically non-relativistic approximation that the kinetic energy KE of the particle is $\mu^2/2m$, and also that $KE = E - V(x)$, the latter the potential function for the problem. Putting the last three relations together gives

$$\frac{4}{\hbar^2}\langle 2m\,KE\rangle = \frac{8m}{\hbar^2}\langle E - V(x)\rangle = \frac{8m}{\hbar^2}\int \mathrm{d}x[E - V(x)]\sum_n \psi_n^*\psi_n \equiv J \,, \tag{17.112}$$

where (17.86) was used. The EPI extremization principle (17.102) is now

$$K \equiv I - J = \int \mathrm{d}x\, k = \text{extrem}\,,$$

$$k = \sum_n \left[4\frac{\mathrm{d}\psi_n^*}{\mathrm{d}x}\frac{\mathrm{d}\psi_n}{\mathrm{d}x} - \frac{8m}{\hbar^2}[E - V(x)]\psi_n^*\psi_n \right]. \tag{17.113}$$

Eq. (17.87) was used.

The Euler–Lagrange equation for the problem is (G.14),

$$\frac{\mathrm{d}}{\mathrm{d}x}\left(\frac{\partial k}{\partial \psi_{nx}^*}\right) = \frac{\partial k}{\partial \psi_n^*}, \quad \psi_{nx}^* \equiv \frac{\mathrm{d}\psi_n^*}{\mathrm{d}x}, \quad n = 1, \ldots, N/2\,. \tag{17.114}$$

Use of the information density k from (17.113) in this equation gives a solution

$$\psi_n''(x|t) + \frac{2m}{\hbar^2}[E - V(x)]\psi_n(x|t) = 0, \quad n = 1, \ldots, N/2\,. \tag{17.115}$$

Finally, (17.111) brings in the time, as

$$-\frac{\hbar^2}{2m}\psi_n''(x|t) + V(x)\psi_n(x|t) = \mathrm{j}\hbar\frac{\partial \psi_n}{\partial t}\,. \tag{17.116}$$

This is the same second-order differential equation for each subscript value n. One function ψ_n suffices to describe its general solution. Hence $N = 2$, $n = 1$, and we may drop the now superfluous subscript on $\psi_n \equiv \psi$ in (17.116). It is the Schroedinger wave equation in one coordinate.

It may be noted that if, instead, *all three space coordinates* are measured the same approach gives the SWE (17.116) with $\psi_n(x|t)$ replaced by $\psi_n(x, y, z|t)$,$\psi_n''(x|t)$ replaced by $\nabla^2\psi_n(x, y, z|t)$, and $V(x)$ replaced by $V(x, y, z)$. See Ex. 17.3.2. Eq. (17.114) is the projection of the latter into one dimension. Analogously, the SWE in $\psi_n(x, y, z|t)$ may be thought of as being the projection of the relativistic, four-dimensional Klein–Gordon equation (17.106) into three dimensions. *EPI gives the correct answer within the dimensionality set by the given measurement.*

There are additional prices paid when measuring with less than full dimensionality. First, the SWE is only a non-relativistic approximation to the fully-dimensioned Klein–Gordon equation (KGE). Second, the SWE is actually a temporally filtered version of the KGE, lacking the frequency $mc^2/\hbar$ [17.12].This amounts to a temporally "coarse-grained" version of the KGE. Thus EPI states that, since the time was not measured and instead was assumed, an effect with less time-resolution is elicited. Third, the derivation of the SWE did not allow us to derive many of the other truths uncovered by EPI in the four-dimensional KGE derivation: neither the mass-energy relation (17.92), the ultimate resolution length (17.93), or the universal constancy of the constants $\hbar$, m and L_c.

17.3.21 Dimensionality, and Plato's Cave

The preceding derivation of the SWE in one coordinate shows that that *EPI gives an answer on the level of accuracy and dimensionality that is demanded by the observer's question.* Then, is there an upper limit to the dimensionality of a space that obeys the SWE or more general wave equations? The EPI viewpoint is that, in principle, there is no upper limit to the number of dimensions *except as we can detect them.* In the above example of deriving the one-dimensional SWE, the observer simply asked a smaller question than nature is equipped to answer.

The Greek philosopher Plato actually anticipated these effects of unknown dimensionality. This is through his fable of the man in the cave. Suppose that a man is born and raised inside a cave, without ever having set foot outside and seeing other humans. It is night time and a person outside the cave builds a fire. The fire casts that person's shadow upon a wall inside the cave. The man in the cave sees the shadow and infers that that is what people look like. A two-dimensional "truth" is all the caveman can make of the three- (and higher-) dimensional person.

The point Plato was making is that man can never know the absolute truth: in his fable, the absolute dimensionality of the truth. Kant picked up on these ideas and further developed them as the concepts of "noumenon" and "phenomenon" (Sect. 17.3.2). As we saw above, EPI agrees with these ideas, but, with an important distinction: Plato and Kant were being *pessimistic,* in believing that man is ultimately limited in his quest for knowledge. By contrast, EPI is an optimistic statement, in that it makes creative use of the very disparity $I - J$ between noumenon and phenomenon to seek the knowledge we lack, within the chosen dimensionality of a measurement space.

17.3.22 Wheeler's "Participatory Universe"

The modern physicist J.A. Wheeler made some qualitative statements [17.16] that predate EPI and yet well describe its overall viewpoint:

"All things physical are information-theoretic in origin and this is a participatory universe ... Observer participancy gives rise to information and information gives rise to physics."

From the standpoint of EPI, "the observer" is the measurer, "the participancy" is the measurement process, "the information" is that of Fisher, and it "gives rise to physics" via the enaction of the EPI principle. The only important component of EPI that is lacking in Wheeler's statements is J, the source of the observed information I.

17.3.23 Exhaustivity Property, and Future Research

Maupertuis actually invented the "Lagrangian" approach, and regarded it as a somewhat mystical route to understanding nature. It remained a mystery until EPI gave it both a prior meaning and a framework for construction. What are the limitations in the use of EPI for finding probability laws?

We found that the unitary transformation, as in Table 17.1, is central to the use of EPI. Each new unitary transformation defines a new probability law of physics. Other invariance principles, such as in Table 17.2, can be used as well in an approximate, non-quantum, sense. On the other hand, EPI appears to have an *exhaustive nature.* All possible solutions it predicts have, so far, been either found in nature or are plausible extensions of what is observed. This exhaustive nature seems to imply, then, that the scope of physics that is comprehensible by man is limited only by his imagination in constructing new invariance principles. A search for other such transformations should yield other laws of physics via EPI. Unknown problems remain in the theories of turbulence and of *complex systems* [17.26]. The former has already been partially resolved using EPI [17.27]. Examples of the latter are problems of modelling the weather, the economy of a country, or the interaction between the environment and technology.

17.3.24 Can EPI be Used in a Design Mode?

These uses of EPI have been in what might be called *estimation mode*, whereby the principle is used to estimate, or model, existing probability laws. Each such law arises out of a corresponding invariance principle (as in Tables 17.1 and 17.2). All invariance principles must be physically realizable, i.e., describe physically meaningful effects such as conservation of flow, or physically meaningful transformation spaces, such as a momentum space. The resulting laws are in a "default" mode, i.e., as we find them to naturally occur.

One wonders if instead EPI can be used in a *design mode*. In this mode, a desired probability law is, instead, *imposed upon nature*. This amounts to a reverse use of EPI. A computer program would be built which subjects EPI to a succession of trial invariance principles, until one is found whose EPI output meets the required law. Although this program is bound to be successful to some degree, the problem remains of finding *an analog method* for enforcing the design invariance principle. This might not always exist.

However, EPI at present appears to be an exhaustive principle, in the sense that all of its mathematical solutions occur physically. Then EPI might also turn out to be reversible. If so, any design invariance principle will exist in nature as well.

Conceivably this approach could usher in an era of ideal effects that are produced by design, and not by default as at present.

17.3.25 EPI as a Knowledge Acquisition Game

Knowledge acquisition is at the heart of EPI and its ability to find the fundamental probability laws. EPI can be modelled as a *mathematical game* of knowledge acquistion [17.12], whereby the observer plays nature (the measured phenomenon) for a finite available amount of Fisher information. The game is zero-sum, so that the information that is gained by the observer is at the expense of nature. This has a counterpart in the Second Law of thermodynamics, whereby the acquisition of Shannon information, measured in "bits", is more than offset by an equal expenditure of Boltzmann entropy [17.19].

The EPI game has a "fixed-point" solution, whereby the winning move for both protagonists is the same. At this point, each player is satisfied that he has optimized his level of Fisher information (or knowledge). This payoff point depends upon the particular invariance principle that applies to the measurement scenario. The payoff represents, as well, the physical law that EPI derives.

For example, as a result of the particle reaction $e^{+} + e^{-} \rightarrow Z^{0} + H^{0}$ where e^{-} is an electron, $e^{+}$is a positron, $Z^{0}$is a vector boson and H^{0} is an observed scalar Higgs boson, the EPI information game is effectively between the two bosons. During this play of the game the information "prize" takes the form of an amount of mass that is acquired by the vector boson at the expense of an equivalent amount of energy from the scalar boson. This is the origin of the famous Higgs mass phenomenon [17.11].

As with the entangled realities previously noted for these two particles (Sect. 17.3.12), the game expresses the fact that they have to share between them a fixed amount of Fisher information, $I = J$.

17.3.26 Ultimate Uses of EPI

In general, the acquisition of knowledge represents a local increase of order, on the level of thoughts and concepts. But since knowledge can only be acquired as the ultimate result of irreversible measurement, each measurement increases the overall level of disorder and entropy. Thus, *each increase in knowledge is accompanied by a larger increase in disorder.* The difference can be measured as a Fisher information deficit ($I - J$). The existence of this deficit is an alternative statement of the Second Law of thermodynamics. See Ex. 17.3.8. Thus, at each measurement, *the measurer always plays an EPI game with the Second Law* as well as with the immediate phenomenon of interest. The presence of a net information deficit means that he always loses the game. Once all available entropy is used up, there can be no further particle interaction and, hence, no further acquired knowledge (see steps (3) and (4) of the EPI process, Sect. 17.3.16.). The Second Law is the ultimate limitor of knowledge (and everything else).

On the other hand, acquired knowledge inspires creative ideas in people. And, ideas about nature lead inevitably to new research and solved problems. The value of an "idea" has never been meaningfully quantified. However, an idea about nature represents new knowledge about nature and hence, by the tenets of EPI, must affect some observable aspect of nature. This means that an idea is a kind of interactive force. *As such, it must interact with the other forces of physics.* This has an important ramification.

If the past is any indicator, most physical limits are illusions, i.e., exist only in the absence of sufficient knowledge. Hence, if civilization is allowed to continue developing so that the level of overall knowledge keeps increasing, critical levels of knowledge will be attained such that many apparently fundamental physical limits are, in turn, overcome by ideas and subsequent EPI games. This includes the possibility of reversing the Second Law and the information deficit. We may note in this regard that the extremum attained by ($I - J$) in (17.102) can be of any type: a minimum, maximum, or point of inflection. Thus, the sense of inequality (17.24) could be reversed in this way.

Such a reversal would allow a net *gain net of information* as the prize in the information game that is played at subsequent observations. These conceivably could be attained by the use of the computer program described in Sect. 17.3.24. Its output would be the required invariance principle for winning this game.

Exercise 17.3

The aims of the following exercises are to work out the details of certain of the above derivations (Exs. 17.3.1–17.3.3); to gain practice with the use of EPI in other derivations (Exs. 17.3.9–17.3.13); and to derive other important properties of EPI and informations I and J (Exs. 17.3.4–17.3.8) and (17.3.14)–(17.3.16).

17.3.1 Carrying through the operations mentioned before (17.77), show that with no absolute origin the defining form (17.76) for Fisher information goes over into the shift-invariant form (17.77).

17.3.2 Show that, if instead *all three space coordinates* are measured, the approach (17.110)–(17.116) gives the three-dimensional version of the SWE (17.116), with $\psi_n(x|t)$ replaced by $\psi_n(x, y, z|t)$, $\psi_n''(x|t)$ replaced by $\nabla^2\psi_n(x, y, z|t)$, and $V(x)$ replaced by $V(x, y, z)$.

17.3.3 Using the Lagrangian k given by (17.105), show that

$$\frac{\partial k}{\partial \psi^*_{nm}} = -4c\psi_{nm}, \quad \frac{\partial k}{\partial \psi^*_{n4}} = \frac{4}{c}\psi_{n4}, \quad \frac{\partial k}{\partial \psi^*_{n}} = -4c\left(\frac{mc}{\hbar}\right)^2\psi_n .$$

Using these derivatives in the Euler–Lagrange equations (17.104), show that the result is the Klein–Gordon equation (17.106).

17.3.4 The preceding derivation presumes that the functions ψ_n^* are varied. Show that if instead the functions ψ_n are varied the corresponding Euler–Lagrange equations give the *complex conjugate* of the Klein–Gordon equation (17.106). Is this equivalent to (17.106) or is it an independent solution?

17.3.5 ***Why the Compton resolution length is an ultimate limit of resolution.*** The Compton resolution length (17.93) is an ultimate limit of x-resolution for a particle of mass m because, by the Heisenberg uncertainty principle (3.89), to accomplish this resolution would excite energy fluctuations that are comparable to the particle's rest energy mc^2. Show this. These energy values would break up the particle.

Hint: The Heisenberg principle (3.89) must be specialized to the case $\omega \equiv \mu$, the momentum. Then it takes the form

$$\langle x^2\rangle\langle\mu^2\rangle \geq \hbar^2/4 . \tag{17.117}$$

Also use $E^2 \sim c^2\mu^2$ to define the energy E imparted to the particle by the momentum μ [see (17.92)].

17.3.6 ***Heisenberg uncertainty principle from Fisher information.*** The Heisenberg uncertainty principle (17.117) may also be derived as purely an effect of measurement, i.e., out of the Cramer–Rao inequality (17.18),

$$e^2 I_0 \geq 1\,. \tag{17.118}$$

This shows reciprocity between the mean-square error e^2 in locating the position of a particle of mass m and the information level I_0 in the measured position value. Furthermore, the information capacity I in measurement of the particle obeys the relation below (17.111), repeated here as

$$I = \frac{4}{\hbar^2}\left\langle \mu^2 \right\rangle\,. \tag{17.119}$$

Since this is an information capacity, it exceeds the corresponding information level I_0,

$$I \geq I_0\,. \tag{17.120}$$

Show that eqs. (17.118)-(17.120) imply the Heisenberg principle (17.117).

17.3.7 Eq. (3.98) states that the Fisher information I is the negative of the cross entropy for a special choice of the reference function $r(x)$,

$$I = \lim_{\Delta x \to 0}\left(\frac{2}{\Delta x^2}\right)\int \mathrm{d}x\, p(x)\log\frac{p(x)}{p(x+\Delta x)}\,. \tag{17.121}$$

What is the reference function here, and why is it not strictly a Bayesian term?

17.3.8 ***Second Law of thermodynamics in terms of*** I. It is known [17.22] that if two probability laws $p(x)$ and $r(x)$ obey a differential equation called the Fokker–Planck equation, then their cross entropy

$$G(t) \equiv -\int \mathrm{d}x\, p\ln(p/r), \quad p \equiv p(x|t), \quad r \equiv r(x|t) \tag{17.122}$$

obeys an inequality

$$\mathrm{d}G(t)/\,\mathrm{d}t \geq 0\,. \tag{17.123}$$

Suppose, then, that the probabilities $p(x|t)$ and $p(x+\epsilon|t)$ obey a Fokker–Planck equation. Using the latter probability as $r(x|t)$ in (17.122), show that the result (17.123) becomes

$$\frac{\mathrm{d}I}{\mathrm{d}t} \leq 0 \tag{17.124}$$

after use of (17.121).

This says that I tends to decrease with time [17.23]. The form (17.78) for I shows decreasing values as the underlying probability amplitude $q(x)$ becomes more spread out and non-localized. This indicates increased disorder in x. Hence, (17.124) says that *disorder tends to increase with time*. This agrees as well with the behavior of entropy in the Second Law.

The EPI measurement process is the transition $J \to I$. This indicates a net change in Fisher information of amount $\Delta I = I - J$. Then the statement (17.24) states that $I - J \leq 0$, and this verifies the model assumption (17.95) of EPI.

Information inequality (17.124) is also obeyed if, in place of the Fokker–Planck equation, the process $p(x)$ obeys a *homogenous Boltzmann transport equation* [17.24]. Fisher information has also recently been related to *topological* entropy as a lowest-order measure of *the complexity* of a system [17.25].

17.3.9 ***EPI derivation of Maxwell's equations of electrodynamics.*** These can be simply derived from the vector wave equation [17.12]. The latter is derived via EPI as follows. A value of the electric- or magnetic field is measured at a required space-time coordinate. A current density j_n and a charge density ρ are present. The actual space-time coordinate suffers random errors $(\boldsymbol{r}, ct)$. The self-consistency EPI solution gives a result

$$\kappa = 1/2, \quad J = 4\pi \int\int \mathrm{d}\boldsymbol{r}\,\mathrm{d}t \sum_{n=1}^{4} q_n j_n\,; \quad j_n \equiv j_n(\boldsymbol{r}, t),\ n = 1, 2, 3;$$

$$j_4 \equiv c\rho(\boldsymbol{r}, t)\,. \tag{17.125}$$

We identify $q_n = A_n(\boldsymbol{r}, t)$, $n = 1, 2, 3$; and $q_4 = \phi(\boldsymbol{r}, t)$, the vector- and scalar potentials, respectively.

(a) Using the coordinates $\boldsymbol{x} = (\boldsymbol{r}, \mathrm{j}ct)$ and $M = N = 4$, show that the information (17.80) becomes

$$I = c \int\int \mathrm{d}\boldsymbol{r}\,\mathrm{d}t \sum_{n=1}^{4} \left[\nabla q_n \cdot \nabla q_n - \frac{1}{c^2}\left(\frac{\partial q_n}{\partial t}\right)^2 \right]. \tag{17.126}$$

(b) Using I and J as given by (17.125) and (17.126), respectively, show that the EPI solution provided by the Euler–Lagrange equation (G.16) of Appendix G is

$$\nabla^2 q_n - \frac{1}{c^2}\frac{\partial^2 q_n}{\partial t^2} = -\frac{4\pi}{c} j_n,\ n = 1 - 4\,. \tag{17.127}$$

Identifying the amplitude functions q_n with potentials $A_n(\boldsymbol{r}, t)$, $\phi(\boldsymbol{r}, t)$ as above, this becomes the vector wave equation as required.

17.3.10 ***EPI derivation of the Maxwell–Boltzmann law.*** An isolated gas consists of a large number of identical molecules of mass m that are moving randomly within a container that is kept at a constant temperature T. The molecules collide elastically with themselves and the container walls.

Suppose that we want to know the joint probability law on the momentum $\boldsymbol{\mu} = (\mu_x, \mu_y, \mu_z)$ of a particle. Accordingly, the momentum value of a randomly selected particle is measured.

The self-consistency EPI approach gives a result [17.12]

$$\kappa = 1, \quad J = \frac{4}{N} \int \mathrm{d}\boldsymbol{\mu} \sum_{n=1}^{N} q_n^2 (A_n + B\mu^2), \quad q_n = q_n(\boldsymbol{\mu}) \tag{17.128}$$

with A_n, B constants.

(a) Show that the physical information obeys

$$K = \frac{4}{N} \int \mathrm{d}\boldsymbol{\mu} \sum_{n=1}^{N} \left[\sum_{m=1}^{3} \left(\frac{\partial q_n}{\partial \mu_m}\right)^2 - q_n^2 (A_n + B\mu^2) \right] \tag{17.129}$$

(b) Show that the Euler–Lagrange equations (G.16) of Appendix G give a differential equation

$$\nabla_\mu^2 q_n + q_n(A_n + B\mu^2) = 0, \quad n = 1, \ldots, N . \tag{17.130}$$

(c) Show that the use of separation of variables $q_n = q_{n1}(\mu_x)q_{n2}(\mu_y)q_{n3}(\mu_z)$ in (17.130) gives a differential equation of the parabolic cylinder type, in each variable. Show that the general solution q_n is a superposition

$$q_n = e^{-\mu^2/4a^2} 2^{-n/2} \sum_{\substack{ijk \\ i+j+k=n}} a_{nijk} H_i(\mu_x/a\sqrt{2}) H_j(\mu_y/a\sqrt{2}) H_k(\mu_z/a\sqrt{2}) \tag{17.131}$$

where the functions H_i are Hermite polynomials (3.54), $a = const.$ and the a_{nijk} are arbitrary weights.

The lowest-order solution, for $n = 0$, is the smoothest since $H_0(x) = 1$ [see (3.54)]. Hence it has the lowest Fisher information level I over all orders. It therefore represents the equilibrium probability law for momentum values.

(d) Show that the equilibrium probability law has the simple form

$$p(\boldsymbol{\mu}) = \frac{1}{\sqrt{2\pi a^2}} e^{-\mu^2/2a^2} . \tag{17.132}$$

This is one form of the Maxwell-Boltzmann distribution. The constant a^2 is the variance, and is usually directly computable as the mean-square momentum value. For example, for a one-dimensional gas, by the principle of "equipartition of energy" the mean-square momentum obeys $\langle\mu^2\rangle/2m = (3/2)kT$, with k the Boltzmann constant.

Solution (17.131) has the form of a superposition of probability amplitudes. Such a superposition is usually considered the hallmark of quantum mechanics, and yet these are purely classical in nature. EPI emphasizes the common roots of diverse phenomena.

The higher-order solutions (17.131) represent subsidiary minima in information I, and therefore unstable equilibrium distributions *en route* to the final one. They have not yet been confirmed experimentally.

17.3.11 ***Probability law on particle speed.*** The Maxwell-Boltzmann law (17.132) is the probability on the joint fluctuation of the three momentum components. It is often of more interest to know the probability law on the magnitude of the momentum. From that one easily gets the probability law on the speed, independent of direction.

The general approach is to Jacobian transform the law (17.132) from its given rectangular coordinates (μ_x, μ_y, μ_z) to the corresponding one in spherical polar coordinates (μ, θ, ϕ) and then integrate out the angular dependencies to get the marginal law on just the magnitude of the momentum. The relation between the two sets of coordinates is of course

$$\mu_x = \mu \sin\theta \cos\phi, \ \mu_y = \mu \sin\theta \sin\phi, \ \mu_z = \mu \cos\theta .$$

(a) Using the one-root Jacobian approach of Sect. 5.6, show that the probability law on the magnitude μ of the momentum is of the form

$$p_M(\mu) = B\mu^2 e^{-\mu^2/2a^2} . \tag{17.133}$$

Find the necessary value of constant B for the law to obey normalization.

(b) This law has $p_M(0) = 0$, the minimum probability, while (17.132) has $p(0) = \frac{1}{\sqrt{2\pi a^2}}$, the *maximum* probability over all momenta. Is this an inconsistency?

Since momentum $\mu = mv$, where v is the speed of the particle, the probability law on speed is of the same form as (17.133). Find this, once again using the Jacobian method.

17.3.12 ***An approximate use of EPI; J as mean kinetic energy.*** As indicated in (17.112) and (17.128), sometimes information J is proportional to the mean kinetic energy KE of the measured object [17.31].That this is not always the case is illustrated by (17.125). Nevertheless, the mean kinetic energy is often a good approximation for J, and can lead to a correct or nearly correct probability law. (This might be considered a Bayesian use of EPI.)

A case in point is the problem of finding the probability law on energy E for a particle in Ex. 17.3.10. There $E = KE$. Hence $p(E)$ can be *approximately* found by EPI by the use of a $J = b \int dx x p(x)$, $b = const.$, $x \equiv E$.

Show that this J when used in the EPI extremum principle (17.102) leads to an Airy differential equation in the probability amplitude $q(x)$, of the form

$$q''(x) - xq(x) = 0 .$$

This has two possible solutions: either a non-normalizable Airy function $Bi(x)$ or a normalizable Airy function $Ai(x)$. Using the latter and squaring to get $p(x)$ yields the indicated curve in Fig. 17.12. The correct answer for these particles, the Boltzmann law

$$p(x) = \frac{1}{\langle x \rangle} \exp\left(-\frac{x}{\langle x \rangle}\right), \quad x \equiv E , \tag{17.134}$$

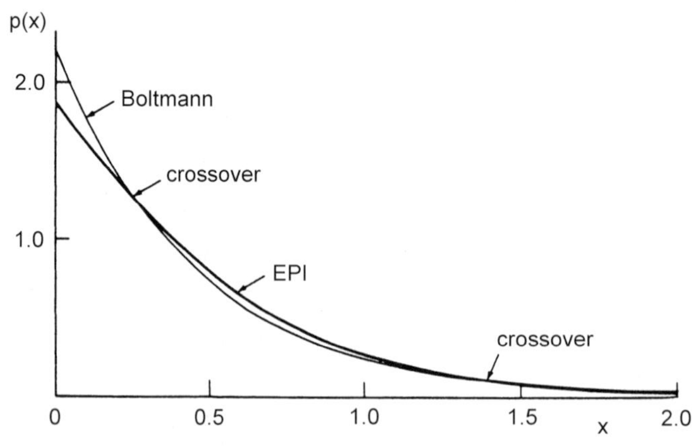

Fig. 17.12. Approximate EPI solution (*bold curve*) compared with correct answer (*thin curve*), the Boltzmann probability law

$\langle x \rangle = (3/2)kT$, is plotted as well for comparison. The two curves are fundamentally different, but numerically agree remarkably well.

By comparison, rigorous use of EPI derives the Boltzmann law. The correct functional form of J turns out to be, not the mean kinetic energy, but a normalization integral [17.12].

17.3.13 ***Equation of genetic change.*** Consider a population of living creatures of genetic types $n = 1, \ldots, N$. Each type, called an "allele", characterizes a sub population that is a fraction $p_n \equiv p(n|t)$ of the total population at the time t. This is also of course a probability p_n. A generation later, at time $t + \mathrm{d}t$ (in this continuous formulation), suppose that the nth allele has produced a relative number w_n of offspring called its "fitness" value. Quantities p_n, w_n have definite values at the time t. A generation later, at time $(t + \mathrm{d}t)$, the "fittest" alleles are expected to have increased fractions p_n. By how much does each probability p_n change over the generation? Since quantity $\dot{p}_n\,\mathrm{d}t \equiv \mathrm{d}p_n$ represents the change, *the* $\dot{p}_n(t)$ *are the unknowns of the problem*. These are called the "genetic changes".

The degree of disorder I in the population at a fixed time t is measured by the variability of its observable allele types n. This suggests that the EPI measurement experiment consists of observing an allele type n (corresponding to y in (17.75)) at a time t which is the unknown (fixed) parameter a. By (17.76), the information in the measurement n about the parameter t is

$$I_F(p;t) \equiv I = \sum_{n=1}^{N} p_n \left(\frac{\partial \log p_n}{\partial t} \right)^2 = \sum_{n=1}^{N} \frac{\dot{p}_n^2}{p_n}, \quad p_n \equiv p(n|t) \, . \tag{17.135}$$

(a) Verify the second and third equalities.

Use of the self-consistency approach to EPI gives an information

$$J = 2 \sum_{n=1}^{N} \dot{p}_n (w_n - \langle w \rangle), \ \langle w \rangle \equiv \sum_{n=1}^{N} w_n p_n \, . \tag{17.136}$$

(b) Show that use of (17.135) and (17.136) in the EPI extremum principle (17.102) gives as a solution

$$\dot{p}_n = p_n (w_n - \langle w \rangle) \, . \tag{17.137}$$

This is the called the "equation of genetic change" since it predicts the changes $\dot{p}_n\,\mathrm{d}t$ in the p_n. *Hints*: Since each $\dot{p}_n$ obviously depends upon p_n the extremization (17.102) must be in one of these sets of variables, but not both. The solution (17.137) results from extremizing in the $\dot{p}_n$. Notice that since the time t is a *fixed* parameter, the $\dot{p}_n$ are here simply *discrete unknowns,* to be found by ordinary differentiation $\partial/\partial \dot{p}_n$. Euler–Lagrange equations are not used.

(c) Show that if instead the extremization is in the p_n the solution is $p_n = 1/N =$ *const.*, quite different from the prediction (17.137). Which one is right? Show that both solutions minimize information K, but the absolute (smallest) minimum is attained by solution (17.137). In general EPI applications, when there are multiple solutions the one giving the absolute minimum value for K is the one that is most often observed.

17.3.14 ***Fisher temperature.*** A temperature value T is *a measure of inertia*. It measures the resistance of the level of disorder to a change in the internal energy of a system. The entropy H is one measure of the level of disorder of a system. Hence there is a corresponding measure T_B of inertia obeying

$$\frac{1}{T_B} \equiv k\frac{\mathrm{d}H}{\mathrm{d}E}, \quad H \equiv -\int \mathrm{d}E\, p(E)\ln p(E) \tag{17.138}$$

with k the Boltzmann constant. T_B is called the Boltzmann temperature.

(a) Show that the entropy $H = H(E)$ of a system obeying a Boltzmann law (17.134) is

$$H = 1 + \ln(\langle E\rangle) . \tag{17.139}$$

(b) Using $E \sim \langle E\rangle$, show that definition (17.138) of the Boltzmann temperature T_B yields a value

$$T_B = \frac{E}{k}, \tag{17.140}$$

which approximates equipartition of energy used below (17.132).

In view of inequality (17.124) and the discussion following it, information I is another measure of the disorder of a system. A Fisher temperature T_F may accordingly be defined in analogy with (17.138). This is [17.12]

$$\frac{1}{T_F} = -k_F\frac{\mathrm{d}I}{\mathrm{d}E} . \tag{17.141}$$

(The minus sign is present because we want T_F to be positive and, by (17.124), I *decreases* with an increase in disorder.) Parameter k_F is a constant. The two temperatures (17.138) and (17.141) are not in general equal or even linear in each other.

(c) Show that, for a Boltzmann law (17.134)

$$I = \frac{1}{\langle E\rangle^2} . \tag{17.142}$$

(b) Again using $E \sim \langle E\rangle$, show that definition (17.141) of the Fisher temperature T_F yields a value

$$T_F = \frac{E^3}{2k_F} . \tag{17.143}$$

Comparing this with (17.140) shows that *for Boltzmann particles* $T_F \sim T_B^3$, a cubic connection with the Boltzmann temperature. The Fisher temperature is, resultingly, the more sensitive of the two measures.

17.3.15 ***Fisher information and Shannon entropy as local and global measures, respectively.*** It is convenient to work here with discrete approximations to Fisher information I and entropy H,

$$I \approx 4\Delta x^{-1}\sum_{n=1}^{N-1}[q_{n+1} - q_n]^2, \quad p_n \equiv q_n^2, \tag{17.144}$$

$$H \approx -\Delta x\sum_{n=1}^{N} p_n \ln p_n .$$

The approximations become equalities in the limit as $\Delta x \to 0$, which will be taken at the end. Draw any curve of discrete points q_n vs n, $n = 1, \ldots, N$. Suppose that certain of the points are interchanged, say, points $(q_1, 1)$ and $(q_3, 3)$. We are interested in the effects of the interchange upon the two measures of disorder.

What is the change in entropy H from before to after the interchange? Show that

(a) the entropy H remains invariant. Notice that this is true even in the limit $\Delta x \to 0$.

What is the change in Fisher information I from before to after the interchange? Show from the graphs that

(b) Unless the curve is a uniform one, the Fisher information I changes drastically. Notice that the change grows unboundedly in the limit $\Delta x \to 0$.

An information measure that is invariant under such an interchange is called a "global measure", while one that changes under the interchange is called a "local measure". *Thus, H is a global measure of disorder while I is a local measure.* The local nature of I traces from its dependence upon the gradient of the curve $q(x)$, i.e. values $q'^2(x)$, as compared with values $q^2(x)$ for H. The student may note that it is precisely this gradient-dependence that leads to differential equations as solutions to the EPI Euler–Lagrange problem (17.114). This is useful, since differential equations describe the fundamental probability laws of physics.

When instead H is extremized, with the absence of any gradient-dependence pure exponential solutions (10.28) instead occur. This is why maximization of H leads to correct physical solutions only when they happen to be of an exponential form, as in statistical mechanics. As indicated at Eq. (17.132), EPI gives these solutions anyhow.

17.3.16 ***Additivity property.*** The information capacity (17.80) obeys a property of additivity if the Fisher coordinates x_m are independent, i.e., if each

$$p_n(\boldsymbol{x}) = \prod_m p_{nm}(x_m), \quad p_{nm}(x_m) \equiv q_{nm}^2(x_m) \,. \tag{17.145}$$

Under this condition (17.80) becomes

$$I = \sum_m I_m, \quad I_m = \frac{4}{N} \int \mathrm{d}x_m \sum_n \left(\frac{\mathrm{d}q_{nm}}{\mathrm{d}x_m} \right)^2 \,. \tag{17.146}$$

Show this.

Hint: Work from the equivalent expression to (17.80),

$$I = \frac{1}{N} \int \mathrm{d}\boldsymbol{x} \sum_{n,m} p_n(\boldsymbol{x}) \left(\frac{\partial \ln p_n}{\partial x_m} \right)^2 \,. \tag{17.147}$$

After substituting in the equalities (17.145), the ln operation becomes a sum over index m, so that the subsequent differentiation $\partial/\partial x_m$ gives just one term. Its square is still one term, and this defines the mth term in the indicated sum. Next, the integrations $\mathrm{d}\boldsymbol{x} \equiv \mathrm{d}x_1 \cdots \mathrm{d}x_M$ are all unit operations because of normalization of each p_{nm}, except for one integration which mates with the mth term in the indicated sum over m. The result is (17.146), after converting back to amplitudes q_{nm} via the definition (17.145).

17.3.17 ***Klein–Gordon equation from the wave representation for ψ.*** Derive the free-field Klein–Gordon equation (17.106) by successive differentiation of the representation (17.88) for ψ. *Hint*: Partially differentiate the latter twice with respect to each space coordinate and add the three resulting equations. Then differentiate (17.88) twice with respect to the time, and subtract $(1/c^2)$ times this equation from the preceding sum. If identity (17.92) is now used inside the integrand, the K–G equation results.

17.3.18 ***Wave-particle duality and EPI.*** The Heisenberg uncertainty principle (17.18) was originally derived in Sect. 3.16 out of the representation (3.83) for the particle's amplitude function ψ. This is a superposition of plane waves, and hence represents the *wave aspect* of the "wavicle" (a name for a wave-particle). By comparison, in this chapter we find the Heisenberg principle to be derived, in Ex. 17.3.6, out of the Cramer–Rao inequality (17.18). This regards the particle as a measured entity with a well-defined position a, thereby emphasizing the *particle nature* of the wavicle. In summary, then, the two approaches to the Heisenberg principle arise out of the well-known wave-particle duality of a wavicle.

We turn next to derivations of the Klein–Gordon (K–G) equation. It was shown in Ex. 17.3.17 that the K–G can be derived by successive differentiations of the wave representation (17.88) for the wavicle. This is a *wave-based* derivation of the K–G equation. Alternatively, the K–G is derived in Sect. 17.3.18 by using EPI, which is an outgrowth of measurement theory, in particular the measurement of the position a of the wavicle. This appears to regard the wavicle *as a particle*.

Does, then, EPI provide a *particle-based* derivation of the K–G equation that is a counterpart to the *wave-based* derivation? Discuss pros and cons of this view. A complicating factor is that at (17.77) the unknown position a drops out of the information expression.

17.3.19 ***The speed of light and information flow.*** A relativistic electron whose Fisher information (17.77) on 3−position is I, and whose rate of change of Shannon entropy is $\partial H/\partial t$, obeys (Note: the derivation in [17.12], Eq. (17.114), is off by a factor of 3)

$$c \geq \frac{\partial H/\partial t}{\sqrt{I}} . \tag{17.148}$$

Verify that the right-hand side indeed has units of length/time.

Notice that there are no explicitly electromagnetic factors on the r.h.s. of (17.148). Instead the speed of light is defined completely by two information flow quantities. Does this mean that, as an alternative to its usual electromagnetic interpretation, the speed of light can be regarded as an information quantity?

Cross-multiplying (17.148) gives $c\sqrt{I}$ as an upper bound to the Shannon *information flow rate* $\partial H/\partial t$. Compare this result with the oft-repeated statement that c per se *is* the ultimate speed with which *information* can flow.

Appendix A. Error Function and Its Derivative [4.12]

x	$\frac{2}{\sqrt{\pi}} e^{-x^2}$	erf x	x	$\frac{2}{\sqrt{\pi}} e^{-x^2}$	erf x
0.00	1.1283791671	0.0000000000	0.50	0.8787825789	0.5204998778
0.01	1.1282663348	0.0112834156	0.51	0.8699515467	0.5292436198
0.02	1.1279279057	0.0225645747	0.52	0.8610370343	0.5378986305
0.03	1.1273640827	0.0338412223	0.53	0.8520434444	0.5464640969
0.04	1.1265752040	0.0451111061	0.54	0.8429751813	0.5549392505
0.05	1.1255617424	0.0563719778	0.55	0.8338366473	0.5633233663
0.06	1.1243243052	0.0676215944	0.56	0.8246322395	0.5716157638
0.07	1.1228636333	0.0788577198	0.57	0.8153663461	0.5798158062
0.08	1.1211806004	0.0900781258	0.58	0.8060433431	0.5879229004
0.09	1.1192762126	0.1012805939	0.59	0.7966675911	0.5959364972
0.10	1.1171516068	0.1124629160	0.60	0.7872434317	0.6038560908
0.11	1.1148080500	0.1236228962	0.61	0.7777751846	0.6116812189
0.12	1.1122469379	0.1347583518	0.62	0.7682671442	0.6194114619
0.13	1.1094697934	0.1458671148	0.63	0.7587235764	0.6270464433
0.14	1.1064782654	0.1569470331	0.64	0.7491487161	0.6345858291
0.15	1.1032741267	0.1679959714	0.65	0.7395467634	0.6420293274
0.16	1.0998592726	0.1790118132	0.66	0.7299218814	0.6493766880
0.17	1.0962357192	0.1899924612	0.67	0.7202781930	0.6566277023
0.18	1.0924056008	0.2009358390	0.68	0.7106197784	0.6637822027
0.19	1.0883711683	0.2118398922	0.69	0.7009506721	0.6708400622
0.20	1.0841347871	0.2227025892	0.70	0.6912748604	0.6778011938
0.21	1.0796989342	0.2335219230	0.71	0.6815962792	0.6846655502
0.22	1.0750661963	0.2442959116	0.72	0.6719188112	0.6914331231
0.23	1.0702392672	0.2550225996	0.73	0.6622462838	0.6981039429
0.24	1.0652209449	0.2657000590	0.74	0.6525824665	0.7046780779
0.25	1.0600141294	0.2763263902	0.75	0.6429310692	0.7111556337
0.26	1.0546218194	0.2868997232	0.76	0.6332957399	0.7175367528
0.27	1.0490471098	0.2974182185	0.77	0.6236800626	0.7238216140
0.28	1.0432931885	0.3078800680	0.78	0.6140875556	0.7300104313
0.29	1.0373633334	0.3182834959	0.79	0.6045216696	0.7361034538
0.30	1.0312609096	0.3286267595	0.80	0.5949857863	0.7421009647
0.31	1.0249893657	0.3389081503	0.81	0.5854832161	0.7480032806
0.32	1.0185522310	0.3491259948	0.82	0.5760171973	0.7538107509
0.33	1.0119531119	0.3592786550	0.83	0.5665908944	0.7595237569
0.34	1.0051956887	0.3693645293	0.84	0.5572073967	0.7651427115
0.35	0.9982837121	0.3793820536	0.85	0.5478697173	0.7706680576
0.36	0.9912210001	0.3893297011	0.86	0.5385807918	0.7761002683
0.37	0.9840114337	0.3992059840	0.87	0.5293434773	0.7814398455
0.38	0.9766589542	0.4090094534	0.88	0.5201605514	0.7866873192
0.39	0.9691675592	0.4187387001	0.89	0.5110347116	0.7918432468
0.40	0.9615412988	0.4283923550	0.90	0.5019685742	0.7969082124
0.41	0.9537842727	0.4379690902	0.91	0.4929646742	0.8018828258
0.42	0.9459006256	0.4474676184	0.92	0.4840254639	0.8067677215
0.43	0.9378945443	0.4568866945	0.93	0.4751533132	0.8115635586
0.44	0.9297702537	0.4662251153	0.94	0.4663505090	0.8162710190
0.45	0.9215320130	0.4754817198	0.95	0.4576192546	0.8208908073
0.46	0.9131841122	0.4846553900	0.96	0.4489616700	0.8254236496
0.47	0.9047308685	0.4937450509	0.97	0.4403797913	0.8298702930
0.48	0.8961766223	0.5027496707	0.98	0.4318755710	0.8342315043
0.49	0.8875257337	0.5116682612	0.99	0.4234508779	0.8385080696
0.50	0.8787825789	0.5204998778	1.00	0.4151074974	0.8427007929

Appendix A (cont.)

x	$\frac{2}{\sqrt{\pi}} e^{-x^2}$	erf x	x	$\frac{2}{\sqrt{\pi}} e^{-x^2}$	erf x
1.00	0.4151074974	0.8427007929	1.50	0.1189302892	0.9661051465
1.01	0.4068471315	0.8468104962	1.51	0.1154038270	0.9672767481
1.02	0.3986713992	0.8508380177	1.52	0.1119595356	0.9684134969
1.03	0.3905818368	0.8547842115	1.53	0.1085963195	0.9695162091
1.04	0.3825798986	0.8586499465	1.54	0.1053130683	0.9705856899
1.05	0.3746669570	0.8624361061	1.55	0.1021086576	0.9716227333
1.06	0.3668443034	0.8661435866	1.56	0.0989819506	0.9726281220
1.07	0.3591131488	0.8697732972	1.57	0.0959317995	0.9736026275
1.08	0.3514746245	0.8733261584	1.58	0.0929570461	0.9745470093
1.09	0.3439297827	0.8768031019	1.59	0.0900565239	0.9754620158
1.10	0.3364795978	0.8802050696	1.60	0.0872290586	0.9763483833
1.11	0.3291249667	0.8835330124	1.61	0.0844734697	0.9772068366
1.12	0.3218667103	0.8867878902	1.62	0.0817885711	0.9780380884
1.13	0.3147055742	0.8899706704	1.63	0.0791731730	0.9788428397
1.14	0.3076422299	0.8930823276	1.64	0.0766260821	0.9796217795
1.15	0.3006772759	0.8961238429	1.65	0.0741461034	0.9803755850
1.16	0.2938112389	0.8990962029	1.66	0.0717320405	0.9811049213
1.17	0.2870445748	0.9020003990	1.67	0.0693826972	0.9818104416
1.18	0.2803776702	0.9048374269	1.68	0.0670968781	0.9824927870
1.19	0.2738108437	0.9076082860	1.69	0.0648733895	0.9831525869
1.20	0.2673443470	0.9103139782	1.70	0.0627110405	0.9837904586
1.21	0.2609783664	0.9129555080	1.71	0.0606086436	0.9844070075
1.22	0.2547130243	0.9155338810	1.72	0.0585650157	0.9850028274
1.23	0.2485483805	0.9180501041	1.73	0.0565789788	0.9855784998
1.24	0.2424844335	0.9205051843	1.74	0.0546493607	0.9861345950
1.25	0.2365211224	0.9229001283	1.75	0.0527749959	0.9866716712
1.26	0.2306583281	0.9252359418	1.76	0.0509547262	0.9871902752
1.27	0.2248958748	0.9275136293	1.77	0.0491874012	0.9876909422
1.28	0.2192335317	0.9297341930	1.78	0.0474718791	0.9881741959
1.29	0.2136710145	0.9318986327	1.79	0.0458070274	0.9886405487
1.30	0.2082079868	0.9340079449	1.80	0.0441917233	0.9890905016
1.31	0.2028440621	0.9360631228	1.81	0.0426248543	0.9895245446
1.32	0.1975788048	0.9380651551	1.82	0.0411053185	0.9899431565
1.33	0.1924117326	0.9400150262	1.83	0.0396320255	0.9903468051
1.34	0.1873423172	0.9419137153	1.84	0.0382038966	0.9907359476
1.35	0.1823699865	0.9437621961	1.85	0.0368198653	0.9911110301
1.36	0.1774941262	0.9455614366	1.86	0.0354788774	0.9914724883
1.37	0.1727140811	0.9473123980	1.87	0.0341798920	0.9918207476
1.38	0.1680291568	0.9490160353	1.88	0.0329218811	0.9921562228
1.39	0.1634386216	0.9506732958	1.89	0.0317038307	0.9924793184
1.40	0.1589417077	0.9522851198	1.90	0.0305247404	0.9927904292
1.41	0.1545376130	0.9538524394	1.91	0.0293836241	0.9930899398
1.42	0.1502255027	0.9553761786	1.92	0.0282795101	0.9933782251
1.43	0.1460045107	0.9568572531	1.93	0.0272114412	0.9936556502
1.44	0.1418737413	0.9582965696	1.94	0.0261784752	0.9939225709
1.45	0.1378322708	0.9596950256	1.95	0.0251796849	0.9941793336
1.46	0.1338791486	0.9610535095	1.96	0.0242141583	0.9944262755
1.47	0.1300133993	0.9623728999	1.97	0.0232809986	0.9946637246
1.48	0.1262340239	0.9636540654	1.98	0.0223793244	0.9948920004
1.49	0.1225400011	0.9648978648	1.99	0.0215082701	0.9951114132
1.50	0.1189302892	0.9661051465	2.00	0.0206669854	0.9953222650

Appendix B. Percentage Points of the χ^2-Distribution; Values of χ^2 in Terms of α and ν [4.12]

$\nu \backslash \alpha$	0.995	0.99	0.975	0.95	0.9	0.75	0.5	0.25
1	(−5)3.92704	(−4)1.57088	(−4)9.82069	(−3)3.93214	0.0157908	0.101531	0.454937	1.32330
2	(−2)1.00251	(−2)2.01007	(−2)5.06356	0.102587	0.210720	0.575364	1.38629	2.77259
3	(−2)7.17212	0.114832	0.215795	0.351846	0.584375	1.212534	2.36597	4.10835
4	0.206990	0.297110	0.484419	0.710721	1.063623	1.92255	3.35670	5.38527
5	0.411740	0.554300	0.831211	1.145476	1.61031	2.67460	4.35146	6.62568
6	0.675727	0.872085	1.237347	1.63539	2.20413	3.45460	5.34812	7.84080
7	0.989265	1.239043	1.68987	2.16735	2.83311	4.25485	6.34581	9.03715
8	1.344419	1.646482	2.17973	2.73264	3.48954	5.07064	7.34412	10.2188
9	1.734926	2.087912	2.70039	3.32511	4.16816	5.89883	8.34283	11.3887
10	2.15585	2.55821	3.24697	3.94030	4.86518	6.73720	9.34182	12.5489
11	2.60321	3.05347	3.81575	4.57481	5.57779	7.58412	10.3410	13.7007
12	3.07382	3.57056	4.40379	5.22603	6.30380	8.43842	11.3403	14.8454
13	3.56503	4.10691	5.00874	5.89186	7.04150	9.29906	12.3398	15.9839
14	4.07468	4.66043	5.62872	6.57063	7.78953	10.1653	13.3393	17.1170
15	4.60094	5.22935	6.26214	7.26094	8.54675	11.0365	14.3389	18.2451
16	5.14224	5.81221	6.90766	7.96164	9.31223	11.9122	15.3385	19.3688
17	5.69724	6.40776	7.56418	8.67176	10.0852	12.7919	16.3381	20.4887
18	6.26481	7.01491	8.23075	9.39046	10.8649	13.6753	17.3379	21.6049
19	6.84398	7.63273	8.90665	10.1170	11.6509	14.5620	18.3376	22.7178
20	7.43386	8.26040	9.59083	10.8508	12.4426	15.4518	19.3374	23.8277
21	8.03366	8.89720	10.28293	11.5913	13.2396	16.3444	20.3372	24.9348
22	8.64272	9.54249	10.9823	12.3380	14.0415	17.2396	21.3370	26.0393
23	9.26042	10.19567	11.6885	13.0905	14.8479	18.1373	22.3369	27.1413
24	9.88623	10.8564	12.4011	13.8484	15.6587	19.0372	23.3367	28.2412
25	10.5197	11.5240	13.1197	14.6114	16.4734	19.9393	24.3366	29.3389
26	11.1603	12.1981	13.8439	15.3791	17.2919	20.8434	25.3364	30.4345
27	11.8076	12.8786	14.5733	16.1513	18.1138	21.7494	26.3363	31.5284
28	12.4613	13.5648	15.3079	16.9279	18.9392	22.6572	27.3363	32.6205
29	13.1211	14.2565	16.0471	17.7083	19.7677	23.5666	28.3362	33.7109
30	13.7867	14.9535	16.7908	18.4926	20.5992	24.4776	29.3360	34.7998
40	20.7065	22.1643	24.4331	26.5093	29.0505	33.6603	39.3354	45.6160
50	27.9907	29.7067	32.3574	34.7642	37.6886	42.9421	49.3349	56.3336
60	35.5346	37.4848	40.4817	43.1879	46.4589	52.2938	59.3347	66.9814
70	43.2752	45.4418	48.7576	51.7393	55.3290	61.6983	69.3344	77.5766
80	51.1720	53.5400	57.1532	60.3915	64.2778	71.1445	79.3343	88.1303
90	59.1963	61.7541	65.6466	69.1260	73.2912	80.6247	89.3342	98.6499
100	67.3276	70.0648	74.2219	77.9295	82.3581	90.1332	99.3341	109.141

$$\alpha = [2^{\nu/2}\Gamma(\nu/2)]^{-1} \int_{\chi^2}^{\infty} e^{-t/2} t^{\nu/2} / dt$$

From E.S. Pearson, H.O. Hartley (eds.): *Biometrika tables for statisticians*, vol. I (Cambridge Univ. Press, Cambridge, England, 1954) (with permission).

Appendix B (cont.)

$\nu \backslash \alpha$	0.1	0.05	0.025	0.01	0.005	0.001	0.0005	0.0001
1	2.70554	3.84146	5.02389	6.63490	7.87944	10.828	12.116	15.137
2	4.60517	5.99147	7.37776	9.21034	10.5966	13.816	15.202	18.421
3	6.25139	7.81473	9.34840	11.3449	12.8381	16.266	17.730	21.108
4	7.77944	9.48773	11.1433	13.2767	14.8602	18.467	19.997	23.513
5	9.23635	11.0705	12.8325	15.0863	16.7496	20.515	22.105	25.745
6	10.6446	12.5916	14.4494	16.8119	18.5476	22.458	24.103	27.856
7	12.0170	14.0671	16.0128	18.4753	20.2777	24.322	26.018	29.877
8	13.3616	15.5073	17.5346	20.0902	21.9550	26.125	27.868	31.828
9	14.6837	16.9190	19.0228	21.6660	23.5893	27.877	29.666	33.720
10	15.9871	18.3070	20.4831	23.2093	25.1882	29.588	31.420	35.564
11	17.2750	19.6751	21.9200	24.7250	26.7569	31.264	33.137	37.367
12	18.5494	21.0261	23.3367	26.2170	28.2995	32.909	34.821	39.134
13	19.8119	22.3621	24.7356	27.6883	29.8194	34.528	36.478	40.871
14	21.0642	23.6848	26.1190	29.1413	31.3193	36.123	38.109	42.579
15	22.3072	24.9958	27.4884	30.5779	32.8013	37.697	39.719	44.263
16	23.5418	26.2962	28.8454	31.9999	34.2672	39.252	41.308	45.925
17	24.7690	27.5871	30.1910	33.4087	35.7185	40.790	42.879	47.566
18	25.9894	28.8693	31.5264	34.8053	37.1564	42.312	44.434	49.189
19	27.2036	30.1435	32.8523	36.1908	38.5822	43.820	45.973	50.796
20	28.4120	31.4104	34.1696	37.5662	39.9968	45.315	47.498	52.386
21	29.6151	32.6705	35.4789	38.9321	41.4010	46.797	49.011	53.962
22	30.8133	33.9244	36.7807	40.2894	42.7956	48.268	50.511	55.525
23	32.0069	35.1725	38.0757	41.6384	44.1813	49.728	52.000	57.075
24	33.1963	36.4151	39.3641	42.9798	45.5585	51.179	53.479	58.613
25	34.3816	37.6525	40.6465	44.3141	46.9278	52.620	54.947	60.140
26	35.5631	38.8852	41.9232	45.6417	48.2899	54.052	56.407	61.657
27	36.7412	40.1133	43.1944	46.9630	49.6449	55.476	57.858	63.164
28	37.9159	41.3372	44.4607	48.2782	50.9933	56.892	59.300	64.662
29	39.0875	42.5569	45.7222	49.5879	52.3356	58.302	60.735	66.152
30	40.2560	43.7729	46.9792	50.8922	53.6720	59.703	62.162	67.633
40	51.8050	55.7585	59.3417	63.6907	66.7659	73.402	76.095	82.062
50	63.1671	67.5048	71.4202	76.1539	79.4900	86.661	89.560	95.969
60	74.3970	79.0819	83.2976	88.3794	91.9517	99.607	102.695	109.503
70	85.5271	90.5312	95.0231	100.425	104.215	112.317	115.578	122.755
80	96.5782	101.879	106.629	112.329	116.321	124.839	128.261	135.783
90	107.565	113.145	118.136	124.116	128.299	137.208	140.782	148.627
100	118.498	124.342	129.561	135.807	140.169	149.449	153.167	161.319

Appendix C. Percentage Points of the t-Distribution; Values of t in Terms of A and ν [4.12]

$\nu \backslash A$	0.2	0.5	0.8	0.9	0.95	0.98	0.99	0.995	0.998	0.999	0.9999	0.99999	0.999999
1	0.325	1.000	3.078	6.314	12.706	31.821	63.657	127.321	318.309	636.619	6366.198	63 661.977	636 619.772
2	0.289	0.816	1.886	2.920	4.303	6.965	9.925	14.089	22.327	31.598	99.992	316.225	999.999
3	0.277	0.765	1.638	2.353	3.182	4.541	5.841	7.453	10.214	12.924	28.000	60.397	130.155
4	0.271	0.741	1.533	2.132	2.776	3.747	4.604	5.598	7.173	8.610	15.544	27.771	49.459
5	0.267	0.727	1.476	2.015	2.571	3.365	4.032	4.773	5.893	6.869	11.178	17.897	28.477
6	0.265	0.718	1.440	1.943	2.447	3.143	3.707	4.317	5.208	5.959	9.082	13.555	20.047
7	0.263	0.711	1.415	1.895	2.365	2.998	3.499	4.029	4.785	5.408	7.885	11.215	15.764
8	0.262	0.706	1.397	1.860	2.306	2.896	3.355	3.833	4.501	5.041	7.120	9.782	13.257
9	0.261	0.703	1.383	1.833	2.262	2.821	3.250	3.690	4.297	4.781	6.594	8.827	11.637
10	0.260	0.700	1.372	1.812	2.228	2.764	3.169	3.581	4.144	4.587	6.211	8.150	10.516
11	0.260	0.697	1.363	1.796	2.201	2.718	3.106	3.497	4.025	4.437	5.921	7.648	9.702
12	0.259	0.695	1.356	1.782	2.179	2.681	3.055	3.428	3.930	4.318	5.694	7.261	9.085
13	0.259	0.694	1.350	1.771	2.160	2.650	3.012	3.372	3.852	4.221	5.513	6.955	8.604
14	0.258	0.692	1.345	1,761	2.145	2.624	2.977	3.326	3.787	4.140	5.363	6.706	8.218
15	0.258	0.691	1.341	1.753	2.131	2.602	2.947	3.286	3.733	4.073	5.239	6.502	7.903
16	0.258	0.690	1.337	1.746	2.120	2.583	2.921	3.252	3.686	4.015	5.134	6.330	7.642
17	0.257	0.689	1.333	1.740	2.110	2.567	2.898	3.223	3.646	3.965	5.044	6.184	7.421
18	0.257	0.688	1.330	1.734	2.101	2.552	2.878	3.197	3.610	3.922	4.966	6.059	7.232
19	0.257	0.688	1.328	1.729	2.093	2.539	2.861	3.174	3.579	3.883	4.897	5.949	7.069
20	0.257	0.687	1.325	1.725	2.086	2.528	2.845	3.153	3.552	3.850	4.837	5.854	6.927
21	0.257	0.686	1.323	1.721	2.080	2.518	2.831	3.135	3.527	3.819	4.784	5.769	6.802
22	0.256	0.686	1.321	1.717	2.074	2.508	2.819	3.119	3.505	3.792	4.736	5.694	6.692
23	0.256	0.685	1.319	1.714	2.069	2.500	2.807	3.104	3.485	3.768	4.693	5.627	6.593
24	0.256	0.685	1.318	1.711	2.064	2.492	2.797	3.090	3.467	3.745	4.654	5.566	6.504
25	0.256	0.684	1.316	1.7.08	2.060	2.485	2.787	3.078	3.450	3.725	4.619	5.511	6.424
26	0.256	0.684	1.315	1.706	2.056	2.479	2.779	3.067	3.435	3.707	4.587	5.461	6.352
27	0.256	0.684	1.314	1.703	2.052	2.473	2.771	3.057	3.421	3.690	4.558	5.415	6.286
28	0.256	0.683	1.313	1.701	2.048	2.467	2.763	3.047	3.408	3.674	4.530	5.373	6.225
29	0.256	0.683	1.311	1.699	2.045	2.462	2.756	3.038	3.396	3.659	4.506	5.335	6.170
30	0.256	0.683	1.310	1.697	2.042	2.457	2.750,	3.030	3.385	3.646	4.482	5.299	6.119
40	0.255	0.681	1.303	1.684	2.021	2.423	2.704	2.971	3.307	3.551	4.321	5.053	5.768
60	0.254	0.679	1.296	1.671	2.000	2.390	2.660	2.915	3.232	3.460	4.169	4.825	5.449
120	0.254	0.677	1.289	1.658	1.980	2.358	2.617	2.860	3.160	3.373	4.025	4.613	5.158
∞	0.253	0.674	1.282	1.645	1.960	2.326	2.576	2.807	3.090	3.291	3.891	4.417	4.892

Appendix D. Percentage Points of the F-Distribution; Values of F in Terms Q, ν_1, ν_2 [4.12]

$Q(F|\nu_1, \nu_2) = 0.5$

$\nu_2 \backslash \nu_1$	1	2	3	4	5	6	8	12	15	20	30	60	∞
1	1.00	1.50	1.71	1.82	1.89	1.94	2.00	2.07	2.09	2.12	2.15	2.17	2.20
2	0.667	1.00	1.13	1.21	1.25	1.28	1.32	1.36	1.38	1.39	1.41	1.43	1.44
3	0.585	0.881	1.00	1.06	1.10	1.13	1.16	1.20	1.21	1.23	1.24	1.25	1.27
4	0.549	0.828	0.941	1.00	1.04	1.06	1.09	1.13	1.14	1.15	1.16	1.18	1.19
5	0.528	0.799	0.907	0.965	1.00	1.02	1.05	1.09	1.10	1.11	1.12	1.14	1.15
6	0.515	0.780	0.886	0.942	0.977	1.00	1.03	1.06	1.07	1.08	1.10	1.11	1.12
7	0.506	0.767	0.871	0.926	0.960	0.983	1.01	1.04	1.05	1.07	1.08	1.09	1.10
8	0.499	0.757	0.860	0.915	0.948	0.971	1.00	1.03	1.04	1.05	1.07	1.08	1.09
9	0.494	0.749	0.852	0.906	0.939	0.962	0.990	1.02	1.03	1.04	1.05	1.07	1.08
10	0.490	0.743	0.845	0.899	0.932	0.954	0.983	1.01	1.02	1.03	1.05	1.06	1.07
11	0.486	0.739	0.840	0.893	0.926	0.948	0.977	1.01	1.02	1.03	1.04	1.05	1.06
12	0.484	0.735	0.835	0.888	0.921	0.943	0.972	1.00	1.01	1.02	1.03	1.05	1.06
13	0.481	0.731	0.832	0.885	0.917	0.939	0.967	0.996	1.01	1.02	1.03	1.04	1.05
14	0.479	0.729	0.828	0.881	0.914	0.936	0.964	0.992	1.00	1.01	1.03	1.04	1.05
15	0.478	0.726	0.826	0.878	0.911	0.933	0.960	0.989	1.00	1.01	1.02	1.03	1.05
16	0.476	0.724	0.823	0.876	0.908	0.930	0.958	0.986	0.997	1.01	1.02	1.03	1.04
17	0.475	0.722	0.821	0.874	0.906	0.928	0.955	0.983	0.995	1.01	1.02	1.03	1.04
18	0.474	0.721	0.819	0.872	0.904	0.926	0.953	0.981	0.992	1.00	1.02	1.03	1.04
19	0.473	0.719	0.818	0.870	0.902	0.924	0.951	0.979	0.990	1.00	1.01	1.02	1.04
20	0.472	0.718	0.816	0.868	0.900	0.922	0.950	0.977	0.989	1.00	1.01	1.02	1.03
21	0.471	0.716	0.815	0.867	0.899	0.921	0.948	0.976	0.987	0.998	1.01	1.02	1.03
22	0.470	0.715	0.814	0.866	0.898	0.919	0.947	0.974	0.986	0.997	1.01	1.02	1.03
23	0.470	0.714	0.813	0.864	0.896	0.918	0.945	0.973	0.984	0.996	1.01	1.02	1.03
24	0.469	0.714	0.812	0.863	0.895	0.917	0.944	0.972	0.983	0.994	1.01	1.02	1.03
25	0.468	0.713	0.811	0.862	0.894	0.916	0.943	0.971	0.982	0.993	1.00	1.02	1.03
26	0.468	0.712	0.810	0.861	0.893	0.915	0.942	0.970	0.981	0.992	1.00	1.01	1.03
27	0.467	0.711	0.809	0.861	0.892	0.914	0.941	0.969	0.980	0.991	1.00	1.01	1.03
28	0.467	0.711	0.808	0.860	0.892	0.913	0.940	0.968	0.979	0.990	1.00	1.01	1.02
29	0.466	0.710	0.808	0.859	0.891	0.912	0.940	0.967	0.978	0.990	1.00	1.01	1.02
30	0.466	0.709	0.807	0.858	0.890	0.912	0.939	0.966	0.978	0.989	1.00	1.01	1.02
40	0.463	0.705	0.802	0.854	0.885	0.907	0.934	0.961	0.972	0.983	0.994	1.01	1.02
60	0.461	0.701	0.798	0.849	0.880	0.901	0.928	0.956	0.967	0.978	0.989	1.00	1.01
120	0.458	0.697	0.793	0.844	0.875	0.896	0.923	0.950	0.961	0.972	0.983	0.994	1.01
∞	0.455	0.693	0.789	0.839	0.870	0.891	0.918	0.945	0.956	0.967	0.978	0.989	1.00

$Q(F|\nu_1, \nu_2) = 0.25$

$\nu_2 \backslash \nu_1$	1	2	3	4	5	6	8	12	15	20	30	60	∞
1	5.83	7.50	8.20	8.58	8.82	8.98	9.19	9.41	9.49	9.58	9.67	9.76	9.85
2	2.57	3.00	3.15	3.23	3.28	3.31	3.35	3.39	3.41	3.43	3.44	3.46	3.48
3	2.02	2.28	2.36	2.39	2.41	2.42	2.44	2.45	2.46	2.46	2.47	2.47	2.47
4	1.81	2.00	2.05	2.06	2.07	2.08	2.08	2.08	2.08	2.08	2.08	2.08	2.08
5	1.69	1.85	1.88	1.89	1.89	1.89	1.89	1.89	1.89	1.88	1.88	1.87	1.87
6	1.62	1.76	1.78	1.79	1.79	1.78	1.78	1.77	1.76	1.76	1.75	1.74	1.74
7	1.57	1.70	1.72	1.72	1.71	1.71	1.70	1.68	1.68	1.67	1.66	1.65	1.65
8	1.54	1.66	1.67	1.66	1.66	1.65	1.64	1.62	1.62	1.61	1.60	1.59	1.58
9	1.51	1.62	1.63	1.63	1.62	1.61	1.60	1.58	1.57	1.56	1.55	1.54	1.53
10	1.49	1.60	1.60	1.59	1.59	1.58	1.56	1.54	1.53	1.52	1.51	1.50	1.48
11	1.47	1.58	1.58	1.57	1.56	1.55	1.53	1.51	1.50	1.49	1.48	1.47	1.45
12	1.46	1.56	1.56	1.55	1.54	1.53	1.51	1.49	1.48	1.47	1.45	1.44	1.42
13	1.45	1.55	1.55	1.53	1.52	1.51	1.49	1.47	1.46	1.45	1.43	1.42	1.40
14	1.44	1.53	1.53	1.52	1.51	1.50	1.48	1.45	1.44	1.43	1.41	1.40	1.38
15	1.43	1.52	1.52	1.51	1.49	1.48	1.46	1.44	1.43	1.41	1.40	1.38	1.36
16	1.42	1.51	1.51	1.50	1.48	1.47	1.45	1.43	1.41	1.40	1.38	1.36	1.34
17	1.42	1.51	1.50	1.49	1.47	1.46	1.44	1.41	1.40	1.39	1.37	1.35	1.33
18	1.41	1.50	1.49	1.48	1.46	1.45	1.43	1.40	1.39	1.38	1.36	1.34	1.32
19	1.41	1.49	1.49	1.47	1.46	1.44	1.42	1.40	1.38	1.37	1.35	1.33	1.30
20	1.40	1.49	1.48	1.47	1.45	1.44	1.42	1.39	1.37	1.36	1.34	1.32	1.29
21	1.40	1.48	1.48	1.46	1.44	1.43	1.41	1.38	1.37	1.35	1.33	1.31	1.28
22	1.40	1.48	1.47	1.45	1.44	1.42	1.40	1.37	1.36	1.34	1.32	1.30	1.28
23	1.39	1.47	1.47	1.45	1.43	1.42	1.40	1.37	1.35	1.34	1.32	1.30	1.27
24	1.39	1.47	1.46	1.44	1.43	1.41	1.39	1.36	1.35	1.33	1.31	1.29	1.26
25	1.39	1.47	1.46	1.44	1.42	1.41	1.39	1.36	1.34	1.33	1.31	1.28	1.25
26	1.38	1.46	1.45	1.44	1.42	1.41	1.38	1.35	1.34	1.32	1.30	1.28	1.25
27	1.38	1.46	1.45	1.43	1.42	1.40	1.38	1.35	1.33	1.32	1.30	1.27	1.24
28	1.38	1.46	1.45	1.43	1.41	1.40	1.38	1.34	1.33	1.31	1.29	1.27	1.24
29	1.38	1.45	1.45	1.43	1.41	1.40	1.37	1.34	1.32	1.31	1.29	1.26	1.23
30	1.38	1.45	1.44	1.42	1.41	1.39	1.37	1.34	1.32	1.30	1.28	1.26	1.23
40	1.36	1.44	1.42	1.40	1.39	1.37	1.35	1.31	1.30	1.28	1.25	1.22	1.19
60	1.35	1.42	1.41	1.38	1.37	1.35	1.32	1.29	1.27	1.25	1.22	1.19	1.15
120	1.34	1.40	1.39	1.37	1.35	1.33	1.30	1.26	1.24	1.22	1.19	1.16	1.10
∞	1.32	1.39	1.37	1.35	1.33	1.31	1.28	1.24	1.22	1.19	1.16	1.12	1.00

Compiled from E.S. Pearson, H.O. Hartley (eds.): *Biometrika tables for statisticians*, vol. I (Cambridge Univ. Press, Cambridge, England, 1954) (with permission).

Appendix D (cont.)

$Q(F|\nu_1, \nu_2) = 0.1$

$\nu_2 \backslash \nu_1$	1	2	3	4	5	6	8	12	15	20	30	60	∞
1	39.86	49.50	53.59	55.83	57.24	58.20	59.44	60.71	61.22	61.74	62.26	62.79	63.33
2	8.53	9.00	9.16	9.24	9.29	9.33	9.37	9.41	9.42	9.44	9.46	9.47	9.49
3	5.54	5.46	5.39	5.34	5.31	5.28	5.25	5.22	5.20	5.18	5.17	5.15	5.13
4	4.54	4.32	4.19	4.11	4.05	4.01	3.95	3.90	3.87	3.84	3.82	3.79	3.76
5	4.06	3.78	3.62	3.52	3.45	3.40	3.34	3.27	3.24	3.21	3.17	3.14	3.10
6	3.78	3.46	3.29	3.18	3.11	3.05	2.98	2.90	2.87	2.84	2.80	2.76	2.72
7	3.59	3.26	3.07	2.96	2.88	2.83	2.75	2.67	2.63	2.59	2.56	2.51	2.47
8	3.46	3.11	2.92	2.81	2.73	2.67	2.59	2.50	2.46	2.42	2.38	2.34	2.29
9	3.36	3.01	2.81	2.69	2.61	2.55	2.47	2.38	2.34	2.30	2.25	2.21	2.16
10	3.29	2.92	2.73	2.61	2.52	2.46	2.38	2.28	2.24	2.20	2.16	2.11	2.06
11	3.23	2.86	2.66	2.54	2.45	2.39	2.30	2.21	2.17	2.12	2.08	2.03	1.97
12	3.18	2.81	2.61	2.48	2.39	2.33	2.24	2.15	2.10	2.06	2.01	1.96	1.90
13	3.14	2.76	2.56	2.43	2.35	2.28	2.20	2.10	2.05	2.01	1.96	1.90	1.85
14	3.10	2.73	2.52	2.39	2.31	2.24	2.15	2.05	2.01	1.96	1.91	1.86	1.80
15	3.07	2.70	2.49	2.36	2.27	2.21	2.12	2.02	1.97	1.92	1.87	1.82	1.76
16	3.05	2.67	2.46	2.33	2.24	2.18	2.09	1.99	1.94	1.89	1.84	1.78	1.72
17	3.03	2.64	2.44	2.31	2.22	2.15	2.06	1.96	1.91	1.86	1.81	1.75	1.69
18	3.01	2.62	2.42	2.29	2.20	2.13	2.04	1.93	1.89	1.84	1.78	1.72	1.66
19	2.99	2.61	2.40	2.27	2.18	2.11	2.02	1.91	1.86	1.81	1.76	1.70	1.63
20	2.97	2.59	2.38	2.25	2.16	2.09	2.00	1.89	1.84	1.79	1.74	1.68	1.61
21	2.96	2.57	2.36	2.23	2.14	2.08	1.98	1.87	1.83	1.78	1.72	1.66	1.59
22	2.95	2.56	2.35	2.22	2.13	2.06	1.97	1.86	1.81	1.76	1.70	1.64	1.57
23	2.94	2.55	2.34	2.21	2.11	2.05	1.95	1.84	1.80	1.74	1.69	1.62	1.55
24	2.93	2.54	2.33	2.19	2.10	2.04	1.94	1.83	1.78	1.73	1.67	1.61	1.53
25	2.92	2.53	2.32	2.18	2.09	2.02	1.93	1.82	1.77	1.72	1.66	1.59	1.52
26	2.91	2.52	2.31	2.17	2.08	2.01	1.92	1.81	1.76	1.71	1.65	1.58	1.50
27	2.90	2.51	2.30	2.17	2.07	2.00	1.91	1.80	1.75	1.70	1.64	1.57	1.49
28	2.89	2.50	2.29	2.16	2.06	2.00	1.90	1.79	1.74	1.69	1.63	1.56	1.48
29	2.89	2.50	2.28	2.15	2.06	1.99	1.89	1.78	1.73	1.68	1.62	1.55	1.47
30	2.88	2.49	2.28	2.14	2.05	1.98	1.88	1.77	1.72	1.67	1.61	1.54	1.46
40	2.84	2.44	2.23	2.09	2.00	1.93	1.83	1.71	1.66	1.61	1.54	1.47	1.38
60	2.79	2.39	2.18	2.04	1.95	1.87	1.77	1.66	1.60	1.54	1.48	1.40	1.29
120	2.75	2.35	2.13	1.99	1.90	1.82	1.72	1.60	1.55	1.48	1.41	1.32	1.19
∞	2.71	2.30	2.08	1.94	1.85	1.77	1.67	1.55	1.49	1.42	1.34	1.24	1.00

$Q(F|\nu_1, \nu_2) = 0.05$

$\nu_2 \backslash \nu_1$	1	2	3	4	5	6	8	12	15	20	30	60	∞
1	161.4	199.5	215.7	224.6	230.2	234.0	238.9	243.9	245.9	248.0	250.1	252.2	254.3
2	18.51	19.00	19.16	19.25	19.30	19.33	19.37	19.41	19.43	19.45	19.46	19.48	19.50
3	10.13	9.55	9.28	9.12	9.01	8.94	8.85	8.74	8.70	8.66	8.62	8.57	8.53
4	7.71	6.94	6.59	6.39	6.26	6.16	6.04	5.91	5.86	5.80	5.75	5.69	5.63
5	6.61	5.79	5.41	5.19	5.05	4.95	4.82	4.68	4.62	4.56	4.50	4.43	4.36
6	5.99	5.14	4.76	4.53	4.39	4.28	4.15	4.00	3.94	3.87	3.81	3.74	3.67
7	5.59	4.74	4.35	4.12	3.97	3.87	3.73	3.57	3.51	3.44	3.38	3.30	3.23
8	5.32	4.46	4.07	3.84	3.69	3.58	3.44	3.28	3.22	3.15	3.08	3.01	2.93
9	5.12	4.26	3.86	3.63	3.48	3.37	3.23	3.07	3.01	2.94	2.86	2.79	2.71
10	4.96	4.10	3.71	3.48	3.33	3.22	3.07	2.91	2.85	2.77	2.70	2.62	2.54
11	4.84	3.98	3.59	3.36	3.20	3.09	2.95	2.79	2.72	2.65	2.57	2.49	2.40
12	4.75	3.89	3.49	3.26	3.11	3.00	2.85	2.69	2.62	2.54	2.47	2.38	2.30
13	4.67	3.81	3.41	3.18	3.03	2.92	2.77	2.60	2.53	2.46	2.38	2.30	2.21
14	4.60	3.74	3.34	3.11	2.96	2.85	2.70	2.53	2.46	2.39	2.31	2.22	2.13
15	4.54	3.68	3.29	3.06	2.90	2.79	2.64	2.48	2.40	2.33	2.25	2.16	2.07
16	4.49	3.63	3.24	3.01	2.85	2.74	2.59	2.42	2.35	2.28	2.19	2.11	2.01
17	4.45	3.59	3.20	2.96	2.81	2.70	2.55	2.38	2.31	2.23	2.15	2.06	1.96
18	4.41	3.55	3.16	2.93	2.77	2.66	2.51	2.34	2.27	2.19	2.11	2.02	1.92
19	4.38	3.52	3.13	2.90	2.74	2.63	2.48	2.31	2.23	2.16	2.07	1.98	1.88
20	4.35	3.49	3.10	2.87	2.71	2.60	2.45	2.28	2.20	2.12	2.04	1.95	1.84
21	4.32	3.47	3.07	2.84	2.68	2.57	2.42	2.25	2.18	2.10	2.01	1.92	1.81
22	4.30	3.44	3.05	2.82	2.66	2.55	2.40	2.23	2.15	2.07	1.98	1.89	1.78
23	4.28	3.42	3.03	2.80	2.64	2.53	2.37	2.20	2.13	2.05	1.96	1.86	1.76
24	4.26	3.40	3.01	2.78	2.62	2.51	2.36	2.18	2.11	2.03	1.94	1.84	1.73
25	4.24	3.39	2.99	2.76	2.60	2.49	2.34	2.16	2.09	2.01	1.92	1.82	1.71
26	4.23	3.37	2.98	2.74	2.59	2.47	2.32	2.15	2.07	1.99	1.90	1.80	1.69
27	4.21	3.35	2.96	2.73	2.57	2.46	2.31	2.13	2.06	1.97	1.88	1.79	1.67
28	4.20	3.34	2.95	2.71	2.56	2.45	2.29	2.12	2.04	1.96	1.87	1.77	1.65
29	4.18	3.33	2.93	2.70	2.55	2.43	2.28	2.10	2.03	1.94	1.85	1.75	1.64
30	4.17	3.32	2.92	2.69	2.53	2.42	2.27	2.09	2.01	1.93	1.84	1.74	1.62
40	4.08	3.23	2.84	2.61	2.45	2.34	2.18	2.00	1.92	1.84	1.74	1.64	1.51
60	4.00	3.15	2.76	2.53	2.37	2.25	2.10	1.92	1.84	1.75	1.65	1.53	1.39
120	3.92	3.07	2.68	2.45	2.29	2.17	2.02	1.83	1.75	1.66	1.55	1.43	1.25
∞	3.84	3.00	2.60	2.37	2.21	2.10	1.94	1.75	1.67	1.57	1.46	1.32	1.00

Appendix D (cont.)

$Q(F|\nu_1, \nu_2) = 0.025$

$\nu_2 \backslash \nu_1$	1	2	3	4	5	6	8	12	15	20	30	60	∞
1	647.8	799.5	864.2	899.6	921.8	937.1	956.7	976.7	984.9	993.1	1001	1010	1018
2	38.51	39.00	39.17	39.25	39.30	39.33	39.37	39.41	39.43	39.45	39.46	39.48	39.50
3	17.44	16.04	15.44	15.10	14.88	14.73	14.54	14.34	14.25	14.17	14.08	13.99	13.90
4	12.22	10.65	9.98	9.60	9.36	9.20	8.98	8.75	8.66	8.56	8.46	8.36	8.26
5	10.01	8.43	7.76	7.39	7.15	6.98	6.76	6.52	6.43	6.33	6.23	6.12	6.02
6	8.81	7.26	6.60	6.23	5.99	5.82	5.60	5.37	5.27	5.17	5.07	4.96	4.85
7	8.07	6.54	5.89	5.52	5.29	5.12	4.90	4.67	4.57	4.47	4.36	4.25	4.14
8	7.57	6.06	5.42	5.05	4.82	4.65	4.43	4.20	4.10	4.00	3.89	3.78	3.67
9	7.21	5.71	5.08	4.72	4.48	4.32	4.10	3.87	3.77	3.67	3.56	3.45	3.33
10	6.94	5.46	4.83	4.47	4.24	4.07	3.85	3.62	3.52	3.42	3.31	3.20	3.08
11	6.72	5.26	4.63	4.28	4.04	3.88	3.66	3.43	3.33	3.23	3.12	3.00	2.88
12	6.55	5.10	4.47	4.12	3.89	3.73	3.51	3.28	3.18	3.07	2.96	2.85	2.72
13	6.41	4.97	4.35	4.00	3.77	3.60	3.39	3.15	3.05	2.95	2.84	2.72	2.60
14	6.30	4.86	4.24	3.89	3.66	3.50	3.29	3.05	2.95	2.84	2.73	2.61	2.49
15	6.20	4.77	4.15	3.80	3.58	3.41	3.20	2.96	2.86	2.76	2.64	2.52	2.40
16	6.12	4.69	4.08	3.73	3.50	3.34	3.12	2.89	2.79	2.68	2.57	2.45	2.32
17	6.04	4.62	4.01	3.66	3.44	3.28	3.06	2.82	2.72	2.62	2.50	2.38	2.25
18	5.98	4.56	3.95	3.61	3.38	3.22	3.01	2.77	2.67	2.56	2.44	2.32	2.19
19	5.92	4.51	3.90	3.56	3.33	3.17	2.96	2.72	2.62	2.51	2.39	2.27	2.13
20	5.87	4.46	3.86	3.51	3.29	3.13	2.91	2.68	2.57	2.46	2.35	2.22	2.09
21	5.83	4.42	3.82	3.48	3.25	3.09	2.87	2.64	2.53	2.42	2.31	2.18	2.04
22	5.79	4.38	3.78	3.44	3.22	3.05	2.84	2.60	2.50	2.39	2.27	2.14	2.00
23	5.75	4.35	3.75	3.41	3.18	3.02	2.81	2.57	2.47	2.36	2.24	2.11	1.97
24	5.72	4.32	3.72	3.38	3.15	2.99	2.78	2.54	2.44	2.33	2.21	2.08	1.94
25	5.69	4.29	3.69	3.35	3.13	2.97	2.75	2.51	2.41	2.30	2.18	2.05	1.91
26	5.66	4.27	3.67	3.33	3.10	2.94	2.73	2.49	2.39	2.28	2.16	2.03	1.88
27	5.63	4.24	3.65	3.31	3.08	2.92	2.71	2.47	2.36	2.25	2.13	2.00	1.85
28	5.61	4.22	3.63	3.29	3.06	2.90	2.69	2.45	2.34	2.23	2.11	1.98	1.83
29	5.59	4.20	3.61	3.27	3.04	2.88	2.67	2.43	2.32	2.21	2.09	1.96	1.81
30	5.57	4.18	3.59	3.25	3.03	2.87	2.65	2.41	2.31	2.20	2.07	1.94	1.79
40	5.42	4.05	3.46	3.13	2.90	2.74	2.53	2.29	2.18	2.07	1.94	1.80	1.64
60	5.29	3.93	3.34	3.01	2.79	2.63	2.41	2.17	2.06	1.94	1.82	1.67	1.48
120	5.15	3.80	3.23	2.89	2.67	2.52	2.30	2.05	1.94	1.82	1.69	1.53	1.31
∞	5.02	3.69	3.12	2.79	2.57	2.41	2.19	1.94	1.83	1.71	1.57	1.39	1.00

$Q(F|\nu_1, \nu_2) = 0.01$

$\nu_2 \backslash \nu_1$	1	2	3	4	5	6	8	12	15	20	30	60	∞
1	4052	4999.5	5403	5625	5764	5859	5982	6106	6157	6209	6261	6313	6366
2	98.50	99.00	99.17	99.25	99.30	99.33	99.37	99.42	99.43	99.45	99.47	99.48	99.50
3	34.12	30.82	29.46	28.71	28.24	27.91	27.49	27.05	26.87	26.69	26.50	26.32	26.13
4	21.20	18.00	16.69	15.98	15.52	15.21	14.80	14.37	14.20	14.02	13.84	13.65	13.46
5	16.26	13.27	12.06	11.39	10.97	10.67	10.29	9.89	9.72	9.55	9.38	9.20	9.02
6	13.75	10.92	9.78	9.15	8.75	8.47	8.10	7.72	7.56	7.40	7.23	7.06	6.88
7	12.25	9.56	8.45	7.85	7.46	7.19	6.84	6.47	6.31	6.16	5.99	5.82	5.65
8	11.26	8.65	7.59	7.01	6.63	6.37	6.03	5.67	5.52	5.36	5.20	5.03	4.86
9	10.56	8.02	6.99	6.42	6.06	5.80	5.47	5.11	4.96	4.81	4.65	4.48	4.31
10	10.04	7.56	6.55	5.99	5.64	5.39	5.06	4.71	4.56	4.41	4.25	4.08	3.91
11	9.65	7.21	6.22	5.67	5.32	5.07	4.74	4.40	4.25	4.10	3.94	3.78	3.60
12	9.33	6.93	5.95	5.41	5.06	4.82.	4.50	4.16	4.01	3.86	3.70	3.54	3.36
13	9.07	6.70	5.74	5.21	4.86	4.62	4.30	3.96	3.82	3.66	3.51	3.34	3.17
14	8.86	6.51	5.56	5.04	4.69	4.46	4.14	3.80	3.66	3.51	3.35	3.18	3.00
15	8.68	6.36	5.42	4.89	4.56	4.32	4.00	3.67	3.52	3.37	3.21	3.05	2.87
16	8.53	6.23	5.29	4.77	4.44	4.20	3.89	3.55	3.41	3.26	3.10	2.93	2.75
17	8.40	6.11	5.18	4.67	4.34	4.10	3.79	3.46	3.31	3.16	3.00	2.83	2.65
18	8.29	6.01	5.09	4.58	4.25	4.01	3.71	3.37	3.23	3.08	2.92	2.75	2.57
19	8.18	5.93	5.01	4.50	4.17	3.94	3.63	3.30	3.15	3.00	2.84	2.67	2.49
20	8.10	5.85	4.94	4.43	4.10	3.87	3.56	3.23	3.09	2.94	2.78	2.61	2.42
21	8.02	5.78	4.87	4.37	4.04	3.81	3.51	3.17	3.03	2.88	2.72	2.55	2.36
22	7.95	5.72	4.82	4.31	3.99	3.76	3.45	3.12	2.98	2.83	2.67	2.50	2.31
23	7.88	5.66	4.76	4.26	3.94	3.71	3.41	3.07	2.93	2.78	2.62	2.45	2.26
24	7.82	5.61	4.72	4.22	3.90	3.67	3.36	3.03	2.89	2.74	2.58	2.40	2.21
25	7.77	5.57	4.68	4.18	3.85	3.63	3.32	2.99	2.85	2.70	2.54	2.36	2.17
26	7.72	5.53	4.64	4.14	3.82	3.59	3.29	2.96	2.81	2.66	2.50	2.33	2.13
27	7.68	5.49	4.60	4.11	3.78	3.56	3.26	2.93	2.78	2.63	2.47	2.29	2.10
28	7.64	5.45	4.57	4.07	3.75	3.53	3.23	2.90	2.75	2.60	2.44	2.26	2.06
29	7.60	5.42	4.54	4.04	3.73	3.50	3.20	2.87	2.73	2.57	2.41	2.23	2.03
30	7.56	5.39	4.51	4.02	3.70	3.47	3.17	2.84	2.70	2.55	2.39	2.21	2.01
40	7.31	5.18	4.31	3.83	3.51	3.29	2.99	2.66	2.52	2.37	2.20	2.02	1.80
60	7.08	4.98	4.13	3.65	3.34	3.12	2.82	2.50	2.35	2.20	2.03	1.84	1.60
120	6.85	4.79	3.95	3.48	3.17	2.96	2.66	2.34	2.19	2.03	1.86	1.66	1.38
∞	6.63	4.61	3.78	3.32	3.02	2.80	2.51	2.18	2.04	1.88	1.70	1.47	1.00

Appendix E. A Crib Sheet of Statistical Parameters and Their Errors

This summary is meant as a quick aid in attacking practical problems of statistical estimation.
All but one of the topics (error in correlation r) are derived and developed in the text, along with examples of their use.

To be estimated	Sampling expression	Rms error or relative error	Assumptions about data
Mean $\langle x \rangle$	$\overline{x} = N^{-1} \sum x_n$	$\sigma_{\overline{x}} = \sigma/\sqrt{N}$	$\{x_n\}$ are i.i.d., any probability law, with variance σ^2.
Variance σ^2	$S^2 = (N-1)^{-1} \sum (\overline{x} - x_n)^2$	$\sigma_{S^2}/\sigma^2 = \sqrt{2/(N-1)}$	$\{x_n\}$ are i.i.d. normal, variance σ^2.
Weighted mean	$\overline{x} = \sum w_n x_n / \sum w_n$	$\sigma_{\overline{x}} = \sigma \sqrt{\sum w_n^2 / (\sum w_n)^2}$	$\{x_n - \langle x_n \rangle\}$ are i.i.d., any probability law. The means $\langle x_n \rangle$ are not necessarily equal. Variance of each x_n is σ^2.
Signal to noise ratio S/N $= \langle x \rangle/\sigma$	$\mathrm{SNR} = \frac{\overline{x}}{S}\sqrt{\frac{2}{N-1}} \cdot \frac{\Gamma[(N-1)/2]}{\Gamma[(N-2)/2]}$	$\frac{\sigma_{\mathrm{SNR}}}{\mathrm{S/N}} = \sqrt{2/(N-3)}$	$\{x_n\}$ are i.i.d. normal
Correlation coefficient ρ	$^*r = \frac{N^{-1}\sum(x_n - \overline{x})(y_n - \overline{y})}{S_1 S_2}$	$\sigma_r/\rho = N^{-1/2}(1-\rho^2)/\rho$	$\{x_n\}$, $\{y_n\}$ are jointly normal.
Probability p	$\hat{p} = m/N$	$\sigma_{\hat{p}}/p = \sqrt{\frac{1-p}{Np}}$	The N events are independent (Bernoulli), with m "successes."

Median x_0	Sample median MW, the $(N+1)/2$ highest value in the sample	$\sigma_{MW} \cong \langle x \rangle / \sqrt{N-3}$	$\{x_n\}$ are i.i.d. RV's obeying an exponential probability law; e.g., laser speckle intensities
Regression coefficients $\{\alpha_n\}$	$\hat{\alpha} = ([X^T][X])^{-1}[X]^T \boldsymbol{y}$ where $[X]$ is the matrix of factors, $\boldsymbol{y}$ is the data	Covariance of errors $\Gamma_{\hat{\alpha}} = \sigma^2([X^T][X])^{-1}$	$\{y_n - \langle y_n \rangle\}$ are i.i.d. Gaussian, same variance σ^2. Means $\langle y_n \rangle$ are not necessarily equal.
Principal component eigenvalues $\{\lambda_n\}$	Solution to eigenvalue equation $[\overline{C}]\boldsymbol{u}_n = \lambda_n \boldsymbol{u}_n$	$\sigma_{\lambda_n}/\lambda_n = \sqrt{2/(N-1)}$	$[\overline{C}]$ is the sample covariance matrix. Errors $\Delta C(i, j)$ are small and multivariate normal.

* In forming S_1^2, S_2^2 use N^{-1} in place of $(N-1)^{-1}$

Appendix F. Synopsis of Statistical Tests

The aim of a statistical test is to arrive at a conclusion about some data, with a certain level of confidence.
The mechanism is rejection of an initial hypothesis, on the grounds that it was simply too improbable to be true.
The test hypothesis is therefore chosen to be the *opposite* of what it is desired to conclude, oddly enough.
The following tests are derived, developed and applied in the text.

Name	Aim or Test Hypothesis	Test Statistic	Assumption about data	How to use Table
Student t-test	To infer the spread $\pm\Delta x$ about the sample mean $\overline{x}$ of probable values of the true mean $\langle x\rangle$. This is accomplished without the need for knowing σ.	$^*t_0 \equiv \sqrt{N-1}\,\Delta x/S_0$	$\{x_n\}$ are i.i.d. normal, σ unknown.	Use Appendix C. For given row $\nu \equiv N-1$, find column for $A = 0.95$ (say), for confidence level $1 - A = 0.05$. Equate tabulated t_0 to t_0 statistic at left, solving for Δx.
Student comparison t-test	To test two sets of data $\{x_n\}$, $\{y_n\}$ for the hypothesis that they have the same mean, based on an observed difference $\overline{x} - \overline{y}$ in their sample means.	$^*t_0 \equiv \frac{\overline{x}-\overline{y}}{\sqrt{N_1 S_1^2 + N_2 S_2^2}} \left(\frac{N_1+N_2-2}{1/N_1+1/N_2}\right)^{1/2}$	$\{x_n\}$, $\{y_n\}$ are independent, and i.i.d. normal, from either the same or different laws (subject of the test),	Use Appendix C. Look along row $\nu \equiv N_1 + N_2 - 2$ to find tabulated item closest to computed t_0. Column value A is one minus the confidence level for rejecting the hypothesis of equal means.

Appendix F (cont.)

Snedecor *F*-test	To test two sets of data $\{x_n\}$, $\{y_n\}$ for the hypothesis that they have the same variance, based upon an observed ratio for their sample variances.	* $f_0 \equiv S_1^2/S_2^2$. Make $f_0 > 1$ by proper naming of the two data sets.	As preceding.	Use Appendix D. Each table is identified by its Q or confidence level for rejecting the hypothesis of equality. For given table Q, find the tabulated F for $\nu_1 \equiv N_1 - 1$ and $\nu_2 \equiv N_2 - 1$. The table whose F approximates f_0 most closely identifies the confidence level Q.
Chi-square test	To test a set of occurances $\{m_i\}$ for consistency with a chance hypothesis $\{P_i\}$. Nonconsistency means "significance."	$\chi_0^2 \equiv \sum_{i=1}^{M}(m_i - NP_i)^2/NP_i$	The N trials are independent and $N \gtrsim 4$, so that the central limit theorem holds for each RV m_i.	Use Appendix B. For a given row $\nu \equiv M - 1$, find the column whose tabulated χ^2 comes closest to the computed χ_0^2 at the left. The column identifies the confidence level α for rejecting the hypothesis of chance $\{P_i\}$.

* In forming the S^2 quantities, use N^{-1} in place of $(N - 1)^{-1}$

Appendix F (cont.)

Name	Aim or Test Hypothesis	Test Statistic	Assumption about data	How to use Table
R-test	To test a regression model for the possibility that its last K factors are insignificant.	$R_0 \equiv \frac{N-M-1}{K}\left(\frac{e_1-e}{e}\right)$	Data $\{y_n\}$ suffer additive noise $\{u_n\}$, where $u_n = N(0, \sigma^2)$.	Use Appendix D, with $\nu_1 \equiv K$ and $\nu_2 \equiv N - M - 1$. Each table is identified by its Q or confidence level for rejecting the hypothesis of insignificance of the factors. For a given table Q, find the tabulated F for the given ν_1, ν_2. The table whose F most closely approximates R_0 identifies the required confidence level Q.
Correlation test	To test two data samples $\{x_n\}, \{y_n\}$ for any degree of correlation. The hypothesis is that RV's $\{x_n\}, \{y_n\}$ do not correlate, even though their sample correlation coefficient r is finite.	$t_0 \equiv \sqrt{N-2}\left(\frac{r^2}{1-r^2}\right)^{1/2}$	$\{x_n\}, \{y_n\}$ are i.i.d. joint normal, with unknown correlation (subject of the test).	Use Appendix C. For given row $\nu \equiv N - 2$, find the column whose tabulated item is closest to computed t_0 at the left. Column value A is one minus the confidence level for rejecting the hypothesis of no correlation.

Appendix G. Derivation of Euler-Lagrange Equations

About 100 years ago, H. Helmholtz unified physics by showing that each known phenomenon of that era has an appropriate Lagrangian. That is, one may work *backwards from a known effect* to find the Lagrangian that implies it. Since then, in similar fashion Lagrangians have been formed for each new known physical effect. In this way, Lagrangians have supplied a unifying *mathematical route to known natural law*. How these Lagrangians mathematically give rise to these laws is the subject treated here.

As preparation for the main derivation, we review the chain rule of differentiation. A quantity $k = k(\epsilon, q(\epsilon), r(\epsilon))$ which depends upon its independent coordinate ϵ both directly and indirectly through intermediary functions $q(\epsilon)$ and $r(\epsilon)$, has a total derivative

$$\frac{dk}{d\epsilon} = \frac{\partial k}{\partial \epsilon} + \frac{\partial k}{\partial q}\frac{dq}{d\epsilon} + \frac{\partial k}{\partial r}\frac{dr}{d\epsilon} \; . \tag{G.1}$$

For example, if $k = \epsilon^2 q(\epsilon)/[r(\epsilon)]^2$ with $q(\epsilon) = \cos^3(A\epsilon)$, $A = \text{const.}$ and $r(\epsilon) = \exp(B\epsilon)$, $B = \text{const.}$, then directly differentiating $k = \epsilon^2 \cos^3(A\epsilon)\exp(-2B\epsilon)$ with respect to ϵ gives the same result as differentiating using the rule (G.1). The student is invited to verify this.

In (G.1), the two functions q and r are completely arbitrary as long as they are differentiable. For example, $r(\epsilon)$ could be *related to* $q(\epsilon)$. This will be our case below.

One-dimensional problem. Consider the problem of finding the function $q(x)$ that satisfies

$$K = \int_a^b \mathrm{d}x\, k[x, q(x), q'(x)] = \text{extrem.}, \quad q'(x) \equiv \mathrm{d}q/\mathrm{d}x \; . \tag{G.2}$$

The integral has fixed limits as indicated. Quantity K is an "objective functional" to be extremized. The integrand k is a known function of the indicated arguments and is called the "Lagrangian". Consider any nearby curve $q_\epsilon(x, \epsilon)$ to the solution $q(x)$, formed as

$$q_\epsilon(x, \epsilon) \equiv q(x) + \epsilon\eta(x) \; . \tag{G.3}$$

Here ϵ is a finite number and $\eta(x)$ is any function that perturbs $q(x)$ except at the endpoints a, b of the integration range, i.e.

$$\eta(a) = \eta(b) = 0 \; . \tag{G.4}$$

The situation is as in Fig. G.1. Notice that all perturbed curves pass through the same endpoints.

Differentiating (G.3) gives a property to be used later,

$$\frac{\mathrm{d}q_\epsilon}{\mathrm{d}\epsilon} = \eta(x) \; . \tag{G.5}$$

Back substituting representation (G.3) for q into (G.2) gives

$$K = \int_a^b dx\, k[x, q_\epsilon(x, \epsilon), q_\epsilon'(x, \epsilon)] \equiv K(\epsilon), \quad q_\epsilon'(x, \epsilon) \equiv \mathrm{d}q_\epsilon/\mathrm{d}x \; . \tag{G.6}$$

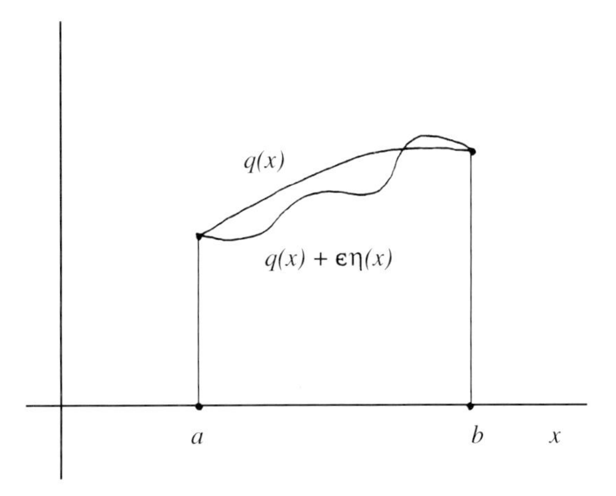

Fig. G.1. A one-dimensional optimization problem. Both the solution $q(x)$ and any perturbation $q_\epsilon(x, \epsilon) \equiv q(x) + \epsilon\eta(x)$ of it must pass through the endpoints $x = a$ and $x = b$.

After the indicated integration over all x, K becomes solely a function of ϵ.

We now use ordinary calculus to find the solution. By (G.3), when $\epsilon = 0$ the perturbed function $q_\epsilon(x, \epsilon) \equiv q(x)$, the extremum solution. Consequently, by (G.6) $K(\epsilon)$ has an extremum at $\epsilon = 0$. As illustrated in Fig. G.2 the one-dimensional curve $K(\epsilon)$ must therefore have zero slope at $\epsilon = 0$,

$$\left(\frac{\mathrm{d}K}{\mathrm{d}\epsilon}\right)_{\epsilon=0} = 0\,. \tag{G.7}$$

This condition will ultimately lead to the solution for $q(x)$.

Taking this road, by (G.6)

$$\frac{\mathrm{d}K}{\mathrm{d}\epsilon} = \int_a^b \mathrm{d}x\,\frac{\mathrm{d}k}{\mathrm{d}\epsilon} = \int_a^b \mathrm{d}x\left(\frac{\partial k}{\partial \epsilon} + \frac{\partial k}{\partial q_\epsilon}\frac{\mathrm{d}q_\epsilon}{\mathrm{d}\epsilon} + \frac{\partial k}{\partial q'_\epsilon}\frac{\mathrm{d}q'_\epsilon}{\mathrm{d}\epsilon}\right), \tag{G.8}$$

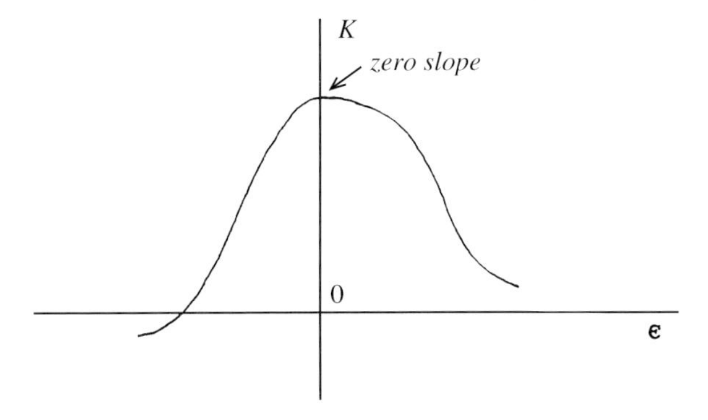

Fig. G.2. Functional K as a function of perturbation size parameter ϵ.

where the first equality is by the usual rule for differentiating an integral, and the second is by identity (G.1) with $q \equiv q_\epsilon$ and $r = q'_\epsilon$. The right-hand terms in (G.8) are evaluated next.

By the form of (G.6), k is not an explicit function of ϵ, so that

$$\frac{\partial k}{\partial \epsilon} = 0\,. \tag{G.9}$$

Also, the third right-hand integral in (G.8) is

$$\begin{aligned}\int_a^b \mathrm{d}x \frac{\partial k}{\partial q'_\epsilon}\frac{\mathrm{d}q'_\epsilon}{\mathrm{d}\epsilon} &= \int_a^b \mathrm{d}x \frac{\partial k}{\partial q'_\epsilon}\frac{d^2 q_\epsilon}{\mathrm{d}x\,\mathrm{d}\epsilon} \\ &= \left(\frac{\partial k}{\partial q'_\epsilon}\frac{\mathrm{d}q_\epsilon}{\mathrm{d}\epsilon}\right)_a^b - \int_a^b \frac{\mathrm{d}q_\epsilon}{\mathrm{d}\epsilon}\frac{\mathrm{d}}{\mathrm{d}x}\left(\frac{\partial k}{\partial q'_\epsilon}\right)\mathrm{d}x\,. \end{aligned} \tag{G.10}$$

The first equality is by definition of q'_ϵ, and the second is by integration by parts (in the usual notation, setting $u = \partial k/\partial q'_\epsilon$ and $\mathrm{d}v = \mathrm{d}^2 q_\epsilon / dx d\epsilon$). By (G.5) and then (G.4), the first far-right term is zero. Then (G.10) is

$$\int_a^b \mathrm{d}x \frac{\partial k}{\partial q'_\epsilon}\frac{\mathrm{d}q'_\epsilon}{\mathrm{d}\epsilon} = -\int_a^b \frac{\mathrm{d}q_\epsilon}{\mathrm{d}\epsilon}\frac{\mathrm{d}}{\mathrm{d}x}\left(\frac{\partial k}{\partial q'_\epsilon}\right)\mathrm{d}x\,. \tag{G.11}$$

Substituting this and (G.9) into (G.8) gives

$$\begin{aligned}\frac{\mathrm{d}K}{\mathrm{d}\epsilon} &= \int_a^b \mathrm{d}x \left[\frac{\partial k}{\partial q_\epsilon}\frac{\mathrm{d}q_\epsilon}{d\epsilon} - \frac{\mathrm{d}q_\epsilon}{\mathrm{d}\epsilon}\frac{\mathrm{d}}{\mathrm{d}x}\left(\frac{\partial k}{\partial q'_\epsilon}\right)\right] \\ &= \int_a^b \mathrm{d}x\,\eta(x)\left[\frac{\partial k}{\partial q_\epsilon} - \frac{\mathrm{d}}{\mathrm{d}x}\left(\frac{\partial k}{\partial q'_\epsilon}\right)\right]\end{aligned} \tag{G.12}$$

after factoring out the common term $dq_\epsilon/d\epsilon$ and using (G.5).

According to plan, we evaluate (G.12) at $\epsilon = 0$, giving

$$\left(\frac{\mathrm{d}K}{\mathrm{d}\epsilon}\right)_{\epsilon=0} = \int_a^b \mathrm{d}x\,\eta(x)\left[\frac{\partial k}{\partial q} - \frac{\mathrm{d}}{\mathrm{d}x}\left(\frac{\partial k}{\partial q'}\right)\right] \equiv 0\,. \tag{G.13}$$

The zero is by requirement (G.7) for attaining an extremum. Also, by (G.3), in this limit $q_\epsilon \to q$ and $q'_\epsilon \to q'$, allowing these replacements to be made in the square brackets. Since this zero is to hold for any perturbing curve $\eta(x)$, the factor in square brackets must be zero at each x, that is

$$\frac{\partial k}{\partial q} - \frac{\mathrm{d}}{\mathrm{d}x}\left(\frac{\partial k}{\partial q'}\right) = 0\,. \tag{G.14}$$

This is the celebrated Euler-Lagrange equation for the problem. Since k is a known function of x, q and q', (G.14) gives a differential equation in $q(x)$. This provides the final solution once particular boundary conditions are imposed.

Example. In classical mechanics the Lagrangian is known to be $k = \frac{1}{2}mq'^2 - V(q)$ where m is the particle mass, $q = q(t)$ defines the particle position at a time coordinate $t \equiv x$ of the preceding theory, and $V(q)$ is a known potential function of position q. Then $\partial k/\partial q = -V'(q)$, $\partial k/\partial q' = mq'$, and consequently the Euler-Lagrange solution (G.14) in q obeys $-V'(q) = mq''$. This is of course Newton's second law, the correct general answer $q(t)$ to the problem. Two imposed initial conditions, such as position and speed at $t = 0$, give the particular solution.

The reader may be curious to know where this Lagrangian of mechanics came from. The answer may be somewhat disappointing: It was chosen merely because "it works", i.e., leads to *the known solution,* which is Newton's second law. Unfortunately, such "reverse engineering" of known theory has been the traditional use of Lagrangians in general. Can Lagrangians be formed that derive new and unknown effects? The aim of Sect. 17.3 of the text is to provide a framework for doing this.

Multiple coordinates and curves. The case of multiple coordinates and multiple curves [17.34] turns out to have the same form of solution as (G.14). Consider a problem

$$K = \int_a^b \mathrm{d}\boldsymbol{x}\, k[\boldsymbol{x}, \boldsymbol{q}(\boldsymbol{x}), \boldsymbol{q}'(\boldsymbol{x})] = \text{extrem.}\,, \tag{G.15}$$

where $\boldsymbol{x} = (x_1, \ldots, x_M)$, $\mathrm{d}\boldsymbol{x} = \mathrm{d}x_1 \cdots \mathrm{d}x_M$, $\boldsymbol{q} = q_1, \ldots, q_N$ and $\boldsymbol{q}'(\boldsymbol{x}) = \partial \boldsymbol{q}/\partial x_1, \ldots, \partial \boldsymbol{q}/\partial x_M$. By the analogous steps to (G.1)–(G.14) the solution is

$$\sum_{m=1}^{M} \frac{\mathrm{d}}{\mathrm{d}x_m}\left(\frac{\partial k}{\partial q_{nm}}\right) - \frac{\partial k}{\partial q_n} = 0, \quad n = 1, \ldots, N\,, \tag{G.16}$$

where $q_{nm} \equiv \partial q_n / x_m$. Solution (G.16) is a system of N Euler-Lagrange equations that are an obvious generalization of the lowest-dimensional solution (G.14). The student should verify that the Eqs. (G.16) collapse to the answer (G.14) in the case $M = N = 1$ of one coordinate x and one unknown function $q(x)$.

Equations (G.16) are generally coupled in the unknown functions q_n. Fortunately, however, in many applications the equations become uncoupled. For example, see (17.106) and (17.130).

Incorporating constraints into the extremization problem. The problem (G.15) is sometimes complemented by knowledge of equality constraints that the unknown functions q_n must obey, that are of the form

$$\int_a^b \mathrm{d}\boldsymbol{x}\, f_{jm}(\boldsymbol{q}(\boldsymbol{x})) = F_{jm}, \quad j = 1, \ldots, J; \;\; m = 1, \ldots, M\,. \tag{G.17}$$

The constraint kernels $f_{jm}(\boldsymbol{q}(\boldsymbol{x}))$ are assumed known. An example is the case $f_{jm}(\boldsymbol{q}(\boldsymbol{x})) = x_m^j p(\boldsymbol{x})$ where $p = \sum_n q_n^2$, the probability law. In this case the numbers F_{jm} are known moments.

General constraints (G.17) may be easily accomodated by the overall Euler-Lagrange approach [17.34]. One simply *adds a term* to the objective functional K for each such constraint, giving a new "constrained functional"

$$K_c \equiv K + \sum_{j,m} \lambda_{jm} \left[\int_a^b d\boldsymbol{x}\; f_{jm}(\boldsymbol{q}(\boldsymbol{x})) - F_{jm}\right] \equiv \text{extrem}\,. \tag{G.18}$$

by hypothesis. The constants λ_{jm} are called "Lagrange multipliers".

Since there are now two sets of unknowns, the λ_{jm} and $\boldsymbol{q}$, we must extremize the constrained objective functional K_c in both sets. That is, K_c is doubly-extremized. Extremizing in the λ_{jm} is by the usual requirements $\partial K_c/\partial\lambda_{jm} = 0$. Operating in this way upon (G.18) gives back the constraint equations (G.17), exactly as we want. This is why the terms were added.

Extremizing in the $\boldsymbol{q}$ is equivalent, by the form of (G.18), to use of a "constrained Lagrangian"

$$k_c = k + \sum_{j,m} \lambda_{jm} f_{jm}(\boldsymbol{q}(\boldsymbol{x})) \,. \tag{G.19}$$

Operationally, the problem is solved by first substituting the constrained Lagrangian k_c of (G.19) into the Euler-Lagrange equation (G.16). The resulting solution $\boldsymbol{q}(\boldsymbol{x})$ will now contain the λ_{jm} as free parameters. Since, as we found, the doubly-extremized solution $\boldsymbol{q}(\boldsymbol{x})$ must also obey the constraint equations (G.17), the parametrized solution $\boldsymbol{q}(\boldsymbol{x})$ is back substituted into (G.17). This results in JM equations in the JM unknowns λ_{jm}. The problem may be nonlinear, but if the functional k_c is either concave or convex, it will have a unique solution. Nonlinear problems may be solved by any convenient search algorithm, such as the Newton-Raphson method [17.34].

References

Chapter 1

1.1 S. Brandt: *Statistical and Computational Methods in Data Analysis,* 2nd rev. ed. (North-Holland, Amsterdam 1976)
1.2 A. Papoulis: *Probability, Random Variables and Stochastic Processes* (McGraw-Hill, New York 1965)
1.3 E. Parzen: *Modern Probability Theory and Its Applications* (Wiley, New York 1966)
1.4 B. Saleh: *Photoelectron Statistics,* Springer Series in Optical Sciences, Vol. 6 (Springer, Berlin, Heidelberg, New York 1978)
1.5 E.L. O'Neill: *Introduction to Statistical Optics* (Addison-Wesley, Reading, MA 1963)
1.6 B. Spinoza: *Calculation of Chances* (Amsterdam 1687)

Chapter 2

2.1 A. Kolmogorov: *Foundations of the Theory of Probability* (Chelsea, New York 1950); *Grundbegriffe der Wahrscheinlichkeitsrechnung* (Springer, Berlin, Heidelberg, New York 1977)
2.2 A. Papoulis: *Probability, Random Variables, and Stochastic Processes* (McGraw-Hill, New York 1965)
2.3 C.E. Shannon: Bell Syst. Tech. J. **27**, 379–423, 623–656 (1948)
2.4 R.A. Fisher: Proc. Cambridge Philos. Soc. **22**, 700–725 (1925)
2.5 R.V.L. Hartley: Bell Syst. Tech. J. **7**, 535 (1928)
2.6 W.H. Richardson: J. Opt. Soc, Am. **62**, 55–59 (1972)
2.7 R.A. Howard: *Dynamic Probability Systems* (Wiley, New York 1971)
2.8 R. Von Mises: *Wahrscheinlichkeit, Statistik und Wahrheit* (Springer-Verlag OHG, Vienna 1936)
2.9 H. Everett III: *Theory of the Universal Wavefunction* (Princeton Univ., Ph.D. dissertation, 1956); also Revs. Mod. Phys. **29**, pp. 454–462 (1957)
2.10 H.L. Van Trees: *Detection, Estimation, and Modulation Theory, Part I* (Wiley, New York 1968)
2.11 C. Sagan, ed: *Communication with Extra-terrestrial Intelligence* (M.I.T. Press, Cambridge, Mass. 1973)
2.12 I. Bialynicki-Birula: in *Prog. in Optics, vol. XXXVI*, ed. E. Wolf (Elsevier, Amsterdam 1996)
2.13 D.J. Watts: *Small Worlds: The Dynamics of Networks Between Order and Randomness* (Princeton U.P., Princeton, N.J. 2000)

Additional Reading

Feller, W.: *An Introduction to Probability Theory and Its Applications,* Vol. I (Wiley, New York 1966)

Moran, P. A. P.: *An Introduction to Probability Theory* (Clarendon, Oxford 1968)
Parzen, E.: *Modern Probability Theory and Its Applications* (Wiley, New York 1966)
Reza, F.: *An Introduction to Information Theory* (McGraw-Hill, New York 1961)
Yu, F. T. S.: *Optics and Information Theory* (Wiley, New York 1976)

Chapter 3

3.1 E.L. O'Neill: *Introduction to Statistical Optics* (Addison-Wesley, Reading, MA 1963)
3.2 C. Dainty (ed.): *Laser Speckle and Related Phenomena,* 2nd ed., Topics in Applied Physics, Vol. 9 (Springer, Berlin, Heidelberg, New York 1982)
3.3 G.N. Plass et al.: Appl. 0pt. **16**, 643–653 (1977)
3.4 C. Cox, W. Munk: J. Opt. Soc. Am. **44**, 838–850 (1954)
3.5 J.W. Goodman: *Introduction to Fourier Optics* (McGraw-Hill, New York 1968)
3.6 B.R. Frieden: J. Opt. Soc. Am. **57**, 56 (1967)
3.7 S.Q. Duntley: J. Opt. Soc. Am. **53**, 214–233 (1963)
3.8 D.M. Green, J.A. Swets: *Signal Detection Theory and Psychophysics* (Wiley, New York 1966)
3.9 J.A. Swets (ed.): *Signal Detection and Recognition by Human Observers* (Wiley, New York 1964)
3.10 C.H. Slump, H.A. Ferwerda: Optik **62**, 98 (1982)
3.11 I.I. Hirschman: Am. J. Math. **79**, 152 (1957)
3.12 W. Beckner: Ann. Math **102**, 159 (1975)
3.13 S. Kullback: *Information Theory and Statistics* (Wiley, New York 1959)
3.14 A. Renyi: *On Measures of Entropy and Information, Proc. of 4th Berkeley Symp. on Math., Stat. and Prob. Theory* (Univ. of California Press, Berkeley 1961)
3.15 W.K. Wootters: Phys. Rev. D **23**, 357 (1981)
3.16 C. Tsallis: J. Stat. Phys. **52**, 479 (1988)
3.17 R.A. Fisher: Phil. Trans. R. Soc. Lond. **222**, 309 (1922)
3.18 B.R. Frieden: J. Opt. Soc. Am. **73**, 927–938 (1983)
3.19 D.G. Simpson, J. Amer. Stat. Assoc. **84**, 107 (1989)
3.20 B.R. Frieden, Founds. of Phys. **29**, 1521 (1999)
3.21 M.C. Alonso and M.A. Gil: in *Advances in Intelligent Computing - IPMU'94*, eds. B. Bouchon-Meunier et. al. (Springer-Verlag, Berlin 1994)

Additional Reading

Clarke, L. E.: *Random Variables* (Longman, New York 1975)
Feller, W.: *An Introduction to Probability Theory and Its Applications,* Vol. II (Wiley, New York 1966)
Papoulis, A.: *Probability, Random Variables and Stochastic Processes* (McGraw-Hill, New York 1965)
Pfeiffer, R. E.: *Concepts of Probability Theory,* 2nd rev. ed. (Dover, New York 1978)

Chapter 4

4.1 R.M. Bracewell: *The Fourier Transform and Its Applications* (McGraw-Hill, New York 1965)
4.2 J.W. Goodman: *Introduction to Fourier Optics* (McGraw-Hill, New York 1968)
4.3 B.R. Frieden: In *Picture Processing and Digital Filtering,* 2nd ed., ed. by T.S. Huang, Topics Appl. Phys., Vol. 6 (Springer, Berlin, Heidelberg, New York 1975)
4.4 A. Einstein: Ann. Phys. **17**, 549 (1905)
4.5 R. Barakat: Opt. Acta **21**, 903 (1974)
4.6 E.L. O'Neill: *Introduction to Statistical Optics* (Addison-Wesley, Reading, MA 1963)

4.7 B. Tatian: J. Opt. Soc. Am. **55**, 1014 (1965)
4.8 H. Lass: *Elements of Pure and Applied Mathematics* (McGraw-Hill, New York 1957)
4.9 E. Parzen: *Modern Probability Theory and Its Applications* (Wiley, New York 1966)
4.10 A.G. Fox, T. Li: Bell Syst. Tech. J. **40**, 453 (1961)
4.11 R.W. Lee, J.C. Harp: Proc. IEEE **57**, 375 (1969)
4.12 M. Abramowitz, I.A. Stegun (eds.): *Handbook of Mathematical Functions* (National Bureau of Standards, Washington, DC 1964)
4.13 J.W. Strohbehn: In *Laser Beam Propagation in the Atmosphere,* ed. J.W. Strohbehn, Topics in Applied Physics, Vol. 25 (Springer, Berlin, Heidelberg, New York 1978)
4.14 N.M. Blachman: IEEE Trans. Inf. Theory **IT-11**, 267 (1965)

Additional Reading

Araujo, A., E. Gine': *The Central Limit Theorem for Real and Banach Valued Random Variables* (Wiley, New York 1980)
Lukacs, E.: *Characteristic Functions,* 2nd ed. (Griffin, London 1970)
Moran, P. A. P.: *An Introduction to Probability Theory* (Clarendon, Oxford 1968)

Chapter 5

5.1 L.I. Goldfischer: J. Opt. Soc. Am. **55**, 247 (1965)
5.2 J.C. Dainty: Opt. Acta **17**, 761 (1970)
5.3 D. Korff: Opt. Commun. **5**, 188 (1972)
5.4 J.C. Dainty (ed.): *Laser Speckle and Related Phenomena,* 2nd ed., Topics in Applied Physics, Vol. 9 (Springer, Berlin, Heidelberg, New York 1982)
5.5 J.W. Goodman: Proc. IEEE **53**, 1688 (1965)
5.6 K. Miyamoto: "Wave Optics and Geometrical Optics in Optical Design," in *Progress in Optics,* Vol. 1, ed. by E. Wolf (North-Holland, Amsterdam 1961)
5.7 J.J. Burke: J. Opt. Soc. Am. **60**, 1262 (1970)
5.8 W. Feller: *An Introduction to Probability Theory and Its Applications,* Vol. 2 (Wiley, New York 1966) p. 50
5.9 J.W. Strohbehn: In *Laser Beam Propagation in the Atmosphere,* ed. J.W. Strohbehn, Topics in Applied Physics, Vol. 25 (Springer, Berlin, Heidelberg, New York 1978)
5.10 R.L. Phillips and L.C. Andrews: J. Opt. Soc. Am. **72**, 864 (1982)
5.11 B.R. Frieden and A. Plastino: "Classical trajectories compatible with quantum mechanics", Phys. Lett. A, under review
5.12 D. Bohm: Phys. Rev. **85**, 166 (1952); see also P.R. Holland: *The Quantum Theory of Motion* (Cambridge Univ. Press, England 1993)
5.13 H.H. Jeffreys: *Scientific Inference, 3rd ed.* (Cambridge Univ. Press, England 1973)
5.14 B.R. Frieden: Found. Phys. **16**, 883 (1986)
5.15 E.T. Jaynes: Found. Phys. **3**, 477 (1973)
5.16 A. Papoulis: *Probability, Random Variables and Stochastic Processes* (McGraw-Hill, New York 1965) pp. 11–12
5.17 V. Volterra: Mem. Acad. Lincei **2**, 31 (1926)

Additional Reading See the readings for Chapters 2 and 3

Chapter 6

6.1 E. L. O'Neill: *Introduction to Statistical Optics* (Addison-Wesley, Reading, MA 1963)
6.2 J.R. Klauder, E.C.G. Sudarshan: *Fundamentals of Quantum Optics* (Benjamin, New York 1968)
6.3 M. Born, E. Wolf: *Principles of Optics* (Macmillan, New York 1959)

6.4 R. Hanbury-Brown, R.Q. Twiss: Proc. R. Soc. London A **248**, 201 (1958)
6.5 R.E. Burgess: Disc. Faraday Soc. **28**, 151 (1959)
6.6 R. Shaw, J.C. Dainty: *Image Science* (Academic, New York 1974)
6.7 S. Chandrasekhar: Rev. Mod. Phys. **15**, 1 (1943)
6.8 J.O. Berger and D.A. Berry: *American Scientist* (Sigma Xi, March-April 1988)

Additional Reading

Barrett, H. H., W. Swindell: *Radiological Imaging,* Vol. 1 (Academic, New York 1981)
Chow, Y. S., H. Teicher: *Probability Theory. Independence, Interchangeability, Martingales* (Springer, Berlin, Heidelberg, New York 1978)
Goodman, J. W: *Statistical Optics* (Wiley, New York, 1985)
Moran, P. A. P.: *An Introduction to Probability Theory* (Clarendon, Oxford 1968)
Saleh, B. E. A., M.C. Teich: Proc. IEEE **70**, 229 (1982)
Thomas, J. B.: *An Introduction to Statistical Communication Theory* (Wiley, New York 1969)

Chapter 7

7.0 N.L. Balazs et. al.: Obituary for Nicholas Constantine Metropolis, in *Physics Today* (Amer. Inst. of Physics, October 2000) pp. 100–101
7.1 J.J. De Palma, J. Gasper: Phot. Sci. Eng. **16**, 181 (1972)
7.2 G.N. Plass, G.W. Kattawar, J.A. Guinn: Appl. Opt. **14**, 1924 (1975)
7.3 J.C. Dainty (ed.): *Laser Speckle and Related Phenomena,* 2nd ed., Topics Appl. Phys. Vol. 9 (Springer, Berlin, Heidelberg, New York 1982)
7.4 H. Fujii, J. Vozomi, T. Askura: J. Opt. Soc. Am. **66**, 1222 (1976)
7.5 R.C. Gonzalez, P. Wintz: *Digital Image Processing* (Addison-Wesley, Reading, MA 1977)
7.6 B.R. Frieden: Appl. Opt. **4**, 1400 (1965)

Additional Reading

Binder, K. (ed.): *Monte Carlo Methods in Statistical Physics,* Topics in Current Physics, Vol. 7 (Springer, Berlin, Heidelberg, New York 1978)
Brandt, S.: *Statistical and Computational Methods in Data Analysis,* 2nd rev. ed. (North-Holland, New York 1976)
Marchuk, G. I., G.A. Mikhailov, M.A. Nazaraliev, R.A. Darbinjan, B.A. Kargin, B.S. Elepov: *Monte Carlo Methods in Atmospheric Optics,* Springer Series in Optical Sciences, Vol. 12 (Springer, Berlin, Heidelberg, New York 1979)
Pratt, W. K.: *Digital Image Processing* (Wiley, New York 1978)

Chapter 8

8.1 R.E. Hufnagel, N.R. Stanley: J. Opt. Soc. Am. **54**, 52 (1964)
8.2 P. Hariharan: Appl. Opt. **9**, 1482 (1970)
8.3 J.C. Dainty (ed.): *Laser Speckle and Related Phenomena,* 2nd ed., Topics Appl. Phys. Vol. 9 (Springer, Berlin, Heidelberg, New York 1982)
8.4 A.M. Schneiderman, P.F. Kellen, M.G. Miller: J. Opt. Soc. Am. **65**, 1287 (1975)
8.5a D. Korff: J. Opt. Soc. Am. **63**, 971 (1973)
8.5b D. Korff: Opt. Commun. **5**, 188 (1972)
8.6 K.T. Knox: J. Opt. Soc. Am. **66**, 1236 (1976)
8.7 A. Papoulis: *Probability, Random Variables, and Stochastic Processes* (McGraw-Hill, New York 1965)
8.8 C.W. Helstrom: J. Opt. Soc. Am. **57**, 297 (1967)

8.9 J.L. Harris: In *Evaluation of Motion-Degraded Images.* NASA Spec. Publ. **193**, 131 (1968)
8.10 P.B. Fellgett, E.H. Linfoot: Philos. Trans. R. Soc. London A **247**, 369 (1955)
8.11 B.R. Frieden: J. Opt. Soc. Am. **60**, 575 (1970)
8.12 J.W. Goodman: *Introduction to Fourier Optics* (McGraw-Hill, New York 1968)
8.13 B. Mandelbrot: *Fractals, Form, Chance, and Dimension* (Freeman, San Francisco 1977)
8.14 T.A. Witten and L.M. Sander: *Phys. Rev. Letts.* **47**, 1400 (1981)

Additional Reading

Barrett, H. H., W. Swindell: *Radiological Imaging,* Vols. 1 and 2 (Academic, New York 1981)
Blanc-Lapierre, A., R. Fortet: *Theory of Random Functions,* Vol. I (Gordon and Breach, New York 1967)
Cramer, H., M.R. Leadbetter: *Stationary and Related Stochastic Processes* (Wiley, New York 1967)
Thomas, J. B.: *An Introduction to Statistical Communication Theory* (Wiley, New York 1969)
Wax, N. (ed.): *Selected Papers on Noise and Stochastic Processes* (Dover, New York 1954)
The journal *Stochastic Processes and their Application* (North-Holland, Amsterdam 1973-) for reference on specific problems

Chapter 9

9.1 B.R. Frieden: In *Picture Processing and Digital Filtering,* 2nd ed., ed. by T.S. Huang, Topics Appl. Phys., Vol. 6 (Springer, Berlin, Heidelberg, New York 1979)
9.2 P.G. Hoel: *Introduction to Mathematical Statistics* (Wiley, New York 1956) pp. 205–207
9.3 I.S. Gradshteyn, I.M. Ryzhik: *Tables of Integrals, Series, and Products* (Academic, New York 1965) p. 337; p. 284
9.4 B.R. Frieden: J. Opt. Soc. Am. **64**, 682 (1974)
9.5 M. Fisz: *Probability Theory and Mathematical Statistics* (Wiley, New York 1963) p. 343
9.6 B.R. Frieden, Fortran program *Aberr.f*, available upon request
9.7 G.D. Boyd and J.P. Gordon: *Bell System Tech. J.* **40**, 489 (1961)

Additional Reading

Justusson, B. I.: In *Two-Dimensional Digital Signal Processing II: Transforms and Median Filters* ed. by T.S. Huang, Topics in Applied Physics, Vol. 43 (Springer, Berlin, Heidelberg, New York 1981)
Kendall, M. G., A. Stuart: *The Advanced Theory of Statistics,* Vol. 1 (Griffin, London 1969)
Mises, R. von: *Mathematical Theory of Probability and Statistics* (Academic, New York 1964)

Chapter 10

10.1 E.T. Jaynes: IEEE Trans. SSC-**4**, 227 (1968)
10.2 N.N. Chenkov: Soviet Math. Dokl. **3**, 1559 (1962); E.L. Kosarev: Comput. Phys. Commun. **20**, 69 (1980)
10.3 J. MacQueen, J. Marschak: Proc. Natl. Acad. Sci. USA **72**, 3819 (1975)
10.4 R. Kikuchi, B.H. Soffer: J. Opt. Soc. Am. **67**, 1656 (1977)
10.5 J.P. Burg: "Maximum entropy spectral analysis," 37th Annual Soc. Exploration Geophysicists Meeting, Oklahoma City, 1967
10.6 S.J. Wernecke, L.R. D'Addario: IEEE Trans. C-**26**, 352 (1977)
10.7 B.R. Frieden: Proc. IEEE **73**, 1764 (1985)

10.8 S. Goldman: *Information Theory* (Prentice-Hall, New York 1953)
10.9 A. Papoulis: *Probability, Random Variables and Stochastic Processes* (McGraw-Hill, New York 1965)
10.10 D. Marcuse: *Engineering Quantum Electrodynamics* (Harcourt, Brace and World, New York 1970)
10.11 L. Mandel, E. Wolf: Rev. Mod. Phys. **37**, 231 (1965)
10.12 B.R. Frieden: J. Opt. Soc. Am. **73**, 927 (1983)

Additional Reading

Baierlein, R.: *Atoms and Information Theory* (Freeman, San Francisco 1971)
Boyd, D. W., J.M. Steele: Ann. Statist. **6**, 932 (1978)
Brillouin, L.: *Science and Information Theory* (Academic, New York 1962)
Frieden, B. R.: Comp. Graph. Image Proc. **12**, 40 (1980)
Kullbach, S.: *Information Theory and Statistics* (Wiley, New York 1959)
Sklansky, J., G.N. Wassel: *Pattern Classifiers and Trainable Machines* (Springer, Berlin, Heidelberg, New York 1981)
Tapia, R. A., J.R. Thompson: *Nonparametric Probability Density Estimation* (Johns Hopkins University Press, Baltimore 1978)

Chapter 11

11.1 A. Papoulis: *Probability, Random Variables and Stochastic Processes* (McGraw-Hill, New York 1965)
11.2 M. Abramowitz, I.A. Stegun (eds.): *Handbook of Mathematical Functions* (National Bureau of Standards, Washington, DC 1964)
11.3 F.B. Hildebrand: *Introduction to Numerical Analysis* (McGraw-Hill, New York 1956)
11.4 G.L. Wied, G.F. Bahr, P.H. Bartels: In *Automated Cell Identification and Cell Sorting,* ed. by G.L. Wied, G.F. Bahr (Academic, New York 1970) pp. 195–360
11.5 C.W. Allen: *Astrophysical Quantities, 3rd. ed.* (Athlone Press, London 1973)

Additional Reading

Breiman, L.: *Statistics: With a View Toward Applications* (Houghton-Mifflin, Boston 1973)
Hogg, R. V., A. T. Craig: *Introduction to Mathematical Statistics,* 3rd ed. (Macmillan, London 1970)
Kendall, M. G., A. Stuart: *The Advanced Theory of Statistics,* Vol. 2 (Griffin, London 1969)
Mises, R. von: *Mathematical Theory of Probability and Statistics* (Academic, New York 1964)

Chapter 12

12.1 Student: Biometrika **6**, 1 (1908)
12.2 M. Abramowitz, I.A. Stegun (eds.): *Handbook of Mathematical Functions* (National Bureau of Standards, Washington, DC 1964)
12.3 C. Stein: *Proceedings Third Berkeley Symposium on Mathematics, Statistics, and Probability* (University of California Press, Berkeley, CA 1955) p. 197

Additional Reading

Brandt, S.: *Statistical and Computational Methods in Data Analysis,* 2nd. rev. ed. (North-Holland, New York 1976)
Hogg, R. V., A.T. Craig: *Introduction to Mathematical Statistics,* 3rd ed. (Macmillan, London 1970)

Mendenhall, W., R.L. Scheaffer: *Mathematical Statistics with Applications* (Duxbury, North Scituate, MA 1973)
Kendall, M. G., A. Stuart: *The Advanced Theory of Statistics,* Vol. 2 (Griffin, London 1969)

Chapter 13

13.1 F. Snedecor: *Statistical Methods,* 5th ed. (Iowa State University Press, Ames 1962)
13.2 M. Abramowitz, I.A. Stegun (Eds.): *Handbook of Mathematical Functions* (National Bureau of Standards, Washington, DC 1964)
13.3 P.H. Bartels, J.A. Subach: "Significance Probability Mapping, and Automated Interpretation of Complex Pictorial Scenes," in *Digital Processing of Biomedical Images,* ed. by K. Preston, M. Onoe (Plenum, New York 1976) pp. 101–114

Additional Reading

Brandt, S.: *Statistical and Computational Methods in Data Analysis,* 2nd rev. ed. (North-Holland, New York 1976)
Breiman, L.: *Statistics: With a View Toward Applications* (Houghton-Mifflin, Boston 1973)
Kempthorne, O., L. Folks: *Probability, Statistics, and Data Analysis* (Iowa State University Press, Ames 1971)

Chapter 14

14.1 L. Breiman: *Statistics: With a View Toward Applications* (Houghton Mifflin, Boston 1973) Chap. 10
14.2 M.G. Kendall, A. Stuart: *The Advanced Theory of Statistics,* Vol. 1 (Charles Griffin, London 1969)

Additional Reading

Bjerhammar, A.: *Theory of Errors and Generalized Matrix Inverses* (Elsevier Scientific, Amsterdam 1973)
Mendenhall, W., R.L. Scheaffer: *Mathematical Statistics with Applications* (Duxbury, North Scituate, MA 1973)

Chapter 15

15.1 H. Hotelling: J. Educ. Psych. **24**, 417, 498 (1933)
15.2 J.L. Simonds: J. Phot. Sci. Eng. **2**, 205 (1958)
15.3 R.C. Gonzalez, P. Wintz: *Digital Image Processing* (Addison-Wesley, Reading, MA 1977) pp. 310–314
15.4 M.G. Kendall, A. Stuart: *The Advanced Theory of Statistics,* Vol. 3 (Charles Griffin, London 1968)
15.5 M.A. Girshick: Ann. Math. Statist. **10**, 203 (1939)

Additional Reading

Wilkinson, J. H.: *The Algebraic Eigenvalue Problem* (Clarendon, Oxford 1965)

Chapter 16

16.1 J. Weber: *Historical Aspects of the Bayesian Controversy* (University of Arizona, Tucson 1973)

16.2 A. Papoulis: *Probability, Random Variables and Stochastic Processes* (McGraw-Hill, New York 1965)
16.3 E. Parzen: *Modern Probability Theory and Its Applications* (Wiley, New York 1960)
16.4 B.R. Frieden: Found. Phys. **16**, 883 (1986)

Additional Reading

Good, I. J.: "The Bayesian Influence, or How to Sweep Subjectivism Under the Carpet," in *Foundations of Probability Theory, Statistical Inference, and Statistical Theories of Science,* Vol. II, ed. by W.L. Harper, C.A. Hooker (Reidel, Boston 1976)
Jaynes, E. T.: "Confidence Intervals vs Bayesian Intervals", in preceding reference

Chapter 17

17.1 B.R. Frieden: Phys. Rev. **A41**, 4265 (1990)
17.2 H.L. Van Trees: *Detection, Estimation and Modulation Theory,* Part I (Wiley, New York 1968)
17.3 B.R. Frieden: Appl. Opt. **26**, 1755 (1987)
17.4 R. Kikuchi, B.H. Soffer: J. Opt. Soc. Am. **67**, 1656 (1977)
17.5 B.R. Frieden: Am. J. Phys. **57**, 1004 (1989)
17.6 B.R. Frieden: Opt. Lett. **14**, 199 (1989)
17.7 T.D. Milster *et al.*: J. Nucl. Med. **31**, 632 (1990)
17.8 A.J. Stam: Inform, and Control **2**, 101 (1959)
17.9 A. Bhattacharyya: *Sankhya* **8**, 1, 201, 315 (1946–1948)
17.10 B.R. Martin and G. Shaw: *Particle Physics* (Wiley, New York 1992)
17.11 B.R. Frieden and A. Plastino: Phys. Letts. A **278**, 299 (2001)
17.12 B.R. Frieden: *Physics from Fisher Information: a Unification* (Cambridge Univ. Press, England 1998)
17.13 I. Kant: *Kritik der reinen Vernunft* (J.F. Hartknoch, Riga 1787). Also, English transl., *Critique of Pure Reason* (Willey Book Co., New York 1900)
17.14 J.D. Jackson: *Classical Electrodynamics, 2nd. ed.* (Wiley, New York 1975)
17.15 M.B. Mensky: *Phys. Rev. D* **20**, 384 (1979); also, *Continuous Quantum Measurements and Path Integrals* (Inst. of Physics Publ., Bristol, England 1993)
17.16 J.A. Wheeler: in *Proceedings of the 3rd International Symposium on Foundations of Quantum Mechanics, Tokyo, 1989,* eds. S. Kobayashi et. al. (Physical Society of Japan, Tokyo 1990), 354; also: in *Physical Origins of Time Asymmetry,* eds. J.J. Halliwell et. al. (Cambridge Univ. Press, England 1994), 1–29
17.17 B.R. Frieden and B.H. Soffer: *Phys. Rev. E* **52**, 2274 (1995)
17.18 B.R. Frieden and A. Plastino: *Physics Letts. A* **272**, 326 (2000)
17.19 L. Brillouin: *Science and Information Theory:* (Academic, New York 1956)
17.20a H. Goldstein: *Classical Mechanics* (Addison-Wesley, Cambridge. Mass. 1956), p. 365
17.20b P.M. Morse and H. Feshbach: *Methods of Theoretical Physics, Part I* (McGraw-Hill, New York 1953), p. 278
17.21 G.V. Vstovsky: *Phys. Rev. E* **51**, 975 (1995)
17.22 H. Risken: *The Fokker-Planck Equation* (Springer-Verlag, Berlin 1984)
17.23 A.R Plastino and A. Plastino: *Phys. Rev. E* **54**, 4423 (1996)
17.24 C. Villani and G. Toscani: *Comm. Math. Phys.* **203**, 667 (1999)
17.25 P.M. Binder: *Phys. Rev. E* **61**, R3303 (2000)
17.26 R. Badii and A. Politi: *Complexity: Hierarchical Structure and Scaling in Physics* (Cambridge Univ. Press, England 1997)
17.27 W.J. Cocke: *Phys. of Fluids* **8**, 1609 (1996)
17.28 E. Giulini et. al.: *Decoherence and the Appearance of a Classical World in Quantum Theory* (Springer-Verlag, Berlin 1996)

17.29 H.D. Zeh: "What is achieved by decoherence?", in: *New Developments on Fundamental Problems in Quantum Physics*, eds. M. Ferrero and A. van der Merwe (Kluwer Academic, Dordrecht, Netherlands 1997)
17.30 R.A. Tapia and J.R. Thompson: *Nonparametric probability density estimation* (Johns Hopkins Univ. Press, Baltimore 1978)
17.31 B.R. Frieden, Am. J. Physics **57**, 1004 (1989); also Phys. Rev. A **41**, 4265 (1990)
17.32 P.H. Cootner, ed: *The Random Character of Stock Market Prices* (MIT Press, Cambridge, Massachusetts 1964)
17.33 R.J. Hawkins, M. Rubinstein and G.J. Daniell: "Reconstruction of the probability density function implicit in option prices from incomplete and noisy data," in: *Proceedings of the Fifteenth International Workshop on Maximum Entropy and Bayesian Methods*, eds. K.M. Hanson and R.N. Silver (Kluwer Academic, Dordrecht, Netherlands 1996)
17.34 G.A. Korn and T.M. Korn: *Mathematical Handbook for Scientists and Engineers, 2nd. ed.* (McGraw-Hill, New York 1968)
17.35 B.R. Frieden, A. Plastino, A.R. Plastino and B.H. Soffer, Phys. Rev. E **60**, 48 (1999)
17.36 J.A. Eisele: *Modern quantum mechanics with applications to elementary particle physics* (Wiley, New York 1969), 37

Additional Reading

J-P. Bouchard and M. Potters: *Theory of Financial Risk: From Statistical Physics to Risk Management* (Cambridge Univ. Press, England 2000)
T. Cover and J. Thomas: *Elements of Information Theory* (Wiley, New York 1991)
I.J. Good: in *Foundations of Probability Theory, Statistical Inference and Statistical Theories of Science, Vol. II*, eds. W.L. Harper and C.A. Hooker (Reidel, Boston 1976)

Index

Printing (Computer to Film): Saladruck, Berlin
Binding: Stürtz AG, Würzburg

Editors: Thomas S. Huang Teuvo Kohonen Manfred R. Schroeder

1 **Content-Addressable Memories**
By T. Kohonen 2nd Edition

2 **Fast Fourier Transform and Convolution Algorithms**
By H. J. Nussbaumer 2nd Edition

3 **Pitch Determination of Speech Signals**
Algorithms and Devices By W. Hess

4 **Pattern Analysis and Understanding**
By H. Niemann 2nd Edition

5 **Image Sequence Analysis**
Editor: T. S. Huang

6 **Picture Engineering**
Editors: King-sun Fu and T. L. Kunii

7 **Number Theory in Science and Communication**
With Applications in Cryptography, Physics, Digital Information, Computing, and Self-Similarity
By M. R. Schroeder
3rd Edition

8 **Self-Organization and Associative Memory**
By T. Kohonen 3rd Edition

9 **Digital Picture Processing**
An Introduction By L. P. Yaroslavsky

10 **Probability, Statistical Optics, and Data Testing**
A Problem Solving Approach
By B. R. Frieden 3rd Edition

11 **Physical and Biological Processing of Images** Editors: O. J. Braddick and A. C. Sleigh

12 **Multiresolution Image Processing and Analysis** Editor: A. Rosenfeld

13 **VLSI for Pattern Recognition and Image Processing** Editor: King-sun Fu

14 **Mathematics of Kalman-Bucy Filtering**
By P. A. Ruymgaart and T. T. Soong
2nd Edition

15 **Fundamentals of Electronic Imaging Systems**
Some Aspects of Image Processing
By W. F. Schreiber 3rd Edition

16 **Radon and Projection Transform-Based Computer Vision**
Algorithms, A Pipeline Architecture, and Industrial Applications By J. L. C. Sanz, E. B. Hinkle, and A. K. Jain

17 **Kalman Filtering with Real-Time Applications**
By C. K. Chui and G. Chen 3rd Edition

18 **Linear Systems and Optimal Control**
By C. K. Chui and G. Chen

19 **Harmony: A Psychoacoustical Approach**
By R. Parncutt

20 **Group-Theoretical Methods in Image Understanding**
By Ken-ichi Kanatani

21 **Linear Prediction Theory**
A Mathematical Basis for Adaptive Systems
By P. Strobach

22 **Psychoacoustics** Facts and Models
By E. Zwicker and H. Fastl 2nd Edition

23 **Digital Image Restoration**
Editor: A. K. Katsaggelos

24 **Parallel Algorithms in Computational Science**
By D. W. Heermann and A. N. Burkitt

25 **Radar Array Processing**
Editors: S. Haykin, J. Litva, and T. J. Shepherd

26 **Discrete H^∞ Optimization**
With Applications in Signal Processing and Control Systems 2nd Edition
By C. K. Chui and G. Chen

27 **3D Dynamic Scene Analysis**
A Stereo Based Approach
By Z. Zhang and O. Faugeras

28 **Theory of Reconstruction from Image Motion**
By S. Maybank

29 **Motion and Structure from Image Sequences**
By J. Weng, T. S. Huang, and N. Ahuja